AF251465

Numerical Methods
in
Aeronautical Fluid Dynamics

The Institute of Mathematics
and its Applications
Conference Series

RECENT TITLES

Computational Methods and Problems in Aeronautical Fluid Dynamics, *edited by* B.L. Hewitt, C.R. Illingworth, R.C. Lock, K.W. Mangler, J.H. McDonnell, Catherine Richards and F. Walkden

Optimization in Action, *edited by* L.C.W. Dixon

The Mathematics of Finite Elements and Applications II, *edited by* J.R. Whiteman

The State of the Art in Numerical Analysis, *edited by* D.A.H. Jacobs

Fisheries Mathematics, *edited by* J.H. Steele

Numerical Software—Needs and Availability, *edited by* D.A.H. Jacobs

Recent Theoretical Developments in Control, *edited by* M.J. Gregson

The Mathematics of Hydrology and Water Resources, *edited by* E.H. Lloyd, T.O'Donnell and J.C. Wilkinson

Mathematical Aspects of Marine Traffic, *edited by* S.H. Hollingdale

Mathematical Modelling of Turbulent Diffusion in the Environment, *edited by* C.J. Harris

Mathematical Methods in Computer Graphics and Design, *edited by* K.W. Brodlie

Computational Techniques for Ordinary Differential Equations, *edited by* I. Gladwell and D.K. Sayers

Stochastic Programming, *edited by* M.A.H. Dempster

Analysis and Optimisation of Stochastic Systems, *edited by* O.L.R. Jacobs, M.H.A. Davis, M.A.H. Dempster, C.J. Harris and P.C. Parks

Numerical Methods in Applied Fluid Dynamics, *edited by* B. Hunt

Recent Developments in Markov Decision Processes, *edited by* R. Hartley, L.C. Thomas and D.J. White

Power from Sea Waves, *edited by* B. Count

Sparse Matrices and their Uses, *edited by* I.S. Duff

The Mathematical Theory of the Dynamics of Biological Populations II, *edited by* R.W. Hiorns and D. Cooke

Floods due to High Winds and Tides, *edited by* D.H. Peregrine

Third IMA Conference on Control Theory, *edited by* J.E. Marshall, W.D. Collins, C.J. Harris and D.H. Owens

Numerical Methods in Aeronautical Fluid Dynamics, *edited by* P.L. Roe

Numerical Methods for Fluid Dynamics, *edited by* K.W. Morton and M.J. Baines

IN PREPARATION

Multi-Objective Decision Making, *edited by* L.C. Thomas

Numerical Methods
in
Aeronautical Fluid Dynamics

*Based on the proceedings of a conference on
Numerical Methods in Aeronautical Fluid Dynamics,
organised by the Institute of Mathematics and its
Applications and held at the University of Reading
on March 30 – April 1, 1981*

Edited by

P. L. ROE

*Aerodynamics Department
Royal Aircraft Establishment
Bedford, UK*

1982

ACADEMIC PRESS

A Subsidiary of Harcourt Brace Jovanovich, Publishers

London   New York
Paris   San Diego   San Francisco   São Paulo
Sydney   Tokyo   Toronto

ACADEMIC PRESS INC. (LONDON) LTD.
24/28 Oval Road
London NW1

United States Edition published by
ACADEMIC PRESS INC.
111 Fifth Avenue
New York, New York 10003

British Library Cataloguing in Publication Data
Numerical methods in aeronautical fluid dynamics.
  (IMA conference series)
  1. Aerodynamics—Data processing—Congresses
I. Roe, P.L.   II. Series
533'.62'02854   TL573

ISBN 0-12-592520-4

LCCCN   81-89583

Printed in Great Britain by
Whitstable Litho Ltd., Whitstable, Kent

LIST OF CONTRIBUTORS

BAKER, T.J. *Aircraft Research Association, Manton Lane, Bedford MK41 8PF.*

BUTTER, D.J. *British Aerospace Aircraft Group, Manchester Division, Chester Road, Woodford, Stockport, Cheshire SK7 1QR.*

CARR, M.P. *Aircraft Research Association, Manton Lane, Bedford MK41 8PF.*

DENTON, J.D. *Whittle Laboratory, University Engineering Department, Madingley Road, Cambridge CB3 0EL.*

DECONINCK, H. *Department of Fluid Mech., Faculty Van De Toegepaste, Wetenschappen, Vrije Universitat Brussel, Pleinlaan, 2, 1050 Brussels, Belgium.*

FIRMIN, M.C.P. *Aerodynamics Department, Royal Aircraft Establishment, Farnborough, Hants. GU14 6TD.*

FORSEY, C.R., AFIMA *Aircraft Research Association, Manton Lane, Bedford MK41 7PF.*

GLOWINSKI, R. *Institut National de Recherche en Information et en Automation, Domaine de Voluceau, Rocquencourt, B.P. 105 – 78150 Le Chesnay, France*

GREEN, J.E. *Ministry of Defence, St Giles Court, London WC2H 8LD.*

HALL, M.G. *Aerodynamics Department, Royal Aircraft Establishment, Farnborough, Hants. GU14 6TD.*

HARGREAVES, G.R. *British Aerospace Aircraft Group, Manchester Division, Chester Road, Woodford, Stockport, Cheshire SK7 1QR.*

HIRSCH, Ch. *Department of Fluid Mech., Faculty Van De Toegepaste, Wetenschappen, Vrije Universitat Brussel, Pleinlaan 2, 1050 Brussels, Belgium.*

HOCKNEY, R.W. *Computer Science Department, University of Reading, Whiteknights, Reading RG6 2AX.*

HUNT, B., AFIMA *Senior Engineer, British Aerospace Aircraft Group, Warton Aerodrome, Preston, Lancashire PR4 1AX.*

JAMESON, A.   *Engineering Quadrangle, Princeton University, Princeton, New Jersey, USA.*

LERAT, A.   *ONERA, 92320 Chatillon, Chatillon-sous-Bagneux, France.*

LOCK, R.C., FIMA   *Aerodynamics Department, Royal Aircraft Establishment, Farnborough, Hants. GU14 6TD.*

ROE, P.L.   *Aerodynamics Department, Royal Aircraft Establishment, Bedford MK41 6AE.*

SIDES, J.   *ONERA, 92329 Chatillon, Chatillon-sous-Bagneux, France.*

SLOOFF, J.W.   *NLR, Anthony Fokkerwej 2, 1859CM, Amsterdam, The Netherlands.*

SMITH, J.H.B., FIMA   *Aerodynamics Department, Royal Aircraft Establishment, Farnborough, Hants, GU14 6TD.*

PREFACE

The papers collected in these proceedings are all concerned with
some aspect of applying numerical methods to solve problems of
aerodynamic analysis or design. Those readers who are already
specialists in this field will no doubt find their way quickly
to those parts of the book which interest them most. This preface is intended to provide some orientation for the more general reader.

To the mathematician who takes a traditional approach to
fluid mechanics the computer may sometimes seem a weapon of last
resort. When the equations which describe some phenomenon do
not yield up the hoped-for closed solution he reports that they
were "solved numerically", and the note of regret which often
accompanies this phrase is understandable. The insight which a
simple result might have revealed has been denied to him. As
second best, tables or graphs have been produced from the machine, possibly by appealing to some library subroutine whose
inner workings remain a dark mystery. By contrast, the authors
of the papers in this book have accepted from the outset the
need for a numerical approach, whose details they are often very
much concerned with. The use of standard subroutines may be
rejected because the problem is too intractable, or because the
general-purpose method cannot match the efficiency of one designed to suit the occasion. However, the skills and knowledge
needed to design and verify the numerical tools have much in
common with traditional fluid mechanics. Loosely, the central
problems may be expressed thus. Can our knowledge of fluid
behaviour be summarised in simple arithmetical rules so as to
create a digital model of a fluid, and for what purposes is such
a model adequate? One purpose of this book will have been well
served if any part of the reservoir of existing talent is diver-
ted to this new and challenging area.

The other group of "near-specialists" to whom a few words may
be addressed are those whose interests already lie in computational fluid mechanics, but with some other application in mind.
The range of such applications is vast, ranging from the evolution of galaxies to the flow of the blood, from the mechanics of
strong explosions to the control of estuaries. Nevertheless,
aerodynamics probably accounts for about half the research being
done on computational methods, and its literature is often

consulted by those seeking ideas which might be more widely
used. Such readers may find the material of this book more
intelligible if they bear in mind certain differences of emphasis which mark off computational aerodynamics from computational
fluid dynamics in general. One distinction is that aerodynamic
shapes are often quite complex, and that even the simpler ones
must be specified with great precision. This leads the aerodynamicist into a preoccupation with geometry which sometimes
makes his programs rather specialised and inflexible, in contrast to the "general hydrocode" approach which is common elsewhere. Also the computational aerodynamicist works alongside
colleagues with a strong experimental tradition, and in the service of a designer who operates within narrow commercial or
military margins of efficiency. There is consequently a very
high demand for accuracy. On the other hand, aerodynamic programs are not usually required to be especially robust. If the
flow is very badly behaved, the shape which produces it is discarded as a failure. For this reason, the non-linearities which
arise in aerodynamics are usually somewhat milder than those
occurring elsewhere. A final distinction, closely related to
the previous one, is the point always emphasised by the late
Professor Dietrich Küchemann*, that the proper activity of the
aerodynamicist is design rather than analysis. He seeks the
geometry which leads to a "healthy flow", rather than the flow
due to an arbitrary geometry.

Taken together, these distinctions make computational aerodynamics a sufficiently separate discipline to merit a conference of its own, but not, I hope, one that is too chauvinist to
communicate with its neighbours. In the past, aerodynamicists
have contributed prominently to the numerical study of mixed
flows, shocked flows, and slightly viscous flows, and to the
development of fast algorithms. All these topics feature
strongly in these proceedings, in papers which can profitably be
read by those with wider interests.

Let us turn now to more specific discussion of the individual
papers. The organisers achieved their intention of covering a
wide variety of methods ranging from those which form the main-
stay of day-to-day industrial design, to those which are as yet
just a gleam in an academic eye. The relationship of these
methods to the phenomena encountered in flight is surveyed in
the introductory paper by J.E. Green. It is emphasised that we
are many decades away (both as regards algorithm design and computer capacity) from the numerical simulation of such a complex
happening as a combat aircraft in mid-manoeuvre. In the mean-
time, the designer's art lies, as it always has, in the synthe-
sis of information from many different sources.

---

*Küchemann, D., "The Aerodynamic Design of Aircraft", Pergamon,
 1976.

One important use of the computer is to assess, quite quickly
and with reasonable accuracy, the effect of radical departures,
such as canard surfaces or slewed wings, from conventional practice. Also the military requirement to pack the maximum of external weaponry into the minimum of space poses problems of
aerodynamic interference which may sometimes be solved by the
ingenious rearrangement of components. The designer's imagina-
tion is less fettered if he no longer has to progress by small
perturbations. The computer gives him the freedom to examine a
great variety of layouts, so that the more promising ones may be
looked at in more detail. However, it is not yet feasible to
use field methods (i.e. those which compute the detailed flow
within some volume of space surrounding the aircraft) for this
purpose. The human effort of specifying a complex coordinate
system, and the machine effort of solving the equations, are
both equally prohibitive. But if the flow is assumed to be
governed by Laplace's equation, then only the surface conditions
need be considered, and both the human and machine effort are
much reduced. The paper by D.J. Butter showed both the flexibility, and the sometimes surprising accuracy of this approach,
which is particularly suitable when the day-to-day needs of the
practical designer require the creation of programs with minimal
input data and execution time. A recent offshoot from this
classical technique is to consider the non-linear effects of
compressibility as source terms superimposed on a Laplacian
field. The hope is that such terms can be evaluated on a rather
coarse grid which does not extend far from the surface. These
integral methods are not yet highly developed, but an indication
of their potential is given in the paper by J. Slooff, who was
also much concerned with the matching of numerical techniques to
industrial requirements.

The main tool used over the past decade for solving two-dimensional problems or simple three-dimensional problems which are
shockfree or have only fairly mild shockwaves has been transonic
small perturbation theory (TSP). The TSP equation has also been
a proving ground for the development of new numerical methods,
as described in the early parts of the articles by M.G. Hall and
T.J. Baker. In practice, it has shown itself to be versatile
and accurate to a far greater extent than might have been expected. Although little is said in these proceedings about its
actual applications, the interested reader can find a recent
comprehensive survey elsewhere*.

Methods based on more complete mathematical models of the
inviscid flow are only now coming into operational service. The
next step is to retain all the non-linear terms in the equation

---

*C.W. Boppe, P.V. Aidala, Complex configuration analysis at trans-
sonic speeds, in AGARD Conference Proceedings 285, "Subsonic/
transonic configuration aerodynamics", Munich 5-7 May, 1980.

of compressible potential flow. These "full potential" methods
are surveyed here by Hall, who is especially concerned with their
adequacy to predict flows with fairly strong shockwaves. It is
important to realise that although the running time of a fully
developed computer code may be roughly proportional to the number of terms retained in the differential equations, the time
taken to develop the method may not follow any simple rule.
Although the experience gained on simple problems can sometimes
be extrapolated to more complicated ones, it may also happen
that the complicated problems expose weaknesses of numerical
technique which take a long time to put right.

Of course there is not only the problem of formulating the
discrete equations, but also the problem of devising a rapid
(usually iterative) method for their solution. In recent years,
very substantial progress has been made in this area, with computing times for some problems being cut by orders of magnitude.
Two popular developments have been Approximate Factorisation,
and the Multigrid Method. Both of these methods are characterised by the fact that although a fairly complete theory exists
for such idealised cases as solving Laplace's equation on a uniform grid, the treatment of more general problems relies heavily
on instinct and experience. In these proceedings, the Approximate Factorisation technique (in which the unwieldly operator
which one would like to invert is replaced by a sequence of
easily invertible operators whose product closely resembles it)
is discussed in detail by Baker. The Multigrid Method (wherein
a set of residual errors evaluated on a fine grid are permitted
to interact through calculations made on a set of progressively
coarser grids) is discussed by no fewer than four of the authors
(Hall, Hirsch, Slooff and J.D. Denton). If one may be allowed
to attempt a generalisation from this work, it could be that
those insights which are most useful in formulating the discrete
equations are also, after a sufficient lapse of time, likely to
be instrumental in their rapid solution.

Another topic dealt with by several authors is that of treating the inviscid flow with even greater realism by going to the
compressible Euler equations. At this level, one is either looking for highly accurate solutions against which to calibrate a
faster method, or else for very robust codes able to deal with
strongly non-linear flows, such as may occur in transonic flight
away from design conditions, or in the inevitably complex environment of turbomachinery. In either case, an even more severe
test of numerical technique has been posed, and it is interesting that the four authors respond in quite different ways. A.
Lerat considers an extremely general class of finite-difference
schemes containing a large number of free parameters, which are
then optimised to meet various criteria. My own contribution
describes recent work on a class of methods which attempt to
reconcile the conservation law and wave propagation aspects of
the problem. Denton proposes a "finite volume" method in which

at each time step, the pressure distribution and momentum distribution are updated by different methods. Denton's method has
been extensively used in turbomachinery blade design for some
time, and has earned a reputation for robustness and reliability;
his present contribution shows a new way of formulating the
method which is tidier, faster and more accurate. The methods
described by Lerat and myself are both too time-consuming for
regular use (in their present form and on present computers) but
each has been developed far enough to be applied to practical
problems. A. Jameson described another "finite volume" method
which seems to be establishing itself as a very promising practical technique. Although Prof. Jameson was too heavily com-
mitted to produce a written paper, he agreed to the inclusion of
a transcript of his lecture which I composed. This somewhat
informal contribution is, I think, especially valuable for the
insight which it gives into the working methods of an exceptionally innovative numerical analyst.

Although the viscous forces which act on the flow around an
aircraft are in general smaller by many orders of magnitude than
the inertial forces, the role which they play in determining
the structure of the flow is of course decisive. A lengthy and
authoritative contribution from R.C. Lock and M.C.P. Firmin
describes those techniques which apply when viscous effects
remain confined within thin boundary layers adjacent to the
surfaces, and thin wakes trailing from them. Although the concepts involved have been part of the aerodynamicist's stock-in-
trade for decades past, there have been many important recent
developments in their practical application. The power of the
computer is utilised here to enable detailed prediction of
three-dimensional turbulent boundary layers, and to match these
with numerical predictions of the inviscid flow. One recent
development to which these authors devote attention is the
second-order boundary layer theory of L.F. East.

When conditions become so severe that the boundary layer sepa-
rates from the surface, the flow cannot yet be calculated by any
method that could be called routine. Nevertheless, progress is
being made, and J.H.B. Smith reviews the current status of the
vortex-sheet model which can be used to represent the separated
flow past strakes, highly-swept wings, and the noses of missiles.
By carefully drawing together the strands of asymptotic analysis,
experimental observation, and numerical techniques, predictions
can be made of flows which are still beyond the reach of so-
called general methods. Nevertheless, as computers continue to
grow bigger and faster, there is more and more incentive to
search for methods which will handle all such complexities automatically by solving the Navier-Stokes equations. The contribution from R. Glowinski describes new and deeply thought-out
numerical methods which can already predict the behaviour of
separated and vortex-type flows, although not yet at the Reynolds numbers which obtain in aircraft flight.

Internal flows in ducts and turbomachinery present their own
special problems. The paper by Denton has already been mentioned but contributions from C. Hirsch and R. Payne also treated
this topic. Hirsch provided a treatment of finite element
methods which have novel and interesting features. His main
application of these methods was to the potential equation, but
he also touched on the Euler equations. No written version of
the paper by Payne was available, but he was able to show the
useful contribution still to be made by quite simple methods.

Finally, there are two contributions devoted to topics which
run like connecting threads through almost all the other material. One of these has already been touched on. It is the aerodynamicist's abiding obsession with geometry; with the need to
define shapes precisely and economically, and to surround them
with computing grids on which a field calculation can be accurately and conveniently carried out. Since the flow field, and
its mathematical description, should in principle be coordinate-
free, the matter of computing grids tends to be thought of as a
minor irritation, almost irrelevant to the main business. However, it has been estimated that in industrial applications of
computational aerodynamics, about half the effort of creating
and running the programs is absorbed by "geometry handling".
Clearly, the problem is of great importance, but few unifying
principles have yet emerged. M.P. Carr and C. Forsey have surveyed the work done so far, classifying and evaluating it as
far as possible, but clearly indicating that "more work remains
to be done".

The second of these universal topics is of course the machine
itself. The use of high level languages does mean in principle
that the programmer can perform his task in complete ignorance
of how the hardware actually operates. But viewing the matter
more realistically, we have to admit that no compiler yet written can perfectly adjust the machine's ability to the programmer's intentions. The authors of large programs need some
awareness of the machine's operating methods if their products
are to be efficient. The conference treated this topic both
formally and informally. The informal treatment took the shape
of an evening discussion on the characteristics of the latest
generation of machines. This was led by Dr J.R. Petersen (on
behalf of the Cray Corporation), Mr S. Fernbach (on behalf of
CDC) and Professor D. Parkinson (on behalf of ICL). The rivalry
was in part softened and in part enlivened by the IMA tradition
of holding such meetings as close as possible to a bar, so it is
perhaps as well that no written record survives. But among many
interesting points made, one (due to Professor Parkinson) was
that every advance in computing equipment is sooner or later
followed by new algorithms which are adapted to the new tools.
This thought was developed in more detail by R.W. Hockney in a
formal contribution, relating specifically to parallel processing machines, but applicable to most available devices. He

attempts to measure the "degree of parallelism" inherent in a
set of numerical operations. Since this concept applies equally
to machines and to algorithms, it provides a means of matching
"horses to courses", which should be very useful.

Ten years ago, there were those who thought that by now the
computer would soon be displacing the wind tunnel, but today most
aerodynamicists would foresee a long partnership between computation and experiment. The brief summaries above perhaps indicate both the scale and the diversity of the enterprise needed
to achieve realistic predictions of aerodynamic flows. Over a
sufficiently long period of time, Darwinian selection may reduce
the numerical methods in use to some small set of basic techniques, but meanwhile I feel sure that all the contributions to
this volume will remain useful and valuable for many years to
come. Assembling such wide-ranging material has been for me a
rewarding and fascinating task, and I would like to thank all
the authors for the efforts they have put into their contribution. I would also like to thank Catherine Richards and the
staff of the IMA for making the conference possible, and for
their help and advice in producing this volume. Liaison with
Academic Press has been through Cath Pettyfer, whose friendly
assistance was invaluable.

P.L. Roe
October, 1982

ACKNOWLEDGEMENTS

The Institute thanks the authors of the papers, the editor,
Mr. P.L. Roe (RAE, Bedford) and also Mrs. S. Hockett, Mrs. J.
Parsons and other typists for typing the manuscript.

They also give thanks to various organisations and individuals
who have given their consent to the inclusion of copyright material. These are: The Controller of Her Majesty's Stationery
Office, for permission to use Crown Copyright material in the
chapters by Green, Hall, Roe, Lock and Firmin, and Smith; The
American Society of Mechanical Engineers, for permission to
reproduce the chapter by J.D. Denton; The American Institute of
Aeronautics and Astronautics, and Professor D.A. Caughey for
permission to include figures 4 and 5 of the chapter by Carr
and Forsey; The United States National Aeronautics and Space
Administration and the Boeing Company for permission to include
figures 10 to 13 of the chapter by Smith.

A challenge to computational aerodynamics;
the Tornado with its load of stores.

CONTENTS

NUMERICAL METHODS IN AERONAUTICAL FLUID DYNAMICS —
AN INTRODUCTION

J.E. GREEN

(RAE, Farnborough, Hants)

SUMMARY

This introductory paper reviews the factors underlying the rapid
development of computational fluid dynamics in recent years, and
considers the present state and future potential of the subject.
It outlines the range of physical phenomena with which aerodynamic design methods must deal and discusses the characteristics, limitations and applications of the various types of
numerical method in current use. Looking ahead, the paper considers the likely rate of development of computer technology
and the advances in computational fluid dynamics that should be
possible in the foreseeable future. The field should provide
mathematicians, numerical analysts and theoretical physicists
with a wide spectrum of challenging and important problems.

1. PREAMBLE

The use of numerical methods in aerodynamics — which I shall
refer to as computational fluid dynamics, CFD for short — has
advanced tremendously over the past ten years. So much so that
CFD is now an essential part of the aircraft designer's armoury
and is recognised as one of the key aeronautical technologies
for the future. I expect the papers at this conference to indicate the great scope there is for advance in CFD: the subject
is still to some degree in its infancy, and I believe few if any
of us appreciate all its potentialities. Nevertheless, we have
already reached the stage at which CFD is in routine use in aircraft design offices for calculating transonic flow fields
about swept winged aircraft. It is salutory to reflect that,
scarcely more than a decade ago, some of the ablest and most
experienced aerodynamicists dismissed such a possibility as an
unrealistic dream.

The factors underlying this flowering of CFD have been the
rapid rate of development of computer technology and, in parallel
with this, a rapid improvement in numerical methods. Chapman
(1979), in his excellent review of the field, shows how the
cost of a given computation has been steadily falling, roughly

by a factor of ten every eight years, as a result of improvements in computer technology. Numerical methods for solving a
given problem have shown a similar improvement. To illustrate
how these effects may be compounded, Chapman cites the example
of calculating the flow over an aerofoil using the Reynolds-
averaged Navier-Stokes equations. On a 1978, Class 6 computer,
using 1978 algorithms, the computation costs less than a thousand
dollars and occupies less than 30 minutes of computer time. Had
the same calculation been attempted in 1958, using the computers
and algorithms then available, it would have cost roughly 10
million dollars in computer time and taken about 30 years to
complete - the calculation would still be running now and it
would take 7 or 8 more years before it was finished!

This amazing increase in the cost effectiveness of CFD has
given it a central role both in aerodynamic design and development and in basic and applied research. In the field of design
it is a powerful means of searching for efficiency, for configurations optimised and well integrated in aerodynamic terms.
As time goes by it will be relied on increasingly to provide
aerodynamic data for the structural designer, the performance
engineer, the designer of the flight control system. It is
particularly useful as a means of investigating possible design
changes and it has an important role to play in support of wind
tunnel testing. In the field of applied research, CFD provides
a means of investigating new design concepts, of studying complex flow interactions in a systematic way and of generating
definitive results against which practical design methods can
be tested. At a more basic level, its use can identify experiments needed to advance our understanding, particularly in the
field of turbulence modelling, and it can make a valuable contribution towards the actual design of the experiment. Numerical
experiments using CFD provide a further way of advancing our
understanding, particularly with respect to turbulent flows, and
work of this nature is likely to expand as the power of CFD
methods increases.

Chapman (1979) puts forward three compelling reasons for us
vigorously to develop CFD. From the aircraft design point of
view, I would offer you four. First, CFD reduces the lead times
for aerodynamic design and development by a very large factor.
Compared with the sequence of design, manufacture of a wind
tunnel model, tunnel testing and re-design, CFD provides virtually an immediate result for a new design and enables a much
larger number of variants on the design to be covered in the
search for an optimum. Secondly, CFD has the potential to deal
with 'real world' phenomena that cannot be reproduced adequately
in model tests. For the aircraft designer these include: full
scale Reynolds numbers; interaction between airframe and power
plant; unsteady aerodynamics in rapid manoeuvres; aeroelastic

distortions; freedom from tunnel interference. Other examples
arise in the fields of turbomachinery and spacecraft. The third
advantage of CFD is that it offers good value for money and,
moreover, is becoming markedly more economical as time goes by
(the example from Chapman cited earlier illustrates this graph-
ically). In contrast, the cost of wind tunnel testing remains
relatively steady. The final point in favour of CFD is its
relatively low energy consumption, an important consideration
in view of the possibility that energy shortages might in the
future lead to restrictions on the operation of the large wind
tunnels currently used in Europe and the USA for aircraft project development.

In this introduction all I aim to do is to set the scene for
what is to follow, leaving later contributors to go into detail
in their individual subjects. I shall cover first of all the
range of physical phenomena with which aerodynamic design methods
must deal, and then go on to outline the characteristics, limitations and main applications of the numerical methods now in use.
I shall end with some remarks on the likely rate of advance of
computer capability over the next decade and the implications
therein for the development of CFD.

2. AIRCRAFT AND THEIR FLOW FIELDS

In order to exploit CFD to the full in aeronautics, immense computing power is needed. The need arises partly from the complexity of the physical phenomena to be modelled, partly from
the complicated geometry of aircraft.

The latter is illustrated in the frontispiece, which shows a
Tornado loaded with stores. Before we can begin to compute the
flow about such a shape, we have to define the shape fully in
space. Thereafter a grid has to be defined suitable for the
numerical calculation of the flow. This must be fine enough
both for the geometry of the aircraft to be represented accurately and for the computed flow to be defined precisely in regions
of rapid variation, such as around leading edges and in the
vicinity of shock waves. For a shape as complicated as the
loaded Tornado, a three-dimensional grid for calculating the
flow field at transonic speeds using present algorithms would
need of the order of 10^6 grid points to achieve reasonable
accuracy.

Even though the geometry of the aircraft may not always be
so complicated, the structure of the flow field around the aircraft is highly complex and contains a number of different
features which it is essential to represent accurately in any
numerical computation. Figure 1 shows the main aspects of the
flow about the wing of a transport aircraft at its design point,

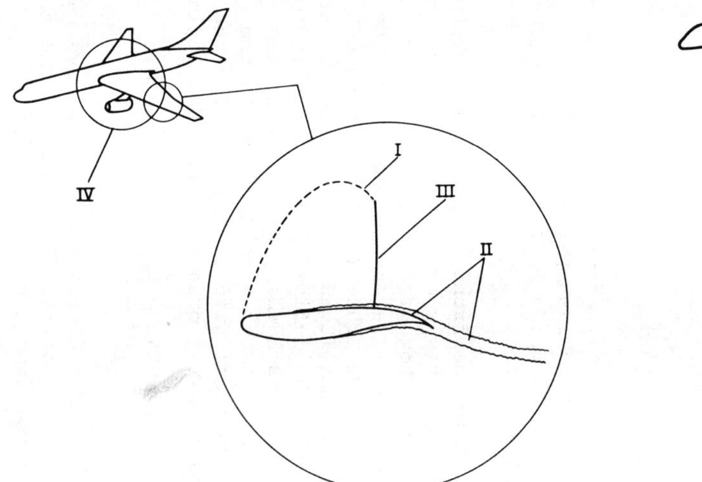

I   Transonic flow

II   Important viscous effects

III   Flow through shock non-isentropic

IV   Strong 3-dimensionality

Fig. 1   Primary features of transonic fluid dynamics 'on design'

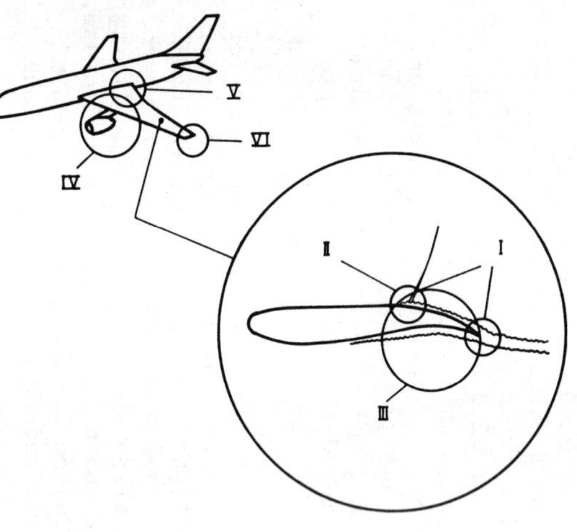

I   Localised strong viscous interactions

II   Shock 'softening' by boundary layer

III   Strong gradients invalidate 1st order BL theory

IV   Interaction with engine flow

V   Viscous effects in wing-body junction

VI   Wing-tip vortices

Fig. 2   Secondary features of transonic fluid dynamics 'on design'

cruising at a high subsonic speed. First, and most important,
the flow about the wing is transonic. The differential equations describing the flow are of mixed type, being elliptic in
the outer subsonic flow but hyperbolic in the localised region
of supersonic flow on the upper (and possibly also the lower)
surface of the wing. A second important feature of the flow is
that viscous effects — the development of the boundary layers
over the wing and the wake downstream — have an important influence on the overall flow. Third, the shock waves on the wing
are not usually of negligible strength and the flow outside the
boundary layer region is therefore not isentropic. Fourth, the
flow has a highly three-dimensional structure: the wing is swept,
tapered, usually supports a double shock system, and is strongly
influenced by the disturbance field due to the fuselage, engines,
etc.

Figure 2 illustrates further complexities of this 'on design'
flow field. In the immediate vicinity of the trailing edge
there is a strong, localised interaction between the boundary
layer and the external flow, and immediately downstream of the
trailing edge the wake is often highly curved: both effects can
have a significant influence on the lift carried by the wing, and
hence on the overall flow field. The interaction between the
shock wave and the boundary layer is another strong, localised
phenomenon with significant overall effects. The effect of the
boundary layer is to 'soften' the shock so that the pressure rise
at the wall is smoothed out rather than being effectively discontinuous: as a result, the entropy rise through the inner part
of the shock can be appreciably less than that through a fully
developed shock giving the same overall pressure rise. Strong
pressure gradients on certain parts of the wing are a further
difficulty, recent evidence indicating that their influence on
the boundary layer cannot be represented adequately by the first-
order boundary layer methods currently in general use. Other
features of this 'on design' flow which are significant are the
engine inlet and exhaust flows, the fuselage boundary layer and
its interaction with the pressure field of the wing to form wing
root vortices, and the complex sweeping of the boundary layer at
the wing tip into the tip vortices.

Figure 3 illustrates some additional, more complex flow
phenomena which are of importance to the designer. At off-design
conditions at transonic speeds, the wing shock wave may be strong
enough to cause separation of the boundary layer, or the angle
of attack may be sufficiently high to cause flow separation to
move forward from the trailing edge: in either case appreciable
flow unsteadiness (buffet) may result. Figure 4 shows the multi-
element configuration of the wing used to generate high lift at
low speeds, which presents us with further complications. The
boundary layers on downstream elements become submerged in the

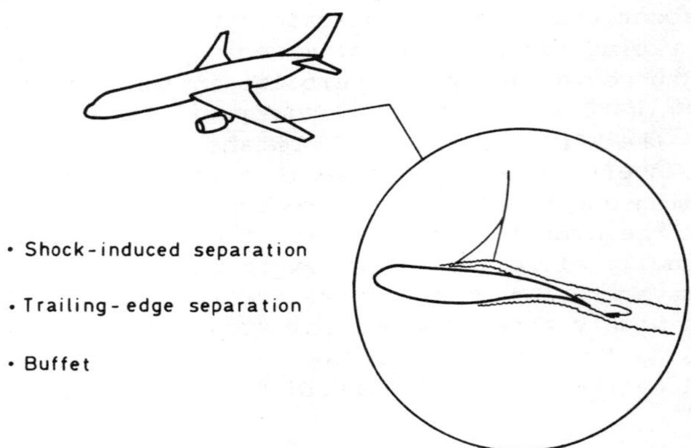

Fig. 3   Transonic fluid dynamics 'off design'

Fig. 4   Fluid dynamics of high lift wings

wakes from upstream elements, there are usually several localised regions of flow separation, and quite often the flow over
the slat (the most upstream element) becomes transonic.

Figure 5 illustrates some important features of flows about
manoeuvring combat aircraft. At high angles of attack, the
vortices formed by flow separation from a long, pointed nose
often lock into an asymmetric pattern which leads to large aerodynamic side forces on the nose. Wing leading-edge strakes and
upstream control surfaces (canards) also generate vortices and
these usually have a significant influence on the flow over the
wing, strake vortices being used specifically to extend the
range of angle of attack over which the aircraft can be flown
controllably when there is extensive flow breakdown over the
outer part of the wing. The forward influence of the engine
exhaust plume on a combat aircraft is another feature which
needs to be represented, particularly where the exhaust is close
to either wing or tail surfaces and can influence the lift,
stability and/or control of the aircraft.

Figures 1–5 have illustrated some of the features of steady flow around fixed wing aircraft. Mention should also be made of the problems of unsteady aerodynamics which arise, coupled with structural dynamics, in the design of helicopter rotors and in the study of flutter and buffet of fixed wing aircraft. Internal flows also provide challenging opportunities for CFD, examples being engine air intakes, diffuser ducts, and turbo-machinery. In general such flows are non-homentropic, highly three-dimensional and unsteady; a compensating feature is that their confined character eases the problem of keeping the number of grid points within manageable bounds.

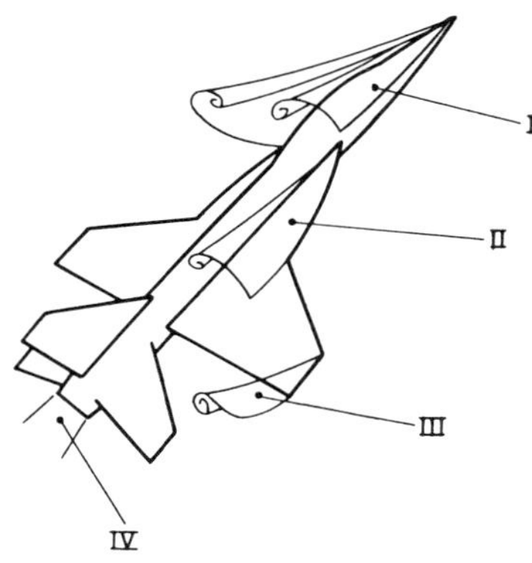

I   Nose vortices, possibly asymmetric

II   Strake or canard vortices

III   Separated flow over wings

IV   Interaction with propulsive jet

    -at all angles of attack

Fig. 5 Fluid dynamics at high angle of attack

3. METHODS

3.1 *Hierarchy*

Figure 6 lays out the hierarchy of calculation methods currently in use in aeronautical fluid dynamics. At the head of the tree stand the full Navier–Stokes equations, which satisfy con-servation of mass, momentum and energy in a viscous, heat-conducting, time-dependent flow. Although it is generally accepted that they provide a complete description of the flows

which concern us, their solution for turbulent flows of practical
interest would seem beyond our reach for several decades. There
are, however, methods for turbulent flows shown in the right-
hand branch of Fig. 6 which are based on some form of averaging
of the full Navier—Stokes equations: these are already providing
results of considerable interest and potentially are of great
importance for the future. The simplest such form of method,
and the one most likely to find practical application in aerodynamic design, uses the Reynolds-averaged form of the equations,
in which all turbulence quantities are averaged with respect to
time and empirical closure formulae are then used to determine

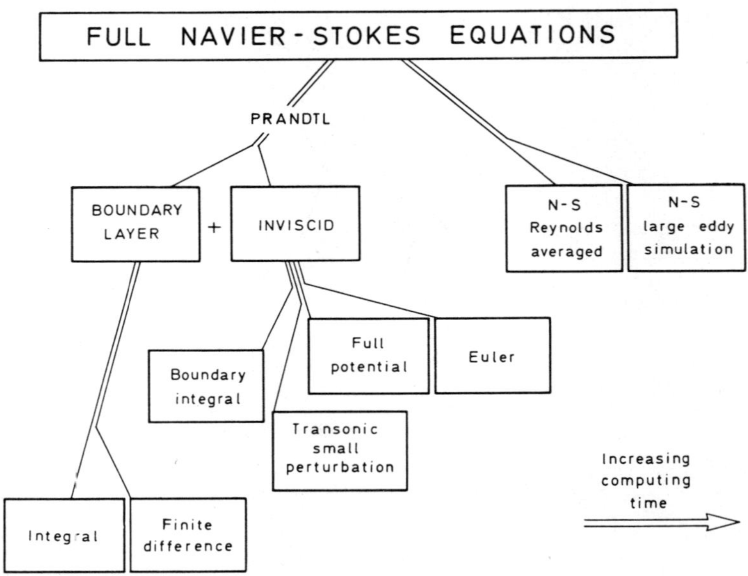

Fig. 6 Hierarchy of computational fluid dynamics

the various Reynolds stresses. A more advanced form of modelling
is the so-called Large Eddy Simulation, in which the full, time-
dependent form of the Navier-Stokes equations is used to calculate the development with time of the large scale turbulent
eddies, which are anisotropic and highly structured, while the
small, sub-grid scale turbulence is modelled by a process
analogous to Reynolds averaging.

The other branch shown in Fig. 6, and the one which is at
present more important to aerodynamic design, adopts the simplifying assumption proposed by Prandtl (1904) that significant
viscous effects are confined to a thin layer, adjacent to and
trailing downstream of a body, and that the entire outer flow
region is effectively inviscid. The sets of equations

describing the two regions are quite different in character and
quite different methods of solving them are required. Indeed,
until fairly recently, the mutual inter-dependence of the two
solutions was not correctly represented in aerodynamic design
methods: whereas the solution for the inviscid flow was used to
provide the boundary conditions for a calculation of the boundary layer, the influence of the boundary layer on the outer flow
was ignored. Over the past decade, however, the need for correct
representation of the coupling between boundary layer and outer
flow has been recognised increasingly and considerable effort is
now going into the problem of viscous-inviscid matching.

Boundary layers on aircraft are, in the main, turbulent and
their prediction requires the same kind of empiricism in modelling the turbulent stresses that is involved in solving the
Reynolds-averaged Navier-Stokes equations. The boundary layer
equations are usually solved by a downstream marching process,
the flow at the outer edge of the boundary layer being defined
by a preceding solution for the outer, inviscid flow. There
are two categories of boundary layer prediction method, the
more general one involving the approximation of the set of differential equations describing the boundary layer by a set of
difference equations which are solved numerically. A widely
used simplification, requiring some additional empiricism, is
to integrate the boundary layer equations with respect to distance normal to the surface. For two-dimensional problems, this
reduces a set of partial differential equations in two variables
to a set of ordinary differential equations, while for three-
dimensional problems a set of partial differential equations in
two variables is obtained.

The methods for solving inviscid flow problems shown in Fig.
6 cover a range of increasing complexity and generality. Boundary integral methods are limited in application to wholly subsonic or wholly supersonic flows. Transonic small perturbation
methods are strictly applicable only to flows in the vicinity
of Mach 1, and involve the application of linearised boundary
conditions. The full potential equation is applicable across
the speed range, but is limited to irrotational flows. It
therefore cannot describe accurately flows with strong shock
waves across which the entropy rise is significant. This limitation is overcome in the Euler equations, which are applicable
generally to all flows in which the effects of viscosity and
thermal conductivity can be neglected. The Euler equations
correctly describe the flow changes across a shock wave, and provide a means of calculating the development of the inviscid but
rotational flow downstream of the shock wave.

In Fig. 6, each successive step to the right involves an increase in the scale of the computation by one or more orders of

magnitude. There are similar consequences in transferring our
interest from two spatial dimensions to three or from steady
to time-varying flows. In the following sections I shall outline
the characteristics of each type of method in ascending order of
complexity, starting with boundary integral methods for inviscid
flows: for simplicity, I shall confine myself to methods for predicting steady flows.

3.2 *Linear Equations — Boundary Integral (Panel) Methods*

The early aerodynamic design methods were derived from classical,
19th century hydrodynamics. They treated the flow as inviscid,
irrotational, incompressible and therefore having a velocity
potential ϕ satisfying Laplace's equation

$$\phi_{xx} + \phi_{yy} + \phi_{zz} = 0. \tag{1}$$

This equation, derived from continuity and irrotationality, is
exact for incompressible flow. The more recent Prandtl—Glauert
equation,

$$(1-M_\infty^2)\,\phi_{xx} + \phi_{yy} + \phi_{zz} = 0, \tag{2}$$

is an analogous linear equation for compressible flow. It is
derived by omitting all non-linear terms from the full potential
equation for a compressible flow, the free stream direction of
which is parallel to the x axis. The linearisation is justified
only for flows which are either subsonic or supersonic in their
entirety and is unreliable for flows in which the Mach number
departs appreciably from its free stream value. For subsonic
flows, solutions of the Prandtl—Glauert equation can be transformed to solutions of Laplace's equation simply by scaling the
x coordinate by $\sqrt{1-M_\infty^2}$, though it should be noted that the scaling
of the exact boundary condition under the transformation is not
straightforward.

Because Laplace's equation is linear, solutions of it can be
added together to provide new solutions: hence flowfields about
complicated shapes can be derived by adding together flowfields
of simpler origin. The building bricks, introduced by Rankine
and still in use, are the flow fields due to line and point singularities — sources, sinks, doublets, vortices. Determining
the flow field about a body then becomes the problem of determining the spatial distribution of singularities which, superimposed on a given free stream, satisfies the boundary condition
of zero normal velocity at all points on the body surface. The
singularity distribution is related to the boundary condition
by an integral equation.

For a given body there are in general an infinite number of
possible solutions to this equation and uniqueness criteria are
needed to isolate those which give the correct, physically
realistic flowfield external to the body. For aerofoils and
wings with subsonic flow over the trailing edge this is achieved
by imposing the Kutta condition, which relies on the argument
that the flow must leave a sharp trailing edge smoothly on both
sides since any alternative would give rise to a physically unreal infinity in the velocity field. Although the Kutta condition is a key element of wing theory, the same physical
reasoning cannot be applied to bodies without sharp trailing
edges and uniqueness criteria for the flowfield around such
bodies are necessarily more empirical and arbitrary.

Uniqueness of the flow external to the body is not in itself
sufficient to yield a unique distribution of singularities.
There are an infinite number of such distributions which produce
the same solution in the domain external to the body but different solutions in the 'non physical' domain inside the body.
Additional constraints have to be imposed to eliminate all but
one of these solutions, but the constraints can be tailored to
suit the particular method being used to solve the integral
equation.

Before the advent of large-scale computing, solutions of the
Laplace equation in three dimensions were practicable only for
fairly simple singularity distributions and linearised boundary
conditions which enabled only approximate representation of an
aircraft configuration. Despite their limitations, these methods
were put to good effect by the aircraft designer and are still
widely used today.

The power of modern computers has enabled the development of
so-called panel methods which overcome these restrictions. The
aim of panel methods is to achieve a numerical solution of the
integral equation which satisfies the boundary condition accurately on the body surface. The three-dimensional geometry
of the aircraft is defined precisely and its surface is subdivided into a large number of panels (1000 or more may be
needed for complex configurations). Each panel is endowed with
a source and doublet singularity of unknown strength and a
control point at which the boundary condition of zero normal
velocity has to be satisfied. The integral equation relating
the singularity distribution to the boundary conditions is thus
reduced to a finite set of linear algebraic equations which can
be solved by a standard procedure. The solution yields the
velocity potential throughout the flow field and thence velocities,
pressures, streamlines, forces and moments.

The most highly developed of modern panel methods provide an
effectively exact solution for the incompressible, inviscid flow

about an aircraft configuration and related methods afford useful estimates of compressible, subsonic or supersonic flows.
Applications in which they have proved particularly effective
include: the detailed study of local flow behaviour in geometrically complicated regions (it is possible to 'panel' the
region of interest in great detail while using a simpler and
more economical panel distribution for remote parts of the aircraft); determining the aerodynamics of complete aircraft configurations (such as shown, for example, in the frontispiece);
mapping streamlines, either as an aid to design (for example,
the integration of engine nacelles and pylons into the flow
field), or as a starting point for further calculations (for
example, store trajectories, boundary layers).

The main limitations of panel methods include their inapplicability to mixed flows and diminishing accuracy as M_∞
approaches unity. Also, they are less suited to supersonic
than subsonic flows, the discontinuities that arise in supersonic flow being difficult to model by panel discretisation.
The principal advantages of panel methods are their economy,
their ability to handle complicated shapes and, at the expense
of some investment in programme development, their adaptability
to a wide range of problems.

3.3 *Non-linear equations — transonic small perturbation (TSP)*
  methods

Although the panel methods developed in the late 1960s demonstrated the power of the computer in solving aerodynamic problems, the transonic flow regime which was of crucial importance
to the aircraft designers remained beyond their reach. This
situation was transformed by Magnus and Yoshihara (1970) and
Murman and Cole (1971) who, for the first time, obtained computer solutions for two-dimensional, inviscid, transonic flows
by use of finite difference methods. The method of Magnus and
Yoshihara used a time-marching procedure to obtain an asymptotic solution at large time as an approximation to the steady-
state solution, that of Murman and Cole used a relaxation procedure to solve the steady form of the flow equations. These
two papers really mark the beginning of computational fluid
dynamics as we in the aeronautical community understand the term.

The method of Murman and Cole, which was an order of magnitude
faster than that of Magnus and Yoshihara, solved the transonic
small perturbation (TSP) equation. In three-dimensional flow,
the classical form of this equation is written

$$[1-M_\infty^2-(\gamma+1)M_\infty^2\phi_x]\phi_{xx} + \phi_{yy} + \phi_{zz} = 0 \qquad (3)$$

where ϕ is the perturbation potential. The crucial difference

between this and the Prandtl Glauert equation is the occurrence
of the non-linear term in $\phi_x\phi_{xx}$, the presence of which allows
the equation to be either elliptic or hyperbolic at different
places in the same flow field. It is this term which enables
the equation to describe transonic flows and which admits jumps,
corresponding to shock waves, in the solution.

Early methods of solving this equation defined a grid in a
cartesian co-ordinate system, approximated the differential
equation by a set of difference equations written at each node
of the computing grid and solved this set of equations by successive line relaxation, marching downstream in the grid and
repeating the cycle until convergence was obtained. Simple coordinate transformations were used to concentrate grid points in
regions of steep gradient in the flow variables and linearised
boundary conditions were applied on simple surfaces, usually
planes of constant y or z.

The key to the success of the method, and to many which have
since followed, was the introduction by Murman and Cole of a
mixed differencing scheme. In regions of subsonic flow the
derivatives in all three directions were approximated by central
differencing: in regions of supersonic flow, upwind or backward
differencing was used for derivatives with respect to x and
central differencing for transverse derivatives. The current
value of ϕ_x was used to test whether the local flow was subsonic
or supersonic and hence what form of difference equation should
be used at that point in space and at that stage of the calculation.

Over the past decade there have been several important improvements in the TSP method, particularly in its applicability to
swept winged aircraft. The classical TSP equation has been
augmented by the inclusion of additional non-linear terms appropriate to transonic flow over an infinite sheared wing

$$[1-M_\infty^2 - (\gamma+1)M_\infty^2\phi_x]\phi_{xx} + \phi_{yy} + \phi_{zz}$$

$$- (\gamma-1)M_\infty^2\phi_x\phi_{yy} - 2M_\infty^2\phi_y\phi_{xy} = 0. \tag{4}$$

The effect of these terms, the last two on the left-hand side,
is to reduce the smearing out of shock-waves on swept wings,
something which afflicted solutions of the classical equation to
an unacceptable degree, and to achieve greater consistency in
approximating the non-linear terms for swept wings.

Further developments have been: empirical 'tuning' of the coefficients of the non-linear terms in the equation to suit particular types of problem; introduction of a non-orthogonal
co-ordinate system to align the grid with the wing planform;

introduction of a 'rotated' difference scheme (Albone (1974)) to
ensure stability of the computation in supersonic regions when
non-orthogonal co-ordinates are used; use of approximate factorisation (Ballhaus et al. (1977)) in place of successive line
relaxation to reduce computing time.

Transonic small perturbation methods suffer from the nature
of the approximations made in deriving the equation. They are
at their best when these approximations are least in error, i.e.
when the free-stream Mach number is close to unity, the wing is
thin, not too highly swept or of too low aspect ratio, and the
lift coefficient is fairly low (though they have proved surprisingly useful for flows in which M_∞ is not close to unity
and perturbations are not small). When the Mach number ahead
of a shock wave appreciably exceeds unity the jump in flow conditions across the shock wave is predicted incorrectly; also,
the linearised boundary conditions limit accuracy in the region
of leading edges and other forward facing surfaces. However,
the simplicity of the boundary conditions and co-ordinate system
has enabled versions of the method to be developed for application to relatively complex configurations; Boppe and Aidala
(1980) report several examples of such applications including
predictions for the launch configuration of Space Shuttle.
Because of their applicability to practical configurations, TSP
methods have made a vital contribution to a number of aerodynamic design studies in recent years and, despite their limitations in accuracy, are likely to remain in wide use for many
years to come.

3.4 Non-linear equations — full potential (FP) methods

The pioneering work of Murman and Cole with the transonic small
perturbation equation was soon followed by numerical solutions
by Garabedian and Korn (1971) of the full potential (FP) equation. In three dimensions this may be written

$$(\rho\phi_x)_x + (\rho\phi_y)_y + (\rho\phi_z)_z = 0 \qquad (5)$$

where ϕ is the velocity potential and

$$\rho = \rho_\infty[1 + \frac{\gamma-1}{2} M_\infty^2 (1 - \phi_x^2 - \phi_y^2 - \phi_z^2)]^{\frac{1}{\gamma-1}} .$$

This equation is exact for the irrotational, isentropic (i.e.
shock free) flow of a perfect gas, is applicable to transonic
flows and admits discontinuous (though inexact) solutions across
shock waves.

To solve the equation in two dimensions, Garabedian and Korn followed Sells (1967) in using a conformal transformation to map the domain exterior to the aerofoil onto the interior of a circle: the resulting differential equation was then solved by a finite-difference procedure which, like that of Murman and Cole, used central or backward differences, depending on whether the flow was subsonic or supersonic, and achieved a converged solution by successive line over-relaxation. The employment of a conformal transformation ensured that the boundary conditions on the aerofoil surface and at infinity were satisfied accurately.

Evolution of the full potential method has included its extension to three-dimensional flows, again making use of a mapping of the external flow onto the interior of a confined domain. Another development has been the emergence of alternative finite difference algorithms to improve agreement between predicted changes across shock waves and the Rankine-Hugoniot relations. The original method of Garabedian and Korn was intended for application to flow through weak shock waves only, and used a non-conservative form of the potential equation. Jameson (1975) subsequently proposed a method for solving the equation in conservation form — i.e. in a form (equation 5) in which conservation of mass (but not momentum) is satisfied across a discontinuity — and thereby improving the accuracy of predicting shock changes. More recently Lock (1980) has reported a 'partially conservative' scheme which gives shock jumps intermediate between the conservative and non-conservative methods. This scheme relies on a difference equation at the shock point which is a weighted combination of the central and backward difference formulae, the weighting factor being derived empirically. Although Lock's evidence favours the use of this intermediate formulation, the issue is still contentious.

A development to reduce computing times is the so-called multi-level adaptive technique, or multi-grid, in which the computation sweeps successively over a sequence of uniform grids in geometrically decreasing mesh size. The process of coarse grid to fine grid interpolation of corrections, followed by fine grid to coarse grid transfer of residuals, is particularly effective in the rapid elimination of numerical error and gives a rapid method of solving algebraic systems of equations. In the review by Brandt (1980) this and several other advantages of the technique are discussed. Solution of the FP equation by the approximate factorisation technique in place of successive line over-relaxation has also reduced computer times substantially.

There are good reasons to believe that numerical methods based on the full potential equation will provide the corner stone of aerodynamic design for the next ten years or so. The main problems to be overcome are the accurate prediction of shock jumps

and the satisfaction of boundary conditions on complex configurations: for the former we need empirically to optimise our
jump algorithms by calibration against more accurate methods,
while for the latter we need to develop flexible and easily
adapted co-ordinate systems which retain the precision of the
present mapping schemes as regards satisfying the boundary
conditions.

3.5 *Inviscid rotational flow — the Euler equations*

The exact equations for an inviscid, non-homentropic flow are
the Euler equations. These satisfy conservation of mass, energy
and momentum, and apply to an inviscid, thermally non-conducting
fluid for which the equation of state is known. To define the
flow at a point, three components of velocity and two thermodynamic properties of the fluid are required, thus we have to
determine five independent variables at every point in the flow
field, as compared with only one in the TSP and FP methods, and
we have five non-linear partial differential equations to solve
instead of one. It will be appreciated that numerically accurate
solutions of the Euler equations require substantially more
computer time and storage than corresponding FP or TSP calculations.

Finite-difference and finite-volume methods have both been
used to solve the Euler equations, solutions for steady flows
usually being obtained as the asymptotic solution at large time
of the unsteady form of the equations. Explicit time-marching
procedures, which provide a straightforward way of solving the
equations, are constrained by the Courant—Friedrichs—Levy
stability criterion to time steps related to the mesh spacing.
Faster, implicit time-marching procedures have been developed,
as also have hybrid schemes.

The high cost of solving these equations is compensated by
the potential accuracy of solutions. If numerical error is
avoided, the Rankine—Hugoniot relations are satisfied across
shock waves and the behaviour of the non-homentropic flow downstream of the shock wave is correctly reproduced. Solutions of
the Euler equations can thus be used to test, and even to calibrate, simpler but less exact methods. Figure 7 shows a comparison, for transonic flow over an aerofoil, between a solution
of the Euler equations and three alternative solutions of the
full potential equation with varying degrees of conservatism.
The Euler calculation was done by the explicit method of Sells
(1980), the potential calculations by Lock: the results support
Lock's proposition that a method intermediate between conservative (Q-C) and non-conservative (N-C) is to be preferred.

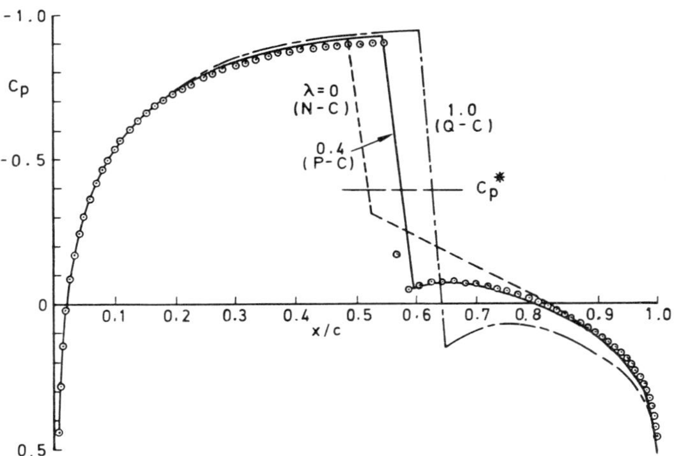

Fig. 7   Comparison of solutions of Euler and potential flow
equations for NACA 0012 aerofoil; $M_\infty = 0.816$, $\alpha = 0^\circ$

3.6 Grids

One avenue which holds out the promise of substantial advance in
the future is the development of new grid systems for flowfield
computation. The prime aims in selecting a computing grid are:

    precise application of boundary conditions;
    accurate and detailed information in regions of rapid variation;
    minimum computation consistent with precision in regions of
        slow variation;
    adaptability to complex geometry;
    high speed of computation;
    low memory requirement, relatively.

This is a diverse field and one which offers scope for a wide
variety of innovative approaches. Co-ordinate transformations
of rather different kinds have played an important part in the
development of the TSP and FP methods and new advances may be
expected, particularly in the application of the FP method to
more complex configurations. The technique of grid imbedding
has been shown by Boppe and Aidala (1980) to be a powerful
means of treating complicated geometries, and there will no
doubt be further extensions of the technique. The development
of a non-aligned mesh for solving the full potential equation
offers greater ease of application to complicated shapes, but
at the expense of greater difficulty in satisfying boundary
conditions accurately. Grid generation by solving a set of
partial differential equations, as Thompson et al (1975) have

done, holds out the prospect of a general method, applicable to
bodies of arbitrary shape, for producing a computing mesh which
is aligned with the body surface so as to minimise the error in
applying boundary conditions. Work on grid systems which adapt
to the changing needs of the calculation as it proceeds should
also be mentioned. In transient, time-dependent calculations,
for example, the use of a grid which evolves so as to maintain
a concentration of grid points in regions of steep gradients
as the flow develops offers both gains in accuracy and savings
in computing time.

3.7 *Viscous-inviscid matching*

For the purposes of aerodynamic design, being able to compute
inviscid flow fields is only half the battle. In virtually every
aspect of practical aerodynamics viscous effects are significant
and in many — above all in modern wing design — they are crucially important. Prandtl's hypothesis, which takes us down the
left-hand branch of Fig. 6, argues that viscosity has a direct
influence only in the narrow, inner regions of boundary layer
and wake, that the outer flow region is effectively inviscid and
that inner and outer regions can be treated separately. It is
now generally appreciated that, whilst this is true, the mutual
influence of the two regions is important: the main factor in
determining the development of the boundary layer and wake is
the pressure field imposed by the external flow, but the growth
of the boundary layer displaces this flow away from the surface
of the body, thereby changing the pressure field. This interaction makes it necessary to couple the separate treatments of
the two regions in such a way that their solutions match at
some notional interface between the regions.

The matching process will be discussed in greater detail in
the later paper by Lock and Firmin (1982) and I shall only outline the principles here. These are sketched in Fig. 8. In the
terminology of Lock and Firmin, the Real Viscous Flow is predicted by a calculation which iterates between solutions for an
Equivalent Inviscid Flow and for the viscous layer until these
two solutions match at all points along the outer edge S of the
viscous layer. When the iteration has converged, the Equivalent
Inviscid Flow will be identical to the Real Viscous Flow everywhere outside S provided the viscous layer prediction is correct.

As the computation proceeds, the boundary conditions for the
viscous layer calculation — pressure and (in three-dimensional
problems) flow direction — are determined from the preceding
solution for the Equivalent Inviscid Flow. The viscous layer
solution in turn provides boundary conditions for the next
calculation of the Equivalent Inviscid Flow. These may be
applied by specifying either a displacement surface, on which

the normal component of velocity is everywhere zero, or a distribution of the velocity component normal to the surface of
the wing itself: this transpiration distribution must be continued downstream along the dividing stream surface in the wake.
Whichever way the displacement effect is represented, a distribution of vorticity is also needed downstream of the trailing
edge in the Equivalent Inviscid Flow in order to represent correctly the pressure difference between upper and lower edges of
the wake in the Real Viscous Flow. From the viscous calculation
in the converged solution the distributions of pressure and
shear stress over the surface of the wing may be determined and
integrated to give overall aerodynamic forces.

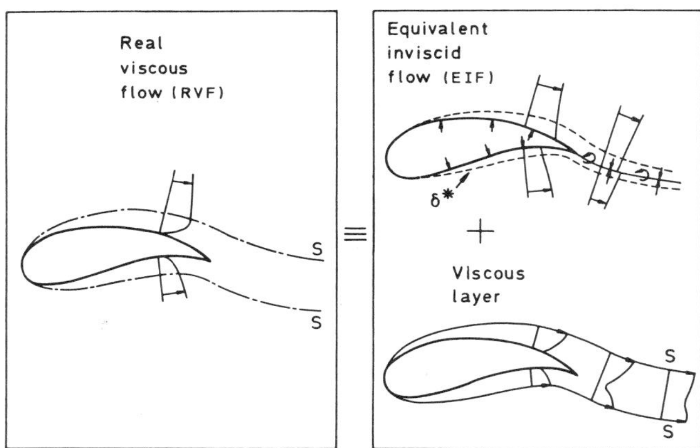

- Boundary conditions for EIF - displacement
  surface or transpiration velocity - given
  by viscous layer calculation

- Boundary conditions for viscous layer from EIF

- Solutions matched on S by iteration give RVF,
  hence surface pressure and shear stress, lift and drag

- First-order boundary-layer theory inadequate
  - higher-order model can include all main N-S terms

Fig. 8 Viscous-inviscid matching

Recent work has uncovered some important flows in which first-
order boundary layer theory is no longer sufficiently accurate,
either for predicting the growth of the viscous layer or for
determining, from the viscous layer calculation, the boundary
conditions for the Equivalent Inviscid Flow. Lock and Firmin
will discuss the inclusion of higher-order terms in the modelling of the viscous layer. What they describe is a self-consistent
framework within which, it is argued, all the significant terms
in the Reynolds-averaged Navier—Stokes equations appear. The

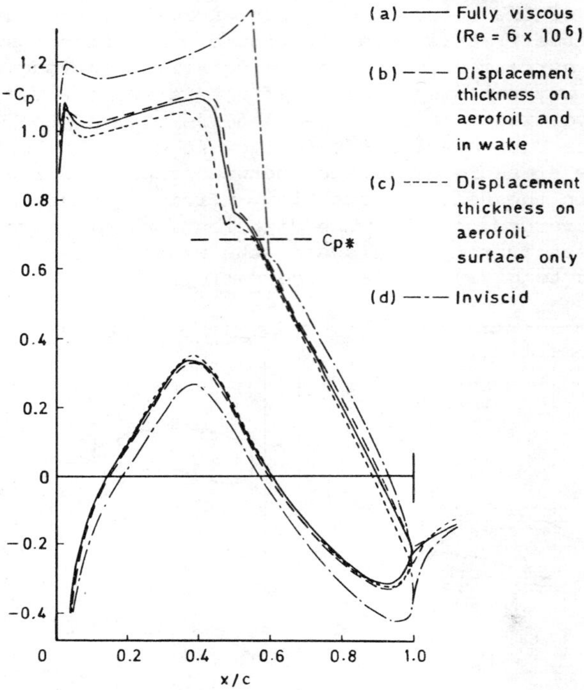

Fig. 9   Comparisons between viscous and inviscid solutions for
RAE 2822 aerofoil at $M_\infty = 0.725$, $\alpha = 2.3^\circ$ (Collyer, 1977)

TABLE 1

Calculated viscous effects on RAE 2822 aerofoil, $M = 0.725$, $\alpha = 2.3^\circ$

	Inviscid (a)	δ^* on aerofoil surface only (b)	δ^* on aerofoil and wake (c)	Full curvature effects included (d)
C_p at t.e.	0.364	0.261	0.211	0.219
C_L	0.922	0.663	0.721	0.699
Change in C_L from previous solution	—	-28%	+9%	-3%

higher order terms introduce upstream influence into the problem
— the parabolic boundary layer equations giving place to the
elliptic Navier—Stokes equations — but, since this influence is
relatively weak, it can be accommodated within the present
iterative procedures for viscous-inviscid matching. This discussion relates to unseparated flows. With some qualification
it applies also to flows with limited regions of separation, this
aspect being discussed further by Lock and Firmin.

The importance of viscous effects on the modern aerofoil, and
the significance of higher order terms, is illustrated in Fig. 9,
showing the results of calculations by Collyer (1977). For a
given Mach number and angle of attack, the pressure distribution
around the aerofoil is plotted as given by calculations (a)
neglecting all viscous effects, (b) including only boundary layer
displacement effects, (c) including boundary layer and wake displacement effects and (d) including second-order terms associated
with streamline curvature (but not all the higher-order terms
referred to above). The calculated aerofoil lift for each of
these cases is given in Table 1.

3.8 Aeroelastic distortion in steady flow

The deformation of wings and control surfaces under aerodynamic
loading is an effect which aircraft designers have faced since
aviation began. It is particularly important for thin-winged
combat aircraft which, during a manoeuvre, experience lift forces
many times the weight of the aircraft.

The effect is in some respects analogous to that of boundary
layer development — the external flow gives rise to forces which
alter the shape of the wing and hence the boundary condition
for the external flow. It can therefore be incorporated into
an aerodynamic calculation in a similar fashion to viscous
effects, by a process of iteration between a solution of the flowfield — giving a distribution of pressure over the wing — and a
calculation of a new shape for the wing from a knowledge of its
unloaded shape, its stiffness and the imposed aerodynamic and
inertial loading (in a 5g turn, the deformation of the wing is
significantly affected by its own weight and that of its stores).
Because of their similarities, calculations of viscous and aeroelastic effects in steady flows may conveniently be combined,
so that in each iterative cycle the boundary conditions for the
Equivalent Inviscid Flow are updated for both effects. When the
inviscid flow is itself calculated by a procedure requiring a
large number of iterations, viscous and aeroelastic calculations
may be interleaved into the procedure so that the overall computation moves progressively to a solution which is converged in
all three aspects.

3.9 *Navier–Stokes equations*

The right-hand branch of Fig. 6, discarding separate treatment
of the viscous and inviscid regions of the flow, offers the
most exciting prospects for the future. However, major advances
in computing power will be needed before Navier–Stokes computations find widespread application in aerodynamic design.
Their primary role, over the next decade at least, is likely
to be in research — particularly into the structure of turbulence
and separated flows — and perhaps in improving the flow modelling
in simpler computational methods. It is worth remarking that
the rate at which the right-hand branch of Fig. 6 catches up with
the left-hand branch depends not only on the rate of progress of
Navier–Stokes methods but also on the rate at which the 'Prandtl'
methods of the left-hand branch advance.

Solution of the Navier–Stokes equations offers the prospect of
predicting flows which, by virtue of their complex structure, are
beyond the scope of simpler treatments. The generality of the
equations is an argument in their favour — i.e. it may be more
economical to develop a reliable procedure for solving the Navier–
Stokes equations than to put time and effort into developing
separate and distinct methods for a range of individual problems.
On the other hand, their generality may tempt us into mental
laziness — for some types of problem they may give us results
without necessarily giving us understanding or insight — and
this is something we have to guard against.

Some recent achievements with numerical solutions of the
Navier–Stokes equations are cited in Chapman's review and I shall
not discuss this aspect in much detail. Figure 10 sketches three
of the most striking applications to date. The first is the
prediction of violent oscillations in transonic flow about an
aerofoil. Levy (1978) found that, for a thick, circular-arc
aerofoil, the Reynolds-averaged Navier–Stokes equations automatically computed an unsteady flow at certain Mach numbers and
a steady one at others. For Mach numbers below a certain value
the computed flow involved trailing edge separation and was
steady; for Mach numbers well above it was also relatively steady
but involved shock-induced separation. In between these two
Mach numbers the flow oscillated violently between trailing-edge
and shock-induced separation on alternate surfaces of the aerofoil. The prediction agreed remarkably with experiment in
respect of the frequency, intensity, Mach number of onset and
variation of pressure with time at fixed positions along the
chord. Chapman cites buffet at high lift and control-surface
flutter as further examples of unsteady transonic flows which
have been accurately predicted by two-dimensional, Reynolds-
averaged methods. The second illustration in Fig. 10 is of
steady, three-dimensional flow separation from a body of revolution at incidence, again predicted using the Reynolds-averaged

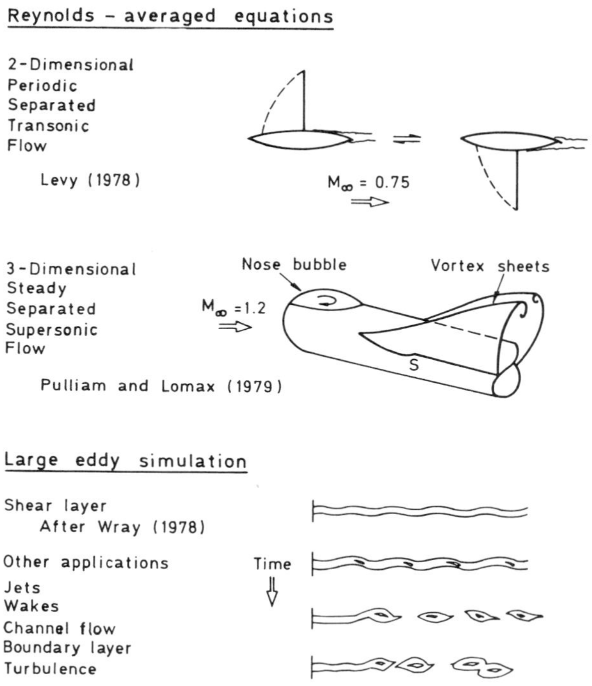

Fig. 10   Examples of types of problem for which Navier—Stokes solutions have been obtained

form of the equations. The computations by Pulliam and Lomax (1979) of this type of flow were in fairly good agreement with experiment, revealing three separate types of flow separation — primary vortex, secondary vortex and nose 'bubble'. The third sketch in Fig. 10 relates to the prediction of turbulent jets and free shear layers by Large Eddy Simulation. Chapman gives the results of unpublished work by Wray, showing the development of the eddy structure to depend strongly on the form of the initial disturbance. Kim and Moin (1979) report related computations of turbulent flow in a two-dimensional channel and draw attention to features in the calculated behaviour of the flow which they suggest provide further insight into the structure of turbulence.

Before leaving the Navier-Stokes equations, we should note the introduction of the 'thin layer' approximation for the Reynolds-averaged form of the equations, as discussed by Baldwin and Lomax (1978). This resolves the principal component of viscous stress

accurately but not the two orthogonal components: it reduces
computing time and programming effort and, for mesh sizes in
current use, gives results which differ little from those
obtained using the full Reynolds-averaged equations.

4.   LOOKING AHEAD

4.1 *Rates of advance*

In his review, Chapman presents data on the development of computer hardware and computational methods over the past thirty
years.   From a wealth of fascinating detail I have extracted
the following table to illustrate, in round numbers, progress in
four key areas.

TABLE 2

Year	1950	1965	1980
COMPUTER SPEED, flops	10^2	10^6	10^8
MAIN MEMORY, words	10^2	10^5	10^7
SCALE OF COMPUTATION (max number of grid points in CFD publications)	270	500 (aerodynamics) 10^4 (meteorology)	5×10^5
RELATIVE EFFICIENCY OF NUMERICAL METHODS		1	100–500

Chapman discusses the factors likely to influence future developments, including technological possibilities, the influence of
the market, requirements of and scope for advance in the CFD
field.   For each of the four areas listed above he conjectures
that advances will continue at a rate of approximately one order-
of-magnitude improvement every seven years — i.e. a twenty-five
fold improvement per decade.   In one sense the picture is encouraging: increases in memory and speed together, coupled with
improvements in algorithms, mean that we should be able to do
computations between one and two orders of magnitude larger than
at present, using the same or less computer time than at present.
In another sense our prospects are decidedly limited: a factor
of twenty five increase in the number of grid points allows us
only a threefold increase in each direction!   It may also be
remarked that, in certain applications, some of the more efficient of current methods may not provide scope for a twenty-five
fold improvement.

4.2 Prospects for 1991

The above extrapolations provide a basis for speculating on
what CFD might offer the aeronautical community in ten years'
time. Since speculation is a subjective business, I shall leave
the aero-engine and helicopter fraternity to do their own and
confine myself to aircraft aerodynamics. The subjects are listed
in ascending order of complexity rather than importance.

(a) *Panel methods plus viscous layer*

Panel methods for inviscid, attached flow have reached virtually
full maturity for steady, incompressible problems and no substantial improvements are foreseen. For compressible flows,
their linearity is a limitation which reduces the incentive to
develop them further, though there is likely to be at least some
refinement of methods for supersonic flow. Methods for unsteady
flow are also fairly well developed. On the other hand, panel
methods matched to viscous layer calculations have as yet realised
only a fraction of their potential. At present, their main such
application is to two-dimensional flow over multiple-aerofoil
high-lift systems (e.g. Butter and Williams (1980)). It was
mentioned in Section 2 that the flow over such systems is often
transonic, and flowfield prediction will require a method based
on the FP equation (or Euler or Navier-Stokes) which can model
the shock wave and its interaction with the boundary layer.
Nevertheless, for the three-dimensional viscous flow over aircraft in the high-lift configuration, with slats and flaps
deployed, 'viscous' panel methods will offer the attraction of
handling the complicated geometry of such configurations in
detail with greater economy than non-linear methods. In ten
years' time I would expect such methods to be in general use in
aircraft design offices as an important aid in working up the
design of high-lift systems in three dimensions. To fulfil this
role the methods will need to be applicable to flows with appreciable regions of separation, and the more general treatment of
three-dimensional separation is another avenue along which
viscous panel methods are likely to evolve. They offer the
prospect of practicable treatments of separated flows intermediate between the highly swept and the two-dimensional varieties
to be discussed respectively by J.H.B. Smith and (briefly) by
Lock and Firmin later in this volume.

(b) *TSP or FP equation plus viscous layer*

Over the next decade these two methods for transonic flows should
become applicable to a wide range of configurations and flow
conditions. Development of transonic small perturbation methods
to widen their applicability is likely to continue in parallel
with but less intensively than similar work on full potential
methods, until such time as the FP methods can be applied econo-

mically to as wide a range of problems as TSP methods. Thereafter, the emphasis is likely to be primarily on continued
development of the FP methods. By 1991 we may expect such
methods to be capable of treating:

> realistic aircraft geometry, in the same sort of detail now
>   achieved by panel methods;
> viscous, aeroelastic, time-dependent problems on practical
>   configurations, including moving controls;
> propulsive jets embedded within aircraft flowfields, including
>   modelling the entrainment of flow into the jet and any
>   shock structure within the jet;
> flows with regions of separation which are in the form either
>   of shallow bubbles or of highly swept vortices.

If these capabilities can be achieved economically, relative
to methods using Euler or Navier—Stokes equations, and if the
representation of shock waves and viscous effects is shown to
be sufficiently accurate 'viscous FP' is likely in ten years time
to be the most heavily used CFD method in aeronautics.

(c) *Euler equations plus viscous layer*

In aircraft aerodynamics, solutions of the Euler equations in
three dimensions should be available for simplified configurations such as a swept wing on a body. It is not clear how
strong the incentive will be over the next decade to extend
their application to more complicated configurations: their
requirement for substantially more memory than FP methods and
their greater computing time and cost are likely to inhibit
their widespread adoption in aircraft design. We may expect to
use them to provide definitive solutions for transonic and supersonic flows over simplified configurations, to serve as test
cases for simpler methods and as a basis for 'tuning' these
methods empirically to reproduce, say, the same pressure distribution as the Euler solution. Methods will be developed for
coupling the Euler solutions with calculations of the viscous
layers and of aeroelastic deformation. Hybrid methods, in which
an Euler solution for the inner, non-isentropic region is embedded within an outer solution of the full potential equation,
merit investigation.

For internal flows, such as air intakes and turbomachinery,
Euler methods should be well developed as the principal means
of computing inviscid flows. Viscous effects are important and
complex in internal flows, however, and limit the value of purely
inviscid calculations. Techniques are likely to be developed
for matching Euler solutions with viscous-layer calculations
for internal flows, but in ten years' time Navier—Stokes methods
are likely to have a strong foothold in this field and may well
have displaced 'viscous Euler' methods almost entirely.

(d) *Reynolds-averaged Navier Stokes methods*

In a decade, solutions of these equations should be possible for
simplified wing-body configurations. It remains to be seen
whether, for steady, unseparated, transonic flows, corresponding
to the design point of an aircraft, the results will be in any
way superior or inferior to those given by 'viscous Euler'
methods: how effectively the two types of method deal with complex viscous effects, at the wing root and tip for example,
will be a factor in this comparison.

There will be a wide range of separated and unsteady flows
which will be beyond the scope of all but Navier—Stokes methods,
wing buffeting at transonic speeds being an important practical
example. Even so, the time and cost of running Navier—Stokes
programs will inhibit their use in aircraft design and their
most important applications in 1991 are likely to be in research
and in underpinning simpler methods. In internal aerodynamics
they may well be more widely used in design, primarily because
the complexity of internal flows limits the value of simpler
methods.

(e) *Large Eddy Simulation in the Navier-Stokes equations*

A twenty-five-fold increase in computer memory and speed will
extend but probably not transform what can be achieved by computations using Large Eddy Simulation. They will play a valuable
part in research into turbulence structure, particularly for flows
in which this structure is in any respect unusual or difficult to
investigate experimentally. They will also assist in the development of improved turbulence models for use in viscous layer calculations and in Reynolds-averaged Navier—Stokes methods. Direct
use in aerodynamic design to any appreciable degree seems unlikely without a further increase of two or more orders of magnitude in computer power beyond what is here suggested for 1991:
even then, applications will be limited to those cases, likely
to be relatively rare, in which simpler and cheaper alternatives
will not suffice. Chapman, looking further ahead, offers detailed justification for a more optimistic view of the role of
Large Eddy Simulation in the long term.

4.3 Avenues for Progress

To reach the general level of capability suggested in the preceding section by 1991 we shall need advances over a wide front,
not all of them in the field of fluid dynamics. First and fore-
most, advances in computer power are foreseen, deriving from
improvements both in component technology and in computer
architecture. These aspects will be the subject of further discussion at this Conference which I shall not pre-empt. Chapman
provides an informative assessment of prospects for the future

and cites a number of relevant references. The main thrust of
his argument is that the driving force behind component technology now appears to be the mass consumer market rather than
the needs of supercomputers, with the result that component
memory will increase more rapidly than speed. This leads to the
conclusion by Best (1978) that supercomputers of the future
can attain their goals only by exploiting all levels of parallelism within a computer structure assembled from conventional
components developed for other end-user requirements (i.e. the
mass market). It is broadly this philosophy which underlies
the proposal by NASA to develop the Numerical Aerodynamic Simu-
lator, the overall capability of which is planned to exceed that
of present Class 6 machines by rather more than an order of
magnitude.

Advances in numerical methods are the second main avenue for
progress. The solution procedures used in the current generation
of methods for steady three-dimensional flow fields require
either a large number of iterations or a large number of time
steps to reach a converged solution. This suggests that there
is still considerable scope for speeding up the calculation and
ample opportunity for innovation on the part of the algorithm
developers. There is also need for an imaginative approach to
the development of new coordinate systems, such as adaptive or
hybrid systems, to enable complicated geometry to be represented
accurately and economically in a flow field computation.

Modelling the physics provides us with a range of more specific problems which need to be addressed. Flow separation in two
or three dimensions, and in steady or time-dependent flows, is
perhaps the greatest challenge: should we rely on Navier–Stokes
methods for all such problems or can we develop simpler methods,
perhaps extensions of the 'viscous FP', to treat many of the
important flows?

Near streamwise vortices, such as arise from canards, strakes,
noses at high angle of attack, must be incorporated into aerodynamic computations. Whilst this has been done already in the
more advanced panel methods, the equally important problem of
incorporating streamwise vortices into a numerical flow field
computation at transonic speeds must also be tackled: this will
eventually have to include modelling the phenomenon of vortex
bursting and also the interaction between a vortex and a shock
wave.

I have already made several references to the improvement of
turbulence modelling, including the prospects of using Large
Eddy Simulation for this purpose: however, in addition to
devising calculation methods which incorporate increasingly
advanced turbulence models, I believe we need also to approach
the problem from the other direction, determining by analysis

and experiment the respects in which our present methods are
most in need of improvement.

   Current uncertainties in the application of the full potential
equation to flows with strong shock waves are in need of resolution: empirical tuning of the shock algorithm is one way forward, but it is worth asking whether a more global analysis of
the flow will teach us anything (the extension of such empirical
treatments to three dimensions introduces further uncertainty
and increases the need to elucidate this aspect of the method).
In Euler methods, accurate prediction of changes across the
shock remains the key to establishing definitive solutions for
inviscid flows. In coupled viscous-inviscid calculations,
satisfactory modelling is needed of the strong interactions
which arise between shock waves and boundary layers, in the trailing edge region, and at flow separation. It remains to be determined whether the use of a high-order viscous layer computation
in the inverse mode, as discussed by Lock and Firmin, is in any
way a less complete treatment than the multi-deck analyses which
seem generally to be accepted as correct models for strong interaction regions.

   For aircraft aerodynamics, incorporation of the hot, high-
velocity jet exhaust into the flow field is necessary and this
will require modelling of the shock structure in the jet, whenever necessary the two-stream structure of the jet, and also
the strong viscous effects within the jet and between the jet
and the external flow. For propeller-driven aircraft, modelling
of the interaction between the propeller and the aircraft flow
field at high subsonic speeds is a further challenge — overall
flow field prediction for a helicopter presents an analogous
problem. Internal flows present many additional problems of
modelling, including strongly three-dimensional viscous effects
and time dependence, but I shall not dwell on these here.

The advances foreseen above in computer hardware, in numerical
methods and in the representation of the physics should lead to
a generation of CFD methods of greater accuracy and appreciably
wider applicability than the present. As the methods become increasingly complex, in respect both of the fluid dynamics they
model and the geometries to which they can be applied, so it
will become increasingly difficult to test them rigorously. But,
as their potential value in aeronautical design increases, so it
will become all the more *necessary* to test them rigorously. Work
to establish a more extensive and reliable set of test cases for
CFD methods must therefore continue in parallel with the development of the methods themselves.

On the experimental front, there will be a need for detailed
studies of a range of turbulent flows, particularly the more

complex types involving separation, strong viscous-inviscid
interaction or marked three-dimensionality in the time-averaged
flow. Careful wind-tunnel experiments on representative aircraft configurations will also be required, including measurements of forces and pressures on the model, boundary conditions
in the wind tunnel and, where practicable, surveys of the
boundary layer and flow field over the model. The report by
AGARD (1979) of Working Group 04 of the Fluid Dynamics Panel,
which tabulates the results of a number of experiments considered
to be suitable at present as test cases, includes detailed recommendations by the Group for future experimental work.

As a complement to experiment, accurate numerical solutions
of the Navier—Stokes and Euler equations can also serve as test
cases and there may, in addition, be a limited role for exact
results derived from analytic solutions of the equations. A new
working group of the AGARD Fluid Dynamics Panel is currently
investigating one aspect of this question, the provision of a
range of numerical or analytical solutions of the Euler equations
to serve as test cases for inviscid methods based on the potential or small perturbation equations.

5. CONCLUSION

Two decades ago, numerical methods provided the means of
solving certain special problems in aerodynamics. Today, they
are at the heart of our aerodynamic design methods and are the
key to future aerodynamic advance. From this introductory paper
and the more detailed contributions which follow one message
above all should emerge: though the development of CFD has revol-
utionised aeronautical fluid dynamics in the past decade, the
subject is still emerging from its infancy and its full poten-
tiality has yet to be appreciated generally. For the mathematician, the numerical analyst and the theoretical physicist the
field of computational fluid dynamics offers a wide spectrum of
challenging and important problems.

6. REFERENCES

AGARD, (1979). 'Experimental data base for computer program
assessment' Report of the Fluid Dynamics Panel Working Group 04
AGARD-AR-138.

Albone, C.M. (1974). 'A finite difference scheme for computing
supercritical flows in arbitrary coordinate systems.' RAE Tech
Report 74090.

Baldwin, B.S. and Lomax, H. (1978). 'Thin layer approximation
and algebraic model for separated turbulent flows.' AIAA Paper
78-257.

Ballhaus, W.F., Jameson, A. and Albert, J. (1977). 'Implicit approximate factorization for the efficient solution of steady transonic flow problems.' AIAA Paper 77-634.

Best, D.R. (1978). 'Technology advances and market forces: Their impact on high speed performance architectures.' NASA CP 2032, pp. 343-353.

Boppe, C.W. and Aidala, P.V. (1970). 'Complex configuration analysis at transonic speeds.' AGARD Fluid Dynamics Panel Symposium, Munich May 1980.

Brandt, A. (1980). 'Multilevel adaptive computations in fluid dynamics.' AIAA J. 13, No. 10, pp. 1165-1172.

Butter, D.J. and Williams, B.R. (1980). 'The development and application of a method for calculating the viscous flow about high lift aerofoils.' AGARD Fluid Dynamics Panel Symposium, Colorado Springs, September 1980.

Chapman, D.R. (1979). 'Computational Aerodynamics; development and outlook.' AIAA J. 17 No. 12, pp. 1293-1312.

Collyer, M.R. (1977). 'An extension of the method of Garabedian and Korn for the calculation of transonic flow past an aerofoil to include the effects of the boundary layer and wake.' RAE TR 77194.

Garabedian, P.R. and Korn, D.G. (1971). 'Analysis of transonic airfoils.' Comm. Pure & Appl. Maths, Vol. 24, pp. 841-851.

Jameson, A. (1975). 'Transonic potential flow calculations using conservation form.' Proc. AIAA 2nd Computational Fluid Dynamics Conf. Hartford, Connecticut, pp. 148-61.

Kim, J. and Moin, P. (1979). 'Large eddy simulation of turbulent channel flow - Illiac IV calculation.' AGARD Fluid Dynamics Panel Symposium, The Hague, September 1979.

Levy, L.L. (1978). 'Experimental and computational steady and unsteady transonic flows about a thick airfoil.' AIAA J. 16, No. 6, pp. 564-72.

Lock, R.C. (1980). 'A review of methods for predicting viscous effects on aerofoils and wings at transonic speeds.' RAE TM Aero No. 1860. AGARD Fluid Dynamics Panel Symposium, Colorado Springs, September 1980.

Lock, R.C. and Firmin, M.C.P. (1982). 'Survey of techniques for estimating viscous effects in external aerodynamics' in these proceedings.

Magnus, R. and Yoshihara, H. (1970). 'Inviscid transonic flow over airfoils.' AIAA J. Vol. 8, pp. 2157-2162.

Murman, E.M. and Cole, J.D. (1971). 'Calculation of plane steady transonic flows.' AIAA J. Vol. 9, pp. 114-121.

Prandtl, L. (1904). 'Ueber Flussigkeitsbewegung bei sehr kleiner Reibung.' Proc. III International Maths. Cong., Heidelberg, pp. 484-491.

Pulliam, T.H. and Lomax, H. (1979). 'Simulation of three-dimensional compressible viscous flow on the Illiac IV computer.' AIAA Paper 79-0206.

Sells, C.C.L. (1968). 'Plane subcritical flow past a lifting aerofoil.' Proc. Roy. Soc. A308, pp. 377-401.

Sells, C.C.L. (1980). 'Solution of the Euler equations for transonic flow past a lifting aerofoil.' RAE TR 80065.

Smith, J.H.B. (1982). 'Achievements and problems in modelling highly swept flow separations', this volume.

Thompson, J.F., Thames, F.C., Mastin, C.M. and Shanks, S.P. (1975). 'Use of numerically generated body-fitted coordinate systems for solution of the Navier—Stokes equations.' Proc. AIAA 2nd Computational Fluid Dynamics Conf. Hartford, Connecticut, pp. 68-80.

Treadgold, D. and Wilson, K.H. (1980). 'Some aerodynamic interference effects that influence the transonic performance of combat aircraft.' AGARD Fluid Dynamics Panel Symposium, Munich, May 1980.

ADVANCES AND SHORTCOMINGS IN THE CALCULATION OF INVISCID
FLOWS WITH SHOCK WAVES

M.G. Hall

(RAE, Aerodynamics Department, Farnborough, Hants)

SUMMARY

An introductory survey is presented of recent developments
in the calculation of inviscid flows with embedded shocks and
mixed subsonic-supersonic flow. It begins with an account of
methods for solving the compressible potential equation, covering the inherent limitations, the approximate factorization
and multigrid techniques, and unsteady as well as steady flows.
A brief review of methods for the Euler equations, which admit
accurate simulations of shock waves, follows. Finally, the advantages and shortcomings of the various methods are compared,
with special attention to the simulation of shock waves. Other
factors considered are overall accuracy, speed and practicability.

1. INTRODUCTION

When a flow contains shock waves and part of the flow is subsonic it would be expected that there would be difficulties in
calculating the flow for non-trivial problems. Even for purely
supersonic flow, any solution containing discontinuities, or
local regions with very large gradients, would present difficulties
if the locations of the discontinuities are not known in advance.
If the flow is partly supersonic and partly subsonic, and so
described in the respective parts by hyperbolic and elliptic
equations, with the boundaries between the two again not known
in advance, the difficulties become considerable. Of course all
real flows of aerodynamic interest are viscous as well but there
is clearly some incentive to avoid the further complication. At
present there is a great deal of effort being spent by aerodynamicists in calculating inviscid transonic flows with shock
waves and the calculation methods are probably the most widely
used tool in computational aerodynamic design. The reason stems
from the practical importance of transonic flow, the behaviour
of the flow frequently being a crucial factor in the performance
of aircraft, missiles and turbomachinery. At the high Reynolds
numbers that characterize practical flows most of the flow field
is essentially inviscid, viscous effects being confined to
relatively thin layers even when there is separation. Thus

reliable calculation of inviscid flow would be a prerequisite
even if the objective is to solve the Navier—Stokes equations.
When there is no large-scale separation inviscid flow calculations can assume a dominant role, because viscous effects
can then be included by combining iteratively an extra inviscid
solution with a boundary layer solution. This type of calculation nowadays has a central role in aerodynamic design.

   While it is recognised that computational fluid dynamics,
which includes boundary layer and Navier—Stokes calculations, has
revolutionized both research and design in aerodynamics, the
critical breakthrough was in inviscid transonic flow calculations.
Beginning perhaps 40 years ago, a widening gap appeared between
theoretical and practical aerodynamics. Theoretical aerodynamics
was restricted to problems which essentially could be reduced to
the solution of linear or one-dimensional equations. This suf-
ficed, for attached flow at least, while speeds were so low that
the effects of compressibility were small. As speeds increased,
and especially with the advent of jet propulsion, even attached
flows could not be described by anything less than a non-linear
partial differential equation. By the 1960s, theoretical and
practical aerodynamics had become virtually unrelated subjects,
the theory being unable to provide as much as a qualitative
description of the flow over the wing, for example, of an
ordinary cruising airliner. The practical aerodynamicist and
designer had to rely mainly on wind tunnel and flight testing.
In 1970 a new, computational, method for calculating inviscid
transonic flows was proposed by Murman and Cole. Now just over
ten years later, aerodynamics is transformed. The equation
solved by Murman and Cole (1971) was indeed a non-linear partial
differential equation of mixed elliptic-hyperbolic type, but it
was only for plane two-dimensional flow and strictly appropriate
only for slightly perturbed flows at near the speed of sound
with weak shock waves. It was quickly noticed, however, that
with minor modifications the same equation could yield useful
results when the perturbations were not small, speeds were not
near the speed of sound and shock waves were not weak. This
was seized on, for example by Bailey and Ballhaus (1975) in USA,
and Albone, Hall and Joyce (1976) at RAE, as the basis of
methods for three-dimensional flows. Thus within a few years
practical methods became available for wings and simple wing-
body combinations. For the first time since the 1930s the flow
over a real aircraft could be usefully predicted. The designers
found these transonic small-perturbation (TSP) methods irresist-
ible and decided that they wanted much more. The methods were unreliable when perturbations were not small or shocks were not
weak. Full potential (FP) methods were developed, notably by
Garabedian and Korn (1971), Jameson (1974) in USA and Forsey
and Carr (1978) here, in which the small-perturbation approximation is no longer a limitation, and these are gradually

displacing the TSP methods. The methods remain unreliable, however, when shocks are not weak. There is moreover a need to extend the methods to deal with flows past the complex shapes of practical interest and, at the same time, to reduce computational costs. Thus there is at present a great deal of work on the Euler equations which admit the correct representation of strong shocks, on new problem formulations for complex shapes, and on new algorithms for greater computational efficiency.

For anyone working in the field of inviscid mixed flows the range of possibilities is large. Perhaps the first issue to be faced is the choice between [solving] the potential flow equation and the Euler equations. The Euler equations in principle admit accurate solutions for rotational flows with shock waves, whatever their strength, but instead of having one non-linear partial differential equation to solve there are in general five, which greatly increases the computational labour. Moreover, the actual achievement in accuracy at present falls far short of that attainable in principle.

One question to be decided is whether to fit shocks explicitly as discontinuities in the flow or to capture them, generally as steep compressions, in the course of the calculation. To fit a shock implies detecting it, tracking it and applying the appropriate Rankine-Hugoniot relations across it. As one would expect, the difficulties are severe, even in two space dimensions. On the other hand while the capture of a shock is relatively easy, shocks can be seriously displaced and diffused in the process of capture. Oblique shocks in particular may be so diffused as to be unrecognizable as shocks. Since the interaction between shock-waves and boundary layers tends to have a strong effect on the overall flow pattern we can define rough criteria for successful capture. The distance a shock is displaced by the capturing process should be small compared with the relevant body length and should not be so great that the associated boundary layer changes by more than a small amount over that distance, and the width over which a shock is diffused should be much less than the boundary layer thickness. Successful capture may be no easier than successful fitting.

Another question is raised by the form to be chosen for discretization of the problem. Finite-difference techniques, in which derivatives in the differential equations of motion are replaced by finite-difference approximations, are the most common. Also in use are finite-volume techniques, where integral forms of the equations of motion are satisfied across individual volume elements. Finite-element techniques, in which the equations of motion are replaced by a variational condition or an equivalent, have yet to gain practical acceptance. This is partly because the advantages in efficiency over finite-difference

methods that hold for systems governed by, say, the Laplace
equation do not seem to carry over to the more complex systems
of interest here. Both finite-element and finite-volume techniques have been claimed to offer a better capability for dealing with flows past complex shapes, because the individual
elements may be of arbitrary shape and the array of elements
need not have a regular structure. However, it has been found
that, unless the variation in shape from one element to the
next is small and the array of elements does have a regular
structure, the elements are difficult to construct, truncation
errors are large and the numerical solution is lengthy.

The means to be adopted to deal with the essentially mixed,
subsonic-supersonic, character of the flow raises yet another
question. For steady flow the governing partial differential
equations are mixed, elliptic-hyperbolic, in type. The difficulties of dealing with mixed elliptic-hyperbolic systems can
be circumvented by treating the steady state as one that is
approached by advancing in time or some pseudo-time, for then the
system is simply hyperbolic, and this is indeed the usual practice with the Euler equations. With the potential flow equations
it is more usual to switch difference, or discretization, schemes
depending on the local nature of the flow. The classical Murman—
Cole proposal was to switch from central differences where the
flow is subsonic to backward or up-wind differences where the
flow is supersonic.

After each of the above issues is decided several subsidiary
choices have to be made. One that arises frequently is whether
to adopt a 'conservative' scheme. For one-dimensional time-
dependent flow Lax (1954) has shown that to capture shocks satisfactorily one should discretize the 'conservation-law' or divergence form of the Euler equations. Otherwise 'shocks' will be
obtained across which mass flux, etc, is not conserved. Of
course, conservative discretization is unnecessary if shocks
are fitted and in that case non-conservative schemes may be preferred for simplicity. Lax's conclusion seems to have been
adopted as a guiding principle for multi-dimensional problems.
The resulting shocks appear to be captured well provided they
are aligned with a coordinate direction and provided care is
taken to ensure that there is no loss of conservation at any
switch of discretization scheme. The capture of shocks that are
oblique to the coordinate directions seems far less successful,
the shocks being smeared or diffused. For potential flows it is
not possible to conserve both mass and momentum across a shock
and there is no clear guiding principle.

The present survey begins with an account of methods for solving the full potential equation for steady flow, with intro-
ductions to the approximate-factorization and multi-grid tech-

niques. Solutions of the potential equation for unsteady flow
are considered next. A brief review of a selection of methods
for the Euler equations follows, with the emphasis on the dual
objectives of accurate capture of shocks and high efficiency
of numerical solution. Finite-element methods are not included
in the above although they have been proposed for transonic
flows, because they do not so far appear to offer practical
advantages over finite-difference and finite-volume methods.
Finally, the comparative advantages and shortcomings of the
various methods are discussed, with special attention to the
simulation of shock waves. Other factors considered are overall
accuracy, speed and practicability.

2.   THE FULL POTENTIAL EQUATION FOR STEADY FLOW

2.1 Background

The full potential equation may take many forms. In mass-
conservation form and Cartesian coordinates we have, for steady
two-dimensional flow,

$$(\rho\phi_x)_x + (\rho\phi_y)_y = 0, \tag{1a}$$

where

$$\rho = \left[1 + \frac{\gamma-1}{2} M_\infty^2 (1 - \phi_x^2 - \phi_y^2)\right]^{\frac{1}{\gamma-1}} . \tag{1b}$$

An alternative, quasi-linear form is

$$(a^2 - \phi_x^2)\phi_{xx} - 2\phi_x\phi_y\phi_{xy} + (a^2 - \phi_y^2)\phi_{yy} = 0, \tag{2a}$$

where

$$a^2 = a_0^2 - \frac{\gamma-1}{2} (\phi_x^2 + \phi_y^2) . \tag{2b}$$

The canonical form referred to below is

$$\left[1 - \frac{q^2}{a^2}\right]\phi_{ss} + \phi_{nn} = 0 \tag{3}$$

where s and n are local streamwise and normal Cartesian coordinates and q is the local speed.

The best known iterative technique for solving the equation
is successive line over-relaxation (SLOR). In any one iterative

step the effects of a local perturbation in the numerical solution are transmitted only one mesh width in the backward direction. This makes rapid convergence impossible.

Since SLOR is very well documented we turn here to two newer techniques, approximate factorization and multigrid, which are appreciably faster. Already they are beginning to supersede line relaxation. Although the latter may well soon be completely outdated some of the associated developments will continue to be of use. Rotated differences and the introduction of artificial viscosity to effect the switch between subsonic and supersonic difference schemes are two such developments.

Rotated differences were proposed independently by Albone (1974) and Jameson (1974) to ensure stable calculation of a locally supersonic flow when, as is sketched in Fig. 1, the coordinate system is not aligned with the local velocity vector. The use of backward differences alone for the second derivative with respect to x would, in Fig. 1b violate the Courant—Friedrichs-Lewy (CFL) condition for stability, because the characteristic directions are such that the domain of dependance of the point P with respect to the difference equation fails to include that with respect to the differential equation. Albone and Jameson remedy this by proposing a mixture of backward and central differences derived by starting with the differential equation in canonical form (3) and then rotating to the desired coordinate direction. It has been noted recently by Albone and Hall (1980) that the original Albone—Jameson scheme satisfies conditions that are in general stronger than the CFL condition, but it remains to be seen whether their proposed improvement, by which central differences are eliminated where they are not required by the CFL condition, is practically feasible.

Artificial viscosity was introduced by Jameson (1975) to ensure that mass was conserved when difference schemes were switched in solving the full potential equation. If the backward and forward difference operators, $\overleftarrow{\delta}_x$ and $\overrightarrow{\delta}_x$, are respectively defined by

$$\overleftarrow{\delta}_x \phi_i = \frac{1}{\Delta x} (\phi_i - \phi_{i-1})$$

$$\overrightarrow{\delta}_x \phi_i = \frac{1}{\Delta x} (\phi_{i+1} - \phi_i),$$

then a suitable difference approximation to the first term $(\rho\phi_x)_x$ in equation (1a) is, for subsonic flow

$$\overleftarrow{\delta}_x \rho_{i+\frac{1}{2},j} \overrightarrow{\delta}_x \phi_{i,j} = \frac{1}{(\Delta x)^2}\left[\rho_{i+\frac{1}{2},j}(\phi_{i+1,j} - \phi_{i,j}) - \rho_{i-\frac{1}{2},j}(\phi_{i,j} - \phi_{i-1,j})\right].$$

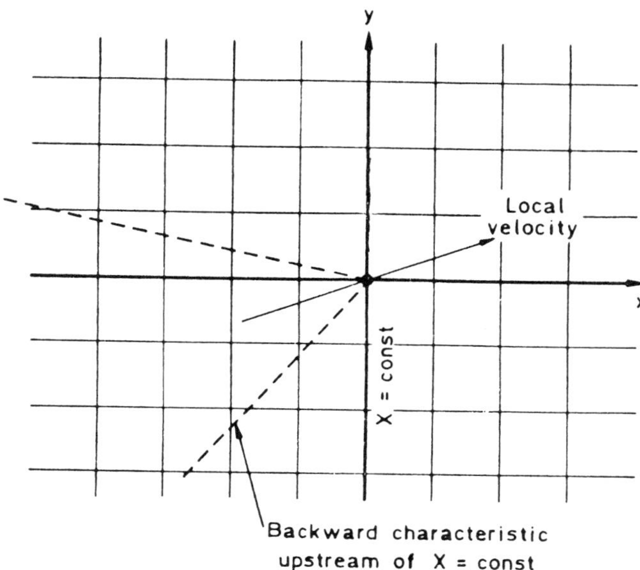

a   Central differences not needed for ϕ_{xx}

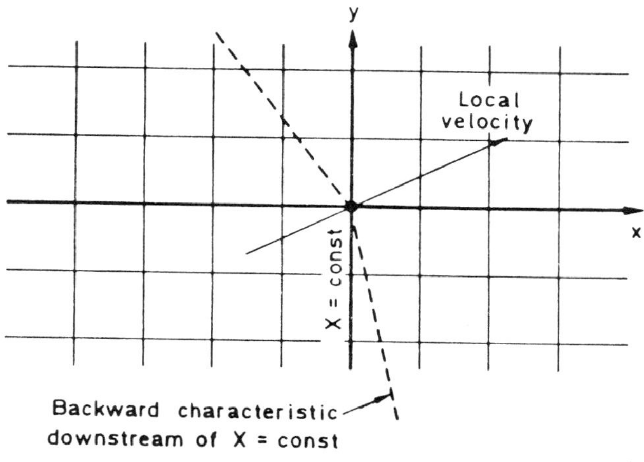

b   Central differences needed for ϕ_{xx}

Fig. 1 Characteristics and the need for central differences

To admit supersonic flow as well, Jameson adds to the above term

$$- \overleftarrow{\delta}_x \mu_{i,j} (\rho_{i+\frac{1}{2},j} - \rho_{i-\frac{1}{2},j}) \overrightarrow{\delta}_x \phi_{i,j} \doteq - (\Delta x \mu \rho_x \phi_x)_x, \qquad (4)$$

where μ is the switching function defined by

$$\mu = \max\left\{0, \ \frac{1-a^2}{q^2}\right\}.$$

A similar term that approximates $-(\Delta y \mu \rho_y \phi_y)_y$ is added to the supersonic difference approximation to $(\rho \phi_y)_y$. Together the additional terms represent an artificial viscosity in divergence or conservation form. If the artificial viscosity were not in divergence form mass would not be conserved across supersonic-subsonic shocks even though it is the mass-conservation form (1a) of the potential equation that is discretized.

Now Holst and Ballhaus (1979) have observed that Jameson's sum

$$\overleftarrow{\delta}_x \rho_{i+\frac{1}{2},j} \overrightarrow{\delta}_x \phi_{i,j} - \overleftarrow{\delta}_x \mu (\rho_{i+\frac{1}{2},j} - \rho_{i-\frac{1}{2},j}) \overrightarrow{\delta}_x \phi_{i,j}$$

may be written

$$\overleftarrow{\delta}_x \tilde{\rho}_{i+\frac{1}{2},j} \overrightarrow{\delta}_x \phi_{i,j}$$

with

$$\tilde{\rho}_{i+\frac{1}{2},j} = (1 - \mu)\rho_{i+\frac{1}{2},j} + \mu \rho_{i-\frac{1}{2},j},$$

so that Jameson's addition of artificial viscosity is equivalent to giving the density an upwind or backward bias for supersonic flow. This makes it clear that only through the density does any backward bias enter this formulation of the problem. The difference approximations for ϕ in (1a) are here all of the central type, even in supersonic flow, and this has been exploited by Holst and Ballhaus to simplify the numerical solution.

It appears to be the usual practice outside UK to discretize the mass-conservation form of the potential equation in a manner such as has been indicated above. Since the flow is isentropic it is not possible, of course, to model physical shock waves correctly: when mass is conserved momentum is not conserved. The resulting shocks are usually found to be stronger, by varying degrees, than the shocks in the corresponding solutions of

the Euler equations. Better agreement with solutions of the Euler
equations has been obtained by Lock (1981) by use of an empirically chosen scheme in which neither mass nor momentum is conserved. This scheme is conveniently written down with the aid of
the operator δ_{xx} defined by

$$\delta_{xx}\phi_{i,j} = \frac{1}{(\Delta x)^2}\,(\phi_{i+1,j}-2\phi_{i,j}+\phi_{i-1,j}).$$

The original non-conservative Murman-Cole switching scheme for
ϕ_{xx} in equation (2a) may then be written

$$\left[(\phi_{xx})_{N\text{-}C}\right]_{i,j} = \delta_{xx}\phi_{i,j} - \mu_{i,j}\left[\delta_{xx}\phi_{i,j}-\delta_{xx}\phi_{i-1,j}\right], \qquad (5)$$

where

$$\mu_{i,j} = \begin{cases} 1, M > 1 \\ 0, M < 1 \end{cases}.$$

Noting that the artificial viscosity on the right-hand side is
not in divergence form, Jameson had suggested the quasi-
conservative alternative

$$\left[(\phi_{xx})_{Q\text{-}C}\right]_{i,j} = \delta_{xx}\phi_{i,j} - \mu_{i,j}\delta_{xx}\phi_{i,j} + \mu_{i-1,j}\delta_{xx}\phi_{i-1,j} \qquad (6)$$

where the artificial viscosity, which is an approximation to

$$- (\Delta x\mu\phi_{xx})_x$$

is not in divergence form. Lock's scheme, termed partially
conservative, is

$$\left[(\phi_{xx})_{P\text{-}C}\right]_{i,j} = (1 - \lambda)\left[(\phi_{xx})_{N\text{-}C}\right]_{i,j} + \lambda\left[(\phi_{xx})_{Q\text{-}C}\right]_{i,j}, \qquad (7)$$

where λ is a constant. Lock finds that if $\lambda \doteq 0.25$ fair agreement is obtained with solutions of the Euler equations for a use-
ful range of cases, provided the comparisons are made for equal
lifts. Two examples are shown in Fig. 2. Note that Jamesons's
quasi-conservative scheme ($\lambda = 1$) differs from his other conservative scheme (4) in retarding ϕ rather than ρ, so the corresponding numerical solutions will also differ, though both exhibit
shocks that appear over-strong for practical purposes. The comparison on the right-hand side shows excellent agreement with a solution of the Euler equations, for equal lifts. It now seems clear,
however, that the angle of incidence required for a given lift

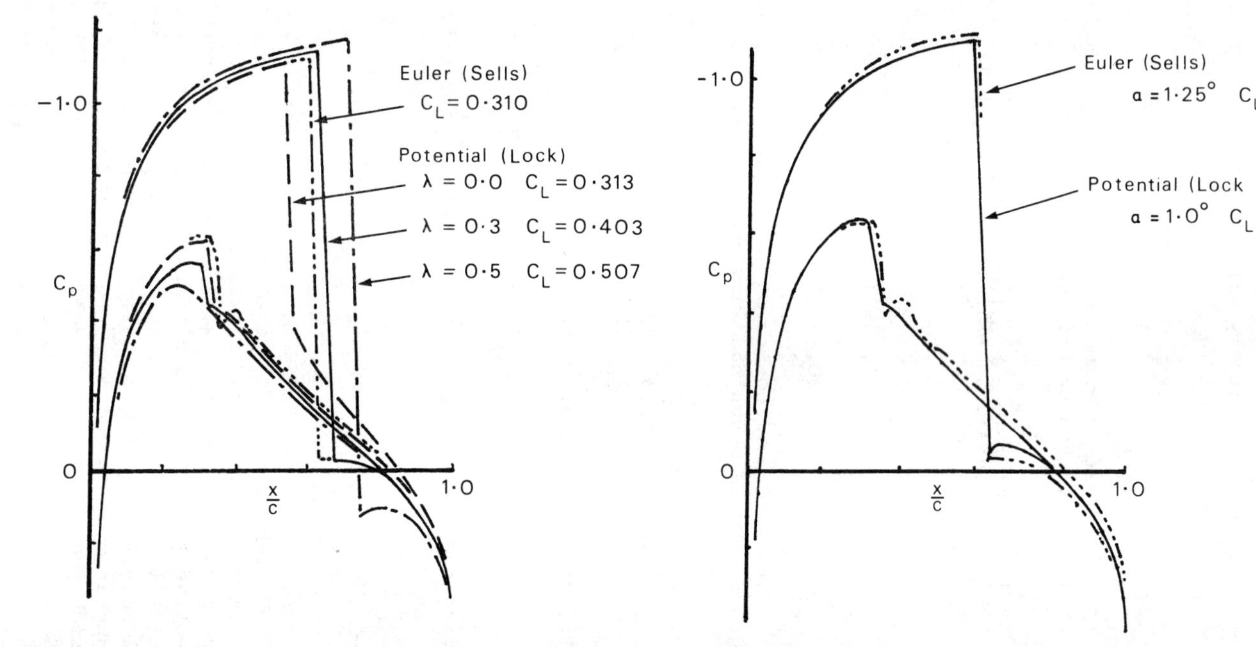

Fig. 2 Comparison of potential (Lock) and Euler (Sells) solutions

is in general appreciably less with the FP equation than with
the Euler equations.

2.2 *Approximation factorization*

Approximate factorization is undergoing rapid development
and we present here only the basic principle, as described by
Ballhaus, Jameson and Albert (1978) and an outline of one of
the more successful schemes (that of Holst and Ballhaus (1979)),
with some final remarks on the course of developments.

Let the difference equation to be solved be denoted by

$$L\phi = 0, \tag{8}$$

and let ϕ^n be the solution vector at the nth iterative level, so
that the residual is $L\phi^n$ and one can define a correction vector

$$C^n = \phi^{n+1} - \phi^n \tag{9}$$

and an error

$$e^n = \phi^n - \phi. \tag{10}$$

Then the iteration schemes under consideration are of the type

$$NC^n = \alpha L\phi^n, \tag{11}$$

where N is a difference operator to be carefully chosen and α
is an acceleration parameter that is changed from one iterative
level to the next. It follows immediately from (8) to (11) that

$$e^{n+1} = (I + \alpha N^{-1}L)e^n, \tag{12}$$

I being the identity matrix. Thus the rate of convergence
depends crucially on the choice of N (and α). Clearly if N was
identical to L and if it were practicable to invert L then, by
setting $\alpha = -1$, an exact solution to $L\phi = 0$ would be obtained
in a single iteration. This suggests that convergence would be
rapid if N could be chosen to resemble L closely. Since N must
be such that $N^{-1}L\phi^n$ is readily computed, it is proposed to construct N as the product of two or more linear factors,
$N = N_1N_2 \ldots$, each of which can be readily inverted. The
operator N should be implicit, in the sense that its factors,
in combination, should ensure that the effects of any local perturbation are transmitted over the entire field at each iterative level. There is considerable scope for optimization in the
choice of factors.

The approximate factorization scheme we outline here is the AF2 scheme of Holst and Ballhaus (1979). As indicated in the previous section (2.1) the mass-conservation form (la) of the full potential equation is replaced by the finite-difference equation

$$L\phi \equiv [\overset{\leftarrow}{\delta}_x \tilde{\rho}_{i+\frac{1}{2}} \overset{\rightarrow}{\delta}_x + \overset{\leftarrow}{\delta}_y \tilde{\rho}_{j+\frac{1}{2}} \overset{\rightarrow}{\delta}_y]\phi_{i,j} = 0 \tag{13}$$

where

$$\tilde{\rho}_{i+\frac{1}{2}} = (1 - \mu_{i,j})\rho_{i+\frac{1}{2},j} + \mu_{i,j}\rho_{i-\frac{1}{2},j}$$

$$\tilde{\rho}_{j+\frac{1}{2}} = (1 - \mu_{i,j})\rho_{i,j+\frac{1}{2}} + \mu_{i,j}\rho_{i,j-\frac{1}{2}} \; .$$

The factored form for N is, following Ballhaus and Steger (1975), chosen to be

$$N = - (\alpha\overset{\leftarrow}{\delta}_x - \overset{\leftarrow}{\delta}_y \tilde{\rho}^n_{j+\frac{1}{2}} \overset{\rightarrow}{\delta}_y)(\alpha - \tilde{\rho}^n_{i+\frac{1}{2}} \overset{\rightarrow}{\delta}_x) \; . \tag{14}$$

Substitution of equations (13) and (14) in the basic iteration (11) leads to the two-step procedure:

Step 1
$$(\alpha\overset{\leftarrow}{\delta}_x - \overset{\leftarrow}{\delta}_y \tilde{\rho}^n_{j+\frac{1}{2}} \overset{\rightarrow}{\delta}_y) f^n_{i,j} = \alpha L\phi^n_{i,j} \tag{15a}$$

Step 2
$$(\alpha - \tilde{\rho}^n_{i+\frac{1}{2}} \overset{\rightarrow}{\delta}_x) C^n_{i,j} = f^n_{i,j} \tag{15b}$$

In step 1 the intermediate result $f^n_{i,j}$ is obtained by solving a tridiagonal matrix for each line x = const. The correction $C^n_{i,j}$ is then obtained in the second step from the $f^n_{i,j}$ by solving a simple bidiagonal matrix equation for each line y = const. Note that this simplicity of the matrix inversions has been achieved by providing for supersonic flow through a backward bias of the density alone and evaluating this at the nth iterative level.

The choice of the values to be given to the acceleration parameter α is guided by an analogy to the time-dependent problem. If the iterations are regarded as an advance in pseudo-time then α can be regarded as the inverse of the time step. Since large time steps would be appropriate for reducing the long-wave components of error, while small time steps would be appropriate for the short-wave components, it is appropriate here to vary α in cycles through a sequence of different values.

Holst and Ballhaus chose to give α eight different values in a geometric sequence over the range

$$2 \leq \alpha \leq \frac{2}{\Delta y} \ .$$

The above factored scheme is much faster than SLOR. Typically, whereas several hundred SLOR iterations are required to reduce the rms error to 1% the factored scheme requires less than 100. Holst (1980) has recently extended the scheme to treat three-dimensional flows past wings. Baker (1981) has derived a related scheme for solving the quasi-linear form (2a) of the full potential equation, and this appears to be appreciably faster than the Holst—Ballhaus scheme. It is described by Baker himself elsewhere in this volume. This also has been extended to three dimensions. Catherall (1981) has devised a hybrid scheme consisting of Baker's scheme for the outer part of the field and SLOR for the inner. This simplifies the treatment of boundary conditions in certain cases and appears to be faster still. In addition, he has shown how to devise 'optimal' approximate factorization schemes for two-dimensional flows; in particular, he has provided guidelines for the choice of coefficients in the individual factors. For three-dimensional problems it has not so far been possible to reproduce the same advantage in speed over SLOR as has been achieved in two; there is much scope for improvement, but the analysis of the efficiency of factored schemes is a much more complex task than in two dimensions.

2.3 Multigrid techniques

The 'multigrid' technique seems to have been first proposed by Federenko (1964) and has been studied by a number of people, notably Brandt (1977). There are many variations and only recently has one been devised that has proved completely successful for solutions of the potential equation for transonic flow, perhaps because the original technique was more naturally suited to the solution of elliptic partial differential equations. The successful technique was devised by Jameson (1979) and is outlined here.

The underlying idea is that an approximate solution of the finite difference (or other discrete) equation for a given co-ordinate grid can be improved by calculating the correction to the solution on a coarser grid. If the solution on each grid effectively reduces error components of wave-length proportional to mesh size, a sequence of calculations on successively coarser grids could yield high accuracy at minimal cost. The idea may be illustrated by considering a simple two-grid process. Let the partial differential equation to be solved be denoted by

$$LU = F, \tag{16}$$

where L is a non-linear differential operator, and let the difference approximation to (16) on the fine grid h be

$$L^h U^h = F^h. \tag{17}$$

This equation is solved to yield an approximation to U^h which we denote by u^h. Put $U^h = u^h + v^h$ where v^h is the required correction. Since L^h is non-linear we do not attempt to derive a difference operator for V but, instead, we write (17) in the form

$$L^h(u^H + v^h) - L^h u^h = r^h, \tag{18}$$

where $r^h (\equiv F^h - L^h u^h)$ is the residual.

To proceed from the fine grid to a coarse grid H, or vice versa, interpolations will be needed, and we take I_h^H to denote the operation of interpolating from the fine to the coarse grid (for example by taking weighted means of neighbouring quantities). Then to calculate the correction v^h on the coarse grid we solve, instead of (18), the corresponding equation

$$L^H(I_h^H u^h + v^H) - L^H(I_h^H u^h) = \bar{I}_h^H r^h, \tag{19}$$

where $\bar{I}_h^H$ may be different from I_h^H. To retain the form of equation (17) and so simplify the numerical work, we re-write (19) as

$$L^H \bar{u}^H = f^H \tag{20}$$

where $\bar{u}^H \equiv I_h^H u^h + v^H$

and $\quad f^H \equiv L^H(I_h^H u^h) + \bar{I}_h^H r^h .$

Equation (20) is solved on the coarse grid to yield an approximation to $\bar{u}^H$ which we denote by u^H. Finally, since $v^H \doteq u^H - I_h^H u^h$ we have an improved new approximation to u^h, namely

$$(u^h)_{new} = u^h + I_H^h(u^H - I_h^H u^h) . \tag{21}$$

In the multigrid process both the algorithms for the approximate solution at each level and the interpolation formulae for transferring results from one level to the next must be chosen with care. Jameson shows that for a practical grid, which is non-uniform or curvilinear in physical space, neither the point nor line relaxation algorithms will damp the high frequency error components as required. He proposes instead a factored scheme of the form

$$(S - A\delta_x^2)(S - B\delta_y^2)C = SLu \tag{22}$$

where C is a correction, and the operator S is given by

$$S \equiv \alpha_0 + \alpha_1 \bar{\delta}_x + \alpha_2 \bar{\delta}_y \, ,$$

where $\bar{\delta}$ denotes a one-sided difference and the αs are constants.

Jameson's solution strategy is to perform a cycle of calculations repeatedly until convergence is achieved. Each cycle begins on the finest grid and succeeding grids have the numbers of intervals halved in each direction. The factored equation is solved once on each grid until the coarsest grid is reached. Then it is solved once again on each grid going back up to the second finest grid, and the cycle terminates with the interpolation of the correction from the second finest grid to the finest grid. It was found, in a test calculation with a 192 × 32 fine grid, that there was no improvement in convergence when the number of grid levels was extended beyond five. Convergence was complete, for any practical purpose, in about 10 cycles. If we note that roughly three-fifths of the computational effort in a cycle is spent on the finest grid we can conclude that the multi-grid technique, even at its very first appearance, is faster than the approximate factorization schemes of the previous section.

It seems too early, however, for a practical choice to be made between multi-grid and approximate factorization. The latter may appear inefficient in not employing coarser grids for iterations at the smaller values of the acceleration parameter α, but there are no interpolations, simpler demands on computer memory and the programming is more straightforward. The final criteria are how readily the methods can be used and how well they perform, for practical three-dimensional flows, in iterative combination with boundary layer calculations. Work on three-dimensional problems has only just begun.

3 SOLUTIONS OF POTENTIAL FLOW EQUATIONS FOR UNSTEADY FLOW

3.1 *The additional difficulties in unsteady flow problems*

The development of methods for unsteady transonic flows has
lagged behind that for steady flows, in spite of the importance
of high aircraft performance in unsteady conditions. Aerodynamic flutter and helicopter rotor blades provide perhaps the
most obvious examples of flows that are essentially unsteady.
The pace of development reflects the difficulty of the subject.
Not only may the body be in motion but it may be undergoing
deformation as well. Thus a body-fitted coordinate system would
vary in time, which complicates the governing equations of
motion. This gives small-perturbation methods, which admit
stationary coordinates, an attraction that ensures that they
will outlast their counterparts for steady flow. A further
complication lies in the implementation of the Kutta condition
at the trailing edge of an aerofoil. Strictly, the flow will
leave the trailing edge at a tangent to either the upper or the
lower surface, depending on the past history of the motion.
Another difficulty frequently met in attempts to follow numerically the variations of a flow in time is a contamination of the
results by reflections of disturbances from the outer part of
the field. The above apply to a wide range of unsteady flows.
For unsteady potential flows the capture of shock waves raises
a special problem. It will be recalled that if, for steady
flow, the mass-conservation form of the potential equation is
solved the captured shocks will be stronger than the corresponding Rankine-Hugoniot shocks and also be displaced in position,
and that this shortcoming has led Lock to choose, by trial, a
partially-conservative scheme for practical calculations. For
unsteady flow the mass-conservation form must be expected to
yield shocks that in general move at a speed that differs from
that of Rankine-Hugoniot shocks. This follows from the displacement of shocks from their correct position in steady flow.
Some such displacement would be expected in unsteady flow and,
as is easily seen from oscillatory flows, the associated shock
speeds must therefore be incorrect. Partially conservative
schemes could again be tried, but the task of matching solutions
of the Euler equations would now be more demanding, and even
for steady flows the pressure distributions over an aerofoil
could be matched well only by making the comparison at a specified value of the lift.

3.2 *An alternating-direction implicit method*

An outline is given here of the best known and probably the
most widely used of numerical methods for unsteady transonic
flows. The method was proposed by Ballhaus and Steger (1975)
for the restricted class of unsteady flows governed by the low-

frequency transonic small-perturbation equation

$$\beta_1 \phi_{xt} = f_x + \phi_{yy} \tag{23}$$

where $\beta_1 = 2kM_\infty^2$ ($k \equiv$ non-dimensional frequency $\ll 1$) and

$$f = \left(1 - M_\infty^2 - \frac{\gamma+1}{2} M_\infty^m \phi_x\right)\phi_x \ .$$

Equation (23) is replaced by the implicit difference approximation

$$\beta_1 \delta_x (\Delta t)^{-1} (\phi_{j,k}^{n+1} - \phi_{j,k}^n) = D_x \bar{f}_{j,k} + \frac{1}{2}(\delta_{yy}\phi_{j,k}^{n+1} + \delta_{yy}\phi_{j,k}^n) \tag{24}$$

where D_x is a mixed difference operator containing a switching
function to ensure stability and maintain conservation form,
and

$$\bar{f} = \frac{1}{2}\left[\left(\frac{\partial f}{\partial \phi_x}\right)_{j,k}^n (\phi_x)_{j,k}^{n+1} + (1 - M_\infty^2)(\phi_x)_{j,k}^n\right] \doteq \frac{1}{2}(f_{j,k}^{n+1} + f_{j,k}^n) \ .$$

Equation (24) is then solved at each time step by the simple
two-sweep, alternating-direction (ADI) algorithm

x-sweep
$$\beta_1 (\Delta t)^{-1}\delta_x(\tilde{\phi}_{j,k}^{n+1} - \phi_{j,k}^n) = D_x\bar{f}_{j,k} + \delta_{yy}\phi_{j,k}^n$$

y-sweep
$$\beta_1 (\Delta t)^{-1}\delta_x(\phi_{j,k}^{n+1} - \tilde{\phi}_{j,k}^{n+1}) = \tfrac{1}{2}\delta_{yy}\phi_{j,k}^{n+1} - \tfrac{1}{2}\delta_{yy}\phi_{j,k}^n \tag{25}$$

For accuracy and stability the step size Δt must be small enough
that shock waves do not move over more than one spatial mesh
interval per time step. This is not a severe restriction and
much longer time steps can be taken than would be possible with
an explicit method. Typically, 50—100 time steps would be
taken per cycle of an oscillatory motion.

The method was first applied to a range of problems by Ballhaus and Goorjian (1977) and has since been improved, for
example by Houwink and van der Vooren (1980), so that it is now
considered applicable in the range of frequencies k ≤ 0.4. It
has also been extended by Borland, Rizzetta
and Yoshihara, (1980) to treat three-dimensional problems.
The method has been found to reproduce the observed qualitative
features of several types of unsteady flow but its accuracy
is difficult to assess. Good quantitative agreement with

solutions of the Euler equations for a case of transonic flow
past an aerofoil with an oscillating flap has been obtained by
appropriate 'tuning' of the kind that is possible in the transonic small-perturbation approximation. But the latter approximation is known to have important limitations in steady flow. It
will become clearer what are the corresponding limitations for
unsteady flow when solutions of the full potential equation are
more readily available.

3.3 Other recent developments

In addition to the above ADI method for the low-frequency
transonic small-perturbation equation there have been several
other developments, in some of which the full potential
equation is solved. To remove the restriction to low-frequency
motions ADI methods have been developed for solving the complete
transonic small-perturbation equation, including the term in ϕ_{tt}
that was omitted in the low-frequency approximation. The
method of Rizzetta and Chin (1979) is, as a result, more complicated, while that of Isogai (1980) is no longer fully implicit,
and in both the time steps must be smaller. Isogai makes
extensive calculations and goes on to apply the results to a
flutter analysis of an aerofoil with two degrees of freedom,
with an outcome that is instructive. He finds, as the Mach
number is increased, a sharp dip in the flutter speed that can
be associated, by reference to the flow solutions, with the
development of a pronounced phase lag in the motion of the shock
wave. This supports the view that shock motion is not negligible
in transonic flutter and has implications for certain flow calculations that involve a linearization of the time-dependent
component of a motion by assuming it to be a small perturbation
of a mean steady flow. Such time-linearizations, and the need
to account for shock motion, are discussed in a recent review
by Tijdeman and Seebass (1980).

The development of methods for solving the full potential
equation for unsteady flow has only just begun but it is clear
that progress will be rapid. Isogai (1977) has presented a
quasi-conservative ADI method in which the small-perturbation,
rather than the exact, boundary conditions were satisfied.
Chipman and Jameson (1979) and Goorjian (1980) have presented
conservative ADI methods for symmetrical non-lifting flows, in
which the exact boundary conditions were satisfied, but which
were otherwise very different. Finally, Steger and Caradonna
(1980) have presented a conservative ADI method which, while
reverting to small-perturbation boundary conditions, is intended
for three-dimensional flows with lift.

The method of Chipman and Jameson differs in essence from
the others in that, instead of dealing with the velocity

potential, the equations of continuity and momentum in the primitive variables — density and the velocity components — are solved. An artificial viscosity is introduced and shocks are captured as sharp compressions but it is not clear whether mass or momentum, or neither, is conserved across a shock. Although there are three equations to solve simultaneously they take relatively simple forms so that solution is straightforward, by means of a simple approximate factorization and the inversion of block tridiagonal matrices. In contrast, the others choose to solve a single, relatively complicated equation for the velocity potential. The complication follows essentially from the need to express the derivative ρ_t in the basic governing equation

$$\rho_t + (\rho\phi_x)_x + (\rho\phi_y)_y = 0$$

in terms of the unknown ϕ whilst retaining the divergence form. Steger and Caradonna solve their implicit difference equations by a simple approximate factorization

$$L_x L_y (\phi^{n+1} - \phi^n) = R$$

where the factors are identical to each other in form and involve only scalar tridiagonal inversions. The simple structure makes the extension to three dimensions readily understood. Goorjian's ADI scheme appears to resemble the above factorization closely. Steger and Caradonna remark that theirs 'is a very poor relaxation algorithm — steady states can require over 1000 time steps'. It is not known how the other methods compare. Goorjian's method may require fewer operations per time step than that of Chipman and Jameson, but the latter may admit larger time steps. At present all of the methods are considerably slower than the method of Ballhaus and Steger for the low-frequency transonic small-perturbation equation, and considerably faster than any of the methods for the Euler equations.

4. SOLUTIONS OF THE EULER EQUATIONS

4.1 Background

The Euler equations may be written in the compact form

$$U_t + F_x + G_y = 0 \tag{26}$$

where U is an unknown vector with components $[\rho, \rho u, \rho v, \rho E]$ and $F(U)$ and $G(U)$ are vectors whose components are simple functions of the components of U. The equations enable shock waves to be represented correctly, because the flow is not assumed to be isentropic. This is a substantial advance over the formulations

based on the velocity potential. No longer is there necessarily
an inaccuracy in the position, strength or speed of shock waves.
This is not achieved without cost. The equations are now much
more complicated, with four dependent variables (five, for problems in three space dimensions) instead of two. A correspondingly large computational effort is required. It is difficult
to give any useful measure of comparative effort, because this
depends on the method adopted. Only the implicit factored
scheme of Warming and Beam (1978) appears applicable to the
potential equations as well as to the Euler equations, and for
this scheme the computational effort is roughly proportional
to the product of the square of the number of dependent variables and the average number of terms per scalar equation in
(26). Thus, if there were no difference in the number of
terms per equation, solution of the Euler equations might take
four times longer than solution of the equations for potential
flow, and in three dimensions the factor would be about 6.

Most of the present methods for solving the Euler equations
for transonic flow are of the initial value type, marching in
real or pseudo-time. Thus even steady flows are calculated by
solving a system of hyperbolic partial differential equations.
There are explicit, implicit and hybrid explicit-implicit methods
in use. Here no attempt is made even to list the many methods.
Instead we outline and discuss perhaps the best known of the
methods, one from each category.

4.2 The explicit time-splitting method of MacCormack

This now classical method with its many variants has had an
extended life even though it is very demanding in computer time,
perhaps because of its relative simplicity. MacCormack and
Paullay (1972) replace the finite-difference approximation to
equation (26), namely

$$U_{i,k}^{n+1} = L(\Delta t) U_{i,k}^{n} \tag{27}$$

by

$$U_{i,k}^{n+1} = L_y\left(\frac{\Delta t}{2}\right) L_x(\Delta t) L_y\left(\frac{\Delta t}{2}\right) U_{i,k}^{n}$$

where the operators L_x and L_y correspond to finite-difference
approximations that advance U in time in accordance with the
equations $U_t + F_x = 0$ and $U_t + G_y = 0$ respectively. In each of
L_x and L_y there are predictor (p) and corrector (c) stages.
For example $L_x(\Delta t) U_{i,k}^{n}$ is

$$U^p_{i,k} = U^n_{i,k} - \frac{\Delta t}{\Delta x} (F^n_{i,k} - F^n_{i-1,k})$$

$$U^c_{i,k} = \frac{1}{2} \left\{ U^n_{i,k} + U^p_{i,k} - \frac{\Delta t}{\Delta x} (F^p_{i+1,k} - F^p_{i,k}) \right\} \tag{29}$$

For numerical stability it is required that

$$\Delta t \leq \left[\frac{\Delta x}{|u| + a + \ldots} , \quad \frac{2\Delta y}{|v| + a + \ldots} \right] . \tag{30}$$

Thus speed is improved for $\Delta y < \Delta x$. However, if one or both of Δx, Δy have to be very small to admit the required accuracy the computing time becomes very large.

Another shortcoming is in the simulation of shock waves. These are captured as steep compressions accompanied on each side by spatial oscillations. To damp these oscillations extra, dissipative, terms are usually included. The 'mean' pressure rise appears to be sufficiently close to the correct Rankine—Hugoniot value for some engineering applications but the details of any viscous interactions must be seriously in error. More-over the calculated entropy rise through the shock is far from correct and this implies that wave drag cannot be expected to be correct. Poor shock capturing is in fact a feature of most of the current methods for the Euler equations, hybrid and implicit as well as explicit, and unless it can be remedied any advantage over potential flow methods is lost.

The above spurious oscillations can be attributed to an assignment of characteristic directions that is inappropriate. Recently Roe (1979) has shown how MacCormack's method can be modified to take proper account of the characteristic directions, for unsteady unidirectional flow, and Sells (1980) has in-corporated Roe's ideas into an explicit time-splitting method of the MacCormack type for transonic two-dimensional flow past aerofoils. Details of the method are provided in the article by Roe in this volume. There are virtually no oscillations and hence there is no longer a need to add dissipative terms for damping. Wider ranging tests have shown, however, that this great improvement in shock capturing is obtained only when the shock is aligned or roughly aligned with a coordinate direction. It seems generally accepted now that the accurate capture of shock waves in more than one space dimension presents severe difficulties. Note that although the Roe-Sells modification greatly reduces the number of time steps required to reach a steady state, the computational effort per step is much in-creased, so that the method remains very time-consuming, with the equivalent of 2—3 hours on a CDC 7600 being required to reach a steady state for the transonic flow past a lifting aerofoil.

4.3 *The hybrid method of MacCormack (1976)*

This method is intended primarily for solving the Navier—
Stokes equations for high Reynolds number flows, where the cross-
wise mesh dimension must be exceedingly small for adequate
accuracy. It has been widely applied, perhaps the most remarkable case being the simulation by Levy (1978) of transonic
dynamic stall. MacCormack begins with a time-splitting formulation similar to equation (28), with L_y further split into
inviscid (hyperbolic) and viscous parts. We are concerned
here only with the inviscid part. For this a stage is introduced in which the pressure terms are separated from the convective terms and evaluated by use of the method of characteristics.
The resulting stability condition is

$$\Delta t \leq \frac{2\Delta y}{|v|} \quad ,$$

which, when $|v|$ is small, allows much larger values of Δt than
the corresponding condition $\Delta t \leq 2\Delta y/(|v| + a)$ in (30). The
main advantage of the method is the very large reduction in
computing time when compared with the corresponding explicit
method, for problems to which it is well suited. However, it is
necessary that the flow is approximately aligned with the x-
direction and that the large flow gradients that require a fine
mesh are confined to the cross-wise coordinate direction. This
may explain why aerofoils which have leading edges with small
radius of curvature still require the order of half an hour's
computing time on a CDC 7600. Also, shock waves are not captured any better than with the basic explicit method, and
the complexity of the hybrid method makes it difficult to improve capturing by a technique such as that employed by Sells.
Indeed, such hybrid methods are so complex that while the
number of applications continues to grow the only computer
programs available appear to be those prepared by the originators
of the algorithms.

4.4 *The implicit factored scheme of Warming and Beam*

When the Euler equations (26) are approximated by use of an
implicit difference scheme the resulting system of algebraic
equations presents a formidable multi-dimensional matrix inversion problem. The basic idea in overcoming this difficulty
is to replace this formidable inversion by a sequence of one-
dimensional inversions derived by approximate factorization.
For a discussion of the process reference may be made to the
work of Warming and Beam (1978), who have been at the centre
of the development of such implicit factored schemes for aerodynamic calculations. There are many variations, the schemes
being applicable also to the Navier—Stokes equations. Particularly

impressive have been the simulations by Pulliam and Steger
(1980) of three-dimensional separation from inclined hemisphere-
cylinder combinations at transonic speeds and high Reynolds
number. The essential features of the Warming and Beam algorithm
may be indicated as follows.

We begin with an implicit difference approximation to (26),
namely

$$U^{n+1} = U^n - \frac{\Delta t}{2}[(F_x + G_y)^n + (F_x + G_y)^{n+1}] + O(\Delta t^3) . \qquad (31)$$

To deal with the non-linear vectors $F^{n+1} = F(U^{n+1})$, etc, local
Taylor expansions are made, namely

$$\left.\begin{aligned}
F^{n+1} &= F^n + A^n(U^{n+1} - U^n) + O(\Delta t^2) \\
G^{n+1} &= G^n + B^n(U^{n+1} - U^n) + O(\Delta t^2)
\end{aligned}\right\} , \qquad (32)$$

where A and B are the known Jacobian matrices $\partial F/\partial U$ and $\partial G/\partial U$
respectively. On setting $U^{n+1} - U^n = \Delta U^n$ and substituting the
expansions into (31) we obtain

$$[I + \frac{\Delta t}{2}(\frac{\partial}{\partial x} A^n + \frac{\partial}{\partial y} B^n)]\Delta U^n = -\Delta t(F_x + G_y)^n + O(\Delta t^3) . \qquad (33)$$

The left-hand side is now easily factorized approximately to
yield

$$(I + \frac{\Delta t}{2} \frac{\partial}{\partial x} A^n)(I + \frac{\Delta t}{2} \frac{\partial}{\partial y} B^n)\Delta U^n = -\Delta t(F_x + G_y)^n + O(\Delta t^3) \qquad (34)$$

which is solved by the sequence

$$\left.\begin{aligned}
(I + \frac{\Delta t}{2} \frac{\partial}{\partial x} A^n)\overline{\Delta U} &= -\Delta t(F_x + G_y)^n \\
(I + \frac{\Delta t}{2} \frac{\partial}{\partial y} B^n)\Delta U^n &= \overline{\Delta U} \\
U^{n+1} &= U^n + \Delta U^n
\end{aligned}\right\} . \qquad (35)$$

There are no stability conditions that restrict the size of the
time step Δt. This algorithm has provided solutions much
faster than any explicit method but no faster so far than the
hybrid method of MacCormack, convergence to a steady state
requiring several hundred time steps. Of course, implicit
methods have a theoretical advantage in speed over explicit
methods only when the gradients in time are small enough for

the time step that is dictated by accuracy to be larger than the
time step dictated by the explicit stability criteria.

5. COMPARISONS OF METHODS

In this final section we attempt to sum up the present status
of the various methods of solution. As an aid to discussion,
an assessment scheme is set out in Table 1, where various aspects
of performance can be given levels in a five-position range,
from 'completely satisfactory' on the extreme left to 'barely
acceptable' on the extreme right. This rating procedure which
is, of course, to some extent subjective is common in other
fields. For reference, we have included the two SLOR methods
that are currently most used in aerodynamic design.

There is clearly much scope for improvement in the simulation
of shock waves. The Euler equations yield better results than
the potential equation, as they should, but more attention to
the details of the shock-capturing process is needed, especially
for multi-dimensional problems. At present only aligned shocks
can be captured well; other shocks tend to be wrong in strength
and excessively diffused or 'smeared'. The 'shocks' captured
in the course of solving a potential equation must be non-
physical and the discrepancies will be most marked for unsteady
flows involving shock movement. For practical potential flow
calculations it appears to be best to conserve neither mass nor
momentum.

In shockfree flow the TSP methods are at a substantial disadvantage because of the basic limitations inherent in the
small-perturbation approximation. In principle, solutions of
the Euler equations should be as accurate as solutions of the
full potential equation but, in practice, this seems difficult
to achieve, partly because of problems in satisfying boundary
conditions accurately.

There is also much scope for improvement in speed of some of
the implicit methods. The methods for solving the full potential
equation for unsteady flow have only just appeared so some
inefficiencies must be expected. But the implicit methods for
the Euler equations have been known for some time to be dis-
appointingly slow. Perhaps only the resolution of the multi-
dimensional shock-capturing problem will provide the remedy.
Note that the high marks for the approximate factorization and
multi-grid schemes can be justified so far only on performance
in two dimensions. Nevertheless it is clear that SLOR is being
superseded.

Finally it is desirable also that methods should be practicable, in the sense of being readily applicable to practical

TABLE 1

Comparison of methods for inviscid transonic flow with shock waves

Method	Shock simulation					Shockfree flow					Speed						Practicability				
TSP steady																					
SLOR				✓						✓			✓				✓				
TSP unsteady																					
low frequency					✓					✓		✓						✓			
all frequencies					✓					✓			✓						✓		
FP steady																					
SLOR				✓		✓							✓					✓			
approximate factorisation				✓		✓						✓						✓			
multigrid				✓		✓					✓								✓		
FP unsteady																					
implicit					✓	✓							✓						✓		
Euler																					
explicit		✓					✓									✓			✓		
hybrid			✓				✓								✓						✓
implicit			✓				✓								✓					✓	
(Finite element)																					

problems and programmable with acceptable effort. TSP methods
have been invaluable because they are highly practicable even
though their simulation capability is limited. At present
there are no rival methods for the more complex aircraft configurations, and the reason for this lies in the relative ease
with which surface boundary conditions can be satisfied in the
TSP approximation. Multigrid seems at present less practicable
than approximate factorization, mainly because provision must be
made for a multiplicity of grids of differing mesh sizes. The
position could be changed if multigrid turns out to have advantages with complex configurations. Methods for the Euler equations are inherently less practicable than potential flow
methods, because the governing equations are more complex, and
must rely upon superior shock simulation for justification.

All the above methods are of the finite-difference type.
There are finite element methods in existence for solving at
least some of the problems we have covered and an attempt could
be made to include in the Table marks for the various aspects
of performance of these methods. On the available evidence, such
a comparison would probably show finite elements to be inferior,
but it seems premature to draw definite conclusions.

REFERENCES

Albone, C.M. (1974) A finite-difference scheme for computing supercritical flows in arbitrary coordinate systems. RAE Technical Report 74090.

Albone, C.M., Hall, M.G. and Joyce Gaynor (1976) Numerical solutions for transonic flows past wing-body configurations. Symposium Transsonicum II, Göttingen, Springer Verlag.

Albone, C.M. and Hall, M.G. (1980) A scheme for the improved capture of shock waves in potential flow calculations. RAE Technical Report 80128.

Bailey, F.R. and Ballhaus, W.F. (1975) Comparisons of computed and experimental pressures for transonic flows about isolated wings and wing-fuselage configurations. NASA SP-347, 1213–31.

Baker, T.J. (1981) The computation of transonic potential flow. VKI Lecture Series "Computational Fluid Dynamics" Von Karman Institute for Fluid Dynamics, Rhode-St-Genese, Belgium.

Ballhaus, W.F. and Steger, J.L. (1975) Implicit approximate factorization schemes for the low frequency transonic equation. NASA TM X-73082.

Ballhaus, W.F. and Goorjian, P.M. (1977) Implicit finite-difference computations of unsteady transonic flows about airfoils. *AIAA J.* **15**, No. 12, 1728-35.

Ballhaus, W.F., Jameson, A. and Albert, J. (1978) Implicit approximate-factorization schemes for steady transonic flow problems. *AIAA J.* **16**, No. 6, 573-9.

Borland, C., Rizzetta, D. and Yoshihara, H. (1980) Numerical solution of three-dimensional unsteady transonic flow over swept wings. *AIAA Pap.* 80-1369.

Brandt, A. (1977) Multi-level adaptive solutions to boundary-value problems. *Math. Comp.* **31**, No. 138, 333–90.

Catherall, D. (1981) Optimum approximate factorisation schemes for 2D steady transonic flows. *AIAA Pap.* 81-1018.

Chipman, R. and Jameson, A. (1979) Fully conservative numerical solutions for unsteady irrotational transonic flow about airfoils. *AIAA Pap.* 79-1555.

Federenko, R.P. (1964) The speed of convergence of one iterative process. *USSR Comp. Math. and Math. Phys.* **4**, 227–35.

Forsey, C.R. and Carr, M.P. (1978) The calculation of transonic flow over three-dimensional swept wings using the exact potential equation. DGLR Symposium Transonic Configurations, Bad Harzburg.

Garabedian, P.R. and Korn, D.G. (1971) Analysis of transonic airfoils. *Comm. Pure and Applied Maths.* XXIV, 841–51.

Goorjian, P. (1980) Implicit computations of unsteady transonic flow governed by the full potential equation in conservation form. *AIAA Pap.* 80–150.

Holst, T.L. (1980) Fast, conservative algorithm for solving the transonic full-potential equation. *AIAA J.* **18**, No. 12, 1431–39.

Holst, T.L. and Ballhaus, W.F. (1979) Fast conservative schemes for the full potential equation applied to transonic flows. *AIAA J.* **17**, No. 2, 145–52.

Houwink, R. and van der Vooren, J. (1980) Improved version of LTRAN2 for unsteady transonic flow computations. *AIAA J.* **18**, No. 8, 1008–10.

Isogai, K. (1977) Calculation of unsteady transonic flow over oscillating airfoils using the full potential equation. *Proc. AIAA Dynamics Specialist Conf*, San Diego.

Isogai, K. (1980) Numerical study of transonic flutter of a two-dimensional airfoil. National Aerospace Laboratory, Chofu, Tokyo, Japan TR-617T.

Jameson, A. (1974) Iterative solution of transonic flows over airfoils and wings, including flows at Mach 1. *Comm. Pure Appl. Math*, **27**, 283–309.

Jameson, A. (1975) Transonic potential flow calculations using conservation form. *Proc. AIAA 2nd Computational Fluid Dynamics* Conference, Hartford, Connecticut, 148–61.

Jameson, A. and Caughey, D.A. (1977) A finite-volume method for transonic potential flow calculations. *Proc. AIAA 3rd Computational Fluid Dynamics* Conference, Albuquerque, N. Mex. 35–54.

Jameson, A. (1979) Acceleration of transonic potential flow calculations on arbitrary meshes by the multiple grid method. *Proc. AIAA 4th Computational Fluid Dynamics* Conference, Williamsburg, Va, 122–46, *AIAA Pap* 79-1458.

Lax, P.D. (1954) Weak solutions of nonlinear hyperbolic equations and their numerical computation. *Comm. Pure and Applied Maths*, **VII**, 159–93.

Levy, L. L. (1978) Experimental and computational steady and unsteady transonic flows about a thick airfoil. *AIAA J.* **16**, No. 6, 564–72.

Lock, R.C. (1981) A modification to the method of Garabedian and Korn. Numerical methods for the computation of inviscid transonic flows with shock waves (eds A. Rizzi and H. Viviand), 116–24. Friedr. Vieweg and Sohn, Braunschweig.

MacCormack, R.W. and Paullay, A.J. (1972) Computational efficiency achieved by time-splitting of finite-difference operators. *AIAA Pap.* 72-154.

MacCormack, R.W. (1976) An efficient numerical method for solving the time-dependent compressible Navier–Stokes equations at high Reynolds number. *Computing in Applied Mechanics,AMD*, **18**, Soc. Mech. Eng. New York.

Murman, E.M. and Cole, J.D. (1971) Calculation of plane steady transonic flows. *AIAA J.* **9**, No. 1, 114–21.

Pulliam, T.H. and Steger, J.L. (1980) Implicit finite-difference simulations of three-dimensional compressible flow. *AIAA J.* **18**, No. 2, 159–67.

Rizzetta, D.P. and Chin, W.C. (1979) Effects of frequency in unsteady transonic flow. *AIAA J.* **17**, No. 7, 779–81.

Roe, P.L. (1979) An improved version of MacCormack's shock-capturing algorithm. (Unpublished MOD(PE) Report).

Sells, C.C.L. (1980) Solution of the Euler equations for transonic flow past a lifting aerofoil. RAE Technical Report 80065.

Steger, J.L. and Caradonna, F.X. (1980) A conservative implicit finite difference algorithm for the unsteady transonic full potential equation. *AIAA Pap.* 80-1368.

Tijdeman, H. and Seebass, R. (1980) Transonic flow past oscillating airfoils. *Ann. Rev. Fluid Mech.* **12**, 181–222.

Warming, R.F. and Beam, R.M. (1978) On the construction and application of implicit factored schemes for conservation laws. *SIAM-AMS Proc.* **11**, 85–129.

CHARACTERISATION OF PARALLEL COMPUTERS AND ALGORITHMS

R.W. Hockney

*(Computer Science Department,
Reading University, U.K.)*

ABSTRACT

The principal current development in computing is the advent
of the parallel computer in all its various forms, for example
pipelined vector computers (CRAY-1 and CYBER 205) and arrays of
processors (ICL DAP). This paper defines a two-parameter
characterization of such computers that measures both the maximum performance and the amount of hardware parallelism. This
allows a rational comparison of the performance of alternative
algorithms on widely differing computers. As an example we
consider the choice of the best algorithm for the solution of
tridiagonal systems of equations.

1. PARALLEL COMPUTERS

The principal current development in computing is the widespread introduction of parallel operation (or parallelism) into
the architecture of computers. This has occurred because
improvements in technology have been insufficient, by themselves,
to produce the arithmetic performance that is required to solve
the large problems in physics and engineering that are the
subject of this conference.

The principal types of parallelism that have been introduced
are *pipelining* and *replication*. In pipelining, the many different sub operations in an arithmetic operation are overlapped
in time, and the required numbers are "manufactured" on the
assembly-line principle that is to be found, for example, in a
car factory. This method of processing is particularly effective when it is applied to a sequence of identical operations,
as occurs during operations on vectors of numbers. Consequently
the computers usually have separate instructions for vector
operations and may be called *vector processors* or *vector
computers*. Examples of current computers using this principle
are the CRAY-1 (Johnson, 1978) and the CDC Cyber 205 (Kascic,
1979). Both these computers are described in the book by

Hockney and Jesshope (1981). A CRAY-1S has been installed by
the Science Research Council at their Daresbury Laboratory for
use by SRC grant holders, and one of the first Cyber 205s is
to be installed at the U.K. Meteorological Office, Bracknell,
for use in weather forecasting.

In a replicated design, on the other hand, an array of identical processing elements (PEs) perform the same operation on
their own individual and different data in their private memories. In such a processor array the PEs are controlled in
synchronism by a single control unit and usually have the ability
to access data also in the memories of their immediate neighbours
in the array. Such computers are called *array processors* or
processor arrays but we shall use the latter name as it is a
more correct use of the English language, and the former term
is sometimes used quite correctly for computers that are good at
processing arrays of numbers even though they are built on the
pipeline principle (e.g. Floating Point Systems AP-120B;
Wittmayer, 1978). The only processor array currently being
marketed is the ICL Distributed Array Processor (DAP; Reddaway,
1973) which is available to U.K users from Queen Mary College,
London. This machine is a 64 x 64 array of one-bit processing
elements. Both the FPS AP-120B and the ICL DAP are also described in Hockney and Jesshope (1981).

Since the description of the CRAY-1, Cyber 205 and ICL DAP
is the subject of a separate session at this conference and
other papers, we shall concentrate here on the general problem
of characterizing the architecture and performance of such widely differing designs. It is our contention that, at least to a
first approximation, it is possible to look upon the pipelined
computer and processor array as members of the same family,
rather than as fundamentally and qualitatively different types
of computer (see sections 2 and 3). We will do this by studying the performance of all computers on vectors of varying
length, and thereby produce a two-parameter description of the
computer. The first parameter is the traditional maximum performance and the new second parameter is a measure of the
parallelism in the computer. In the world of parallel computers
we find that both parameters are needed if a proper assessment
of the relative performance of two computers is to be made, and
that it is the second parameter - which has not received significant attention in the literature up to now - that determines
the choice of the best algorithm on a particular computer (see
section 4). We illustrate the latter choice by considering in
sections 5 and 6 the solution of single and multiple sets of
tridiagonal systems of equations.

2. TWO-PARAMETER DESCRIPTION

In our approach all computers are characterized by observing the time, t_n, required to perform n elemental operations and fitting this to the generic form (Hockney 1981a)

$$t_n = r_\infty^{-1} (n + n_{\frac{1}{2}}) \qquad (2.1)$$

where

r_∞ - is the *maximum or asymptotic performance* in megaflops (Mflops) and occurs in the limit of infinite vector length.

$n_{\frac{1}{2}}$ - is the *half-performance length,* or vector length that is necessary in order to achieve half the maximum performance. We shall see that this parameter is a measure of the parallelism in the hardware.

In equation (2.1) the number of elemental operations is the number of arithmetic operations between individual elements of the vectors concerned. For a simple dyadic operation like $\underline{z} = \underline{x} + \underline{y}$, it is the length of the vectors $\underline{x}$, $\underline{y}$ and $\underline{z}$. For a triadic operation such as $\underline{z} = \underline{x} + \underline{y}*\underline{w}$, the number of elemental operations is twice the length of the vectors.

The parameters r_∞ and $n_{\frac{1}{2}}$ describe the performance of a computer (be it serial or parallel) on data organized as vectors. We shall see that there is also a useful derived parameter

$\pi = (r_\infty/n_{\frac{1}{2}})$ - the *specific performance* or performance per unit parallelism.

The above parameters are regarded as empirical quantities that are most reliably obtained by measuring t_n as a function of n using simple test programs such as the following FORTRAN sequence testing the multiply operation:

```
            CALL SECOND (T1)
            CALL SECOND (T2)
            TO = T2 - T1

                                              (2.2)

            DO 20 N = 1, NMAX
            CALL SECOND (T1)
            DO 10 I = 1, N
         10 A(I) = B(I)*C(I)
            CALL SECOND (T2)

         20 TN = T2 - T1 -TO
```

in which SECOND(T) is a subprogram delivering the CPU time T in
seconds at the time of the CALL, TN is the time for N elemental
operations and NMAX is the maximum value of N to be tested.
Fig. 1 shows how the three parameters r_∞, $n_{\frac{1}{2}}$ and π can be
obtained directly from the timing curve.

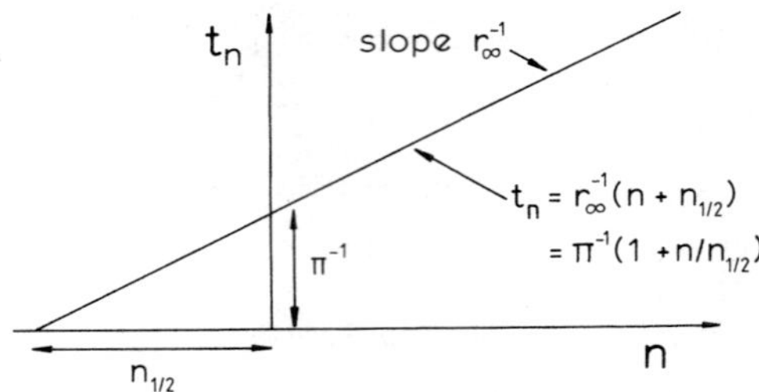

Fig. 1　The generic timing curve, showing how the parameter r_∞,
$n_{\frac{1}{2}}$ and π may be obtained.

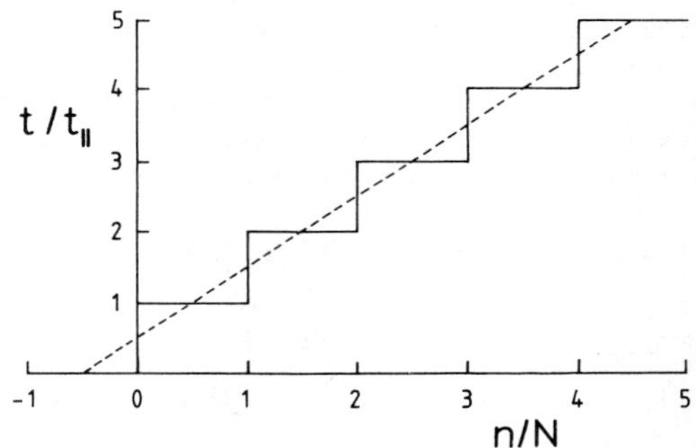

Fig. 2　The timing curve (solid line) for a vector operation of
length n on an array of N processors. $t_\|$ is the time
for one parallel operation of all N processors. The
dotted line is the generic curve for $r_\infty = N/t_\|$ and
$n_{\frac{1}{2}} = N/2$.

We will now consider the timing formulae that are expected
for serial, pipelined and array-like architectures, in order to
determine the values of r_∞ and $n_{\frac{1}{2}}$ for these different architectures. For a serial computer we should expect the execution
time to be proportional to the number of arithmetic operations
hence

$$t_n = t_1 n \tag{2.3}$$

where t_1 is the time for a single arithmetic operation.

Comparing equations (2.3) with (2.1) we obtain for the serial computer

$$r_\infty = t_1^{-1}; \quad n_{\frac{1}{2}} = 0 \qquad (2.4)$$

The timing formulae that are quoted by manufacturers for most pipelined computers are of the form

$$t_n = \{s + \ell + (n-1)\}\tau \qquad (2.5)$$

where τ is the clock period

   s is a fixed startup time (in clock periods), which is
      usually mostly the time to fetch the data from the
      memory or registers to the top of the pipelined
      arithmetic unit

 and

   ℓ is the number of overlapped sub-operations (or segments)
      into which the arithmetic operation is divided. This
      will usually differ slightly depending on the type of
      arithmetic operation.

One may see from equation (2.5) that once the pipeline is full
a new result appears at the output of the pipeline every clock
period. Hence the maximum rate of producing results is one
elemental operation per clock period. However it takes
$(s + \ell - 1)$ clock periods to fill the pipeline. Comparison
with the generic form (2.1) shows that

$$r_\infty = \tau^{-1}; \quad n_{\frac{1}{2}} = s + \ell - 1 . \qquad (2.6)$$

The determination of r_∞ and $n_{\frac{1}{2}}$ is less obvious in an array

of N PEs because the timing curve is steplike, as shown in
Fig. 2. If the number of elemental operations is greater than
the number of PEs then the dotted line represents the best
average description of the operation, from which we obtain

$$r_\infty = N/t_\parallel , \quad n_{\frac{1}{2}} = N/2, \quad n > N \qquad (2.7)$$

where $t_\parallel$ is the time for one parallel operation on the array of

N elements. If, on the other hand, the number of PEs is greater
or equal to the number of elemental operations to be performed,
then the array acts as though it had an infinite number of PEs,
like the *paracomputer* of Schwartz (1980), then

$$\pi = t_\parallel^{-1}, \quad n_{\frac{1}{2}} = \infty, \quad n \leqslant N . \qquad (2.8)$$

In this case the parameter r_∞, turns out not to be relevant.

The significance of the parameters r_∞, $n_{\frac{1}{2}}$ and π can be judged by considering the short and long vector limits. If the vector length n is less than the parallelism of the computer $(n < n_{\frac{1}{2}})$ then we find from equation (2.1)

$$t_n = r_\infty^{-1} n_{\frac{1}{2}} = \pi^{-1} \qquad (2.9)$$

and the average performance

$$r = n/t_n = \pi n \qquad (2.10)$$

Thus the parameter π, rather than r_∞, characterizes the performance on short vectors. Furthermore, in this limit, the time of operation is independent of vector length and the average performance is proportional to vector length. The computer thus acts like an infinite processor array even though it might have quite limited parallelism.

In the long vector limit, when the vector length is much greater than the parallelism of the computer $(n > n_{\frac{1}{2}})$, we find from equation (2.1)

$$t_n = r_\infty^{-1} n \qquad (2.11)$$

$$r = n/t_n = r_\infty \qquad (2.12)$$

Thus the time of operation is proportional to the vector length, and the average rate is equal to the maximum. In this case, the computer "looks" like a serial computer to the vector being processed, even though the computer may have substantial parallelism.

3. SPECTRUM OF COMPUTERS

The results of the last section can be summarized as follows:

$$
\begin{aligned}
n_{\frac{1}{2}} &= 0 && \text{serial computer} \\
&= s+\ell-1 && \text{pipelined computer} \\
&= N/2 && \text{processor array } n > N \\
&= \infty && \left\{ \begin{array}{l} \text{paracomputer, } N = \infty \\ \text{processor array, } n \leqslant N \end{array} \right.
\end{aligned}
$$

It is apparent that $n_{\frac{1}{2}}$ is a measure of the amount of parallelism in a computer, and varies from zero to infinity as computer architectures range from the serial computer to the infinitely

parallel computer. We do not therefore regard computers as
being either serial or parallel, or pipelined or array-like.
Rather we regard computer architectures as forming a continuous
spectrum of designs with parallelism, as measured by $n_{\frac{1}{2}}$, varying
from zero to infinity.

Since the performance of a computer on vectors is determined,
not by r_∞ alone, but by the parameter pair $(n_{\frac{1}{2}}, r_\infty)$, it is
appropriate to plot computers as points on the $(n_{\frac{1}{2}}, r_\infty)$ parameter plane as is done in Fig. 3, which can be described as a
two-dimensional spectrum of computers. Quite frequently a computer may be used in several different modes of operation (e.g.
fixed or floating-point, 32-bit or 64-bit operation) and may
appear as a series of connected points on such a diagram.

As examples of pipelined computers we show the CDC 7600,
CRAY-1 and CDC Cyber 205 which have widely differing values of
$n_{\frac{1}{2}} \simeq 1$, 10,100 respectively. Constant values of specific performance π are diagonal lines at 45 degrees to the axes, and we
note that whereas the CRAY-1 (register mode) has a large value
of π (i.e. better performance on short vectors) than the Cyber
205, the 4-pipe version of the latter machine has a larger
value of r_∞ (i.e. a better performance on long vectors). The
ICL DAP is an interesting case, as its different modes of
operation - scalar, vector, array, floating-point and short
integers - span the full extent of the diagram.

4. PARALLELISM IN ALGORITHMS

An algorithm may be described as a sequence of vector operations of varying length. The total time for the algorithm may
be expressed as

$$T = \sum_{\ell=1}^{\ell \text{max}} r_\infty^{-1} q_\ell \, (p_\ell + n_{\frac{1}{2}}) \tag{4.1}$$

where there are q_ℓ operations with vector length p_ℓ at each
stage $\ell = 1,2,\ldots, \ell\text{max}$ of the algorithm. The total number of
vector operations is

$$q = \sum_{\ell=1}^{\ell \text{max}} q_\ell \tag{4.2}$$

which are equivalent to s elemental operations where

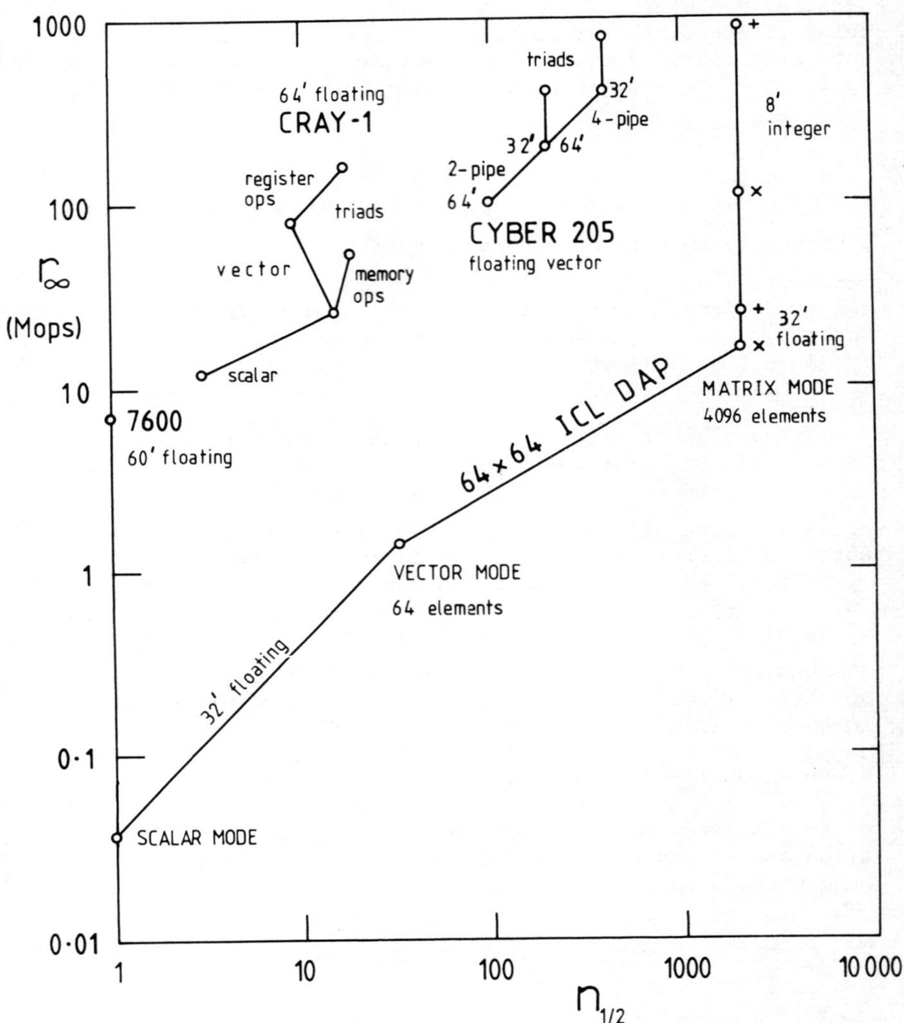

Fig. 3 The two-dimensional spectrum of computers obtained by plotting the performance of computers on the ($n_{1/2}$, r_∞) parameter plane. Constant values of specific performance $\pi = (r_\infty/n_{1/2})$ are diagonal lines at 45 degrees to the axes.

$$s = \sum_{\ell=1}^{\ell max} q_\ell p_\ell \qquad (4.3)$$

We can therefore define the average parallelism of the algorithm as

$$\bar{p} = s/q \qquad (4.4)$$

and the total execution time can be re-expressed as

$$T = r_\infty^{-1} q \; (\bar{p} + n_{\frac{1}{2}}) \qquad (4.5)$$

As with the case of the single vector operations discussed in section 2 the execution time will depend on the ratio of the parallelism of the computer $(n_{\frac{1}{2}})$ to the parallelism of the algorithm $(\bar{p})$.

If we define the performance, P, of an algorithm on a computer as the inverse of the time taken for its execution (i.e. the number of executions possible per second), then we can compare the performance of two competing algorithms (a) and (b) by forming the ratio

$$\frac{P^{(a)}}{P^{(b)}} = \frac{T^{(b)}}{T^{(a)}} = \frac{q^{(b)} (\bar{p}^{(b)} + n_{\frac{1}{2}})}{q^{(a)} (\bar{p}^{(a)} + n_{\frac{1}{2}})} \qquad (4.6)$$

in which we note that the value of r_∞ cancels. When comparing two possible algorithms for use on the same computer, the only relevant property of the computer is, therefore, its half performance length. The quantities q and $\bar{p}$ are likely to be non-linear functions of the vector length, n.

5. SINGLE TRIDIAGONAL SYSTEMS

We illustrate the choice of algorithms and the method outlined in section 4 by considering the solution of single and multiple sets of tridiagonal systems of equations, such as arise frequently in the solution of partial differential equations by difference methods. For a single tridiagonal system we compare two variants of the scalar cyclic reduction algorithm. The latter algorithms are described fully by Hockney (1981b) and we give here only a summary of the results. Let us consider a single tridiagonal system of n equations, then the methods are

SERICR - serial variant of cyclic reduction in which the vector length decreases at each stage of the reduction, and there is a subsequent fill-in phase

$$\ell\text{max} = 2\log_2 n, \quad q = 17 \log_2 n, \quad s = 17n, \quad \bar{p} = n/\log_2 n$$

$$(5.1)$$

PARACR - parallel variant of cyclic reduction in which the vector length is kept at n throughout, so that there is no need for a fill-in phase

$$\ell\text{max} = \log_2 n, \quad q = 12 \log_2 n, \quad s = 12 \, n\log_2 n, \quad \bar{p} = n \quad (5.2)$$

Comparing equations (5.1) and (5.2) we see that SERICR has fewer elemental operations, s, than PARACR; but that PARACR has fewer vector operations, q, than SERICR. Consequently SERICR will be the better on serial computers and PARACR on the paracomputer, which explains the naming of the algorithms. For actual computers, however, which lie somewhere in the spectrum between the serial computer and the paracomputer, we must use the general formula (4.6) to determine the best algorithm. In this way the concept of $n_{\frac{1}{2}}$ provides a rational method of interpolation between the two extremes.

Thus we find that PARACR will have better or equal performance to SERICR if

$$P_{PARACR} \geq P_{SERICR} \quad \text{or} \quad T_{SERICR} \geq T_{PARACR}$$

i.e. when

$$(n_{\frac{1}{2}}/n) \geq 2.4(1 - 1.42/\log_2 n)$$

$$\geq 2.4 \qquad n \to \infty$$

This result is plotted on the $(n_{\frac{1}{2}}/n, n)$ plane in Fig. 4. These coordinates are chosen to make the horizontal axis vary from the serial computer on the left to the paracomputer on the right, as in the spectrum of computers. As expected we find SERICR the better if the $n_{\frac{1}{2}}$ of the computer is small enough and PARACR in the opposite circumstance.

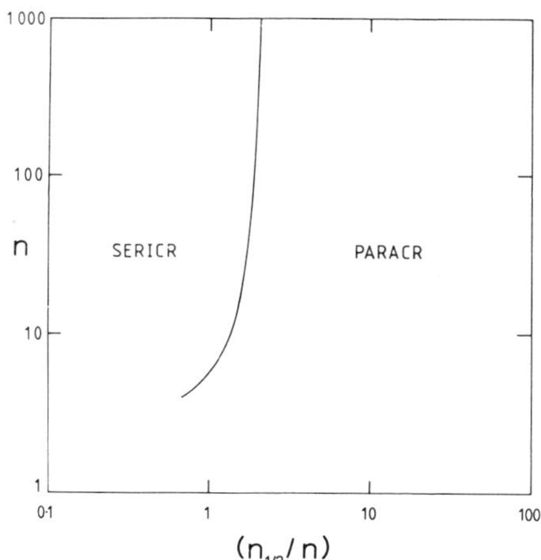

Fig. 4 Regions of the $(n_{\frac{1}{2}}/n, n)$ plane in which either the SERICR or PARACR algorithm has the higher performance for the solution of a single tridiagonal system of n equations. The horizontal coordinate is proportional to the parallelism of the computer and varies from serial computers on the left to increasingly parallel computers to the right.

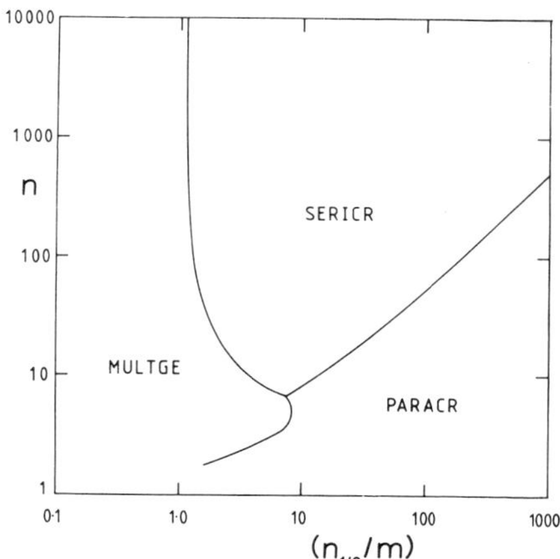

Fig. 5 "Phase diagram" showing the regions of the $(n_{\frac{1}{2}}/m, n)$ plane in which either the MULTGE, SERICR or PARACR algorithm has the highest performance in the solution of m independent tridiagonal systems, each of n equations. Note the "triple-point" at (7.7, 7) when all algorithms have the same performance.

6. PHASE DIAGRAM FOR MULTIPLE SYSTEMS

If the problem is to solve a set of m independent tridiagonal systems each of length n, then we consider three alternatives:

MULTGE - Gauss-elimination, as formulated by Varga (1962), applied to the m systems in parallel

$$T_{MULTGE} = 8n(m+n_{\frac{1}{2}}), \quad \bar{p}_{MULTGE} = m \qquad (6.1)$$

SERICR - the serial variant of cyclic reduction applied to the m systems in parallel

$$T_{SERICR} = 17(mn + n_{\frac{1}{2}}\log_2 n), \quad \bar{p}_{SERICR} = mn/\log_2 n \qquad (6.2)$$

PARACR - the parallel variant of cyclic reduction applied to the m systems in parallel

$$T_{PARACR} = 12(mn+n_{\frac{1}{2}})\log_2 n, \quad \bar{p}_{PARACR} = mn \qquad (6.3)$$

The choice of the best algorithm is now most conveniently displayed on the $(n_{\frac{1}{2}}/m,n)$ plane as in Fig. 5. This shows the regions of the plane for which each of the alternatives has the least execution time. Again serial computers lie to the left and increasingly parallel computers to the right. We find that, for a given n, MULTGE is always the best for computers with small enough $n_{\frac{1}{2}}$ and PARACR for computers with large enough $n_{\frac{1}{2}}$. For n greater than about 8, there is an intermediate range of $n_{\frac{1}{2}}$ for which SERICR is the best algorithm. This diagram plays somewhat the same role as a chemical phase diagram, in that it shows the regions of a parameter plane that favour alternative algorithms. There is even a "triple-point" at $(n_{\frac{1}{2}}/m) = 7.7$ and $n \sim 7$ at which all three algorithms have the same performance.

7. CONCLUSION

We have shown how a two-parameter description of a computer can be derived from the measured performance of the computer when operating on vectors of numbers. The first parameter is the traditional maximum performance in megaflops and the new second parameter, $n_{\frac{1}{2}}$, measures the parallelism of the hardware, and answers the question "How parallel is my computer". It also takes into account the effect of vector length on the execution time of an algorithm, and thus determines the choice of the best algorithm on a particular computer. The example taken of the

solution of tridiagonal systems of equations shows the utility
of $n_{\frac{1}{2}}$, and parameter-plane (or phase-diagram) representations
of the resulting conclusions. In the new world of parallel computers it is evident that the traditional one-parameter description of computer performance is inadequate. Thus we conclude
that $n_{\frac{1}{2}}$ is an essential second parameter to include in any discussion of either parallel-computer performance, or algorithm
performance on parallel computers.

8. REFERENCES

Hockney, R.W. (1981a) "Characterization of Parallel Computer",
IEEE Trans. Comput., submitted.

Hockney, R.W. (1981b) "Characterization of Parallel Algorithms",
IEEE Trans. Comput., submitted.

Hockney, R.W. and Jesshope, C.R. (1981) Parallel Computers -
Architecture, Programming and Algorithms, Adam Hilger, Bristol.

Johnson, P.M. (1978) "An Introduction to Vector Processing",
Comput. Design **17**(2), pp. 89-97.

Kascic, M.J. Jr. (1979) "Vector Processing on the Cyber 200" in
Infotech State of the Art Report: Super-computers (Eds. C.R.
Jesshope and R.W. Hockney), *Infotech Intl. Ltd. Maidenhead* **2**,
pp. 237-270.

Reddaway, S.F. (1973) "DAP - A Distributed Array Processor",
1st Annual Symp. on Comput. Architecture (IEEE/ACM), Florida.

Schwartz, J.T. (1980) Ultracomputers, *ACM Trans. Prog. Lang.
and Systems* **2**, pp. 484-521.

Varga, R.S. (1962) Matrix Iterative Analysis, Prentice-Hall,
Inc., Englewood Cliffs, New Jersey, pp. 195.

Wittmayer, W.R. (1978) Array Processor Provides High Throughput
Rates, *Comput. Design* **17**(3), pp. 93-100.

DEVELOPMENTS IN COORDINATE SYSTEMS

FOR FLOW FIELD PROBLEMS

M.P. Carr and C.R. Forsey

(Aircraft Research Association Ltd., Bedford)

The generation of computing grids for use in the numerical
solution of partial differential equations, such as those
describing fluid motion, is a subject growing in importance as
such methods are applied to ever more complex problems. A general survey of grid generation techniques is presented which
covers grids for specific problems, grids for more general problems based on algebraic or numerical techniques, and some other
approaches. Emphasis is placed on techniques which have already
been used successfully on practical problems or appear to have
such potential. In particular, the methods being developed at
various sites in the UK, for use in aeronautical flow field
problems, are described.

1. INTRODUCTION

There have been major advances in recent years in the use of
field methods to solve fluid mechanics problems. The flow can
be defined by one or more non-linear partial differential equations which in general cannot be solved in closed form and
require an iterative solution. Hence, in order to calculate the
flow, it is necessary to discretise the problem so that one satisfies a numerical approximation to the partial differential
equations at a finite number of points. The adequacy of the
solution of this numerical analogue depends not only on the
numerical (e.g. finite difference) approximation to the partial
differential equation but also on the position of the points at
which the equation is satisfied since any numerical approximation depends not only on the specific point but also on the position of neighbouring points.

If we consider the one-dimensional example of the usual finite
difference approximation to a first derivative $\partial\phi/\partial x$ at a point
i, of a grid of uniform spacing Δx then

$$\frac{\partial\phi_i}{\partial x} = \frac{\phi_{i+1} - \phi_{i-1}}{2\Delta x} - \frac{\Delta x^2}{6}\frac{\partial^3\phi_i}{\partial x^3} \cdots$$

$\partial\phi_i/\partial x$ is usually approximated by the first term on the right
hand side which is $(\phi_{i+1} - \phi_{i-1})/2\Delta x$ and the second and subsequent terms considered negligible; in other words the function ϕ
is assumed to be quadratic in x in the local region of interest.
Clearly, the assumption that the truncation error is small, will
be valid if either ϕ is no more than quadratic in x or the spacing Δx is small. We cannot guarantee the former so one has to
consider the latter. This creates a problem immediately since
in order to have Δx small enough in regions of high flow acceleration, a uniform grid would require very many points and computing times are a function of the number of points. Moreover,
for external aerodynamic problems the flow field is infinite
and so this leads us to providing a transformation from an
infinite physical space to a finite computing space. Hence, we
require a transformation which maps an unequally spaced grid in
physical space, thus allowing for the facility of packing many
points in regions of high flow variation, into an equally spaced
computing space.

   Following the earlier discussions two questions arise, how
can we predict where the flow gradients are large, thus requiring more points, and can the numerical approximations be changed
so that fewer points are required for the same accuracy? In
most, if not all of the flow field methods described at this
conference, the areas of high grid density are prescribed in
advance on the assumption that these coincide with areas of high
flow gradients but there has been some work on adaptive grids
(e.g. Dwyer, Kee and Sanders, 1979) where the grid size is part
of the solution. As regards a reduction in the number of grid
points, this can be achieved by using higher order approximations to the partial derivatives.

   Obviously, the analysis of truncation errors is more complicated when one has coordinate transformations in more than one
dimension and Thompson and Mastin (1980) have produced a detailed discussion of this topic. However, in the final analysis we
are interested in solving fluid mechanics problems and the computing grid is only acceptable if it enables us to do this
successfully.

   In the development of flow field methods in the past decade,
many of the methods which have made an impact were based on an
ad hoc approach to grid generation. In other words, having
decided to tackle a specific problem such as the flow around an
aerofoil, then a grid system was produced for that particular
configuration. Section 2 describes some of these methods.
However, although methods developed by this approach have been
very successful, it is clear that as we progress to more complex
configurations a more general method of grid generation is

required. In deciding on a general grid generation method, one
major decision to make is whether to develop a global system for
complete space or whether to segment the problem with individual
grids in each region, which are then connected. Another is in
the basic form of grid generation and two of the more popular
and promising areas, algebraic and numerical, are discussed in
sections 3 and 4. Alternative techniques, including grids which
are not aligned with the surfaces are dealt with in section 5.

We complete this introduction with a comment on the tools
which one requires to consider grid generation. Obviously, a
computer of some sort is needed although it need not necessarily
be very powerful (the power required for the solution of the
following flow field problem is, of course, another matter).
However, of overriding importance is an adequate graphics system,
both hardware and the back-up software to handle the complex
three-dimensional geometries. Without these facilities, anyone
engaged in grid generation for three-dimensional problems is at
a definite disadvantage.

2. SPECIAL GRIDS

Not unnaturally, the initial calculations of transonic flow
fields were for relatively simple configurations with the
emphasis on the solution algorithms rather than grid generation.
Accordingly, for these problems specific grids were developed
and whilst these are numerous, a limited number will be described. These are either typical of types of grids or relate to
methods which have received wide acceptance. It is interesting
to note that all these grids are aligned with the surface boundaries.

2.1 Aerofoil - Garabedian and Korn (1971)

Probably the most widely used method of calculating compressible, including transonic, flows is that due to Garabedian and
Korn which deals with the case of a single aerofoil. Not only
has it been used in its own right as an inviscid code but has
been extended to include an inverse technique (Tranen, 1974) and
also had viscous effects included (Bauer et al, 1975), (Collyer
and Lock, 1979). The grid which has the aerofoil surface as a
coordinate line and which is the result of mapping the exterior
of the aerofoil into the interior of a unit circle is due to
Sells (1968). A more general and efficient algorithm for producing such a grid was proposed by Jameson (Bauer et al, 1975).

The initial mapping is from the exterior Z of the aerofoil,
to the exterior σ of the circle followed by an inversion with
respect to the circle boundary. The transformation function is

given by

$$\frac{dZ}{d\sigma} = \left(1 - \frac{1}{\sigma}\right)^{1-E/\pi} \exp\left(\sum_{n=0}^{N} \frac{C_n}{\sigma^n}\right)$$

where E is the included angle at the trailing edge.
$(1-1/\sigma)^{1-E/\pi}$ is the Schwarz-Christoffel term which allows for a
corner or a cusp at the trailing edge. The trailing edge thickness $\bar{c}$ is given by

$$2\pi i\bar{c} = 2\pi i \; (E/\pi - 1 + C_1)$$

The values of the series coefficients C are obtained using a
Fast Fourier Transform technique. An example of the grid around
an aerofoil produced by this method is given in Fig. 1. Note
that grid lines are packed radially near the aerofoil and cir-
cumferentially in the region of the leading and trailing edge
points.

2.2 Axisymmetric Nacelle - Baker (1975)

The derivation of a grid suitable for a ducted body, even
axisymmetric, is more difficult than for a single closed body
such as an aerofoil. A major problem is that not only does one
require the body to be a coordinate line but also the axis of
symmetry. Arlinger (1975) derived an orthogonal grid by using
a series of conformal mappings. Baker produced a grid by dividing the region into three subregions and specifying the grid
coordinates in each region by simple analytic expressions. A
schematic representation of the grid is given in Fig. 2. The
full region extends to a large but finite distance upstream and
above the body. The body surface is the line $\xi^1 = 0$ while the
axis and external boundary are $\xi^1 = 1$. ξ^2 varies from -1 at
downstream infinity inside the body to zero at the lip and +1 at
downstream infinity outside the body.

The interfaces of the subregions are lines emanating from the
crest and throat and the analytic functions used are:

1. <u>Downstream of Throat</u>

$$x = X; \; r = R - \lambda\eta + (\lambda - R)\eta^2$$

2. <u>Downstream of Crest</u>

$$x = X; \; r = R + \lambda\eta + (r_\infty - \lambda - R)\eta^2$$

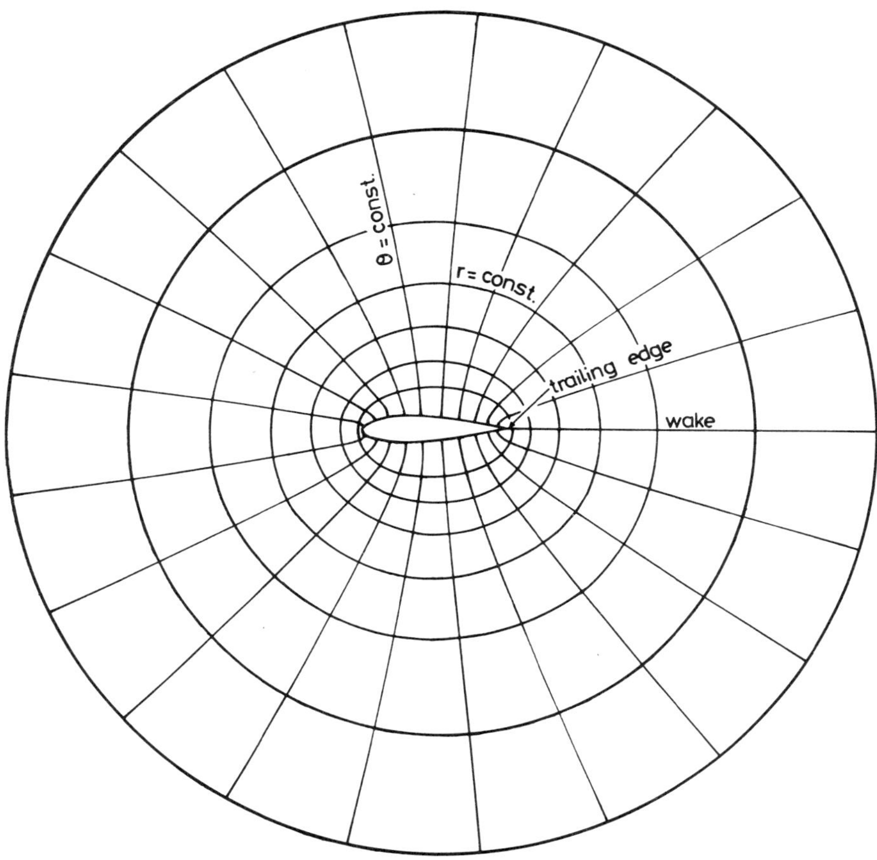

Fig. 1 Computational mesh for two dimensional aerofoil

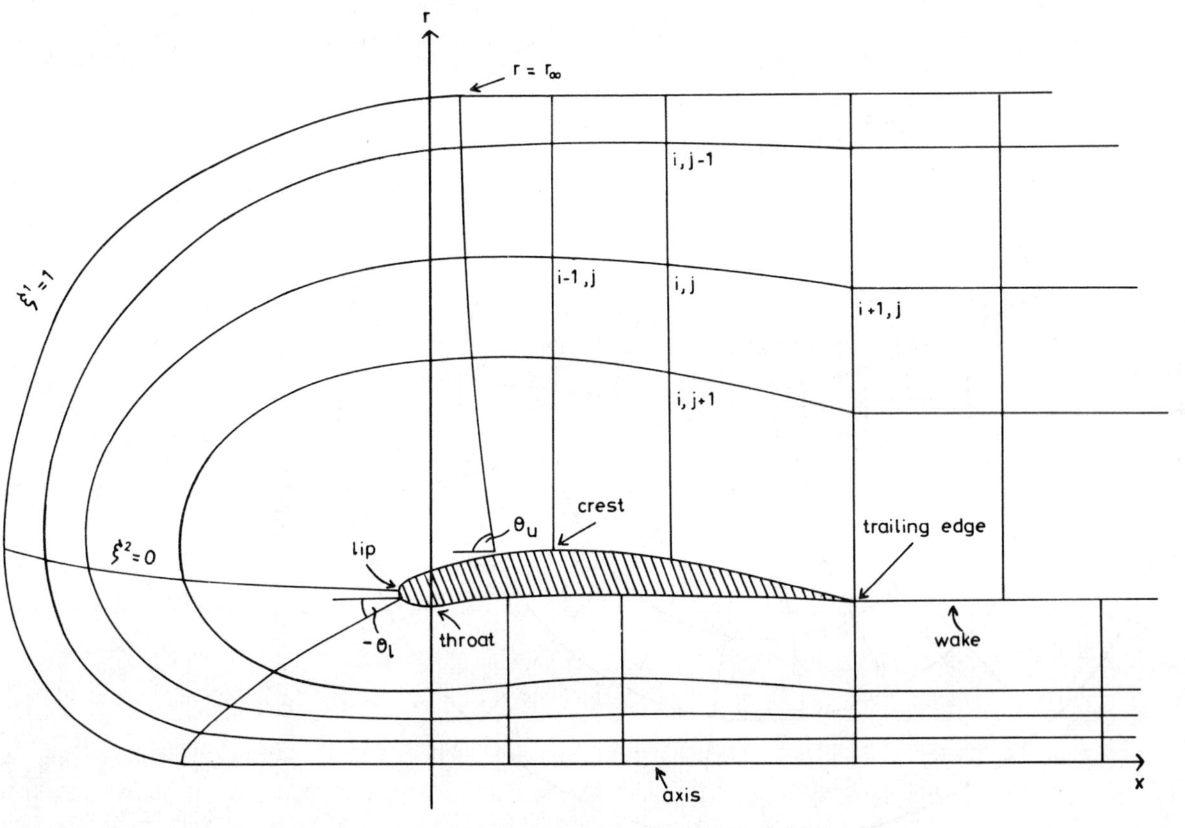

Fig. 2 Computing mesh for ducted body

3. <u>Forward Region</u>

$$x = X + a\eta + (c_1 - a - 3c_2)\frac{\eta^2}{2} + c_2\eta^3$$

$$r = R + b\eta + (r_b - b - R)\eta^2$$

where

(X,R) are the body ordinates; λ is a constant; c_1, c_2 are functions of ξ^2; $a = -\lambda\cos\theta$, $b = \lambda\sin\theta$ with θ a function of ξ^2 and r_b the value of r on the $\xi^1 = 1$ boundary; that is zero on the axis and r_∞ at the crest interface. η is a function of ξ^1 used to concentrate grid lines near the body surface.

This method has been used successfully in the design and analysis of a wide range of cowls.

2.3 Wing-Fuselage - Forsey (1978, 1981)

The previous two examples were two-dimensional in concept and when one considers a three-dimensional problem, grid generation becomes much more difficult. However, if one restricts the complexity of the configuration, then some of the problems are alleviated. The case considered by Forsey is a wing connected to a circular fuselage. The overall problem of calculating transonic flow over a wing-fuselage configuration is conveniently split into two parts; transforming the complex wing-fuselage geometry in physical space into a simple geometry in transform space and solving the potential flow equation in this transform space. These two parts are implemented as separate programs. The overall transform maps the exterior of the wing-fuselage into the interior of a circular cylinder. This transformation as shown in Fig. 3 can be described by the following steps:

1) Analytically map the circular fuselage into a flat plate. This reduces the problem to that of a wing alone with a centre line plane of symmetry.

2) The leading and trailing edges of the planform are extended to infinity using analytic curves to produce a pseudo planform.

3) An analytic spanwise transformation $\eta = \eta(y)$ is performed. This packs constant y planes near the wing root and tip and maps the plane $y = \infty$ to the plane $\eta = 1$, thus producing a finite spanwise computing space.

4) For each constant y plane, corresponding to equal intervals
in η, the region exterior to the aerofoil section on that plane
is conformally mapped to the inside of a unit circle. For sections inboard of the wing tip the mapping is performed numerically using the Fast Fourier Transform technique of Jameson.
Sections outboard of the wing tip are slits and may be mapped
analytically to a unit circle.

5) The spanwise transform derivatives are calculated using spanwise quadratic curve fits through the coordinates.

6) Using all the first and second derivatives calculated in the
steps above the various geometric coefficients appearing in the
flow equation are evaluated at each point in the field (28
values at each point) and written to a disk file ready for use
by the flow calculation program.

The flow program which uses this grid has proved most effective
despite the restriction on fuselage shape. However, when one
considers configurations of more complexity, more general
methods of grid generation appear attractive.

2.4 Limited generalisation - Caughey (1978)

   The grids described so far in this section have been specific
to one configuration and the work described in the following
sections is aimed at general configurations. Caughey's method
can be regarded as a bridge between these two ideas since it is
concerned with generating grids for a limited range of configurations, namely two-dimensional. The basis of the method is conformal mapping which reduces the geometry to a nearly-rectangu-
lar domain. A further shearing transformation is then applied
to produce a nearly-orthogonal grid system with its boundaries
corresponding to grid lines. This technique has been used for a
number of configurations and in particular, the case of an
axisymmetric nacelle (Caughey and Jameson, 1976). Fig. 4 demonstrates how the combination of conformal and shearing transformations are used and Fig. 5 gives an indication of the form of
the grid in physical space.

3.   ALGEBRAIC GRID GENERATION

3.1 Simple analytic functions

   Many grid generation methods, whatever their basic nature,
have features which may be described as algebraic. Such common
features include packing/stretching transformations and shearing
transformations. A packing/stretching transformation is a
transformation of the form $x = f(\xi)$ where equal intervals in ξ
lead to unequal intervals in x so that the grid is finely

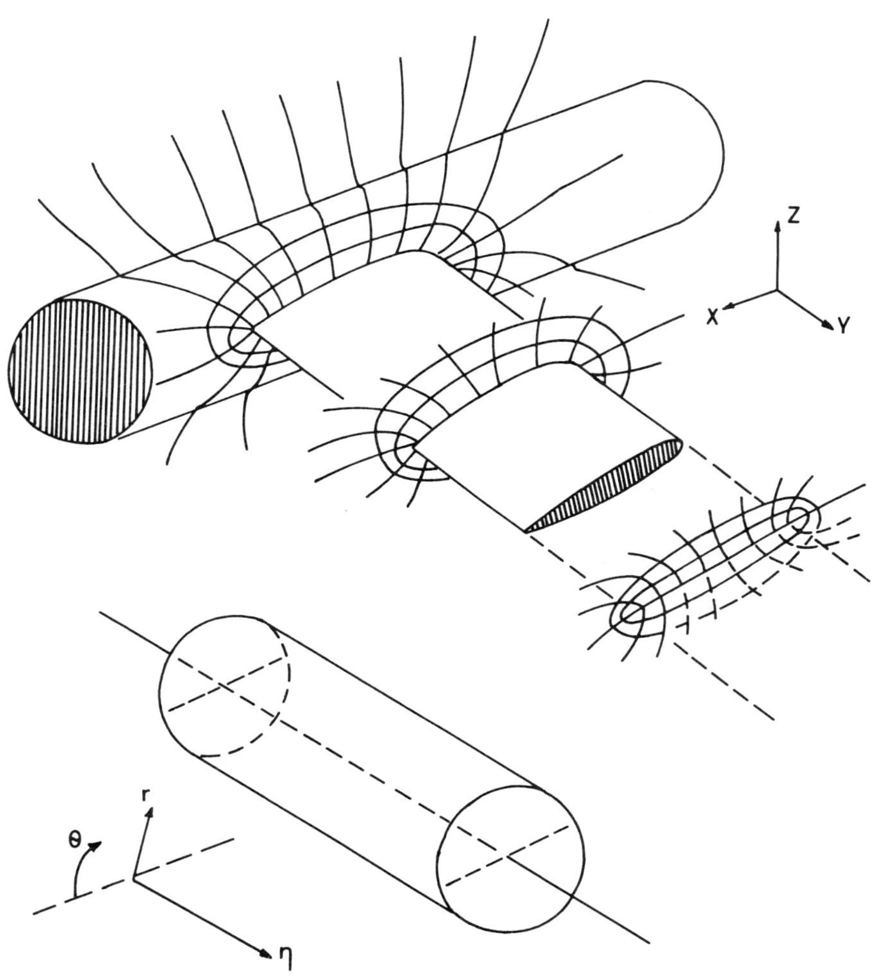

Fig. 3 Forsey's wing/body grid-schematic

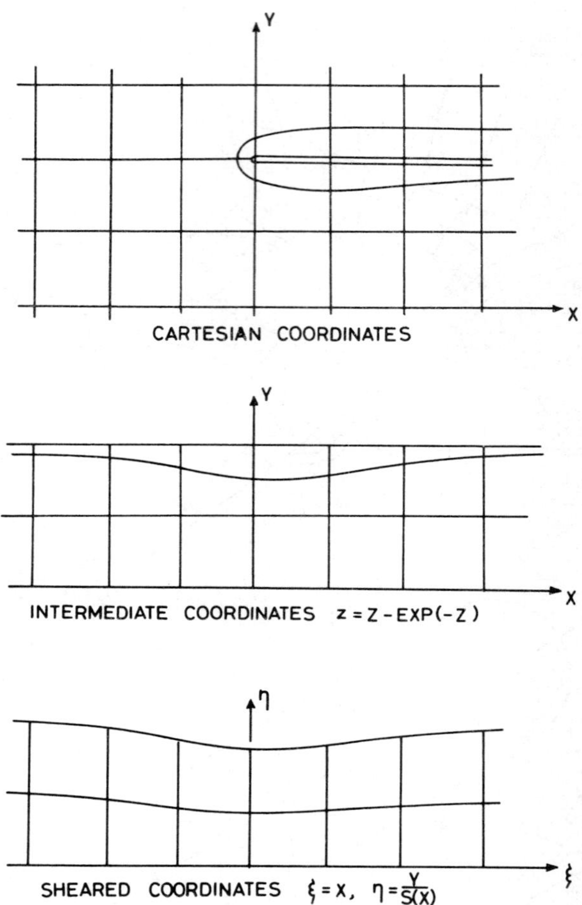

Fig. 4   Nacelle grid (Caughey and Jameson)

packed in some regions and sparse in other regions. Furthermore, by suitable choice of the function $f(\xi)$ it is also
possible to map infinity to a finite point or vice versa. Such
packing/stretching transformations may be simple (Smith and
Weigel, 1980) or rather complex (Oh, 1980) depending on the
amount of grid control required. A shearing transformation is a
linear transformation typically used to nondimensionalise the
distance between two physical boundaries (one of which may be
infinity) without necessarily changing the relative spacing of
grid points between the boundaries. For example, in the nozzle
program of Baker (1979) a shearing transformation

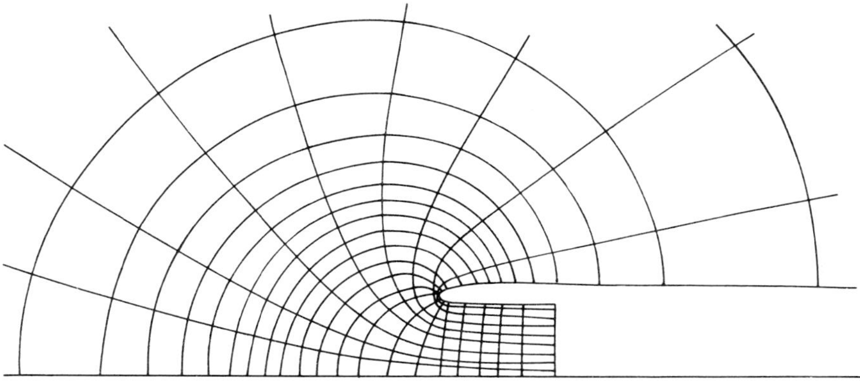

Fig. 5   Nacelle grid lines in physical space
(Caughey and Jameson)

$\eta = (y-y_1)/(y_u-y_1)$ is used to nondimensionalise the distance
between the upper (y_u) and lower (y_1) walls of the nozzle. While
primarily used in conjunction with more general grid generation
techniques, a combination of such simple analytic transformations may be used by themselves for quite complex three-dimensional problems.

One such example is the method of Walkden (1979) for supersonic flows over wing/body combinations. The three-dimensional
grid is generated as a series of two-dimensional grids on cross
section planes normal to the body axis. Fig. 6 shows a typical
cross section plane. Three regions are defined by two large
circles centred on the body axis. A shearing transformation
with packing is used in the region between the wing/body surface
and the first circle, whilst a cylindrical polar system is used
outside the second circle. In the region between the two
circles, cubic curves are used to join the inner and outer grids.
Further flexibility in controlling the grid distributions round
the circumference of the wing/body cross section is obtained by
allowing an unlimited number of regions p_i, i = 1 ... around
this circumference.

Another example of the way simple functions may be built up
into quite complex and flexible grid generation procedures is
the multisurface technique of Eiseman (Eiseman and Smith, 1980).
In this approach, a grid is obtained by joining together corresponding points on inner and outer configuration boundaries by
polynomial curves. Depending on the order of the polynomial,
one or more non-physical surfaces between the two bounding surfaces are used to control such things as grid spacing and the

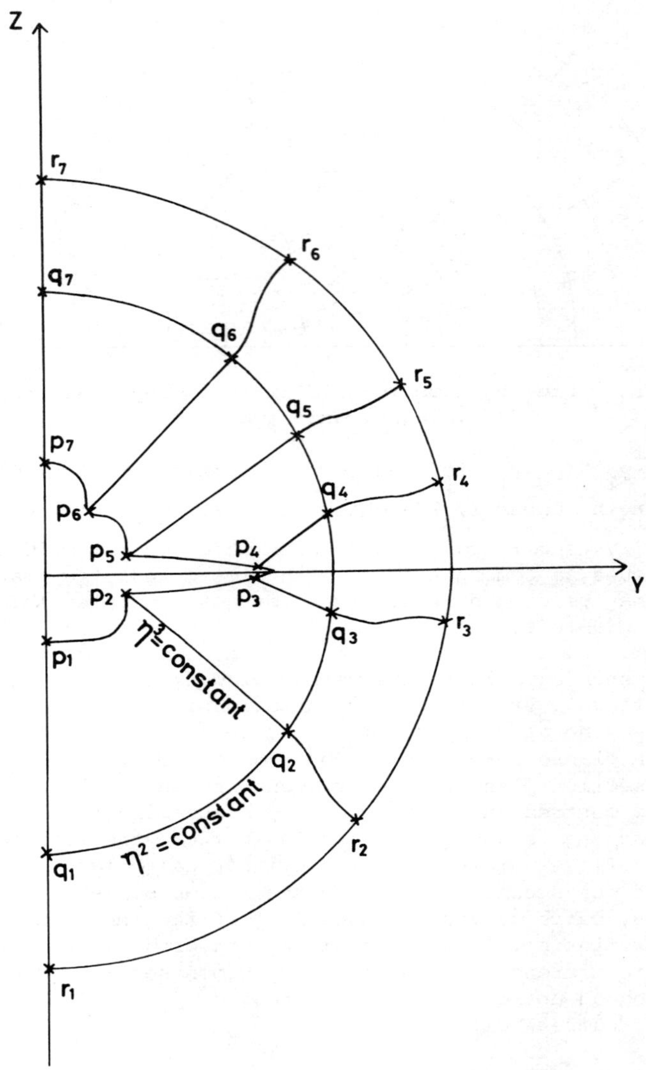

Fig. 6   Walkden's wing/body grid - schematic

direction in which the grid lines meet the boundaries. Although
the method has been used for a number of two-dimensional problems
(Eiseman and Smith, 1980) and at least one three-dimensional
problem (Eiseman, 1979), Eiseman suggests that it is not really
applicable to general three-dimensional problems and favours a
quasi-two-dimensional approach in which two-dimensional grids
are generated on cross sections through a general three-dimensional configuration.

3.2 Blending function methods

Blending function methods describe a class of algebraic grid
generation methods based on isoparametric transformations. Such
transformations were first used in finite element work to derive
specific types of finite element approximations. Zienkiewicz
and Phillips (1971) showed how such isoparametric mappings could
be used to generate grids for configurations whose boundaries
consisted of a few simple analytic curves while Gordon and Hall
(1973) generalised the method to deal with complex boundaries
and showed how extra degrees of freedom could be introduced to
allow some control over the form of the grid produced by the
mapping.

The basic principle behind isoparametric transformations can
best be described by an example. Consider mapping a curvilinear
quadrilateral in physical (x,y) space into a unit square in
parametric (s,t) space as shown in Fig. 7a. s = 0,1 and t = 0,1
correspond to the sides of the quadrilateral where it is assumed
that the distributions of x and y as functions of s, t are known.
The x and y coordinates of points within the quadrilateral are
defined as functions of s and t (and the boundary distributions)
so that a cartesian grid in (s,t) space produces a curvilinear
aligned grid in (x,y) space. The simplest isoparametric transform which does this is

$$f(s,t) = (1-s)f(0,t) + sf(1,t) + 1-t)f(s,0) + tf(s,1) -$$

$$- (1-s)(1-t)f(0,0) - (1-s)tf(0,1) - s(1-t)f(1,0) - stf(1,1)$$

where f = x or y and f(0,t), f(1,t), f(s,0), f(s,1) define the
known distributions along the sides. In effect, this function
consists of bilinear blendings of the point distributions on
opposing sides, so that the functions (1-s),s,(1-t),t are sometimes known as blending functions. The extension to three
dimensions using trilinear blending functions is straightforward.
If polynomials higher than first order are used as blending
functions, extra degrees of freedom are introduced. These can
be used to define the shape and position of some interior grid
lines thus giving control over grid spacing or alternatively to
control the angles at which the grid lines meet the boundaries.

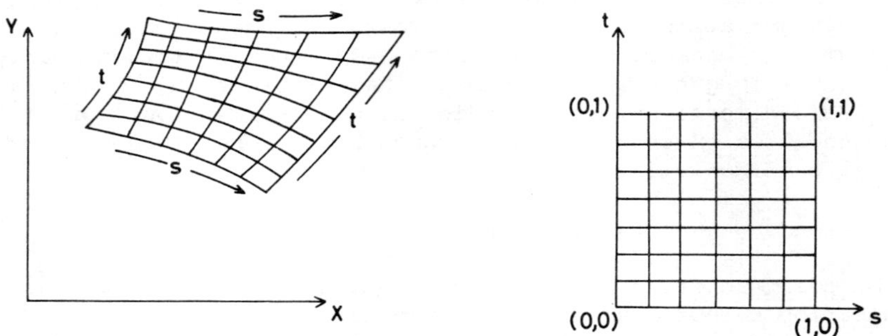

Fig. 7a   Basic isoparametric mapping

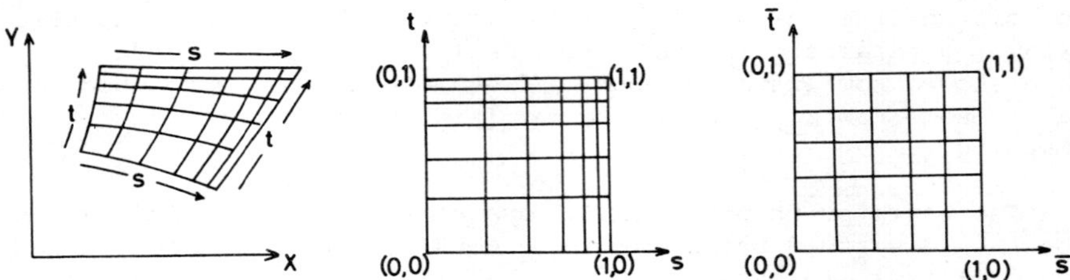

Fig. 7b   Stretched isoparametric mapping

The most general form of the isoparametric transformation is
given by

$$f(s,t) = \sum_{i=0}^{M} \phi_i(s)f(s_i,t) + \sum_{j=0}^{N} \psi_j(t)f(s,t_j)$$

$$- \sum_{i=0}^{M} \sum_{j=0}^{N} \phi_i(s)\ \psi_j(t)f(s_i,t_j)$$

where $f(s_i,t)$, $f(s,t_j)$ define the sets of grid lines $s = s_i$
and $t = t_j$ and where the blending functions $\phi_i(s)$, $\psi_j(t)$ must
satisfy the cardinality conditions

$$\phi_i(s_k) = \delta_{ik}$$
$$\psi_j(t_l) = \delta_{jl} \qquad \text{with } \delta_{ij} = \left.\begin{array}{l} 1 \\ \\ 0 \end{array}\right\} \begin{array}{l} \text{if } i = j \\ \\ \text{if } i \neq j \end{array}$$

Again the extension to three dimensions is straightforward.

Although the use of higher order polynomials as blending
functions is one way of controlling grid spacing, it is perhaps
better suited to controlling the behaviour of grid lines near
boundaries. This is because the use of higher order polynomials
increases the likelihood of grid lines crossing over (i.e. a
non-univalent mapping) and because of the need to specify some
internal grid lines as well as boundaries.

An alternative grid control technique which may be used with
isoparametric transformations is to apply preliminary packing/
stretching transformations to the parametric coordinates s and
t to give new parameters $\bar{s}$ and $\bar{t}$. Taking equal intervals in $\bar{s}$
and $\bar{t}$ leads to unequal intervals in s and t and hence to unequally spaced grid lines in physical space (see Fig. 7b). In
general, separate packing/stretching transformations would be
applied in each coordinate direction, however, greater flexibility may be obtained by allowing the parameters controlling
the packing/stretching in one coordinate direction to be functions of the other coordinate and vice versa.

As an example of both the complexity and the quality of grids
that can be obtained with careful use of the isoparametric
mapping approach, Eriksson (1980) has used the technique to
develop a grid around a wing/body combination. Fig. 8 shows
how the basic cube ($u_1 \leqslant u \leqslant u_2$, $v_1 \leqslant v \leqslant v_2$, $w_1 \leqslant w \leqslant w_2$) in
parametric space is effectively wrapped round the wing/body
combination so that the cube's faces correspond to the configuration surfaces or to branch cuts representing wakes etc. The
isoparametric transformation used appears to consist of a quartic blending in the w direction with specified point distributions on $w = w_1$ and $w = w_2$ together with specified 1st, 2nd and
3rd parametric derivatives on $w = w_1$, followed by a linear blend
in the v direction with specified point distributions on $v = v_1$.
Grid packing is controlled by choice of blending function rather
than by separate packing/stretching transformations. The ultimate aim is to use the grid in conjunction with a solution procedure for the Euler equations, however, as a check on the grid
quality Eriksson shows results for the solution of Laplace's
equation using both finite volume and high order finite element
codes. The same case was also computed by a panel method and
the agreement between the three codes is a measure of the
quality of the grid, particularly in view of the coarse grid
used.

Although the basic isoparametric mapping technique can generate grids round quite complex configurations, the wing/body
grid described above is perhaps at the limit of this ability.
More complex configurations such as wings with stores, nacelles,

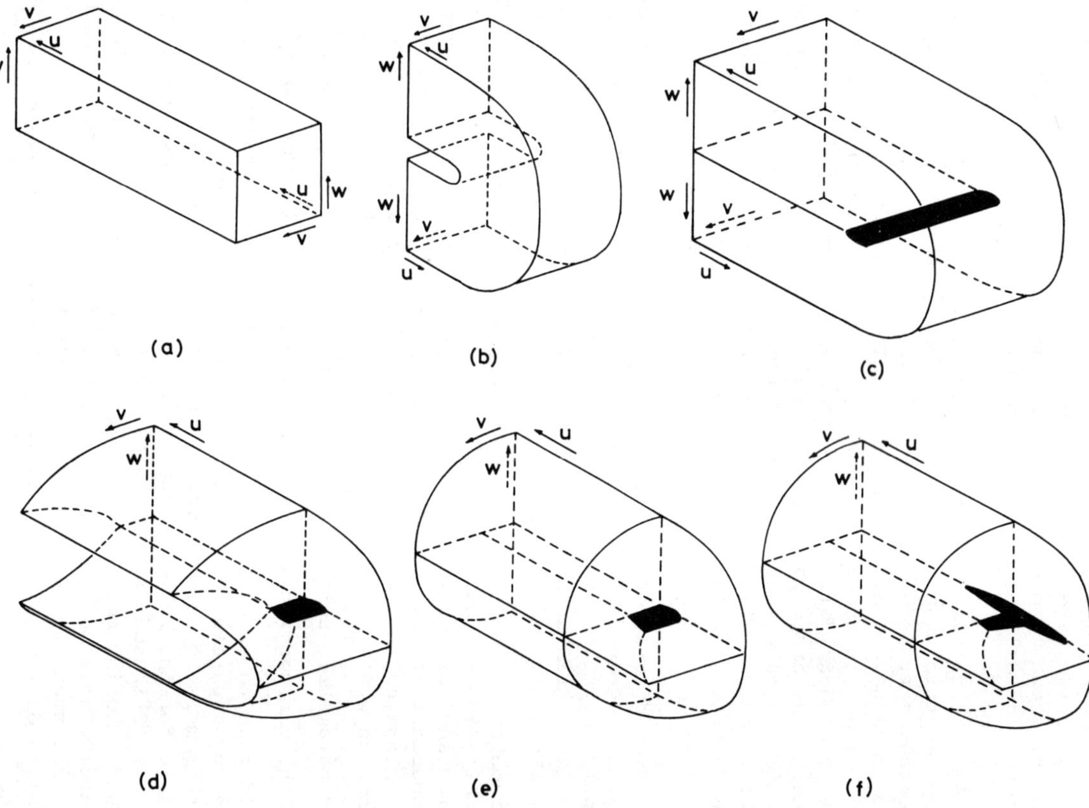

Fig. 8 Eriksson's wing/body grid - schematic

winglets etc. do not have a natural single block topology so
that generally they cannot be mapped successfully into a single
cube. This suggests a multiblock approach in which the region
of interest is artificially divided into a number of blocks each
of which is mapped separately to a cube in parametric space.
Fig. 9 shows how this might be done in two dimensions for an
intake with centre body.

This approach has been widely applied and isoparametric
transformations are frequently used as a convenient and simple
means of performing the individual transformations for each
block.

However, there are certain problems in such an approach.
One problem is that the non-physical boundaries dividing the
region of interest into blocks have to be specified in some
fashion, generally together with point distributions on these
boundaries. A second problem which occurs when low order (e.g.
linear) blending functions are used is that although grid lines
in adjacent blocks meet on their common boundary, the grid lines
usually kink as they pass through the boundary. Higher order
blending functions, which allow control of grid directions at
common boundaries, could be used to alleviate this problem but
at the expense of the other problems associated with higher
order functions. Alternatively, kinks in the grid lines may be
acceptable if the associated flow calculation method is insensitive to such kinks or if special treatment is used at points
where the kinks occur.

Clearly, finite difference methods, which are based on Taylor
series expansions, will be sensitive to kinks in the grid lines.
However, it is possible that finite volume methods, which are
based on flux balance in cells, or finite element methods, which
are based on minimisation procedures, may be less sensitive.
This is perhaps particularly true when the flow equations being
solved are written in terms of the primitive flow variables
(velocity, density etc.) as in the Euler or Navier-Stokes
equations. Although finite volume/element methods have also
been devised for the potential equation, there still remains
the need to calculate derivatives of the potential in order to
obtain the velocity components used in the flux balance terms
and the calculation of these derivatives will probably still be
sensitive to kinks in the grid lines.

One example of the use of the multi-block isoparametric
mapping approach is the General Interpolants Method (GIM) of
Anderson and Spradley (1980). This method has been used to
construct grids for use with a Navier-Stokes code for such complex configurations as the Space Shuttle main engine. Simple
trilinear isoparametric transformations are used in each block

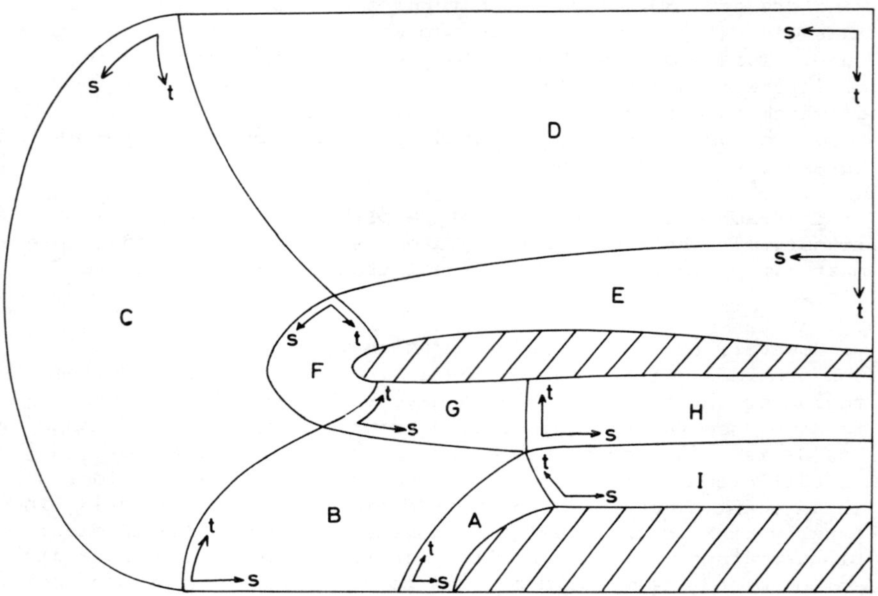

Fig. 9   Multi-block schematic for cowl + centre body

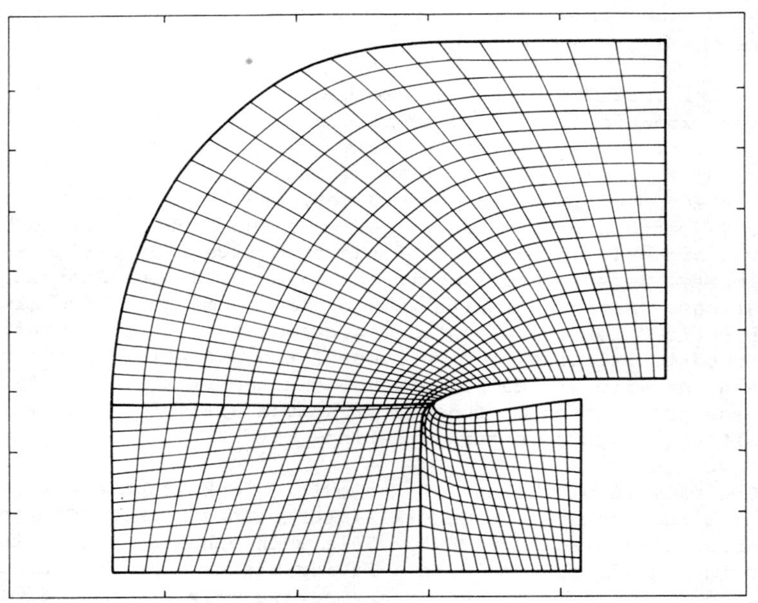

Fig. 10   Paganos   cowl grid - cross section

and the problem of defining the non-physical boundaries between
blocks is simplified by only allowing a small class of analytic
shapes for the faces and edges of each block, (e.g. planar,
cylindrical etc.). Apparently, a finite difference approach is
used for the flow solution with finite element blending across
common boundaries to avoid sensitivity.

A similar approach, for use with an Euler code, is under
development in the UK by Pagano (1980). The grid point distributions in the interior of each block are derived from the
distributions along the twelve edges of each block. Currently,
these edge distributions are represented by cubics. Thus, the
approach can be viewed as an isoparametric mapping with auxiliary packing/stretching transformations. The configuration
surfaces are defined by bicubic (Coons) patches while the non-
physical interior boundaries are generated semi-automatically as
planar surfaces bounded by the edges of already defined adjoin-
ing surfaces. Fig. 10 shows a cross section through part of a
cowl grid generated by this method. The grid is made up of
three blocks (whose faces are defined by the heavier lines) and
the kinks in the grid lines at these boundaries can clearly be
seen. Since the flow solution is to be computed by a finite
volume method, based on MacCormack's second order finite difference algorithm, no special treatment for these boundary points
is contemplated.

The alternative approach, in which special treatment is
applied at common boundary points in conjunction with a finite
difference code, has been investigated by Forsey, Edwards and
Carr (1980) for the solution of the full potential equation.
This investigation has been carried out in two dimensions
although the technique, if successful, should be equally applicable to three dimensions. The grid generation scheme uses
bilinear isoparametric transformations with grid spacing controlled by auxiliary packing/stretching transformations chosen
interactively by the user via an interactive graphics package.
At all interior grid points the potential equation is differenced in the standard fashion while at common boundary points
use is made of the concept of dummy rows of grid points outside
the boundaries. Thus, two separate approximations to the flow
equation are made at each boundary point, each approximation
involving points from one block and one dummy row. A third
equation, which is a discretised form of the condition of continuity of velocity normal to the common boundary involving
points from both blocks and both dummy rows, is solved simultaneously with the previous two equations to eliminate the two
dummy rows of points. When applied at each point on the common
boundary this leads to a tridiagonal system of equations which
may be solved for the potential on the boundary. For boundary
points where the flow is supersonic, further conditions are

required to cope with the effect of backward differences.

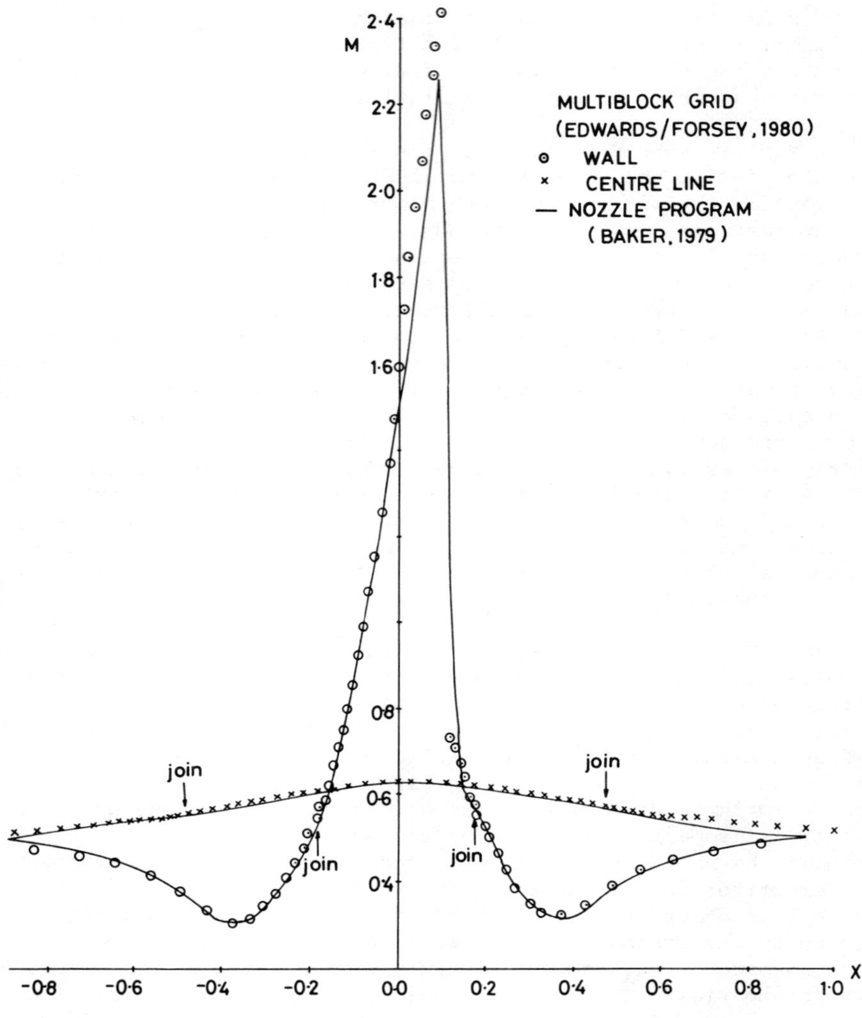

Fig. 11   Nozzle results M_∞ = 0.5

This approach showed early promise and Fig. 11 illustrates
the results obtained for the distribution of Mach number
through a convergent-divergent nozzle using a three block grid
compared with results obtained with the standard transonic
nozzle flow program of Baker (1979).   However, more recent
experience with other configurations has shown problems.   These
appear to be related to the form of stretching function used in

the grid generation rather than to the special treatment at
boundaries and work is now in progress to obtain further understanding of the problems.

In general, an algebraic approach to grid generation is very
attractive. The computer time required to evaluate even fairly
complex algebraic transformations is quite small so that interactive grid generation with user control of grid behaviour via a
graphics terminal becomes quite feasible. Thus, a number of
grids for each problem can be created and compared cheaply in
order to find the most suitable grid. However, when applied to
complex configurations using the multi-block approach, there
may be problems at the non-physical interior boundaries where
the grid lines are generally not smooth. Furthermore, other
problems may arise due to the use of separate grid control
functions in each block leading to discontinuities in grid
spacing near internal boundaries.

4. NUMERICAL GRID GENERATION

4.1 Generation using systems of elliptic equations

In the sense that most grid generation methods ultimately
require some numerical approximations, even if only in defining
the surfaces with which the grid is to be aligned, practically
all methods could be described as numerical. However, this term
has come to be applied to a class of grid generation techniques
based on the numerical solution of sets of partial differential
equations in which the grid coordinates in physical space are
the dependent variables.

Amongst the many possible choices for the partial differential equations to be solved in the grid generation process most
investigations to date have used sets of elliptic equations
derived from Laplace's or Poisson's equation. The use of elliptic equations, particularly those whose solutions only have
extrema on boundaries, such as Laplace's equation, has several
attractive features. Firstly, boundary data must be specified
on all boundaries, thus this method is particularly suited to
constrained (e.g. internal) flows or flows where an outer 'free
stream' boundary can be specified, while the extremum principle
ensures that the mapping is univalent (i.e. grid lines do not
cross over). Furthermore, since Laplace's equation describes a
range of physical phenomena, physical considerations may sometimes help in choosing suitable grid control parameters. For
example, with suitable boundary conditions the grid obtained
consists of streamlines and equipotential lines and the addition
of extra source or sink terms (i.e. replacing Laplace's equation
by Poisson's equation) allows some control over the shape of
these lines. Finally, it should be noted that conformal mapping

in two dimensions is a special case of grid generation using
Laplace's equation.

Although some work had been reported previously, the major
investigation and development of grid generation using Laplace's
and Poisson's equations has been done by Thompson (1979) and
his co-workers.

To describe the basic Thompson technique using Poisson's
equation consider the case shown in Fig. 12a in which a general
quadrilateral in physical space is mapped to a unit square in
transform space. The coordinates in physical space are (x,y)
and in transform space are (ξ,η) and it is assumed that x and y
are known functions of ξ and η on all the boundaries (i.e. the
grid point distributions are specified along the boundaries).
Assume that ξ and η individually satisfy Poisson's equation
(expressed in terms of x and y) inside the unit square subject
to the known Dirichlet conditions on the boundaries, i.e.

$$\frac{\partial^2 \xi}{\partial x^2} + \frac{\partial^2 \xi}{\partial y^2} = P$$

$$\frac{\partial^2 \eta}{\partial x^2} + \frac{\partial^2 \eta}{\partial y^2} = Q$$

where the source terms P and Q are both functions of x and y.

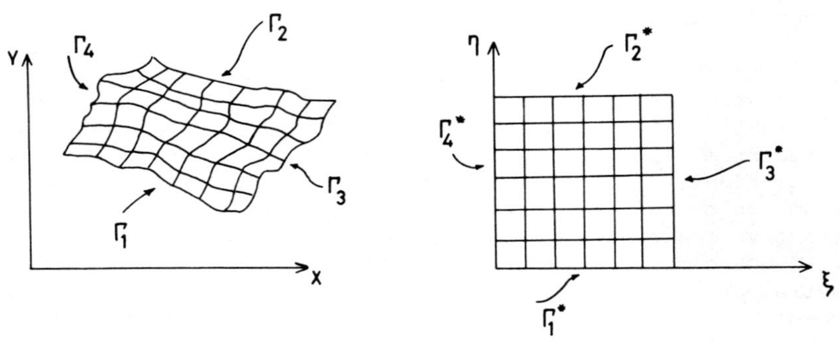

Fig. 12(a)   Basic mapping using elliptic equations

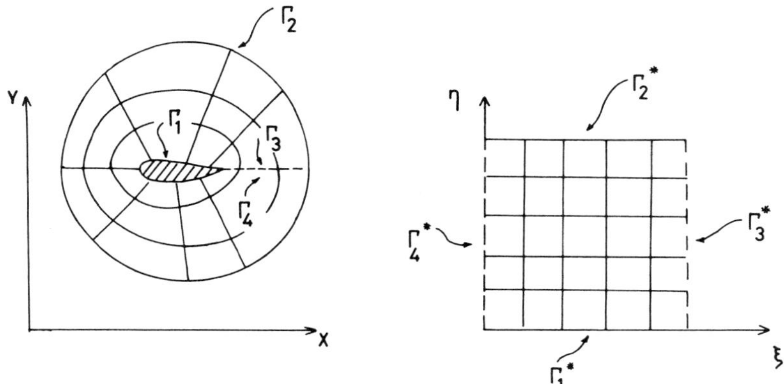

Fig. 12(b) Aerofoil mapping using elliptic equations

In order to obtain a grid in physical space, it is necessary
to interchange the dependent and independent variables so that
the equations may be solved numerically for x and y on an
equally spaced grid in transform space. This interchange is
most easily carried out by defining the vectors

$$x^i = (x,y), \quad \xi^i = (\xi,\eta), \quad p^i = (P,Q).$$

In standard tensor notation the interchanged (or inverted)
equations become

$$g^{ij} \frac{\partial x^k}{\partial \xi^i \partial \xi^j} + p^i \frac{\partial x^k}{\partial \xi^i} = 0 \quad \text{for } k = 1,2$$

where g^{ij} is the contravariant metric tensor defined by
$g_{ij} g^{jk} = \delta^k_i$ with

$$g_{ij} = \frac{\partial x^k}{\partial \xi^i} \frac{\partial x^k}{\partial \xi^j}.$$

The Dirichlet boundary conditions for the inverted equations are
the known values of x and y at each point on the boundary of the
unit square in transform space. Note that in tensor form the
extension to three dimensions is immediate.

The inverted equations can, in principle, be solved by any
convenient method. However, in practice the usual approach has
been to discretise the equations using three point central difference approximations for all derivatives and solve the resulting set of difference equations using a line relaxation scheme.

Recently some of the new fast solver algorithms such as multi-grid or approximate factorisation have been used in place of the relaxation scheme. Now it might be thought that a finite element solution would produce more accurate results because of its higher order accuracy. However, it has been shown that the use of three point finite difference approximations, which are consistent with the approximations used in solving the flow equation on the resulting grid, leads to smaller truncation errors in the flow solution than those obtained using nominally more accurate approximations in the grid generation scheme.

Although this grid generation technique has been illustrated by a simple example, that of mapping a general quadrilateral into a unit square, the approach can be easily applied to other cases with appropriate choice of boundary conditions. For example, Fig 12 shows how a grid can be generated between an aerofoil and an outer boundary by introducing a branch cut downstream of the trailing edge and applying periodic boundary conditions across the cut in transform space.

The spacing of the grid lines in physical space generated by an equally spaced grid in transform space depends strongly on the actual equations being solved and weakly on the boundary point distributions. Hence, changing the point distributions on the boundaries has only a local effect on grid spacing while changing the equations has a more global effect. The actual equations being solved can be effectively changed by varying the values of the source terms P and Q at each grid point, and this is the primary means of grid control in this grid generation technique. Thompson (1979) gives one physical interpretation of grid control using these terms by analogy with a stretching membrane with applied surface pressure. Another physical interpretation might be to consider the grid lines as streamlines (and equipotentials) and the P and Q terms as source or sink distributions modifying the shape of the streamlines. With this concept it is fairly clear that grid lines can be attracted towards particular grid points or grid lines by choosing appropriate values of P and Q at such points or on such lines. Similarly, changing the sign of P and Q will repel rather than attract the grid lines.

This is the basis of the method originally suggested by Thompson for grid control using the P and Q terms. He suggests setting

$$P(\xi,\eta) = \sum_{i=1}^{n} a_i \, \mathrm{sgn}\,(\xi-\xi_i) e^{-c_i |\xi-\xi_i|}$$

$$+ \sum_{j=1}^{m} h_j \, \mathrm{sgn}\,(\xi-\xi_j) e^{-d_j \sqrt{(\xi-\xi_j)^2 + (\eta-\eta_j)^2}}$$

$$Q(\xi,\eta) = \sum_{i=1}^{n} a_i \, \text{sgn} \, (\eta-\eta_i) e^{-c_i|\eta-\eta_i|}$$

$$+ \sum_{j=1}^{m} b_j \, \text{sgn} \, (\eta-\eta_j) e^{-d_j\sqrt{(\xi-\xi_j)^2 + (\eta-\eta_j)^2}}$$

where the first terms in each case control attraction towards the grid lines $\xi = \xi_i$, $\eta = \eta_i$ and the second terms control attractions towards the points (ξ_j,η_j). The amplitudes a_i, b_j control the packing density while the decay factors c_i, d_j control the rate of change of packing density.

For two-dimensional problems this is a powerful and flexible technique. An example of its use is shown in Fig. 13 where two grids for a multi-aerofoil case are shown, one generated using Laplace's equation (P = Q = O) and the other with grid attraction towards the aerofoil surfaces and the region between the aerofoils. However, a certain amount of skill is required in choosing appropriate control parameters which becomes increasingly difficult as the complexity of the configurations increases or for three-dimensional problems. Hence, a number of techniques have been developed for calculating appropriate values of P and Q automatically. Generally, these involve specifying some features of the grid at or near boundaries and choosing P and Q so that these features are spread smoothly through the rest of the grid. Further details will be given later when discussing individual methods.

Of course, grid control using P and Q is not the only possibility. An alternative method is to use auxiliary packing/stretching transforms $\xi = \xi(\bar{\xi})$, $\eta = \eta(\bar{\eta})$ in transform space. One possible way of implementing this would be to modify the metrics appearing in the transformed equations to include the extra (analytic) transform derivatives $d\xi/d\bar{\xi}$, $d\eta/d\bar{\eta}$. An alternative way suggested by Sorenson and Steger (1977) is to produce a temporary grid by solving Laplace's equation (P = Q = O) and then use curve fits along each row of grid points in each direction to interpolate another set of grid points whose distribution is controlled by a packing/stretching function.

Many authors have presented examples of grids generated by using the basic Thompson technique. However, there is much more to grid generation than the development of a good technique, almost equally important is its implementation in a reliable, flexible, easy-to-use code. Two such codes are the TOMCAT code of Thompson (1979) and the GRAPE code of Sorenson and Steger (1980), both of which are two-dimensional codes. Also notable

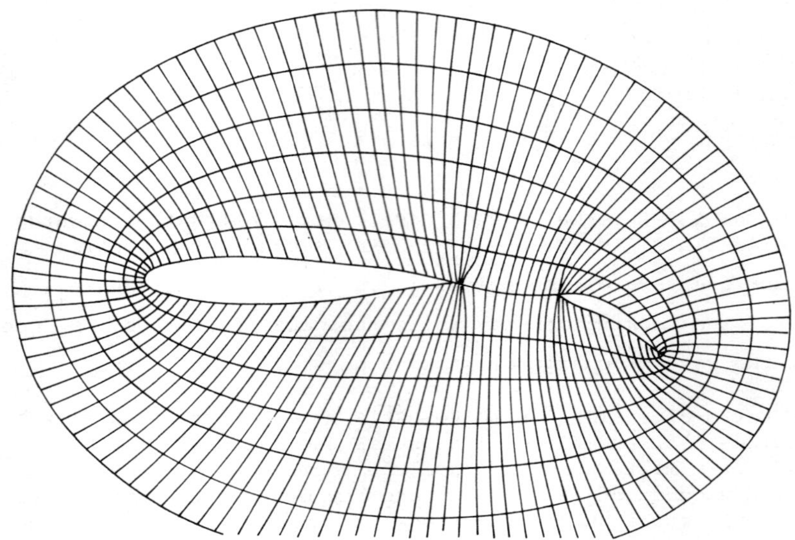

Fig. 13(a)   Multi-aerofoil grid without grid control

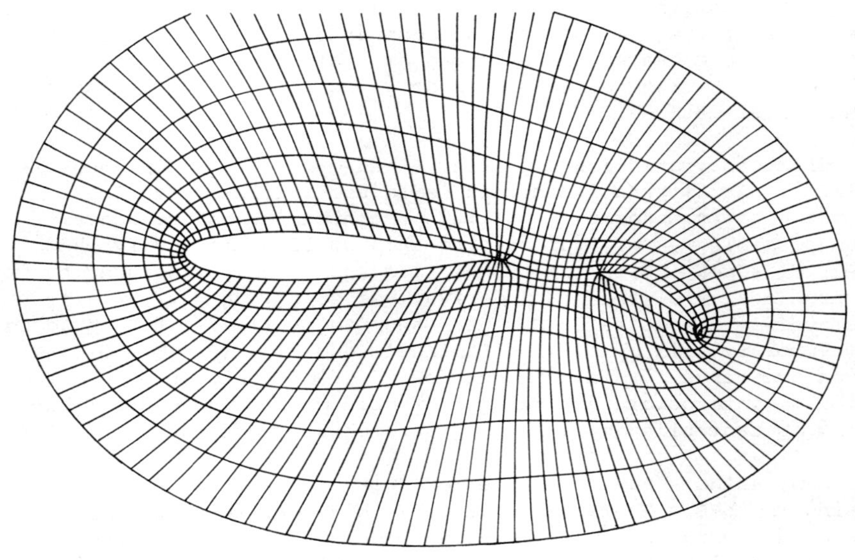

Fig. 13(b)   Multi-aerofoil grid with grid control

is the three-dimensional wing/body code of Yu (1980).

TOMCAT is designed to deal with general two-dimensional configurations of simple or multiple bodies. A multiple-body
region is converted to a simply-connected region by the use of
branch cuts and different grids generated by sliding the physical boundaries corresponding to each body around the single
rectangle in transform space. Alternatively the multiplicity
of the region is left unchanged by transforming bodies in the
interior of the region to slabs or slits in transform space.
Since each approach produces a different form of grid, it is
possible to choose the one most appropriate to the problem concerned. Grid control appears to be by the use of the exponential forms for P and Q described earlier.

GRAPE is designed to generate O-type or C-type grids around
aerofoils and other simple bodies with a variety of outer boundary shapes. The different types of grid are obtained by applying different boundary conditions on the rectangular boundary in
transform space. Two important types of automatic grid control
are implemented. The first is control of the grid spacing
between each boundary and the first interior grid line of the
same family. The second is control of the angle with which grid
lines of the opposite family intersect the boundaries. These
two controls are particularly useful round the leading edges of
aerofoils. The automatic grid control is implemented by assuming P and Q have the form

$$P(\xi,\eta) = p(\xi)e^{-a(\eta-\eta_{min})} + r(\xi)e^{-c(\eta_{max}-\eta)}$$

$$Q(\xi,\eta) = q(\xi)e^{-b(\eta-\eta_{min})} + s(\xi)e^{-d(\eta_{max}-\eta)}$$

where $\eta = \eta_{min}$, $\eta = \eta_{max}$ correspond to the inner and outer
boundaries respectively. On these boundaries $P(\xi,\eta) = p(\xi)$ or
$r(\xi)$, $Q(\xi,\eta) = q(\xi)$ or $s(\xi)$. Also on these boundaries the
derivatives x_η, y_η can be found from the geometric constraints
on grid spacing and grid angles while the derivatives x_ξ, $x_{\xi\xi}$,
y_ξ, $y_{\xi\xi}$, $x_{\xi\eta}$, $y_{\xi\eta}$ can be found by differencing the known boundary data. Thus, Poisson's equation on the boundary may be back
solved to give $p(\xi)$, $r(\xi)$, $q(\xi)$, $s(\xi)$ in terms of $x_{\eta\eta}$ and $y_{\eta\eta}$
and an iterative scheme set up which alternates between solving
for $p(\xi)$ etc with current estimates of $x_{\eta\eta}$, $y_{\eta\eta}$ and solving for
x,y (and hence $x_{\eta\eta}$, $y_{\eta\eta}$) with current estimates of $p(\xi)$ etc.
At convergence a grid with the required geometric constraints is
obtained.

The three-dimensional wing/body method of Yu is somewhat like the algebraic approach of Eriksson in that the rectangular cube in transform space is effectively wrapped round the wing so that the wake becomes a branch cut while the body is modelled by deformation of the plane of symmetry. However, the grid does not wrap round the wing tip in the same way but rather the wing planform is extended outboard of the wing tip as a branch cut. Wing and body coordinates are applied as boundary conditions on appropriate parts of the surface of the cube in transform space with continuity conditions applied across the branch cuts. Grid control is semi-automatic based on the ideas of Middlecoff and Thomas (1979) in which the grid behaviour near all boundaries is defined and used to obtain values of the source control terms (P, Q etc.) on the boundaries. These boundary values are then smoothly interpolated across the region between the boundaries before solving the Poisson equations to generate the grid. The particular grid behaviour chosen is that grid point distributions are specified along all the edges of the cube in transform space together with the condition that all derivatives normal to the edge are zero. So that, for example, on the edge defined by the intersection of the planes ξ^2 = constant, ξ^3 = constant all derivatives involving $\dfrac{\partial}{\partial\xi^2}$ and $\dfrac{\partial}{\partial\xi^3}$ are zero and the inverted Poisson equation becomes

$$g^{11}\left(\frac{\partial^2 x^k}{\partial\xi^1\partial\xi^1} + \frac{p^1}{g^{11}}\frac{\partial x^k}{\partial\xi^1}\right) = 0 \quad k = 1,2,3 \, .$$

Hence $(p^1/g^{11}) = -\left(\dfrac{\partial^2 x^k}{\partial\xi^1\partial\xi^1}\Big/\dfrac{\partial x^k}{\partial\xi^1}\right)$ where the derivatives involving

$\dfrac{\partial}{\partial\xi^1}$ may be calculated by differencing the edge data.

Similar conditions apply on the other edges and when values of p^1/g^{11}, p^2/g^{22}, p^3/g^{33} have been evaluated for all edges, linear interpolation is used to calculate their values at interior points. (Some further constraints are placed on the interior values of p^1/g^{11} etc. to avoid excessive changes in grid spacing between adjacent points.)

The grid generation method has been coupled with a transonic flow code based on the Caughey-Jameson (1977) fully conservative finite volume method and the results presented for a typical transport type wing/body combination are quite impressive. Agreement with experiment is generally good, including pressures

on the body as well as the wing, except perhaps around the nose
of the body where the grid is very sparse.

Although most methods based on solving sets of elliptic
equations perform the actual solution by a finite difference
line relaxation approach, this can be rather slow particularly
for three-dimensional problems and faster methods may be
possible. In the UK, Colbourne and Burton (1979) have developed
a method for three-dimensional ducts in which each cross section
through the duct is mapped to a circle. Clearly, by specifying
the point distribution on the circumference of the circle it
would be possible to solve Poisson's equation in polar coordinates (r, θ) to obtain a grid. Instead, Colbourne and Burton use
the fact that a harmonic function defined on the circumference
of a circle can be represented at interior points by Poisson's
integral. By approximating the boundary point distribution by
a Fourier series in θ, they evaluate the Poisson's integral
analytically in terms of the Fourier coefficients so that internal grid points can be rapidly calculated. Together with a
packing/stretching transform applied to r this yields a fast,
simple grid generation method for this problem. Unfortunately,
however, it is not clear whether such an approach could be generalised to more complex configurations.

When considering complex three-dimensional configurations a
natural approach, as for algebraic grid generation, is to divide
the region of interest into a series of blocks, each of which is
mapped to a separate cube in transform space. The mapping for
each block may, of course, be performed by solving a set of
partial differential equations and several methods using this
approach are currently under development. Of these, the method
which appears to have progressed furthest is that due to Lee et
al (1980) which is closely related to the single block approach
of Yu. The grids are generated by solving a sequence of elliptic equations: firstly in one dimension to generate the point
distributions on block edges, secondly in two dimensions to generate point distributions on block faces and finally in three
dimensions to generate point distributions in the interior of
the blocks. At each level the point distributions from the preceding level serve as boundary conditions for the solution at
the current level. Grid spacing is controlled by the boundary
point distributions in the same semi-automatic way used by Yu
for the single block case. Two sets of elliptic partial differential equations have been investigated. One set is the usual
non-linear inverted Poisson equation, the other is a linear
system which the authors say produces grids of comparable quality
with order of magnitude less cost but with greater possibility
of ill-conditioning. Both sets of equations may be written as

$$g^{ij} \frac{\partial^2 x^k}{\partial \xi^i \partial \xi^j} + p^i \frac{\partial x^k}{\partial \xi^i} = 0$$

where for the non-linear case the g^{ij} are the usual coupling terms (metric components) and the p^i are the grid control (source) terms and for the linear case both g^{ij} and p^i are linear grid control functions of ξ^i with $g^{ij} = 0$ for $i \neq j$.

The multiblock approach to grid generation, although very powerful and flexible, leads to several algorithm issues when solving the flow equation. The problem of dealing with grids which kink across block boundaries has already been mentioned in the section on algebraic grid generation and the same problems may occur with numeric grid generation. This is certainly true in the method of Lee et al and appears to be treated by using a finite volume solution technique which is apparently insensitive to such behaviour. There are other problems however, concerned with the relationship between corners in physical space and transform space. Two such cases which the authors term lost corners and fictitious corners are illustrated in Fig. 14. They are essentially particular cases of the more general mapping singularity where other than four grid lines meet at a point. Lost corners occur when a physical corner is mapped out in transform space and a fictitious corner occurs when a corner in transform space is mapped out in physical space. Such mapping singularities occur naturally in the multiblock approach and are fundamental to the generation of grids for complex shapes. In fact, it is the natural way that such singularities occur which makes the multiblock technique such a powerful tool for complex grid generation. Even so, how to deal with such grid singularities in the solution of the flow equation is by no means clear. In the work of Lee et al no special treatment is mentioned and they appear to suggest that the finite volume approach copes adequately. Although flow solutions have not been reported for three-dimensional configurations, three-dimensional grids have been generated and flow solutions presented for two-dimensional cases using the same technique.

In the UK, Forsey is following a similar approach to grid generation specifically for wing/pylon/nacelle configurations but hopefully with application to other configurations as well. A multiblock Thompson type mapping is used but as far as possible, grid singularities are placed in the field rather than on the boundaries. Furthermore, since a finite difference flow algorithm is to be employed, the problem of kinks in the grid lines at boundaries is avoided by using grid smoothness as the boundary condition on block interfaces rather than specifying

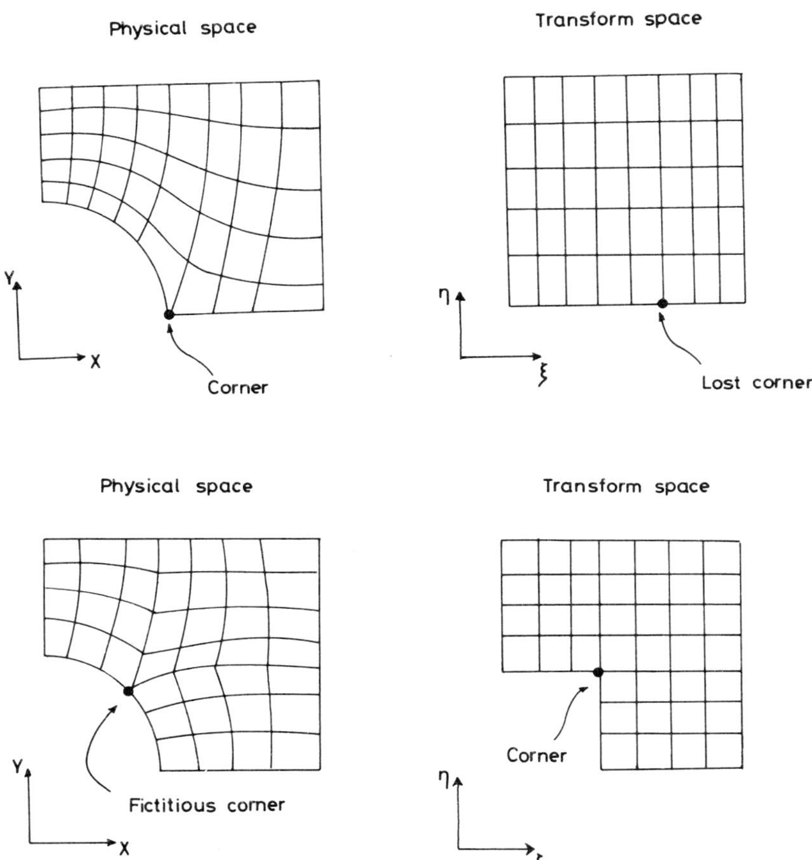

Fig. 14   Types of singular points in multi block mapping

point distributions on such faces.   This leads to somewhat
greater problems with grid control and as yet this problem has
not been solved satisfactorily although several ideas are to be
investigated.   Fig. 15 shows cross sections through two grids
for a wing in a wind tunnel generated with this approach.   The
block boundaries are shown as heavier lines to illustrate the
way the grid is built up.   It should be pointed out that the
second grid is rather artificial being presented solely to show
how the grid singularities mentioned earlier may be generated.
In both cases the grids are generated as solutions of Laplace's
equation without additional grid control via source terms.   Some
flow solutions have been produced for a simple nozzle flow

problem but these are not yet satisfactory due to grid spacing
problems. Note also that in solving both the grid generation
and flow equations, approximate factorisation algorithms are
used, instead of line relaxation, to obtain more rapid convergence.

One other approach based on numeric grid generation is under
development in the UK by Roberts (1980). This approach is both
comprehensive in scope and innovative in nature and it is hard
to do it justice in a short paragraph. Essentially, a multi-
block approach with grid singularities in the field and continuity of grid lines across block interfaces is used. However,
grid lines are to be generated normal to surfaces rather than to
satisfy surface point distributions and all the geometry handling and setting up of the multiblock topology is fully automated. Furthermore, the grid generation equations are not
approximated by finite difference formulae, rather the solutions
are built up as weighted sums of B-spline modes with the weights
chosen so that the spline derivatives satisfy the grid generation equations at each spline node. This leads to greater
accuracy than that obtained with finite differences and is also
particularly effective in conjunction with the multigrid solution algorithm in obtaining very rapid convergence. The method
is being developed to generate grids for wing/body configurations. Later a transonic flow field code will be developed to
use the resulting grid.

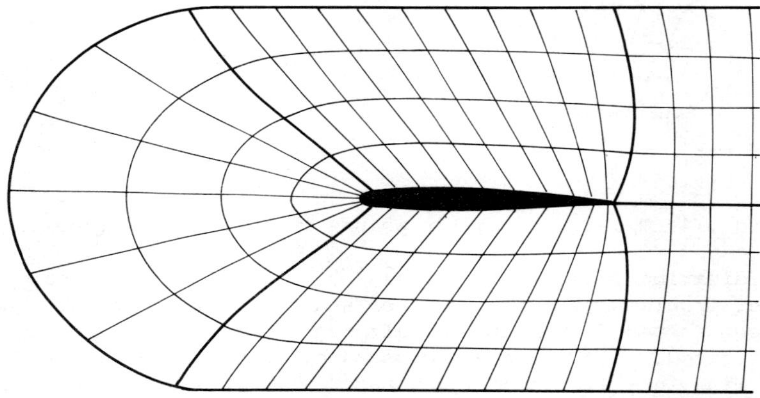

Fig. 15(a)   Wing in wind tunnel - cross section

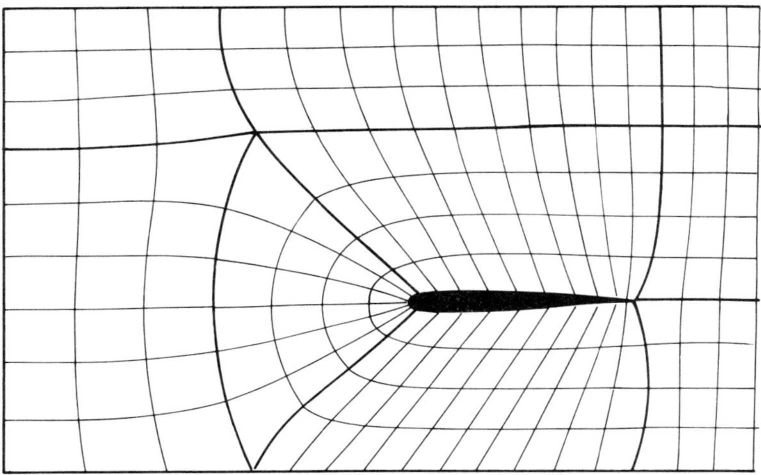

Fig. 15(b)   Wing in wind tunnel - cross section - with
             singular points

4.2 *Generation using other systems*

Although most approaches to numerical grid generation have
been based on solving sets of elliptic equations, other equations may be more appropriate for certain cases. For example,
Steger and Sorenson (1980) have investigated the use of pairs
of hyperbolic equations for certain two-dimensional problems.
Such equations are particularly appropriate where there is no
true outer boundary other than infinity since the hyperbolic
equations may be solved by a marching procedure starting at the
inner boundary and only requiring boundary conditions on the
inner boundary. Furthermore, since the marching procedure is
non-iterative, the solution is obtained rapidly. There remains
the choice of which hyperbolic equations to use and two schemes
were reported. In the arc length/orthogonality scheme:

$$x_\xi^2 + y_\xi^2 + x_\eta^2 + y_\eta^2 = (\Delta s)^2 \qquad \text{(arc length)}$$

$$x_\xi x_\eta + y_\xi y_\eta = 0 \qquad \text{(orthogonality)}$$

the spacing Δs between η = constant grid lines is specified at
each point together with the condition that grid lines of different families are orthogonal. (Note that η = 0 is the inner
boundary.) In the volume/orthogonality scheme:

$$x_\xi y_\eta - x_\eta y_\xi = V \qquad \text{(volume)}$$

$$x_\xi x_\eta + y_\xi y_\eta = 0 \qquad \text{(orthogonality)}$$

the volume V of each grid cell is specified again with the
orthogonality condition.

Of course, there still remains the problem of specifying Δs
or V and one possible way to do this is described by the authors,
however, this appears to be one drawback to the approach.
Another is that discontinuous boundary data are transmitted wave-
like through the grid rather than diffused as they would be with
an elliptic system. Finally, the extension to three dimensions
is not obvious: Steger and Sorenson suggest one scheme based on
two orthogonality conditions and a volume condition but also
point out that a system based on three orthogonality conditions
is neither hyperbolic nor elliptic.

Various other methods using different sets of partial differential equations have also been reported in the literature.

In general, grid generation using a numerical approach offers
great flexibility and relatively straightforward extension to
three dimensions, at least in principle, but frequently requires
significant computational resources comparable to those needed
for the solution of the flow equation. While this may be
acceptable for a series of flow calculations on a fixed configuration, it is less satisfactory for use in a design situation.
Even so, it is probably one of the most hopeful approaches for
calculating flows over complex configurations, at least in the
near future.

5. OTHER APPROACHES

5.1 Non-aligned grids

All the methods described so far have one thing in common,
the lines or surfaces of the configuration are grid lines or
surfaces of the coordinate system. If this constraint is
removed then more options are available. The concept of non-
aligned grids has been used by a number of workers with the
first transonic flow method being developed by Carlson (1975).
He produced a method of calculating transonic flow over an
aerofoil using the full potential equation and a cartesian grid
system. The method of solution is similar to those of the
aligned grid methods with the main difference being in the
application of boundary conditions on the aerofoil surface. One
can perhaps argue that this is no longer a problem of grid generation but rather one of the solution of the flow equation
with more complex boundary conditions. However, as stated in
the introduction, we are concerned with the solution of flow
problems with grid generation as one of the building bricks
required and so non-aligned grids are worthy of inclusion. Also,
particularly in three dimensions, although the grid may be

cartesian, there is still some flexibility in locating it relative to the body and depending on the solution algorithm, this
may produce a variation in solution.

The two-dimensional boundary condition on the surface is the
usual one that the flow is tangential to the surface so that

$$\left(\frac{dz}{dx}\right)_{body} = \left(\frac{w}{u}\right)_{body}$$

where u, w are the velocities in the x,z directions respectively.
The technique is to utilise dummy points within the surface of
the body in order to discretise the equations. Fig. 16 shows
the grid in the neighbourhood of the surface. The boundary
condition can then be expressed, for a uniform grid, in terms
of the derivatives of the total potential Φ, so that

$$\left(\frac{dz}{dx}\right)_b = \left(\frac{\Phi_z}{\Phi_x}\right)_b = \frac{\sin\alpha + (\phi_z)_b}{\cos\alpha + (\phi_x)_b}$$

where ϕ is a perturbation potential, α the angle of attack of
the aerofoil and subscript b denotes body.

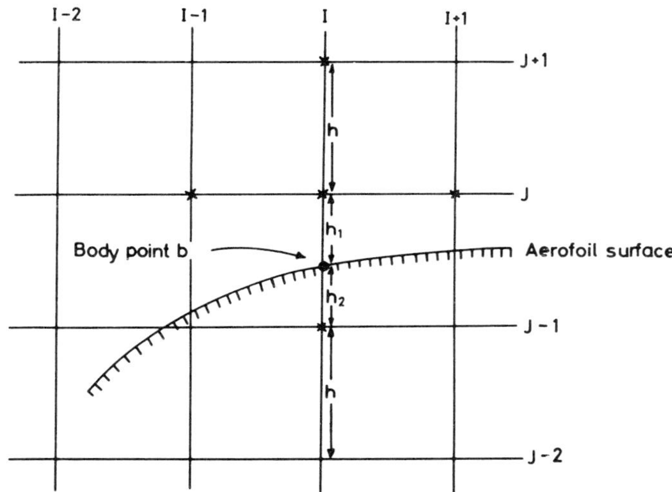

Fig. 16   Non-aligned grid - schematic

$(\phi_x)_b$, $(\phi_z)_b$ are then expressed in terms of a truncated
Taylor series which includes the values of ϕ on the j - 1 row.
These values are thus defined and used as Dirichlet boundary
conditions in the solution of the flow equation on the field
external to the body.

The method appears to be a most useful one and indeed, at
the GAMM workshop on Numerical Methods for the Computation of
Inviscid Transonic Flows with Shock Waves (1979), Carlson was
one of only a few participants who successfully calculated the
majority of the test cases. Moreover, his results agreed well
with the results of Baker who solved the same non-conservative
equation with an aligned grid.

More recently Catherall (1980) has considered the same problem but has attempted to deal with the boundary conditions
more accurately. Care is taken to ensure that the truncation
error is not only small but is continuous along the aerofoil
surface even when the surface crosses grid lines in the free
stream direction. The method has been used for the case of a
wing and elsewhere Morrison (Carr and Morrison, 1981) has
extended the aerofoil program to calculate the flow over an
aerofoil in a wind tunnel.

A non-aligned grid technique has also been used by Reyhner
(1980) to calculate flows about three-dimensional inlets, ducts
and bodies. Comparisons with experiment are shown for one
axisymmetric and two asymmetric geometries containing centre
bodies and at various angles of attack. Overall the results
look impressive and it would appear this program is a very
valuable asset in aerodynamic assessment of a range of configurations.

5.2 *Grid overlapping*

If we accept the requirement for an aligned grid but opt for
local rather than global grid generation, an alternative to the
patching techniques described earlier is grid overlapping.
Here local grids are extended beyond the interface with adjacent
grids and interpolation used at the interface. This was used a
few years ago by Magnus and Yoshihara (1972) and a more recent
method, component-adaptive grid embedding, has been developed
by Atta (1980). The example he shows is an aerofoil with an
inner grid generated by an elliptic grid-generator and a cartesian outer grid. This is successful but it would be surprising
if it were not. The interface between the two grid regions is
situated where the flow gradients are small and we will have to
await more complex cases before a rational assessment can be
made. Nevertheless, the method is promising and could make a
useful contribution.

6. CONCLUSIONS

Although flow field methods have a major part to play in future aerodynamic design work, their application to realistic configurations is heavily dependent on the development of suitable grid generation methods. Current achievements are encouraging but there is still much work to be done. Whilst the recent NASA Workshop on Numerical Grid Generation (1980) is indicative of the importance of grid generation, at the present time no one approach is obviously superior. Also, we should not ignore alternative techniques, not discussed in this paper, such as adaptive grids. Whatever future developments are achieved, it is probable that all aspects of complex configurations, particularly geometry handling, will be labour intensive. Effort must be placed on automating as much as possible using computers and graphics facilities.

7. REFERENCES

Anderson, P.G. and Spradley, L.W. (1980) Finite difference grid generation by multivariate blending function interpolation, paper included in 'Numerical Grid Generation Techniques', NASA Conference Publication 2166.

Arlinger, B.G. (1975) Calculation of transonic flow around axisymmetric inlets, AIAA Paper 75-80.

Atta, E.H. (1980) Component-adaptive grid embedding, paper included in 'Numerical Grid Generation Techniques', NASA Conference Publication 2166.

Baker, T.J. (1975) A numerical method to compute inviscid transonic flow around axisymmetric ducted bodies, IUTAM Symposium Transsonicum II, Göttingen.

Baker, T.J. (1979) Transonic nozzle flow analysis by a relaxation method, ARA Report 51.

Baker, T.J. and Forsey, C.R. (1981) A fast algorithm for the calculation of transonic flow over wing/body combinations, AIAA Paper 81-1015 to be presented at AIAA 5th Computational Fluid Dynamics Conference, Palo Alto.

Bauer, F., Garabedian, P., Korn, D. and Jameson, A. (1975) Supercritical wing sections II, *Lecture notes in Economics and Mathematical Systems,* **108**, Springer-Verlag.

Carlson, L.A. (1975) Transonic airfoil flowfield analysis using Cartesian coordinates, NASA Report CR-2577.

Carr, M.P. and Morrison, J.I. (1981) Use of flowfield methods
to calculate wind tunnel corrections for two-dimensional data in
supercritical flow, paper presented at AGARD Fluid Dynamics
Panel Sub-Committee on Wind Tunnels and Testing Techniques
Convenors' Group Meeting on Integration of Computers and Wind
Tunnel Testing, Bedford.

Catherall, D. (1980) Optimum approximate factorisation schemes
for two-dimensional steady transonic flow, RAE report in
preparation.

Caughey, D.A. and Jameson, A. (1976) Accelerated iterative calculation of transonic nacelle flowfields, AIAA Paper 76-100.

Caughey, D.A. (1978) A systematic procedure for generating
useful conformal mappings, *Int. J. Num. Meth. Engng.* **12**, pp.
1651-1657.

Colbourne, D.E. and Burton, C.G. (1979) A fast method for grid
generation, NGTE Report 79001.

Collyer, M.R. and Lock, R.C. (1979) Prediction of viscous
effects in steady transonic flow past an aerofoil, *Aero Qu.* **30**,
pp. 485-505.

Dwyer, H.A., Kee, R.J. and Sanders, B.R. (1979) An adaptive
grid method for problems in fluid mechanics and heat transfer,
AIAA Paper 79-1464.

Eiseman, P.R. (1979) Three dimensional coordinates about wings,
AIAA Paper 79-1461.

Eiseman, P.R. and Smith, R.E. (1980) Mesh generation using
algebraic techniques, paper included in 'Numerical Grid Generation Techniques', NASA Conference Publication 2166.

Eriksson, L-E. (1980) Three-dimensional spline-generated
coordinate transformations for grids around wing-body configurations, paper included in 'Numerical Grid Generation Techniques',
NASA Conference Publication 2166.

Forsey, C.R. and Carr, M.P. (1978) The calculation of transonic
flow over three-dimensional swept wings using the exact potential equation, DGLR Symposium Transonic Configurations, Bad
Harzburg.

Forsey, C.R., Edwards, M.G. and Carr, M.P. (1980) An investigation into grid patching techniques, paper included in 'Numerical
Grid Generation Techniques', NASA Conference Publication 2166.

Garabedian, P.R. and Korn, D.G. (1971) Analysis of transonic airfoils, *Comm. Pure App. Math.* **24**, pp. 841-851.

Gordon, W.J. and Hall, C.A. (1973) Construction of curvilinear coordinate systems and applications to mesh generation, *Int. J. Num. Meth. Engng.* **7**, pp. 461-477.

Jameson, A. and Caughey, D.A. (1977) A finite volume method for transonic potential flow calculations, AIAA Paper 77-635.

Lee, K.D., Huang, M., Yu, N.J. and Rubbert, P.E. (1980) Grid generation for general three-dimensional configurations, paper included in 'Numerical Grid Generation Techniques', NASA Conference Publication 2166.

Magnus, R. and Yoshihara, H. (1972) Steady inviscid transonic flows over planar airfoils - a search for a simplified procedure, NASA Report CR-2186.

Middlecoff, J.F. and Thomas, P.D. (1979) Direct control of the grid point distribution in meshes generated by elliptic equations, AIAA Paper 79-1462.

Oh, Y.H. (1980) An analytical transformation technique for generating uniformly spaced computational mesh, paper included in 'Numerical Grid Generation Techniques', NASA Conference Publication 2166.

Pagano, A. (1980) Solution of 3D mixed, supersonic and subsonic propulsion problems, A brief description of the Filton programs, BAe internal document.

Reyhner, T.A. (1980) Transonic potential flow computation about three-dimensional inlets, ducts and bodies, AIAA Paper 80-1364.

Roberts, A. (1980) Development of 3D field solution methods, MVDS principles and instruction format, BAe internal document.

Sells, C.C.L. (1968) Plane subcritical flow past a lifting aerofoil, *Proc. Roy. Soc. London* **308A**, pp. 377-401.

Smith, R.E. and Weigel, B.L. (1980) Analytical and approximate boundary-fitted coordinate systems for fluid flow simulation, AIAA Paper 80-192.

Sorenson, R.L. and Steger, J.L. (1977) Simplified clustering of nonorthogonal grids generated by elliptic partial differential equations, NASA Tech. Memo 73252.

Sorenson, R.L. and Steger, J.L. (1980) Numerical generation of two-dimensional grids by the use of Poisson equations with grid control at boundaries, paper included in 'Numerical Grid Generation Techniques', NASA Conference Publication 2166.

Steger, J.L. and Sorenson, R.L. (1980) Use of hyperbolic partial differential equations to generate body fitted coordinates, paper included in 'Numerical Grid Generation Techniques', NASA Conference Publication 2166.

Thompson, J.F. (1978) Numerical solution of flow problems using body-fitted coordinate systems, paper included in *VKI Lecture Series 'Computational Fluid Dynamics'*, **1**.

Thompson, J.F. and Mastin, C.W. (1980) Grid generation using differential systems techniques, paper included in 'Numerical Grid Generation Techniques', NASA Conference Publication 2166.

Tranen, T.L. (1974) A rapid computer aided transonic airfoil design method, AIAA Paper 74-501.

Walkden, F. (1979) SOCSPAC suite of programs, University of Salford internal document.

Yu, N.J. (1980) Grid generation and transonic flow calculations for three-dimensional configurations, AIAA Paper 80-1391.

Zienkiewicz, O.C. and Phillips, D.V. (1971) An automatic mesh generation scheme for plane and curved surfaces by isoparametric coordinates, *Int. J. Num. Meth. Engng.* **3**, pp. 519-528.

APPROXIMATE FACTORISATION METHODS

T.J. Baker

(Aircraft Research Association Ltd., Bedford)

ABSTRACT

Alternating Direction Implicit or Approximate Factorisation
methods were originally developed for parabolic and elliptic
equations. The recent application of these ideas to equations
of mixed elliptic and hyperbolic type has led to significant
improvements in the iterative techniques for computing transonic
potential flow. This paper discusses some of the theory on
which Approximate Factorisation methods are based and illustrates their application to various transonic flow problems.

1. INTRODUCTION

Ever since the pioneering work of Murman and Cole (1971),
successive line over-relaxation (SLOR) has been the predominant
method for solving mixed flow problems. The technique is robust
and stable but its convergence rate is generally slow. A
typical transonic flow computation, for example, can take
several hundred iterations before the pressure distribution is
completely frozen. This deficiency has long been recognised and
several attempts have been made to improve the convergence rate.
Eigenvalue extrapolation (Hafez and Cheng, 1975) has been
successful in accelerating the convergence of SLOR for some flow
problems. Methods based on the Poisson Solver (Martin and
Lomax, 1974) have also been proposed but likewise have achieved
only limited success.

It was the application of Alternating Direction Implicit
(ADI) or approximate factorisation techniques to transonic flow
problems that eventually produced a worthy alternative to SLOR.
These fast implicit methods were originally developed for equations of parabolic and elliptic type (Peaceman and Rachford,
1955 and Yanenko, 1971). The first application to transonic
flow problems was reported by Ballhaus and Steger (1975) who
used these techniques to remove the time step limitation inherent in a relaxation solution of the unsteady TSP equation. The
extension to steady TSP came quickly (Ballhaus et al, 1977) and
their AF2 scheme formed the basis for further developments aimed
at solving the full potential equation. This followed two

independent but similar lines (Holst 1978, 1979 and Baker 1979, 1981) and approximate factorisation algorithms are now replacing SLOR in many existing full potential codes.

Our aim in this paper is to explain the superior performance of approximate factorisation and to illustrate its application to the full potential equation. We start by examining the typical convergence behaviour of point and line iterative techniques. This discussion provides the motivation for selecting implicit schemes of the approximate factorisation or ADI type. The application of these ideas to compute transonic potential flow is described and in the final section we outline a recent application of approximate factorisation to compute viscous transonic flow over an aerofoil.

2.   CONVERGENCE OF EXPLICIT METHODS

To examine the typical convergence properties of point and line iterative methods, we consider a line Gauss Seidel scheme applied to Laplace's equation,

$$\phi_{xx} + \phi_{yy} = 0 \tag{2.1}$$

On a uniform grid ($\Delta x = \Delta y = h$) with central difference approximations for ϕ_{xx} and ϕ_{yy}, we obtain the difference scheme

$$\phi^{n}_{j+1,k} + \phi^{n+1}_{j-1,k} + \phi^{n+1}_{j,k+1} + \phi^{n+1}_{j,k-1} - 4\phi^{n+1}_{j,k} = 0 \tag{2.2}$$

where $\phi^{n}_{j,k}$ is the approximate solution after n iteration cycles. The error between $\phi^{n}_{j,k}$ and the exact solution $\phi_{j,k}$ is defined by

$$e^{n}_{j,k} = \phi^{n}_{j,k} - \phi_{j,k}$$

Let the error be expressed as a Fourier sum, viz

$$e^{n}_{j,k} = \sum_{p,q} \rho^{n}(p,q) \, e^{ipx} e^{iqy} \tag{2.3}$$

We define the amplification factor as

$$G(p,q) = \frac{\rho^{n+1}}{\rho^{n}}$$

and for the line Gauss Seidel scheme (2.2) this is

$$G(p,q) = \frac{e^{i\xi}}{2 - e^{-i\xi} + 2(1-\cos\eta)} \qquad (2.4)$$

where $\xi = ph$, $\eta = qh$ and the wave numbers are

$$p,q = 1 \ \dots\ \frac{\pi}{h}$$

The amplification factor measures the extent to which each frequency component of the error spectrum is increased or reduced after one iteration cycle. The von Neumann condition states that

$$\left| G(p,q) \right| \leqslant 1$$

is a necessary condition for convergence. If the amplification factor is less than but close to the value one, then the iterative scheme may be stable but only converge at a very slow rate. It is therefore desirable to select a method for which the amplification factor is significantly less than one throughout the range of frequencies in the error spectrum. In the above example, it can be seen that expression (2.4) is small for high frequencies. For example, if $\xi \sim \pi$ and $\eta \sim \pi$ then

$$\left| G \right| \sim 1/7$$

At the low frequency end of the error spectrum where $\xi \sim h$ and $\eta \sim h$ we have

$$G \sim 1 - 2h^2$$

which implies a very slow convergence rate. For optimum SLOR this can be improved so that

$$G = 1 - O(h)$$

for low frequency components. Nevertheless, the convergence rate of a relaxation scheme can be expected to become progressively worse as the number of grid points is increased since this causes a corresponding reduction in the size of the grid increment h. In order to alleviate this difficulty it is usual to obtain a relaxation solution on a coarser grid and then interpolate this result onto the fine grid for use as a starting solution for the fine grid calculation. The coarse grid calculation is faster, first because the grid increment h is larger and second since the computation time per iteration cycle is less. This is a useful strategy for decreasing the overall computation time but the convergence rate of a relaxation method still leaves much to be desired. To illustrate this

we first show convergence histories for the calculation of subcritical lifting flow over an aerofoil by relaxation.

Note that all the computed results presented here have been
obtained from methods which use a grid generated by the circle
plane mapping (i.e. by conformally mapping the region exterior
to the aerofoil onto the interior of a circle). By taking equal
increments Δr in the radial direction and $\Delta\theta$ in the circumferential direction one produces a grid that is nonuniform in physical space. The relaxation code mentioned here is very similar
to the familiar Garabedian and Korn method (Garabedian and Korn,
1971) which also uses the circle plane mapping. The approximate
factorisation codes which produced the results shown later use
the same grid.   The subcritical case is that shown in Fig. 1
and the convergence histories for relaxation with and without
grid refinement are presented in Fig. 2.   In each case it can be
seen that the residual drops rapidly at first, corresponding to
the removal of high frequency components. Eventually, the residual is dominated by low frequency errors and the asymptotic
rate of decay is seen to be very slow.

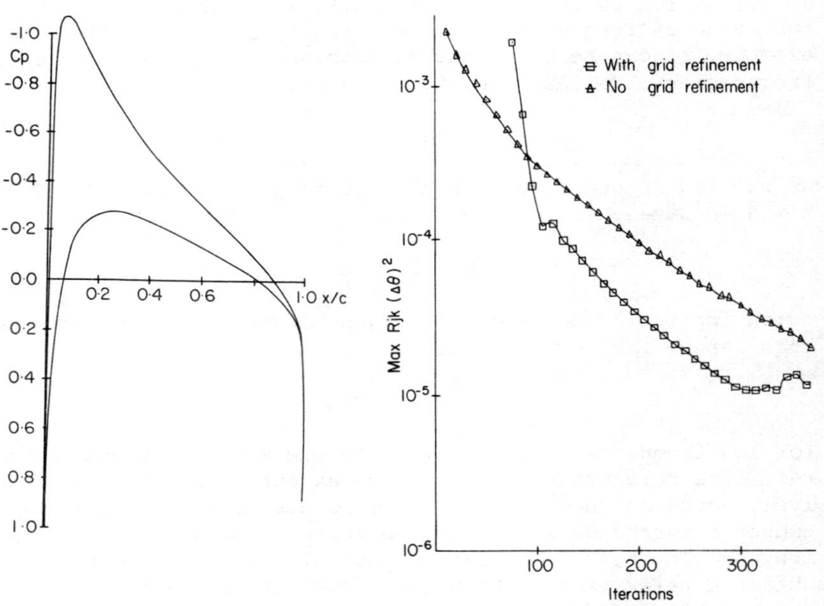

Fig. 1 Pressure distribution   Fig. 2 Convergence history

Figs. 1 & 2 NACA 0012 Aerofoil, M = 0.63, $\alpha = 2^\circ$

Loosely speaking, high frequency errors are associated with localised deviations from the converged solution. On the other hand, low frequency errors reflect global discrepancies, for example, an incorrect circulation. It is therefore not surprising that a good starting solution, such as that provided by a coarse grid calculation, is effective in reducing the initial low frequency content. This behaviour is reflected in the convergence histories which show that grid refinement leads to a lower overall residual even though the asymptotic rate of decay is the same. In practice, one continues the iterative solution until the residual or some other measure of convergence reaches a sufficiently small level. The convergence parameter that is plotted in Fig. 2 is the maximum residual multiplied by h^2. A value of 10^{-5} is usually required to achieve adequate convergence and it can be seen that the slow convergence rate of relaxation prevents this level from being reached until a great many iterations have been carried out.

When computing a transonic flow in which strong shocks are present the convergence rate of the relaxation scheme can be much slower and the manner in which the solution converges can be quite misleading. This is well illustrated by the example shown in Fig. 3. The convergence history with grid refinement is presented in Fig. 4 and the circulation growth is shown in Fig. 5. Between about 200 and 400 iterations the circulation remains almost constant and only reaches the final converged level much later. The final change in circulation and the associated blip in the residual at around 420 iterations correspond with a shift in the shock position by one grid point (see Fig. 3). In view of the essentially unchanging but incorrect pressure distribution and circulation that is predicted prior to this shift, it is quite probable that one would normally have stopped the run before 400 iterations and considered the result to be converged. Admittedly, this example has a much stronger shock than one would usually consider but similar misleading behaviour has been observed with other cases, for example the KORN aerofoil near its shock free design condition.

To gain further insight into the problem it is useful to consider a time dependent analogy. This idea has been successfully exploited by several people (Garabedian, 1956, Jameson, 1974). The difference scheme for the line Gauss Seidel method applied to Laplace's equation on a uniform grid is given by equation (2.2) and we now rearrange this equation into the form

$$N\Delta_{jk}^{n} = L\phi_{jk}^{n}$$

where
$$\Delta_{jk}^{n} = \phi_{jk}^{n+1} - \phi_{jk}^{n}$$
(2.5)

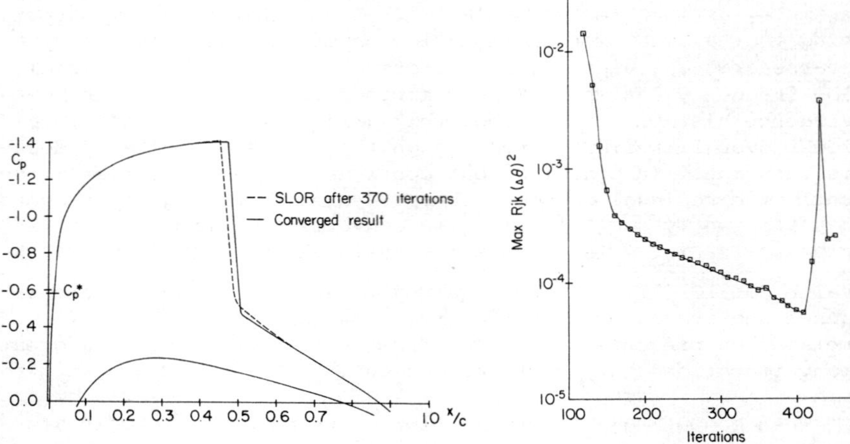

Fig. 3   Pressure distribution   Fig. 4   Convergence history

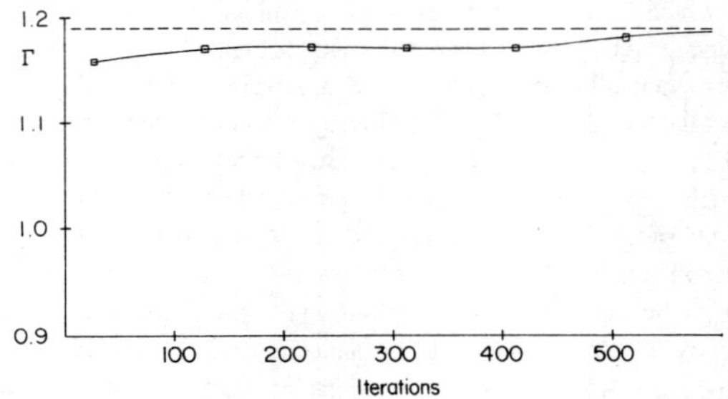

Fig. 5   Circulation growth

Figs. 3-5   NACA 0012 Aerofoil, $M = 0.75$, $\alpha = 3^\circ$

Results from SLOR with grid refinement

The residual on the right hand side is

$$L\phi^n_{jk} = \left(\frac{\delta_{xx}}{h^2} + \frac{\delta_{yy}}{h^2}\right) \phi^n_{jk}$$

$$= \frac{1}{h^2}\left(\phi^n_{j+1,k} + \phi^n_{j-1,k} + \phi^n_{j,k+1} + \phi^n_{j,k-1} - 4\phi^n_{jk}\right)$$

$$(2.6)$$

where the second order difference operator δ_{xx} is defined as

$$\delta_{xx}\, \phi_{j,k} = \phi_{j+1,k} - 2\phi_{j,k} + \phi_{j-1,k}$$

with δ_{yy} defined in a similar way. We also introduce the first order forward and backward difference operators defined by

$$\overrightarrow{\delta}_x\, \phi_{j,k} = \phi_{j+1,k} - \phi_{j,k}$$

$$\overleftarrow{\delta}_x\, \phi_{j,k} = \phi_{j,k} - \phi_{j-1,k}$$

and note that

$$\delta_{xx} = \overrightarrow{\delta}_x\, \overleftarrow{\delta}_x$$

In the above form equation (2.5) can be regarded as a finite difference approximation to a time dependent equation in which the final steady state solution represents the required result. Thus, if one iteration cycle corresponds to a step Δt in artificial time and we take

$$\Delta t = h^2,$$

then to first order in h equation (2.5) approximates the parabolic equation

$$\phi_t = \phi_{xx} + \phi_{yy} \qquad\qquad (2.7)$$

Seen from this viewpoint it seems plausible that the poor attenuation of low frequency error components is related to the small time step that is taken. Ideally we would like to take large time steps in order to move more quickly towards the steady state solution. For a point or line iterative scheme, however, the requirement of numerical stability places a restriction on the maximum size of the effective time step Δt. This is typical of explicit schemes and reflects the observation that several iteration cycles are required before changes at every point are felt at all other points in the flow field.

3. IMPLICIT METHODS

The remedy lies in the adoption of an implicit scheme which will remove the time step limitation. We again take the parabolic equation (2.7) as our prototype and following Mitchell (1969), develop a finite difference approximation by using the formal relationship

$$\phi(x,y,t+\Delta t) = \left[1 + \Delta t \frac{\partial}{\partial t} + \frac{\Delta t^2}{2} \frac{\partial}{\partial t^2} + \ldots\right] \phi(x,y,t)$$

$$= \exp\left(\Delta t \frac{\partial}{\partial t}\right) \phi(x,y,t)$$

Next replace the continuous function $\phi(x,y,t)$ by the finite difference form ϕ_{jk}^n and the differential operation $\partial/\partial t$ by the residual operator (2.6) to get

$$\phi_{jk}^{n+1} = \exp\left[r(\delta_{xx} + \delta_{yy})\right] \phi_{jk}^n \qquad (3.1)$$

where

$$r = \frac{\Delta t}{h^2}$$

If one expands the exponential to first order in r, an explicit formula is obtained with the stability bound

$$r \leq \frac{1}{4}$$

If, however, equation (2.1) is rearranged into the form

$$\exp\left[-\frac{r}{2}(\delta_{xx} + \delta_{yy})\right] \phi_{jk}^{n+1} = \exp\left[\frac{r}{2}(\delta_{xx} + \delta_{yy})\right] \phi_{jk}^n \qquad (3.2)$$

then expanding the exponential to first order in r leads to the following implicit scheme

$$\left[1 - \frac{r}{2}(\delta_{xx} + \delta_{yy})\right] \phi_{jk}^{n+1} = \left[1 + \frac{r}{2}(\delta_{xx} + \delta_{yy})\right] \phi_{jk}^n$$

This difference scheme is convergent for all positive r but, owing to the complexity of the left hand side, a large matrix inversion is required to obtain values at the new time level. An alternative procedure that leads to a set of straightforward tridiagonal inversions follows from the formula (cf equation (3.2))

$$\exp\left(-\frac{r}{2}\delta_{xx}\right) \exp\left(-\frac{r}{2}\delta_{yy}\right) \phi_{jk}^{n+1} = \exp\left(\frac{r}{2}\delta_{xx}\right) \exp\left(\frac{r}{2}\delta_{yy}\right) \phi_{jk}^n$$

This can be expanded to give the ADI scheme

$$(1 - \frac{r}{2} \delta_{xx})(1 - \frac{r}{2} \delta_{yy}) \; \phi_{jk}^{n+1} = (1 + \frac{r}{2} \delta_{xx})(1 + \frac{r}{2} \delta_{yy}) \; \phi_{jk}^{n}$$

$$(3.3)$$

A modal analysis results in the following amplification factor for the residual error

$$G(p,q) = \left\{ \frac{1 - r(1 - \cos\xi)}{1 + r(1 - \cos\xi)} \right\} \left\{ \frac{1 - r(1 - \cos\eta)}{1 + r(1 - \cos\eta)} \right\} \qquad (3.4)$$

where ξ = ph and η = qh as before. It follows that

$$|G| \leqslant 1 \quad \text{for all p and q,}$$

and hence the ADI method is convergent for all positive r.

Rearranging equation (3.3) into the form (2.5) and, for convenience, replacing the parameter r by α where

$$\alpha = \frac{2}{rh^2}$$

leads to the difference scheme

$$\left(\alpha - \frac{\delta_{xx}}{h^2} \right) \left(\alpha - \frac{\delta_{yy}}{h^2} \right) \Delta_{jk}^{n} = 2\alpha L\phi_{jk}^{n} \qquad (3.5)$$

This is the AF1 scheme that was proposed by Ballhaus et al (1977). On examining the amplification factor (3.4) it can be seen that the choice

$$\alpha = \frac{2}{h^2}(1 - \cos ph)$$

will reduce the pth frequency component to zero. Thus the finite sequence

$$\alpha_p = \frac{2}{h^2}(1 - \cos ph), \; p = 1 \ldots \frac{\pi}{h} \qquad (3.6)$$

should remove all error components and hence produce a converged result. The αs are called acceleration parameters and a sequence such as (3.6) is equivalent to the application of a series of different time steps. Large values of α correspond to small time steps which are effective against high frequency error components. On the other hand, small values of α which represent

large time steps reduce the low frequency end of the error
spectrum.

The above analysis is only strictly valid for the solution of
an equation with constant coefficients on a uniform grid with
periodic boundary conditions. In practice, we require the solution of a non-linear equation on a nonuniform grid and so the
above conclusion must be regarded as offering only a guideline
for solving the full potential equation. If the equation to be
solved is

$$C_1 \phi_{xx} + C_2 \phi_{xy} + C_3 \phi_{yy} + D = 0 \qquad (3.7)$$

then the iterative scheme (3.5) can be generalised to

$$(\alpha - C_1 \frac{\delta_{xx}}{\Delta x^2})(\alpha - C_3 \frac{\delta_{yy}}{\Delta y^2}) \Delta_{jk}^n = \sigma \alpha L \phi_{jk}^n \qquad (3.8)$$

where L is now the residual operator corresponding to equation
(3.7). The difference equations (3.8) are solved in two steps.
First we obtain the intermediate values of F_{jk} by inverting

$$(\alpha - C_1 \frac{\delta_{xx}}{\Delta x^2}) F_{jk} = \sigma \alpha L \phi_{jk}^n$$

on each constant y line. The second step requires a similar
series of tridiagonal inversions on constant x lines

$$(\alpha - C_3 \frac{\delta_{yy}}{\Delta y^2}) \Delta_{jk}^n = F_{jk}$$

whence

$$\phi_{jk}^{n+1} = \phi_{jk}^n + \Delta_{jk}^n$$

In practice it is best to use a small sequence of acceleration
parameters running between the minimum and maximum given by
(3.6). The minimum value $\alpha_L = 1$ occurs for $p = 1$, while the
maximum

$$\alpha_H = \frac{4}{h^2}$$

is obtained by taking $ph \sim \pi$. The following geometric sequence
(Ballhaus et al, 1977), which is repeated in a cyclic fashion,

has proved very effective

$$\alpha_k = \alpha_H \left(\frac{\alpha_L}{\alpha_H}\right)^{\cdot k-1/M-1} , \quad k = 1 \ldots M \qquad (3.9)$$

where M typically has a value of around six to eight. The parameter σ in equation (3.8) is usually taken to be two (cf (3.5)).

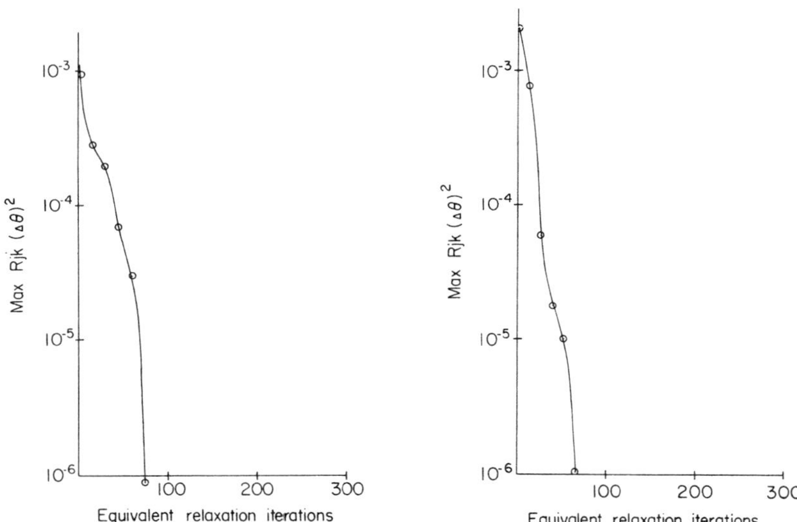

Fig. 6 Convergence history when Fig. 7 Convergence history when
     using AF1   using AF3

Figs. 6-7   NACA 0012 Aerofoil, M = 0.63, $\alpha = 2^{\circ}$

The improvement in convergence rate that can be obtained with
the ADI or AF1 scheme is readily apparent if one considers its
application to the example presented in Fig. 1. The convergence
history for this case is shown in Fig. 6 and it can be seen the
residual drops much more quickly than for SLOR (cf Fig. 2).
Each iteration of the AF1 scheme takes about 1½ times as long as
a relaxation iteration and so the convergence history presented
in Fig. 6 has been plotted against equivalent relaxation iterations. The AF1 scheme clearly offers a significant improvement
over SLOR. The AF1 or ADI scheme described above is a fast
procedure for solving elliptic problems. An important example
is the set of elliptic equations that arise in numerical grid
generation by Thompson's method (Thompson et al, 1974).
Holst (1978, 1979) for example, uses an AF1 algorithm to generate two-dimensional grids by the Thompson technique.

4. THE APPROXIMATE FACTORISATION SCHEME AF2

Although very effective for solving elliptic equations, the
ADI or AF1 factorisation is not well suited to equations of
mixed type. This problem has been carefully examined by Jameson (1974, 1976) who shows that the reason lies in the form of
the time like terms in the equivalent time dependent equation.
In terms of the time dependent analogy the solution of an equation such as

$$(1 - M^2)\ \phi_{xx} + \phi_{yy} = 0$$

by an AF1 scheme is equivalent to solving the equation

$$\phi_t = A\phi_{xx} + \phi_{yy}$$

where $A = 1 - M^2$. When the flow is locally subsonic A is positive and the above equation is parabolic. In this case the ϕ_t
term introduces an effective damping that ensures a stable iterative scheme. When the flow is locally supersonic, A becomes
negative and the ϕ_t term no longer provides the required damping.

Its presence can have a destabilising effect since the equation
then admits solutions that grow exponentially. It is possible
to stabilise the calculation by introducing terms like ϕ_{xt} and
ϕ_{yt} (Ballhaus et al, 1977) but a more satisfactory approach is
the construction of an iterative scheme for which the equivalent
time dependent equation contains no ϕ_t term. We therefore take
the equation

$$\phi_{yt} = A\phi_{xx} + \phi_{yy}$$

as a prototype and seek a factored difference approximation of
the form (2.5). One possibility is to use the identity

$$\delta_{yy} = \vec{\delta}_y\ \overleftarrow{\delta}_y$$

and alter the AF1 factorisation (3.5) so that the difference
operator δ_{yy} is split between the two factors. For example

$$\left(-\alpha\,\frac{\vec{\delta}_y}{h} - A\frac{\delta_{xx}}{h^2}\right)\left(\alpha + \frac{\overleftarrow{\delta}_y}{h}\right)\Delta_{jk}^n = \sigma\alpha L\phi_{jk}^n \qquad (4.1)$$

A modal analysis shows that this scheme is stable and the parameter sequence

$$\alpha_p = \frac{2 \sin ph/2}{h} \, , \ p = 1 \ldots \frac{\pi}{h} \tag{4.2}$$

approximately minimises the amplification factor. Again it is usual to employ a smaller sequence of parameters such as (3.9) and repeat this in a cyclic manner. For the AF2 factorisation, however, we find that

$$\alpha_L = 1, \ \alpha_H = \frac{2}{h} \tag{4.3}$$

and the parameter σ is given a value nearer one. The presence of the $\vec{\delta}_y$ operator in the first factor imposes a sweep restriction. To show this we write out the finite difference equation for the first step of the inversion

$$\left(- \alpha \, \frac{\vec{\delta}_y}{h} - A \, \frac{\delta_{xx}}{h^2} \right) F_{jk} = \sigma \alpha L \phi_{jk}^n \tag{4.4}$$

which expands to give

$$- \frac{F_{j+1,k}}{h^2} + \left(\frac{\alpha}{h} + \frac{2A}{h^2} \right) F_{jk} - \frac{F_{j-1,k}}{h^2} = \sigma \alpha L \phi_{jk}^n + \alpha \, \frac{F_{j,k+1}}{h} \tag{4.5}$$

When solving the tridiagonal system on the line $y = kh$ we require values of $F_{j,k+1}$. Hence we must sweep in the direction of decreasing k. Note that the minus sign in front of the $\vec{\delta}_y$ operator in (4.1) is imposed by the requirement of diagonal dominance. A different factorisation which allows the first factor to be inverted by sweeping in the direction of increasing k is

$$\left(\alpha \, \frac{\overleftarrow{\delta}_y}{h} - A \, \frac{\delta_{xx}}{h^2} \right) \left(\alpha - \frac{\vec{\delta}_y}{h} \right) \Delta_{jk}^n = \sigma \alpha L \phi_{jk}^n \tag{4.6}$$

When A is negative corresponding to locally supersonic flow the term ϕ_{xx} in the residual is backward differenced. The factorisation (4.1) is then replaced by

$$\left(- \alpha \, \frac{\vec{\delta}_y}{h} - A \, \frac{\overleftarrow{\delta}_x \overleftarrow{\delta}_x}{h^2} \right) \left(\alpha + \frac{\overleftarrow{\delta}_y}{h} \right) \Delta_{jk}^n = \sigma \alpha L \phi_{jk}^n \tag{4.7}$$

Apart from minor differences, the factorisations (4.1) and (4.7) are those used in the AF2 scheme proposed by Ballhaus et al (1977). In their paper they show several comparisons of AF2 and relaxation convergence rates for the TSP equation. The improvement they obtained is impressive and encouraged the expectation that a similar improvement would be forthcoming for the full potential equation.

5. HOLST'S SCHEME

The extension of these ideas to the full potential equation is not obvious and there appear to be no sound theoretical criteria for deciding what should be a good factorisation. It is probably fair to say that the two schemes that have proved particularly successful, and which we describe in sections 5 and 6, owe their development more to inspired guesswork than theoretical insight. In order to extract some kind of guideline we note that there is a feature common to both the AF1 factorisation (3.5) and the AF2 factorisation (4.1). The terms in the iteration operator N which are linear in α correspond to the residual operator L. For want of anything better, we adopt the same general guideline when developing a factorisation for the full potential equation. In other words, on expanding the iteration operator as a polynomial in α, we require those terms which are linear in α to resemble closely the residual operator L. On a Cartesian grid the residual for the full potential equation in conservation form is given by

$$L\phi^{n}_{jk} = \vec{\delta}_{x} \left. \frac{(\tilde{\rho}u)^{n}}{\Delta x} \right|_{j-\frac{1}{2},k} + \vec{\delta}_{y} \left. \frac{(\bar{\rho}v)^{n}}{\Delta y} \right|_{j,k-\frac{1}{2}} \qquad (5.1)$$

where

$$\tilde{\rho}_{j+\frac{1}{2},k} = [(1-\nu)\rho]_{j+\frac{1}{2},k} + \nu_{j+\frac{1}{2},k} \, \rho_{j+r+\frac{1}{2},k}$$

$$\bar{\rho}_{j,k+\frac{1}{2}} = [(1-\nu)\rho]_{j,k+\frac{1}{2}} + \nu_{j,k+\frac{1}{2}} \, \rho_{j,k+s+\frac{1}{2}}$$

where

$$r = \begin{cases} -1, & u_{j+\frac{1}{2},k} > 0 \\ 1, & u_{j+\frac{1}{2},k} < 0 \end{cases}$$

$$s = \begin{cases} -1, & v_{j,k+\frac{1}{2}} > 0 \\ 1, & v_{j,k+\frac{1}{2}} < 0 \end{cases}$$

The parameter ν controls the artificial viscosity and is set to zero in subsonic regions of the flow. The form which it should take in supersonic regions is discussed by Holst (1979). In view of the remarks about the AF1 scheme in section 4, it appears best to restrict our attention to a factorisation of the AF2 type (cf equation 4.1). These considerations lead to Holst's AF2 scheme (Holst, 1978, 1979),

$$\left[-\alpha \frac{\overrightarrow{\delta}_y \, \overline{\rho}^{\,n}_{j,k-\frac{1}{2}}}{\Delta y} - \frac{\overrightarrow{\delta}_x \, \tilde{\rho}^{\,n}_{j-\frac{1}{2},k}}{\Delta x^2} \, \overleftarrow{\delta}_x \right] \left(\alpha + \frac{\overleftarrow{\delta}_y}{\Delta y} \right) \Delta^n_{jk} = \sigma \alpha L \phi^n_{jk} \qquad (5.2)$$

This is a very neat factorisation and has some particularly attractive features. First we note that the form taken by the modified densities always maintains the correct upwind influence in the difference scheme when the flow is locally supersonic. Thus the effect of a rotated difference scheme is included in the factorisation as well as the residual. In addition the inversion of the first factor is accomplished by a straightforward tridiagonal solution while the second factor requires only a very simple bidiagonal inversion. It is, however, necessary to evaluate the modified densities $\tilde{\rho}$ and $\overline{\rho}$ at the previous iteration level n which suggests a possible loss of implicitness. As a result one might anticipate some stability problems. This does not appear to cause any difficulty in practice although, when strong shocks are present, Holst augments the factorisation by adding the term

$$-\alpha \frac{\beta \overrightarrow{\delta}_x}{\Delta x} \quad \text{or} \quad +\alpha \frac{\beta \overleftarrow{\delta}_x}{\Delta x} \qquad (5.3)$$

to the first factor. This effectively introduces a ϕ_{xt} term into the equivalent time dependent equation and therefore has a stabilising influence on the iteration scheme. The parameter β is a user defined constant which is set equal to zero in regions of subcritical flow. The choice of a forward or backward difference operator in (5.3) is of course dictated by the local flow direction.

Full details of this scheme can be found in Holst (1979) which shows several impressive examples. These demonstrate that his method is both reliable and very fast for all kinds of flow conditions including cases with extensive regions of supersonic flow.

6. THE APPROXIMATE FACTORISATION SCHEME AF3

In principle it should be possible to use Holst's version of the AF2 scheme to solve the potential equation in nonconservative form. However, an alternative factorisation has been proposed by Baker (1979, 1981) and for the nonconservative equation this is a more suitable extension of the original AF2 method. The name AF3 has been used to identify this particular scheme although it has much in common with Holst's AF2 algorithm.

In Cartesian coordinates the quasilinear or nonconservative form of the potential equation can be written as

$$A \, \phi_{xx} + B \, \phi_{xy} + C \, \phi_{yy} = 0 \qquad (6.1)$$

where

$$A = 1 - \frac{u^2}{a^2}, \quad B = \frac{-2uv}{a^2} \text{ and } C = 1 - \frac{v^2}{a^2}$$

When the flow is subcritical all ϕ derivatives are replaced by central difference formulae and equation (6.1) is approximated by the residual

$$L\phi_{jk}^{n} = \left(A \, \frac{\delta_{xx}}{\Delta x^2} + B \, \frac{\mu_x \mu_y \delta_x \delta_y}{\Delta x \Delta y} + C \, \frac{\delta_{yy}}{\Delta y^2} \right) \phi_{jk}^{n} \qquad (6.2)$$

where μ_x, μ_y are averaging operators, for example

$$\mu_x \, f_{j,k} = \tfrac{1}{2}[f_{j+\frac{1}{2},k} + f_{j-\frac{1}{2},k}]$$

An obvious extension of the AF2 scheme (4.1) is the following

$$\left(- \alpha \, C \, \frac{\vec{\delta}_y}{\Delta y} - A \, \frac{\delta_{xx}}{\Delta x^2} \right) \left(\alpha + \frac{\overleftarrow{\delta}_y}{\Delta y} \right) \phi_{jk}^{n} = \sigma \alpha L \phi_{jk}^{n} \qquad (6.3)$$

We observe that no term corresponding to the cross derivative $B\phi_{xy}$ has been included in the factorisation. This omission does not appear to slow down the convergence rate of the scheme. To illustrate this and demonstrate the algorithm's effectiveness we show the AF3 convergence history in Fig. 7 for the case presented previously in Fig. 1. A cycle of the iteration scheme (6.3) requires one set of tridiagonal and one set of bidiagonal inversions in place of the two sets of tridiagonal inversions required by an AF1 iteration. One therefore finds that each AF2 or AF3 iteration takes slightly less computational effort than an AF1 iteration. In fact, one iteration of the AF3 scheme is equivalent in computational time to about 1.3 relaxation iterations. We have again plotted the convergence history in terms of equivalent relaxation iterations in order to get a fair comparison between convergence rates of the different solution

algorithms. Figs. 2 and 6 when compared with Fig. 7 show that
the AF3 scheme, like the AF1 scheme, is a considerable improvement over SLOR.

When regions of supersonic flow are present, we construct a
residual that contains a combination of upwind and centrally
differenced terms. For convenience we now write

$$A = A_u + A_c, \quad B = B_u + B_c \quad \text{and} \quad C = C_u + C_c$$

where the subscript u refers to the contribution of a coefficient to the upwind differenced term and the subscript c denotes
the centrally differenced contribution. Computational stability
requires the adoption of a rotated difference scheme (Jameson,
1974) which leads to the values

$$A_c = \frac{v^2}{q^2}, \quad B_c = \frac{-2uv}{q^2}, \quad C_c = \frac{u^2}{q^2} \tag{6.4}$$

$$A_u = \left(1 - \frac{q^2}{a^2}\right)\frac{u^2}{q^2}, \quad B_u = 2\left(1 - \frac{q^2}{a^2}\right)\frac{uv}{q^2}, \quad C_u = \left(1 - \frac{q^2}{a^2}\right)\frac{v^2}{q^2}$$

When the two velocity components u and v are both positive, the
residual can be written as

$$L\phi_{jk}^n = \left(A_c \frac{\delta_{xx}}{\Delta x^2} + A_u \frac{\overleftarrow{\delta}_x \overleftarrow{\delta}_x}{\Delta x^2} + B_c \frac{\mu_x \mu_y}{\Delta x \Delta y} \delta_x \delta_y \right.$$
$$\left. + B_u \frac{\overleftarrow{\delta}_x \overleftarrow{\delta}_y}{\Delta x \Delta y} + C_c \frac{\delta_{yy}}{\Delta y^2} + C_u \frac{\overleftarrow{\delta}_y \overleftarrow{\delta}_y}{\Delta y^2} \right) \phi_{jk}^n \tag{6.5}$$

The factorisation (6.3) can then be modified as follows

$$\left(-\alpha C_c \frac{\overrightarrow{\delta}_y}{\Delta y} - A_c \frac{\delta_{xx}}{\Delta x^2} - A_u \frac{\overleftarrow{\delta}_x \overleftarrow{\delta}_x}{\Delta x^2} \right)\left(\alpha + \frac{\overleftarrow{\delta}_y}{\Delta y} \right) \Delta_{jk}^n = \sigma\alpha L\phi_{jk}^n \tag{6.6}$$

When u is negative the backward difference operator $\overleftarrow{\delta}_x \overleftarrow{\delta}_x$ in
equations (6.5 and 6.6) is replaced by the corresponding forward
difference operator $\overrightarrow{\delta}_x \overrightarrow{\delta}_x$. In addition the cross derivative
term $B_u \frac{\overleftarrow{\delta}_x \overleftarrow{\delta}_y}{\Delta x \Delta y}$ in equation (6.5) is replaced by $B_u \frac{\overrightarrow{\delta}_x \overleftarrow{\delta}_y}{\Delta x \Delta y}$

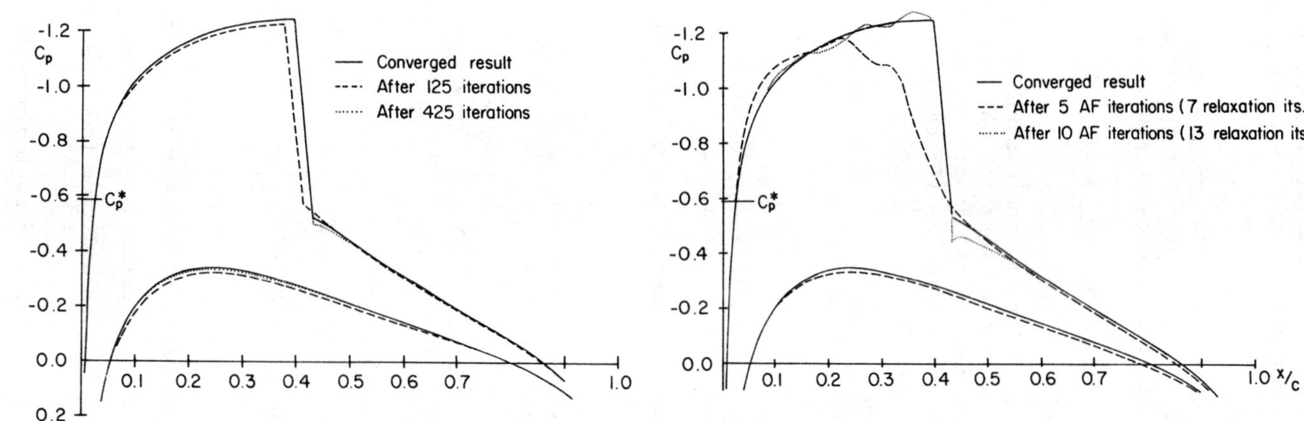

Fig. 8 Pressure distribution computed by
relaxation

Fig. 9 Pressure distribution computed
by AF3

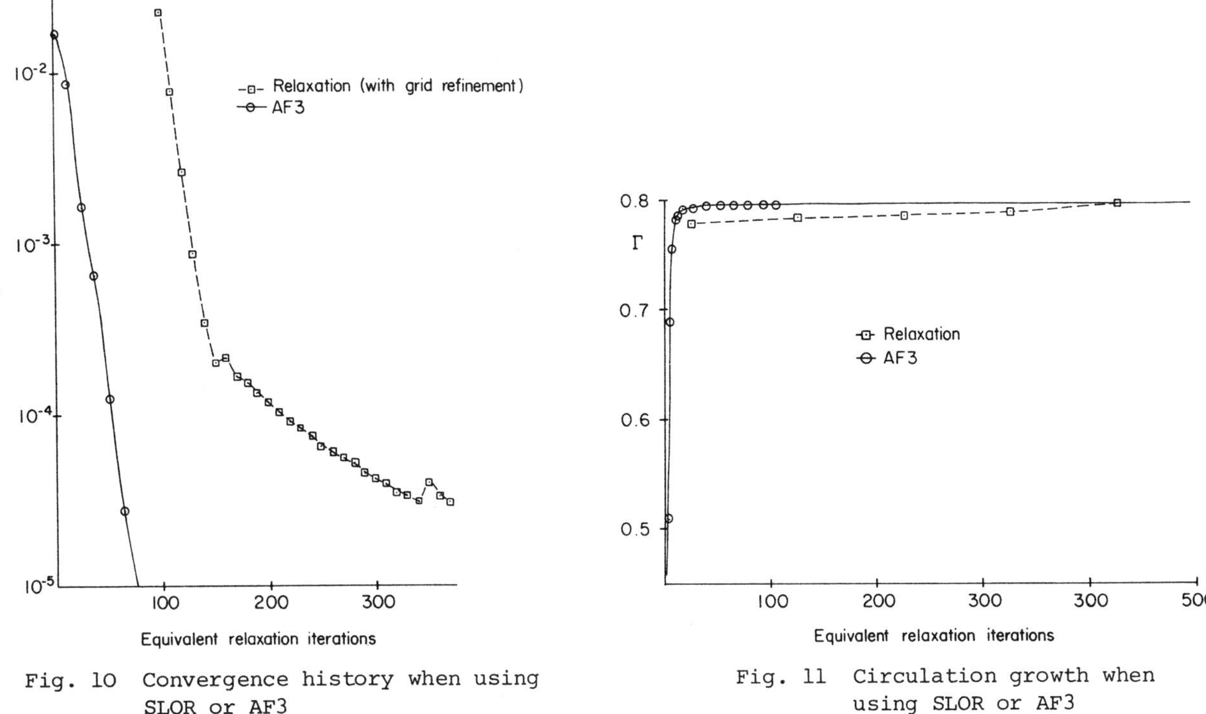

Fig. 10 Convergence history when using SLOR or AF3

Fig. 11 Circulation growth when using SLOR or AF3

Figs. 8-11 NACA 0012 Aerofoil, M = 0.75, $\alpha = 2°$

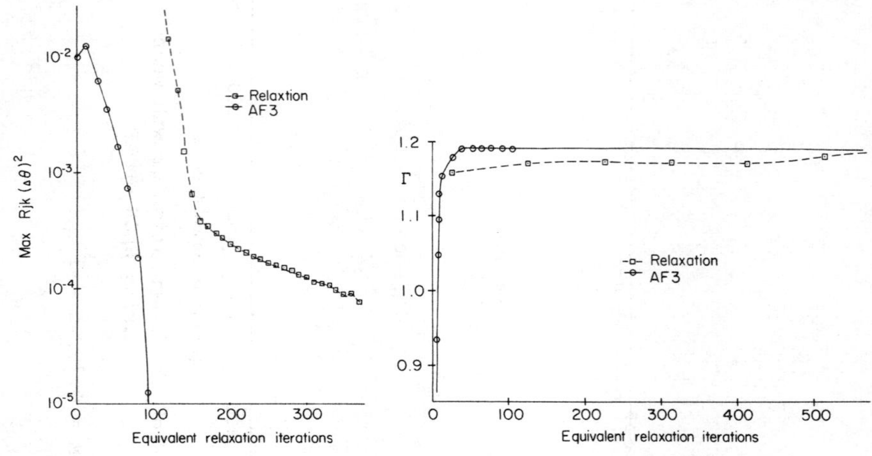

Fig. 12   Convergence history   Fig. 13   Circulation growth
           when using SLOR or   when using SLOR or AF3
           AF3

Figs. 12-13   NACA 0012 Aerofoil, M = 0.75, $\alpha = 3^\circ$

The upwind differencing complicates the algorithm, and allowing
for the possibility of both backward and forward differencing
in the x direction, this requires the inversion of a pentadiagonal matrix. If we drop one of the backward difference operators
$\overleftarrow{\delta}_x$, we obtain the alternative factorisation

$$\left\{ -\alpha\, C_c\, \frac{\overrightarrow{\delta}_y}{\Delta y} - A_c\, \frac{\delta_{xx}}{\Delta x^2} - A_u\, \frac{\overleftarrow{\delta}_x}{\Delta x^2} \right\}\left(\alpha + \frac{\overleftarrow{\delta}_y}{\Delta y} \right) \Delta_{jk}^n = \sigma\alpha L\phi_{jk}^n$$

$$(6.7)$$

which only requires a tridiagonal inversion for the first fac-
tor. Removing one of the $\overleftarrow{\delta}_x$ operators has the effect of replacing the upwind formula

$$\Delta x^2\, D_{xx}\, \phi = \phi_{jk}^{n+1} - 2\phi_{j-1,k}^{n+1} + \phi_{j-2,k}^{n+1}$$

for ϕ_{xx} by the following

$$\Delta x^2\, D_{xx}\, \phi = \phi_{jk}^{n+1} - \phi_{j-1,k}^{n+1} - \phi_{j-1,k}^n + \phi_{j-2,k}^n$$

This alternative form is stable and a comparison of the two factorisations (6.6) and (6.7) for a number of numerical examples
confirms that the convergence rates are similar. The upwind
differenced term in (6.7) now corresponds to a ϕ_{xt} term in the
associated time dependent equation and therefore has a stabilising influence on the iterative scheme. When u is negative the
factorisation (6.7) is replaced by the following

$$\left(- \alpha \, C_c \, \frac{\overset{\rightarrow}{\delta}_y}{\Delta y} - A_c \, \frac{\delta_{xx}}{\Delta x^2} + A_u \, \frac{\overset{\rightarrow}{\delta}_x}{\Delta x^2} \right) \left(\alpha + \frac{\overset{\leftarrow}{\delta}_y}{\Delta y} \right) \Delta^n_{jk} = \sigma \alpha L \phi^n_{jk}$$

To illustrate the application of this scheme to compute transonic lifting flow we first consider computed pressure distributions at various stages. Our example is the NACA 0012 aerofoil at a Mach number of 0.75 and an incidence of 2 degrees.
Fig. 8 compares the converged pressure distribution with results
computed by SLOR after 100 coarse grid iterations (equivalent in
computing time to 25 fine grid iterations) followed by a further
100 and 400 fine grid iterations. It is evident that the result
after an effective computing time of 125 fine grid iterations
is some way off convergence and even after 425 iterations the
shock strength has still not reached its final converged value.
The results can be contrasted with those shown in Fig. 9 for
AF3 after 5 and 10 AF iterations (equivalent in computing time
to 7 and 13 relaxation iterations respectively). After 10 AF
iterations the lower surface distribution is correct and the
final shock position has been reached. After a further 10 AF
iterations there is no plottable difference between the computed
pressure distribution and the converged result. Fig. 10 compares
the convergence histories of SLOR and AF3 for this case and in
Fig. 11 we present the corresponding circulation growth. If the
incidence is increased to 3 degrees we obtain the pressure
distribution shown in Fig. 3; the convergence history and circulation growth for SLOR have already been presented in Figs. 4
and 5 respectively. These are now reproduced in Figs. 12 and 13
together with the corresponding convergence information for AF3.
In sharp contrast to the behaviour of the relaxation scheme, it
can be seen that the AF3 algorithm quickly locks on to the converged solution. This rapid and assured convergence is a particularly gratifying feature of the approximate factorisation
methods.

7. VISCOUS TRANSONIC FLOW OVER AEROFOILS

In all the aerofoil calculations presented so far we have
assumed the flow to be inviscid. In reality the boundary layer
generated by a lifting flow over an aerofoil can be quite thick
and the actual pressure distribution can differ considerably

from the corresponding inviscid solution. Provided the flow
remains attached it is possible to allow for viscous effects by
combining the inviscid algorithm with a boundary layer calculation. The boundary layer can be thought of as a displacement to
the original aerofoil shape. It follows that the inviscid solution corresponding to the modified profile will depend on the
properties of the boundary layer. Likewise, the calculated
boundary layer will depend on the pressure gradients along the
aerofoil surface. It is therefore necessary to link the inviscid
calculation and boundary layer calculation in an iterative
manner in order to arrive at a converged solution of the viscous
flow. A number of such codes have been developed (Bauer et al,
1975, Melnik et al, 1977, Collyer and Lock, 1979) and the
reader is referred to the literature for further details. In
all the codes the inviscid part of the calculation is based on
SLOR. In this section we mention briefly a recent application
of the AF3 algorithm to the method of Collyer and Lock (1979).

   Morrison (1981) has replaced the SLOR routine in the VGK code
(Collyer and Lock (1979)) by an AF3 algorithm. In every other
respect, the program is unchanged and the new VAF (Viscous
Approximate Factorisation) method produces identical results.
However, the benefit of an approximate factorisation scheme
carries over even though the overall convergence for a viscous
computation is somewhat slower than an inviscid case. Fig. 14
shows the pressure distribution over an aerofoil designed about
ten years ago, just prior to the advent of transonic flow field
methods. In Fig. 15 we present convergence histories for this
case obtained from VGK and VAF. Note that the degree of convergence is now measured by the maximum change in ϕ rather than
the maximum residual. This has been done since the VGK code
does not print out values of the residual. The two convergence
indicators are, however, roughly comparable in numerical value.
Over the first 30 to 50 iterations the boundary layer and inviscid solutions are interacting strongly and so initially the VAF
method shows little sign of converging. Once the boundary layer
has stabilised, the VAF code achieves a fast convergence rate in
sharp contrast to VGK.

   Our previous remarks about the reliable, assured convergence
of approximate factorisation, compared with the sometimes misleading behaviour of relaxation, apply equally well to viscous
flow computations. In other respects approximate factorisation
and in particular, the VAF method have proved superior to relaxation. For example, a VAF run can be started from an arbitrary
initial solution. When computing a difficult case by the VGK
method, however, it is usually necessary to obtain a converged
result at an easier condition before proceeding on the final run.
Furthermore, VAF has sometimes been found to converge even
though convergence difficulties have prevented the corresponding

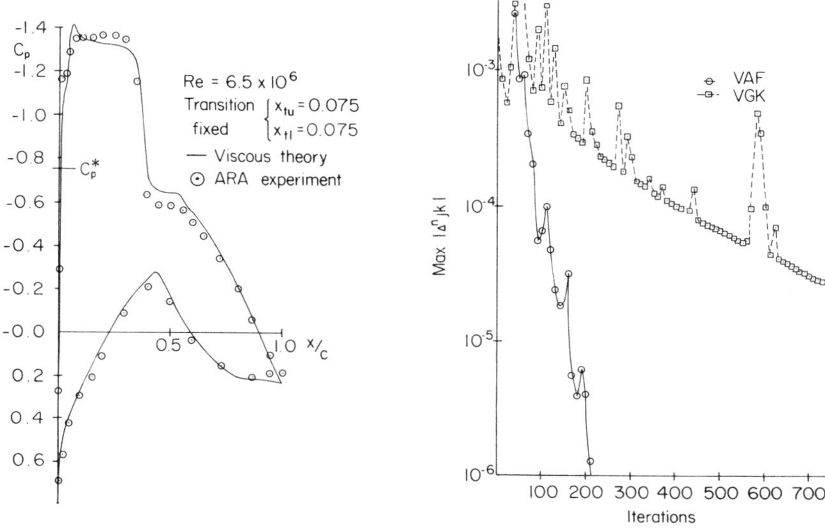

Fig. 14 Pressure distribution at
M = 0.706, C_L = 0.78

Fig. 15 Convergence histories
for VAF and VGK (M =
0.706, C_L = 0.78)

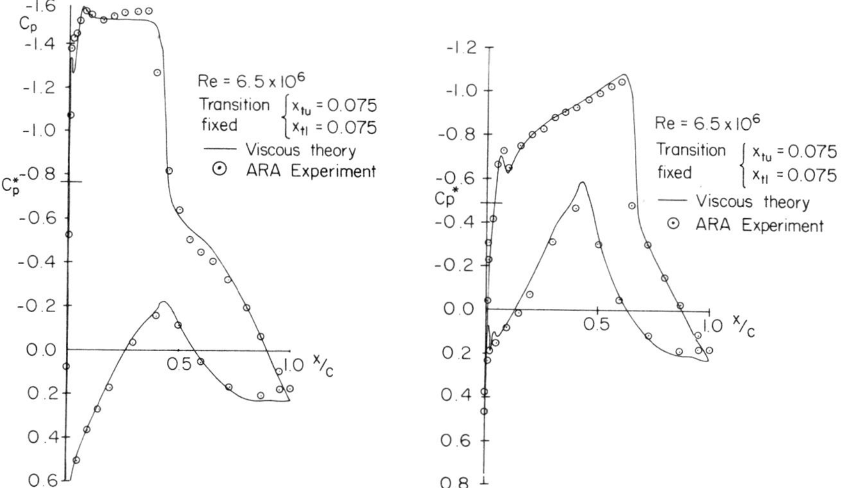

Fig. 16 Pressure distribution at
M = 0.703, C_L = 0.934

Fig. 17 Pressure distribution
at M = 0.782, C_L =
0.524

Figs. 14-17   Results for a Research Aerofoil
(Transition fixed at x/c = 0.075 on both surfaces)

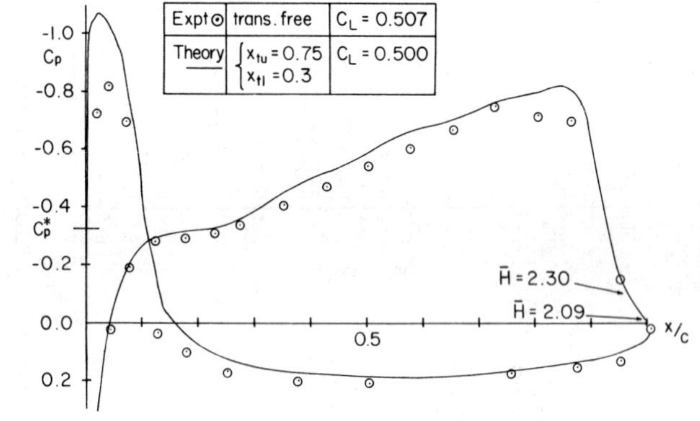

Fig. 18   Pressure distribution over section A
(4% t/c) M = 0.84 Re = 3.0×10⁶

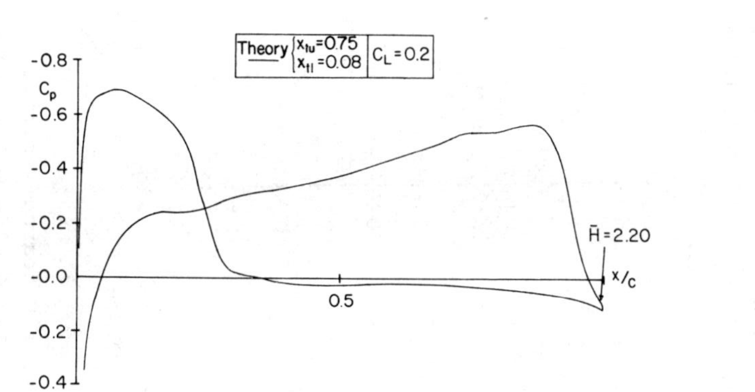

Fig. 19   Pressure distribution over section B
(4% t/c) M = 0.9, Re = 3.0×10⁶

Fig. 18-19   Results for two propeller aerofoils

relaxation solution from being obtained.

In Figs. 16 and 17 we show two more comparisons between theory
(VAF or VGK) and experiment for conditions that are very close
to the aerofoil separation boundary. The agreement between
theory and experiment is good and similar agreement has been
obtained for several aerofoils over a wide range of flow conditions.

Two further pressure distributions are presented in Figs. 18
and 19 for rather different types of aerofoil. These belong to
a family of aerofoil sections designed in about 1975 by Bocci
(1977). They were developed as part of a joint venture between
the Aircraft Research Association Limited and Dowty Rotol Limited
as advanced sections suitable for propellers. Fig. 18 compares
theory and experiment for section A at a freestream Mach number
close to the limit M $\simeq$ 0.85 at which reliable data can be
obtained from the ARA two-dimensional tunnel. The good agreement provided confidence in the adequacy of the viscous theory
and recently calculations using VAF have been carried out to
expand information on the performance of the ARA-D propeller
sections (Bocci, 1981). One example of these computed pressure
distributions is shown in Fig. 19 for a section with lower
camber. It can be seen that large regions of supersonic flow
are present with an upper surface shock wave close to the trailing edge. The experience of this exercise suggests that VAF is
indeed a robust and reliable technique for providing a blanket
investigation of aerofoil performance.

8. REFERENCES

Baker, T.J. (1979) 'A fast implicit algorithm for the nonconservative potential equation', Open Forum Presentation at the
AIAA 4th Computional Fluid Dynamics Conference, Williamsburg,Va.

Baker, T.J. (1981) 'Potential flow calculation by the approximate factorisation method', to appear in *J. Comp. Phys.*

Ballhaus, W.F., Jameson, A. and Albert, J. (1977) 'Implicit
approximate factorisation schemes for the efficient solution of
steady transonic flow problems', AIAA 3rd Computational Fluid
Dynamics Conference, Albuquerque, N. Mex, AIAA Paper 77-634.

Ballhaus, W.F. and Steger, J.L. (1975) 'Implicit approximate
factorisation schemes for the low frequency transonic equation',
NASA TM X-73082.

Bauer, F., Garabedian, P., Korn, D. and Jameson, A. (1975)
'Supercritical wing sections II', *Lecture notes in Economics
and Mathematical Systems*, **108**, pub. Springer.

Bocci, A.J. (1977) 'A new series of aerofoil sections suitable for aircraft propellers', *Aero Qu.*, **28**, pp. 59-73.

Bocci, A.J. (1981) 'Transonic calculations on ARA-D propeller aerofoils', ARA Model Test Note E.21/1.

Collyer, M.R. and Lock, R.C. (1979) 'Prediction of viscous effects in steady transonic flow past an aerofoil', *Aero Qu.*, **30** pp. 485-505.

Garabedian, P.R. (1956) 'Estimation of the relaxation factor for small mesh size', *Math. Tables Aids Comp.*, **10**, pp. 183-185.

Garabedian, P.R. and Korn, D.G. (1971) 'Analysis of transonic airfoils', *Comm. Pure Appl. Math.*, **24**, pp. 841-851.

Hafez, M. and Cheng, H.K. (1975) 'Convergence acceleration and shock fitting for transonic flow computations', AIAA Paper 75-51.

Holst, T.L. (1978) 'An implicit algorithm for the conservative, transonic full potential equation using an arbitrary mesh', AIAA Paper 78-1113.

Holst, T.L. (1979) 'A fast conservative algorithm for solving the transonic full potential equation', AIAA 4th Computational Fluid Dynamics Conference, Williamsburg, Va, AIAA Paper 79-1456.

Jameson, A. (1974) 'Iterative solution of transonic flows over airfoils and wings, including flows at Mach 1', *Comm. Pure Appl. Math.*, **27**, pp. 283-309.

Jameson, A. (1976) 'Transonic flow calculations', VKI Lecture Series 87 on Computational Fluid Dynamics, von Karman Institute for Fluid Dynamics, Rhode-St-Genese, Belgium.

Martin, E.D. and Lomax, H. (1974) 'Rapid finite difference computation of subsonic and transonic aerodynamic flows', AIAA Paper 74-11.

Melnik, R.E., Chow, R. and Mead, H.R. (1977) 'Theory of viscous transonic flow over aerofoils at high Reynolds number', AIAA Paper 77-680.

Mitchell, A.R. (1969) 'Computational methods in partial differential equations', pub. Wiley.

Morrison, J. (1981) 'A fast algorithm for calculating transonic viscous flow over an aerofoil', ARA Memo to appear.

Murman, E.M. and Cole, J.D. (1971) 'Calculation of plane steady transonic flows', *AIAA Journal,* **9**, pp. 114-121.

Peaceman, D.W. and Rachford, H.H. (1955) 'The numerical solution of parabolic and elliptic differential equations', *J. SIAM*, **3**, pp. 28-41.

Thompson, J.F., Thomas, F.C. and Martin, C.W. (1974) 'Automatic numerical generation of body fitted curvilinear coordinate system for field containing any number of arbitrary two-dimensional bodies', *J. Comp. Phys.*, **15**, pp. 299-319.

Yanenko, N.N. (1971) 'The method of fractional steps', pub. Springer.

A SURVEY OF FINITE ELEMENT METHODS FOR TRANSONIC FLOWS

Ch. Hirsch and H. Deconinck

(*Vrije Universiteit Brussel, Department of Fluid Dynamics*)

ABSTRACT

A review is presented of the work performed at the VUB over
the last years in the field of transonic flow calculation using a
Finite Element formulation for the discretized flow equations.
The goal of this research is essentially directed towards cascade
flow computations but many results will be shown for isolated
airfoils.

After an initial attempt with an Optimal Control approach for
the potential equation, much effort has been devoted to an approach incorporating artificial compressibility and iterative
methods. Highly efficient codes have been developed including
higher order elements, SLOR or ADI iteration schemes and body
fitted meshes. Convergence accelerating techniques like grid
refining and finally a multi-grid method have been implemented
for an arbitrary grid.

With regard to the Euler equations, a new method has been
developed, based on "orthogonal shape functions" in order to
allow non-implicit treatments of the time-dependent terms. A
family of possible schemes is obtained which contains as a subclass certain finite difference and finite volume schemes.

INTRODUCTION

Finite Element Methods have been faced initially with several
difficulties when applied to transonic flow problems, mainly due
to the intrinsic "elliptic" character of the finite element
approach which does not introduce naturally the non-symmetric
properties typical of supersonic flows.

With the potential flow approximations, various ways have
been initially presented to circumvent these difficulties with
various degrees of success (Chan and Bradshears, 1975; Glowinski
et al. 1976; Hafez and Murman, 1978). Of these earlier attempts,
the method of optimal control of Glowinski et al. has been
developed with the most success with regard to applications in
external aerodynamics (Periaux 1979) and cascade flows (Deconinck
and Hirsch, 1979a).

However, formulations based on the artificial compressibility concept and iterative solution methods have brought the Finite Element Method for transonic flows to the same level of efficiency as the Finite Difference Methods. The first attempt in this direction came from Eberle (1977) introducing an upwind correction for the density in a Finite Element formulation together with line relaxation (SLOR). A more systematic development was presented by Deconinck and Hirsch (1979b) where next to SLOR, several ADI schemes were developed for Finite Element discretizations on arbitrary meshes. In subsequent work, higher order elements were investigated (Deconinck and Hirsch, 1980) such as the second order Lagrangian quadrilateral element and more recently a general multigrid procedure has been developed for arbitrary body-fitted meshes and applied to transonic flows (Deconinck and Hirsch, 1981).

In the field of Euler equations, very little work has been done up to now and the approach proposed by Hirsch and Warzee (1979) will be summarized. Clearly much more work has to be done in this direction in order to assess the possibilities of the FE methodology within this frame work.

1. POTENTIAL FLOW EQUATION

We consider the potential flow equation

$$\vec{\nabla}[\rho \ \bar{\nabla}\phi] = 0 \tag{1.1}$$

which will be limited to its two-dimensional form in the following.

The specific mass is connected to the velocities for a perfect gas (with specific heat ratio γ and gas constant r) by

$$\rho = \rho_t[1 - k(\bar{\nabla}\phi)^2]^\alpha \tag{1.2}$$

with

$$k = \frac{\gamma - 1}{2 \ \gamma r \ T_t} \quad \text{and} \quad \alpha = \frac{1}{\gamma - 1}$$

where ρ_t and T_t are stagnation values of density and temperature respectively.

1.1 OPTIMAL CONTROL METHOD

The optimal control method developed by Glowinski et al. is actually a form of a least squares approach, since the solution of the potential flow problem is obtained through the minimization of the functional

$$E(\xi) = \| \phi(\xi) - \xi \|^2 = \int \| \nabla [\phi(\xi) - \xi] \|^2 \, ds \qquad (1.1.1)$$

where the relation $\phi(\xi)$ is chosen to be obtained through the solution of a Poisson equation

$$\Delta\phi(\xi) = \Delta\xi + \bar{\nabla}[\rho(\xi)\bar{\nabla}\xi] \qquad (1.1.2)$$

the same boundary conditions being valid for ϕ and ξ.

For the solution of this minimization problem a conjugate gradient method is applied namely the Polak-Ribière algorithm (Polak, 1971), which is an extension of the following fundamental gradient algorithm:

. step (1) : n = O

   calculate an initial solution for the control function assuming incompressible flow: $\Delta\xi^{(o)} = 0$

. step (2) : n = n + 1

. step (3) : calculate $\phi^{(n)}$ $(\xi^{(n-1)})$ using equation (1.1.2)

$$\Delta\phi^{(n)} = \bar{\nabla}[\rho(\xi^{(n-1)})\bar{\nabla}\xi^{(n-1)} + \bar{\nabla}\xi^{(n-1)}]$$

. step (4) : define $E^{(n)} = \| \phi^{(n)} - \xi^{(n-1)} \|^2$

. step (5) : calculate $\chi^{(n)}$ i.e. the derivate of $E^{(n)}$ (ξ) with respect to ξ (see below)

. step (6) : define $\xi_\lambda = \xi^{(n-1)} - \lambda\chi^{(n)}$

   λ being a positive real number; calculate $\lambda*$ for the value of λ for which

$$E(\lambda) = \| \phi(\xi_\lambda) - \xi_\lambda \|^2$$

   is minimal

. step (7) : $\xi^{(n)} = \xi^{(n-1)} - \lambda*\chi^{(n)}$

. step (8) : go to (2).

The Polak-Ribière conjugate gradient algorithm is obtained through a modification of step (7) in the following way:

step (7) $\qquad \xi^{(n)} = \xi^{(n-1)} - \lambda* \, \psi^{(n)}$

where

$$\psi^{(n)} = \chi^{(n)} + \psi^{(n-1)} \frac{\| \chi^{(n)} \|^2 - \langle \chi^{(n)}, \chi^{(n-1)} \rangle}{\| \chi^{(n-1)} \|^2}$$

Step (3) involves the weak solution of a Poisson equation i.e.:

$$\int_S \bar{\nabla}\phi\ \bar{\nabla}W\ dS = \int_S \left[1 + \frac{\rho}{\rho_t}(\xi)\right] \bar{\nabla}\xi\ \bar{\nabla}W\ dS \qquad (1.1.3)$$

where W is an arbitrary weight function.

The derivate of the cost function, step (5), is computed in the following way:

From the definition one has, for any $\delta\xi$

$$E(\xi + \delta\xi) - E(\xi) = \int_S \frac{\partial E(\xi)}{\partial\xi}\ \delta\xi\ dS + O(\delta\xi)$$

and this equation defines a unique element χ such that:

$$<\chi,\delta\xi> = \int_S \bar{\nabla}\chi\ \bar{\nabla}\delta\xi\ dS = \int_S \frac{\partial E(\xi)}{\partial\xi}\ \delta\xi\ dS \qquad (1.1.4)$$

So χ is the function representing the derivate of $E(\xi)$.

Making equation (1.1.4) explicit leads to an equation for χ which is a weak formulation of a Poisson equation:

$$\int_S \bar{\nabla}\chi\bar{\nabla}WdS = -2 \int_S \left\{ \frac{\rho(\xi)}{\rho_t}\left[\frac{2k\alpha[\bar{\nabla}(\phi-\xi)\bar{\nabla}\xi][\bar{\nabla}\xi\bar{\nabla}W]}{1-k|\bar{\nabla}\xi|^2}\right] - \frac{\rho(\xi)}{\rho_t}\bar{\nabla}(\phi-\xi)\bar{\nabla}W \right\}dS$$

$$(1.1.5)$$

for any weight function W.

1.1.1. Finite Element Discretization

The continuous problem is replaced by a discrete problem using a finite element approximation with biquadratic shape functions N_i and quadrilateral isoparametric elements of the same type for the representation of the geometry.

Within an element with node values ϕ_i, χ_i, ξ_i we have:

$$\phi = \Sigma\ \phi_i\ N_i(x,y)$$

$$\xi = \Sigma\ \xi_i\ N_i(x,y)$$

$$\chi = \Sigma\ \chi_i\ N_i(x,y)$$

A Galerkin method is used, replacing the weight functions in equation (1.1.3) and (1.1.5) by the shape functions N_i giving the following discrete equations for each element (with surface S_e and Neumann boundary s_e if the element lies along a Neumann boundary).

For ϕ:

$$\phi_j \int_{S_e} \bar{\nabla} N_j \bar{\nabla} N_i \, dS = \int_{S_e} \left[1 + \frac{\rho(\xi)}{\rho_t}\right] \bar{\nabla}\xi \bar{\nabla} N_i \, dS - \int_{s_e} \frac{\rho}{\rho_t} \frac{\partial \xi}{\partial n} N_i \, ds$$

$$(1.1.6)$$

For χ:

$$\chi_j \int_{S_e} \bar{\nabla} N_j \bar{\nabla} N_i \, dS = -2 \int_{S_e} \frac{\rho(\xi)}{\rho_t} \left[\frac{2k\alpha \bar{\nabla}(\phi-\xi)\bar{\nabla}\xi}{1 - k|\bar{\nabla}\xi|^2} \bar{\nabla}\xi - \bar{\nabla}(\phi-\xi) \right] \bar{\nabla} N_i \, dS$$

$$(1.1.7)$$

1.1.2. Entropy Condition

The potential equation (1.1) is fully isentropic and therefore the solution may contain non-physical expansion shocks which do not appear in real flows due to the viscous effects i.e. due to an entropy increase. There are two ways to eliminate these expansion shocks, first by adding an artificial viscosity term to the potential equation, secondly by including an inequality constraint in the minimization problem (Bristeau, 1978).

In the second method a non-physical shock will be excluded if:

$$\Delta\phi \ < \ +\infty$$

which is practically replaced by:

$$\Delta\phi \ \leq \ M$$

or written in weak form for any weight function $W \geq 0$

$$\int_S -(\bar{\nabla}W \ \bar{\nabla}\xi + MW)\,dS \ < \ 0 \qquad (1.1.8)$$

This inequality is introduced in the problem using a penalty functional

$$P(\xi) = \underset{\forall W}{\Sigma} \ \text{Max}\left[\left\{\int_S -(\bar{\nabla}W \ \bar{\nabla}\xi + MW)\,dS, \ 0\right\}^2\right] \qquad (1.1.9)$$

which is to be added to the cost function $E(\xi)$.

The equations (1.1.6) and (1.1.7) are solved by a direct
method and an example is shown for a nozzle flow.

A divergent nozzle is calculated with a supersonic inlet Mach
number of 1.21 showing a compression shock spread over two mesh
spaces, see Figs 1 and 2. In this calculation no artificial

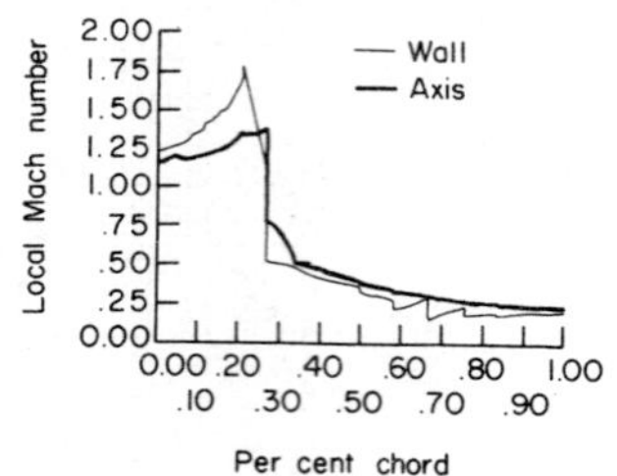

Fig. 1 Transonic nozzle; optimal control method; distribution
        of Mach number

viscosity nor penalization is used and therefore in cases with
strong non-symmetric boundary conditions and with decreasing
velocity (e.g. compressor cascades), no expansion shocks seem
to appear.

The plots of Fig. 1 show also the velocity discontinuities
at the inter-element boundaries.

The introduction of penalisation or artificial viscosity slows
down the convergence rate quite appreciably and it is our experience that the methods described in the following based on the
artificial compressibility are more efficient.

1.2. ARTIFICIAL COMPRESSIBILITY METHODS

In order to introduce in the potential equations the irrever-
sible character of transonic flow with shocks, an artificial
viscosity term is added in the supersonic flow regions. This
is usually done in an implicit way with F.D. methods using upwind differencing in the supersonic regions. In classical finite
element methods this technique is not applicable since the
discretization is always of the elliptic central differencing
type.

However, the artificial viscosity can be introduced explicitly.
Jameson (1975) adds a viscosity term to the equation which can
be shown to be equivalent to adding:

$$-\left[(\nu\, \phi_x\, \rho_{\bar{x}}\, \Delta x)_x + (\nu\, \phi_y\, \rho_{\bar{y}}\, \Delta y)_y \right] \qquad (1.2.1)$$

where:

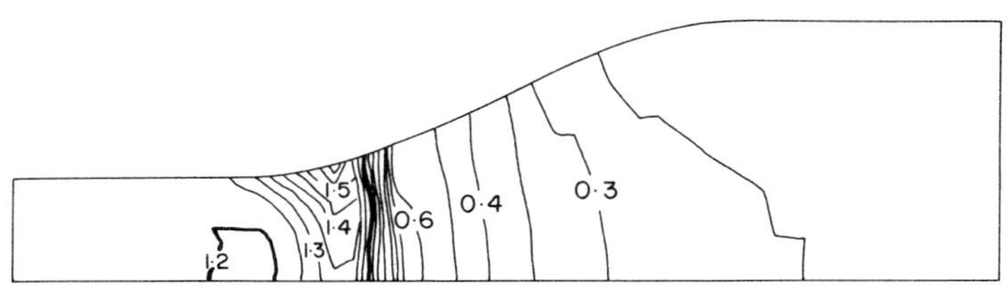

Fig. 2   Transonic nozzle; optimal control method. Isomach lines
at intervals 0.05

$$\nu = \max(0, 1 - \frac{1}{M^2})$$
(1.2.2)

is the switching function setting the artificial viscosity to
zero in the subsonic flow regions, and $\rho^{\leftarrow}_x$ and $\rho^{\leftarrow}_y$ are upwind
differenced.

Expression (1.2.1) is particularly interesting since it offers
the possibility to treat the equation as an elliptic one.

The potential equation corrected with the artificial viscosity
formulation is given by:

$$(\tilde{\rho}\ \phi_x)_x + (\bar{\rho}\ \phi_y)_y = 0$$
(1.2.3)

$$\tilde{\rho} = \rho - \nu\ \rho^{\leftarrow}_x \Delta x$$
(1.2.4)

$$\bar{\rho} = \rho - \nu\ \rho^{\leftarrow}_y \Delta y$$
(1.2.5)

The non-linear coefficients $\bar{\rho}$ and $\tilde{\rho}$ are updated at each iteration, while the equation is solved and discretized in an elliptic
way. So equation (1.2.3) could be discretized with standard
Galerkin finite element techniques.

A further simplification can be obtained by replacing equations (1.2.4) and (1.2.5) by

$$\bar{\rho} = \tilde{\rho} = \rho - \nu\ \rho^{\leftarrow}_s \Delta s$$
(1.2.6)

and with $\rho^{\leftarrow}_s$ the upwind differencing of ρ along a streamline
(Hafez et al., 1978) and the potential equation has now again
the usual form

$$(\tilde{\rho}\ \phi_x)_x + (\tilde{\rho}\ \phi_y)_y = 0$$
(1.2.7)

where the viscosity terms are included in the artificial compressibility $\tilde{\rho}$.

1.2.1. Finite Element Successive Line Overrelaxation Method (SLOR)

The potential equation (1.2.7) with modified density is equivalent to the following weak form:

$$- \int_S \tilde{\rho}(\phi_x W_x + \phi_y W_y)\, dS + \int_s \tilde{\rho}\, \frac{\partial\phi}{\partial n}\, Wds = 0 \qquad (1.2.8)$$

for any weight function W and where S is the flow domain and s
the part of the boundary where the Neumann boundary condition
$\rho(\partial\phi/\partial n)$ is specified.

Applying Galerkin method using the shape functions as weight
functions, the discrete equation for iteration (n) is obtained:

$$\sum_J \phi_J^{(n)} \int_S \tilde{\rho}^{(n-1)} \nabla N_J \nabla N_I\, dS - \int_s (\tilde{\rho}\, \frac{\partial\phi}{\partial n})^\circ N_I\, ds = 0 \qquad (1.2.9)$$

Here the non-linear equation is linearized as usual using the
values of $\tilde{\rho}$ of the previous iteration.

It is important to note that this discretization technique is
fully elliptic, the distinction between the supersonic and
subsonic points being included in the artificial compressibility
expression.

In finite difference notation with row and column indices
(Fig. 3) one can write equation (1.2.9) as

$$\sum_{k,l} \phi_{kl}^{(n)} K_{ij}^{kl} = f_{ij} \qquad (1.2.10)$$

or:

$$\sum_{k,l} \delta\phi_{kl} K_{ij}^{kl} = - R_{ij}^{(n-1)} \qquad (1.2.11)$$

with:

$$\delta\phi = \phi^{(n)} - \phi^{(n-1)}$$

and the residual $R^{(n-1)}$ defined by

$$-R^{(n-1)} = f^{(n-1)} - K^{(n-1)} \phi^{(n-1)} \qquad (1.2.12)$$

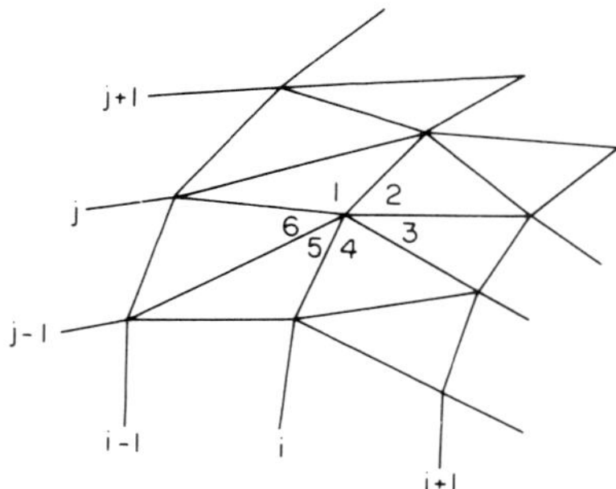

Fig. 3   Linear triangular elements

Due to the FE approach the boundary conditions are included explicitly in the residual since f contains the contribution from
the boundaries.

In iterative methods equation (1.2.11) is replaced by an
approximate form

$$C^n \, \delta\phi = -\tau\omega R^n \tag{1.2.13}$$

where C^n is an approximate operator and ω a relaxation constant.

Successive line relaxation (SLOR) is the standard device for
transonic FD calculations due to the implicit presence of upwind
differences in stream-wise direction in the iteration operator
C^n which acts in a stabilizing way in supersonic points if the
proper sweep direction is applied.   For bilinear quadrilaterals
or linear triangles in the configuration of Fig. 3, the SLOR
leads to the following scheme:

$$\sum_{\ell=j-1}^{j+1} \delta\phi_{i\ell} \, K_{ij}^{i\ell} = -\omega \left[R_{ij}^n + \sum_{\ell=j-1}^{j+1} \delta\phi_{i-1,\ell} \, K_{ij}^{i-1,\ell} \right] \tag{1.2.14}$$

where the coefficients and the residual for the triangles are
identical to those obtained with a classical 5-point difference
star except for boundary contributions and as far as the grid
is uniform.   For instance, Laplace equation with Dirichlet
boundary conditions on a uniform mesh yields for linear triangles the same scheme as a 5-point difference star;

$$(-1\ 4\ -1)\begin{pmatrix}\delta\phi_{i,j+1}\\ \delta\phi_{i,j}\\ \delta\phi_{i,j-1}\end{pmatrix} = -\omega\left[R_{ij} + (0\ -1\ 0)\begin{pmatrix}\delta\phi_{i-1,j+1}\\ \delta\phi_{i-1,j}\\ \delta\phi_{i-1,j-1}\end{pmatrix}\right]$$

where

$$R_{ij} = \begin{array}{ccc} & -1 & \\ -1 & 4 & -1 \\ & -1 & \end{array}$$

The bilinear elements lead to a similar scheme containing however contributions from points $(i\pm1, j\pm1)$ in the residual and in the r.h.s. of the equation

$$(-1\ 8\ -1)\begin{pmatrix}\delta\phi_{i,j+1}\\ \delta\phi_{i,j}\\ \delta\phi_{i,j-1}\end{pmatrix} = -\omega\left[R_{ij} + (-1\ -1\ -1)\begin{pmatrix}\delta\phi_{i-1,j+1}\\ \delta\phi_{i-1,j}\\ \delta\phi_{i-1,j-1}\end{pmatrix}\right]$$

and

$$R_{ij} = \begin{array}{ccc} & -1 & \\ -1 & 4 & -1 \\ & -1 & \end{array} + \begin{array}{ccc} -1 & & -1 \\ & 4 & \\ -1 & & -1 \end{array}$$

In all these cases tridiagonal systems along coordinate line i have to be solved.

For 9 node biquadratic Lagrange elements, Fig. 4, shape functions extend over three nodes in one dimension giving the following pentadiagonal system for each i-line:

$$\sum_{\ell=j-2}^{j+2}\delta\phi_{i\ell}\,K_{ij}^{i\ell} = -\omega\left[R_{ij} + \sum_{\ell=j-2}^{j+2}K_{ij}^{i-1,\ell}\,\delta\phi_{i-1,\ell}\right.$$

$$\left. + \sum_{\ell=j-2}^{j+2}K_{ij}^{i-2,\ell}\,\delta\phi_{i-2,\ell}\right] \qquad (1.2.15)$$

The residual involves now contributions from 9 surrounding meshpoints for the central node up to 25 for a corner node.

For coordinate lines consisting of midside and central mesh-

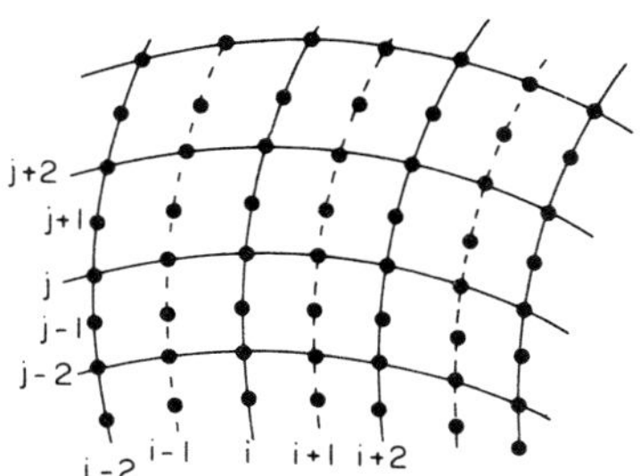

Fig. 4   Biquadratic quadrilateral Lagrange Elements

points (dotted lines) the terms in $K_{ij}^{i-2,\ell}$ are zero.   Equations
written for mid-side or centre nodes are tridiagonal namely
$K_{ij}^{ij\pm2}$ = O, resulting in considerable reduction of the cost for
inversion of the pentadiagonal system.   Application of the 8
node biquadratic serendipity element obtained by dropping the
centre node leads to alternating pentadiagonal and tridiagonal
systems for the coordinate lines consisting of midside nodes.

The use of quadratic elements does not increase the calculation
time in a substantial way as was known from subsonic calculation
experience, the number of elements being four times lower than in
the bilinear case for the same number of meshpoints.

The previous schemes can be modified in order to control the
amount of implicitness by introducing Hopscotch or red and black
ordering of the mesh-points, e.g. with the bilinear elements a
vectorizable scheme very similar to Hafaz' Zebra-SLOR (Hafez and
South, 1979) is obtained by colouring the mesh-points red and
black respectively on odd and even columns and sweeping first
over the red and afterwards over the black points.   In a similar
way, the pentadiagonal systems can be reduced to tridiagonal.

1.2.2. Galerkin FE-ADI Methods

The introduction of factorized forms for the full potential
equation in transonic flows has met with success when formulated
with FD on an arbitrary mesh and with the artificial compressibility form of the basic equation (Holst, 1978).   Here the capability of a FE formulation along these lines is investigated.

In the mathematical literature, some theoretical formulations
can be found for ADI Galerkin methods for elliptic, parabolic or
hyperbolic problems on rectangular meshes (Douglas and Dupont,
1975; Dendy and Fairweather, 1975).

In these papers, convergence proof and error bounds are given even for some classes of non-linear problems, but no application to practical problems is known to us. Neither has the formulation of an ADI Galerkin method on an arbitrary mesh been derived and proven theoretically. These aspects will be developed and presented in the following sections.

1.2.2.1. Galerkin-ADI method on a rectangular mesh Considering the parabolic equation in time:

$$\phi_t - \vec{\nabla}(\rho \, \vec{\nabla} \, \phi) = f \qquad\qquad (1.2.16)$$

Douglas and Dupont define a Galerkin approximation of the form

$$\phi_t = \frac{\delta\phi}{\tau} = \frac{\phi^{(n+1)} - \phi^{(n)}}{\tau} \qquad\qquad (1.2.17)$$

$$\left(\frac{\delta\phi}{\tau}, \, W\right) + \lambda(\vec{\nabla}\delta\phi, \, \vec{\nabla}W) + \lambda^2\tau\left(\frac{\partial^2 \delta\phi}{\partial x \partial y}, \, \frac{\partial^2 W}{\partial x \partial y}\right) = (f^n, \, W) - (\rho^n \, \vec{\nabla} \, \phi^n, \, \vec{\nabla}W)$$

$$(1.2.18)$$

for any weight function $W \in H^1_O$.

If the form function $N_I(x,y)$ factorizes in the form

$$N_I(x,y) = N_i(x) \, N_j(y) \qquad\qquad (1.2.19)$$

where I is the node number and i,j the corresponding column and line numbers, and if a Galerkin approximation is applied, namely W = N, then the following matrix form is obtained

$$[[M] + \lambda \, \tau[K] + \lambda^2 \, \tau^2 \, K_x \otimes K_y] \, \{\delta\phi\} = -\tau R \qquad (1.2.20)$$

where R is the residual

$$R = -(f^n, N) + (\rho^n \, \vec{\nabla} \, \phi^n, \, \vec{\nabla}N) \qquad\qquad (1.2.21)$$

$$M = M_x \otimes M_y = \int N_i(x) \, N_k(x) \, dx \otimes \int N_j(y) \, N_\ell(y) \, dy \quad (1.2.22)$$

$$K = \int \vec{\nabla}N_I \, \vec{\nabla}N_J \, dS \qquad\qquad (1.2.23)$$

or:

$$K = K_x \otimes M_y + K_y \otimes M_x \qquad\qquad (1.2.24)$$

with:

$$K_x = \int \partial_x N_i(x) \, \partial_x N_K(x) \, dx \qquad (1.2.25)$$

$$K_y = \int \partial_y N_j(y) \, \partial_y N_\ell(y) \, dx \qquad (1.2.26)$$

The expression (1.2.20) can be factorized as follows:

$$(M_x + \lambda\tau K_x) \otimes (M_y + \lambda\tau K_y)\delta\phi = -\tau R \qquad (1.2.27)$$

and introducing the intermediate vector g, the following system has to be solved.

$$(M_x + \lambda\tau K_x)g = -\tau R \qquad (1.2.28)$$

$$(M_y + \lambda\tau K_y)\delta\phi = g \qquad (1.2.29)$$

This form of ADI-Galerkin method corresponds to the approximate operator C in the equation

$$C \, \delta\phi = -\tau R \qquad (1.2.30)$$

given by:

$$C = 1 - \lambda L + \lambda^2 \, \partial_{xx} \, \partial_{yy} \qquad (1.2.31)$$

where L is the Laplacian.

The same relations could be obtained in the inverse way, namely by applying first a splitting or factorizing of the equations, following Marchuck (1975) and then a FE-Galerkin formulation. Writing the operator A as a sum of two terms:

$$A = A_x + A_y \qquad (1.2.32)$$

or more specifically

$$-A = \vec{\nabla} \rho \, \vec{\nabla} = \partial_x \rho \, \partial_x + \partial_y \rho \, \partial_y \qquad (1.2.33)$$

a factorized form of the equation is:

$$(1 + \sigma A_x)(1 + \sigma A_y)\delta\phi = -\tau\sigma R \qquad (1.2.34)$$

which corresponds to the approximate operator

$$C = 1 - \sigma A + \sigma^2 A_x A_y \qquad (1.2.35)$$

This could be a better approximation than (1.2.31), where A is replaced by the Laplacian operator.

Equation (1.2.34) is then split as:

$$(1 + \sigma A_x)\, g = -\, \tau \sigma R \qquad (1.2.36)$$

$$(1 + \sigma A_y)\, \delta\phi = g \qquad (1.2.37)$$

and applying a FE-Galerkin method, one obtains

$$(M_x + \sigma K_x)\, g = -\, \sigma\tau R \qquad (1.2.38)$$

$$(M_y + \sigma K_y)\, \delta\phi = g \qquad (1.2.39)$$

where M_x and M_y are defined by (1.2.22) while

$$(K_x)_{ik} = \int \rho^{(n)}\, \partial_x N_i \cdot \partial_x N_k\, dx \qquad (1.2.40)$$

$$(K_y)_{j\ell} = \int \rho^{(n)}\, \partial_y N_j \cdot \partial_y N_\ell\, dy \qquad (2.2.41)$$

1.2.2.2. Application: triangular, bilinear and biquadratic elements

Considering the more general equation

$$(\partial_x\, p\, \partial_x + \partial_y\, q\, \partial_y)\, \phi = f \qquad (1.2.42)$$

and the factorized form

$$(1 + \sigma\partial_x\, p\, \partial_x)\, g = -\, \sigma\tau R \qquad (1.2.43)$$

$$(1 + \sigma\partial_y\, q\, \partial_y)\, \delta\phi = g \qquad (1.2.44)$$

the partial stiffness matrices K_x, K_y become

$$K_x = -\int p\, \partial_x N_i\, \partial_x N_k\, dx \qquad (1.2.45)$$

$$K_y = -\int q\, \partial_y N_j\, \partial_y N_\ell\, dy \qquad (1.2.46)$$

Triangular elements

On an arbitrary mesh, only the 6 nodes surrounding (i,j) will contribute to the equation in this point (Fig. 3).

However, it turns out in the case of an orthogonal mesh that only the nodes (i,j-1), (i,j+1), (i-1,j) and (i+1,j) will contribute and give for the operator the same expression as the one obtained from a central, conservative finite difference discretization:

$$(\delta_x\, p^n_{i+1/2}\, \delta_x + \delta_y\, q^n_{j+1/2}\, \delta_y)^n \qquad (1.2.47)$$

where:

$$p_{i+1/2} = \frac{1}{\Delta x^2} \int_{4+5} p \; dS, \qquad q_{j+1/2} = \frac{1}{\Delta x^2} \int_{3+4} q \, dS \qquad (1.2.48)$$

The mass matrix, as well as the K_x, K_y matrices are tridiagonal and one obtains for the first equation (x equ.) along line j:

$$\left[\frac{1}{6}(1\;4\;1) - \frac{\sigma}{\Delta x}[p_{i-1/2} - (p_{i-1/2} + p_{i+1/2})p_{i+1/2}]]\right\} \left\{\begin{array}{c} g_{i-1} \\ g_i \\ g_{i+1} \end{array}\right\} = -\sigma\tau R_i$$

$$(1.2.49)$$

In an FD scheme, the first matrix is diagonal (0 1 0) but the other terms are identical except that here, the r-h-s of the basic equation ,f, contributes to the residual through the integral:

$$\int f \; N_i(x) \; N_j(y) \; dx \; dy \qquad (1.2.50)$$

over the hexagon of Fig. 3, and that the boundary conditions are automatically included in the residual since:

$$R_j = -\sum_I \phi_I \int_S (p\partial_x N_I \cdot \partial_x N_J + q\partial_y N_I \cdot \partial_y N_J)\,dS + \int_{S_n} (p\frac{\partial\phi}{\partial n_x} + q\frac{\partial\phi}{\partial n_y})N_J \, ds$$

$$- \; f \; N_J \; dS \qquad (1.2.51)$$

over a hexagon around node J.

For the Laplace equation, one would obtain the equation $(\Delta x = \Delta y = 1)$:

$$\frac{1}{6}(1\;4\;1) - \sigma(1\;-2\;1) = -\sigma\tau R = -\sigma\tau \begin{bmatrix} & 1 & \\ 1 & -4 & 1 \\ \hline & 1 & \end{bmatrix}$$

which is identical to the FD form except for the mass matrix.

<u>Bilinear elements</u>

In this case, 9 nodes around (i,j) contribute to the residual and compared with the triangular elements, the mass matrix remains unchanged as well as the K_x and K_y matrices while the residuals will give rise to an integration over the 4 elements surrounding the node (i,j). Hence the ℓ-h-s of equation

(1.2.49) will remain unchanged while the residual will have
another numerical approximation, involving a larger amount of
averaging. This is best illustrated for the Laplace equation
where one obtains the scheme:

$$\frac{1}{6}\ (1\ \ 4\ \ 1)\ -\ \sigma\ (1\ -2\ \ 1)\ =\ -\ \frac{\sigma\tau}{6}\
\begin{array}{ccc} 2 & 2 & 2 \\ 2 & -16 & 2 \\ 2 & 2 & 2 \end{array}$$

$$=\ -\ \frac{\sigma\tau}{3}\ \left[\ \begin{array}{ccc} & 1 & \\ 1 & -4 & 1 \\ & 1 & \end{array}\ +\ \begin{array}{ccc} 1 & & 1 \\ & -4 & \\ 1 & & 1 \end{array}\ \right]$$

Biquadratic elements

Biquadratic Lagrange elements (the Serendipity shape functions
do not factorize) lead to pentadiagonal systems composed of
alternating pentadiagonal equations along odd numbered coordinate
lines and tridiagonal equations along even numbered ones:

$$\begin{array}{l} i\ =\ \text{odd:}\ \ (-4\ \ 8\ \ 32\ \ 8\ -4) \\[4pt] i\ =\ \text{even:}\ \ (0\ \ 8\ \ 64\ \ 8\ \ \ 0) \end{array}\ \ \begin{pmatrix} g_{ij-2} \\ g_{ij-1} \\ g_{ij} \\ g_{ij+1} \\ g_{ij+2} \end{pmatrix}$$

$$+\ \sigma\ \begin{array}{l} (10\ -80\ \ 140\ -80\ \ 10) \\[4pt] (\ 0\ -80\ \ 160\ -80\ \ 0\ ) \end{array}\ \ \begin{pmatrix} g_{ij-2} \\ g_{ij-1} \\ g_{ij} \\ g_{ij+1} \\ g_{ij+2} \end{pmatrix}\ =\ -\ 60\ \sigma\tau R_{ij}$$

where the residual contains again contributions of 9 to 25 surrounding nodes as discussed before.

Formally identical systems are solved for the second equation.
The result is a fully implicit third order accurate scheme containing the exact boundary treatment in the residual.

It can generally be seen that the FE equations lead to a

larger amount of implicitness in the discretized equations since
more surrounding mesh points will contribute to the value of the
solution in a given point.

Hence, better convergence might be expected compared to the
FD form of the equation. This is indeed the case, as illustrated
on a test problem (Deconinck and Hirsch, 1979c).

1.2.2.3. <u>Galerkin-ADI method on an arbitrary mesh</u> In order to
apply the Galerkin-ADI method on arbitrary mesh one can either
transform the equation from the physical plane to a computational
plane which is rectangular, or work directly on the Galerkin
form of the equation. Both methods lead to the same results,
as will be seen in the following.

As is well known, the basic equation (1.2.7), written in
tensor notation:

$$\nabla_i \, \rho \, \nabla^i \, \phi = 0 \qquad\qquad (1.2.52)$$

transforms, in a general coordinate transformation:

$$x^i = x^i(\xi^1, \, \xi^2) \quad \text{or} \quad \xi^i = \xi^i(x^1, \, x^2) \qquad (1.2.53)$$

with metric tensor $g^{i,j}$ defined as:

$$g^{ij} = (J^T J)^{ij} = \begin{pmatrix} \xi_x^2 + \xi_y^2 & \xi_x \eta_x + \xi_y \eta_y \\ \xi_x \eta_x + \xi_y \eta_y & \eta_x^2 + \eta_y^2 \end{pmatrix} \equiv \begin{pmatrix} A_1 & A_2 \\ A_2 & A_3 \end{pmatrix}$$

$$(1.2.54)$$

where:

$$J^i_j = \frac{\partial \xi^i}{\partial x^j} \qquad\qquad (1.2.55)$$

is the Jacobian matrix of the transformation to

$$\partial_\xi \left[\frac{\rho}{|J|} (A_1 \partial_\xi + A_2 \partial_\eta) \phi \right] + \partial_\eta \left[\frac{\rho}{|J|} (A_2 \partial_\xi + A_3 \partial_\eta) \phi \right] = 0 \quad (1.2.56)$$

or:

$$\left(\partial_\xi \frac{\rho}{|J|} U + \partial_\eta \frac{\rho}{|J|} V \right) = 0 \qquad\qquad (1.2.57)$$

which will be written also as:

$$\left[A_{\xi\xi} + A_{\xi\eta} + A_{\eta\xi} + A_{\eta\eta} \right] \phi = 0 \qquad\qquad (1.2.58)$$

It is to be noted that this equation is precisely the form

which is implicitly calculated in the classical finite element approach with isoparametric transformation between the physical and the local coordinates ξ, η.

Within an element, one has the isoparametric transformation and the FE representation of equation (1.2.7) is, after partial integration and for Dirichlet boundary conditions, given by the first term of equation (1.2.9).

This equation is numerically calculated as follows with $\vec{\nabla}'$ being the gradient operation in the ξ, η coordinates:

$$\sum \phi_I \int \rho \, \vec{\nabla}' \, N_I \, (\xi, \eta) \, (J^T J) \, \vec{\nabla}' \, N_J \, (\xi, \eta) \; . \; \frac{1}{|J|} \, d\xi \, d\eta = 0$$

which can be written:

$$\sum \phi_I \int \frac{\rho}{|J|} \, (A_1 \, \partial_\xi \, N_I + A_2 \, \partial_\eta \, N_I) \partial_\xi \, N_J \, d\xi \, d\eta$$

$$+ \sum \phi_I \int \frac{\rho}{|J|} \, (A_2 \, \partial_\xi \, N_I + A_3 \, \partial_\eta \, N_I) \partial_\eta \, N_J \, d\xi \, d\eta = 0$$

and with a reverse partial integration

$$\int \left[\frac{\partial}{\partial \xi} \, \frac{\rho}{|J|} \, (A_1 \phi_\xi + A_2 \phi_\eta) + \frac{\partial}{\partial \eta} \, \frac{\rho}{|J|} \, (A_2 \phi_\xi + A_3 \phi_\eta) \right] \, N_J \, d\xi \, d\eta = 0$$

This is the formulation of equation (1.2.7) which would be obtained by applying directly the Galerkin method to equation (1.2.56) in the (ξ, η) plane.

That the form (1.2.56) is contained in the standard FE-Galerkin method explains the flexibility of the FE approach for arbitrary meshes. Moreover the type of discretization in FE methods, namely the area average of the residual calculation as discussed in the previous section, allows the FE-approach to maintain the order of accuracy of the orthogonal mesh also in the arbitrary mesh calculation. Of course, certain restrictions are imposed in order to maintain this property, which require the mesh not to be too distorted.

The ADI method requires the operator to be written under the form of a sum of two terms

$$A = A_1 + A_2$$

which contain only one type of derivative.

Writing equation (1.2.56) under the form:

$$\{1 - \sigma(A_{\xi\xi} + A_{\xi\eta} + A_{\eta\xi} + A_{\eta\eta})\} \delta\phi = -\tau\sigma R \qquad (1.2.59)$$

and neglecting the mixed operators $A_{\xi\eta}$ and $A_{\eta\xi}$ (which corresponds to approximating the metric tensor g^{ij}, equation (1.2.54) by a diagonal matrix) one obtains

$$(1 - \sigma A_{\xi\xi})(1 - \sigma A_{\eta\eta})\delta\phi = -\sigma\tau R \qquad (1.2.60)$$

This form is identical to the AF1 scheme (Holst and Ballhaus, 1978).

Application of the FE-Galerkin approach to this equation follows the same line as in the previous section.

Explicitly, equation (1.2.60) is written as:

$$[1 - \sigma\partial_\xi \, (\frac{\rho A_1}{|J|} \, \partial_\xi)][1 - \sigma\partial_\eta \, (\frac{\rho A_3}{|J|} \, \partial_\eta)]\delta\phi = -\sigma\tau[(\frac{\rho U}{|J|})_\xi^{(n-1)} + (\frac{\rho V}{|J|})_\eta^{(n-1)}]$$

$$(1.2.61)$$

with:

$$\delta\phi = \phi^{(n)} - \phi^{(n-1)}$$

Applying the FE-Galerkin method, one obtains after splitting and for factorizing shape functions in the (ξ,η) plane, the system

$$(M_\xi + \sigma K_\xi)g = -\sigma\tau R \qquad (1.2.62)$$

$$(M_\eta + \sigma K_\eta)\delta\phi = g \qquad (1.2.63)$$

where:

$$(K_\xi)_{ik} = - \int \partial_\xi [\frac{\rho A_1}{|J|} \, \partial_\xi \, N_i(\xi)] \cdot N_k(\xi) \, d\xi \qquad (1.2.64)$$

$$(K_\eta)_{j\ell} = - \int \partial_\eta [\frac{\rho A_3}{|J|} \, \partial_\eta \, N_j(\eta)] \cdot N_\ell(\eta) \, d\eta \qquad (1.2.65)$$

and the residual R is given by equation (1.2.12).

Hence, the residual contains automatically the boundary conditions and no separate treatment has to be introduced. This maintains one of the strong points of the FE method where the Neumann conditions $\partial\phi/\partial n = 0$ is a natural one and leads to a zero value for the first term in the expression of the residual.

1.3. A FINITE ELEMENT MULTIGRID METHOD ON AN ARBITRARY MESH

For Poisson-type equations, multigrid methods have proven to

be the most efficient solvers available today and recent experience with the transonic potential equation confirms its high
potential value for accelerating the convergence (South and
Brandt, 1977; Jameson, 1979).

In the multigrid framework the previously described iterative
solution methods such as SLOR and ADI merely serve as a smoothing
device to eliminate the high frequency error components present
in the approximate solution, while the low frequency content of
the solution is solved on the coarser grids where the convergence is faster and the computational effort much lower.

1.3.1. *Multigrid algorithm*

Considering the system of non-linear equations (1.2.12) constructed on the finest mesh with characteristic spacing h:

$$R^h(\phi^h) = K^h(\phi^h) - f^h = O \qquad (1.3.1)$$

which may be written in correction form with respect to a known
approximate solution ϕ_n^h:

$$\hat{K}^h(\delta\phi^h) = K^h(\phi_n^h + \delta\phi^h) - K^h(\phi_n^h) = -R^h(\phi_n^h) \qquad (1.3.2)$$

where the unknowns are now the corrections $\delta\phi^h$ given by

$$\phi^h = \phi_n^h + \delta\phi^h \qquad (1.3.3)$$

Supposing that the high frequency errors have been eliminated
effectively by means of a smoothing operation such as SLOR or
ADI, the correction $\delta\phi^h$ and residual $R^h(\phi^h)$ may be considered
as smoothly varying quantities for which an approximation on a
coarser grid makes sense. This mesh with typical spacing 2h
is obtained by dropping the odd numbered coordinate line of the
mesh S^h and an updated approximation ϕ_{n+1}^h can be calculated
according to (1.3.3) by interpolating the coarse grid approximation $\delta\phi^{2h}$ for $\delta\phi^h$ back to the original mesh

$$\phi_{n+1}^h = \phi_n^h + I_{2h}^h \, \delta\phi^{2h} \qquad (1.3.4)$$

where I_{2h}^h is the coarse to fine grid function interpolation
operator called "prolongation". The coarse grid approximation
$\delta\phi^{2h}$ for the fine grid correction is the solution of the following equation on the coarse mesh:

$$\hat{K}^{2h}(\delta\phi^{2h}) = - \; {}^{R}I_{h}^{2h} \; R^{h}(\phi_{n}^{h}) \tag{1.3.5}$$

The fine to coarse residual restriction operator ${}^{R}I_{h}^{2h}$ constructs a meaningful approximation of the coarse grid residual R^{2h} based on the smoothly varying fine grid residuals.

By defining a coarse grid solution ϕ^{2h} as the approximation of ϕ^{h} on the coarse grid:

$$\phi^{2h} = I_{h}^{2h} \; \phi_{n}^{h} + \delta\phi^{2h} \tag{1.3.6}$$

where I_{h}^{2h} is the function restriction, equation (1.3.5) takes again the usual form of (2.3.1)

$$K^{2h}(\phi^{2h}) = f^{2h} \tag{1.3.7}$$

where the right hand side is a known function of the fine grid approximate solution:

$$f^{2h} = - \; {}^{R}I_{h}^{2h} \; R^{h}(\phi_{n}^{h}) + K^{2h}(I_{h}^{2h} \; \phi_{n}^{h}) \tag{1.3.8}$$

and $\delta\phi^{2h}$ can be eliminated from the updating formula (1.3.4) by means of (1.3.6)

$$\phi_{n+1}^{h} = \phi_{n}^{h} + I_{2h}^{h}(\phi^{2h} - I_{h}^{2h} \; \phi_{n}^{h}) \tag{1.3.9}$$

The solution ϕ^{2h} in turn can be approximated on the mesh S^{4h} when it is sufficiently smooth i.e. the whole procedure can be applied in a recursive way to equation (1.3.7). This non linear algorithm (F.A.S. scheme) is due to Brandt, (1977)who describes an adaptive strategy for the transition to a coarser or finer grid depending on the convergence level and speed on a particular grid. A more simple fixed strategy has been used in the present work. Starting on the finest mesh with spacing h one line over-relaxation sweep is performed followed by the transition to the next coarser grid by means of equations (1.3.7) and (1.3.8) until the coarsest grid is reached. On the coarsest grid some additional relaxation sweeps are performed and the solution of the next finer grid is updated by means of (1.3.9) followed by one relaxation step until the second finest grid is reached. The cycle terminates with the updating of the finest grid approximate solution with help of (1.3.9).

As distinct from F.D. approaches the interpolation operators I_{h}^{2h}, I_{2h}^{h} and ${}^{R}I_{h}^{2h}$ are not arbitrary but based on the F.E. interpolation spaces S^h and S^{2h}. They are considered in some more detail in the following sections.

1.3.2. *The coarse to fine grid function prolongation: operator I_{2h}^h*

The only natural choice for the interpolation of a coarse mesh function ϕ^{2h} to a fine mesh location (x_{ij}^h, y_{ij}^h) is to use the value of ϕ^{2h} in the location (x_{ij}^h, y_{ij}^h) given by the F.E. approximation in space S^{2h}:

$$[I_{2h}^h \phi^{2h}]_{ij} = \phi^{2h}(x_{ij}^h, y_{ij}^h) = \sum_{k,\ell} I_{k\ell}^{ij} \phi_{k\ell}^{2h} \qquad (1.3.10)$$

where the matrix $I_{k\ell}^{ij}$ is given by

$$I_{k\ell}^{ij} = N_{k\ell}^{2h}(x_{ij}^h, y_{ij}^h) \qquad (1.3.11)$$

On an arbitrary mesh this results in non-uniform interpolation coefficients I_k^{ij} and for instance with bilinear elements uniform interpolation is only obtained if the fine grid meshpoints are situated in the middle of the coarse grid element sides and in the centre giving only in this case the simple formula (Fig. 5a):

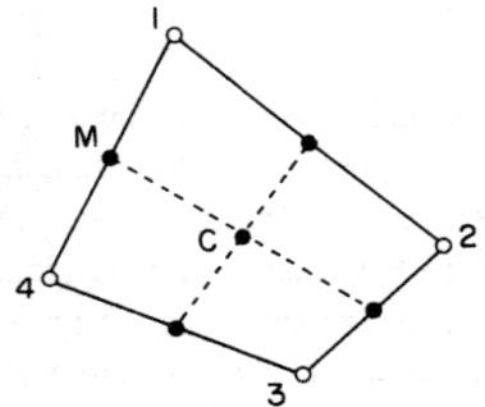

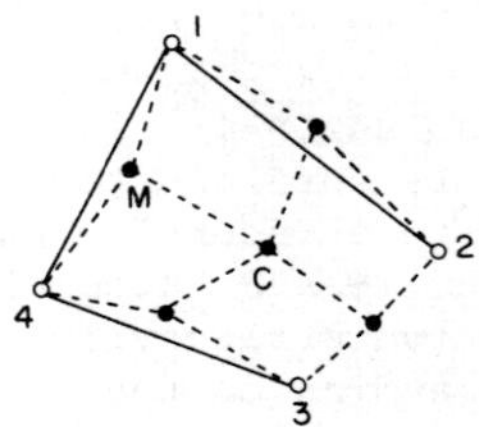

Fig. 5a Uniform interpolation   Fig. 5b Non uniform interpolation

$$[I_{2h}^h \phi^{2h}]_C = \frac{1}{4}(\phi_1^{2h} + \phi_2^{2h} + \phi_3^{2h} + \phi_4^{2h}) \quad \text{for the centre node}$$

$$[I_{2h}^h \phi^{2h}]_N = \frac{1}{2}(\phi_i^{2h} + \phi_j^{2h}) \quad \text{for the midside node i-j}$$

$$[I_{2h}^h \phi^{2h}]_i = \phi_i^{2h} \quad \text{for the corner nodes (identity)}$$

$$(1.3.12)$$

It follows that simple uniform interpolation is only possible for uniform refinements of the coarsest mesh which could be chosen arbitrarily. In the general case with bilinear elements (Fig. 5b) four coefficients are needed for each fine grid meshpoint not coinciding with a coarse grid meshpoint.

1.3.3. *Fine to coarse grid restriction operators* $(I_h^{2h}$ *and* $^R I_h^{2h})$

Function restriction

The value of ϕ^h in the coarse mesh location calculated with the F.E. approximation in S^h leads to the identity since the coarse gridpoints belong also to the fine grid

$$[I_h^{2h} \phi^h]_{ij} = \phi^h_{ij} \qquad (1.3.13)$$

This type of restriction is sometimes called injection.

Integral restriction: operation $^R I_h^{2h}$

As distinct from the F.D. case the residual is an integral quantity which is scaled differently on different grids. It cannot be represented in the spaces S^h and S^{2h} and the previous interpolation rules are inapplicable. In the Galerkin approach the volume integrals are always of the form:

$$[R^{2h}(\phi^{2h})]_{ij} = \int_S N^{2h}_{ij} g(\phi^{2h}) \, dS \qquad (1.3.14)$$

For instance the residual can be written in this form with

$$g(\phi^{2h}) = \nabla[\rho(\phi^{2h}) \nabla\phi^{2h}] \qquad (1.3.15)$$

A consistent representation of R^{2h}_{ij} by means of fine grid quantities is found by approximating the coarse mesh functions in the integrant of (1.3.14) with fine mesh interpolations in the space S^h, namely

$$[^R I_h^{2h} R^h]_{ij} = \int_{S^{2h}_{ij}} I_h^{2h} N^{2h}_{ij} g(I_h^{2h} \phi^h) \, dS \qquad (1.3.16)$$

where S^{2h}_{ij} is the coarse mesh residual integration domain, i.e. the part of S where $N^{2h}_{ij} \neq 0$ (Fig. 6).

On a uniformly subdivided coarse mesh (Fig. 6a) this general expression reduces to

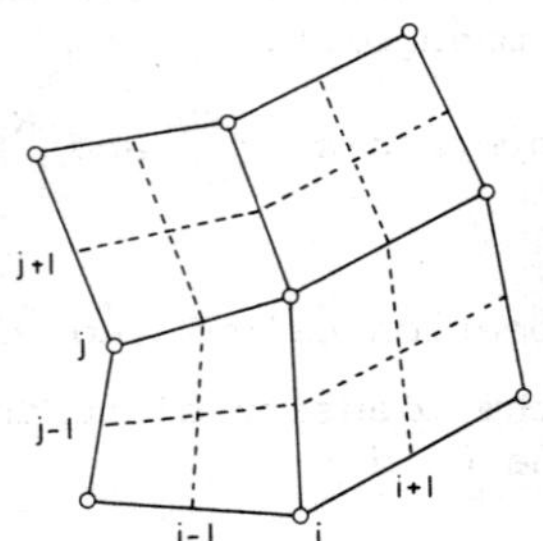

Fig. 6a S^{2h}_{ij} uniformly sub-
         divided mesh

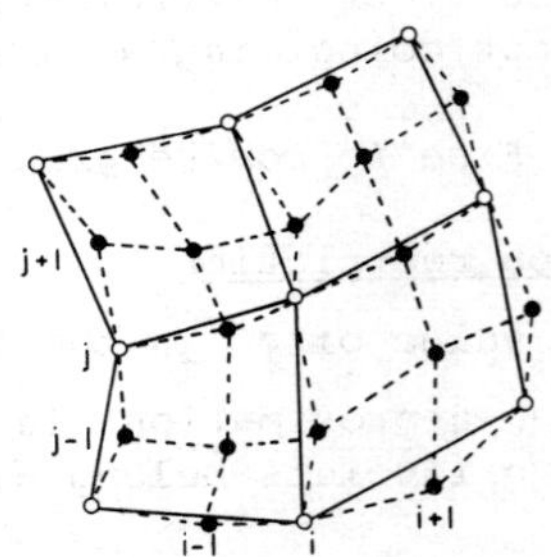

Fig. 6b arbitrarily subdivided
         mesh

$$[^R I^{2h}_h \, R^h]_{ij} = \sum_{k,\ell} I^{k\ell}_{ij} \, R^h_{k\ell} \qquad (1.3.17)$$

where the summation extends only over the 9 inner points in the
domain S^{2h}_{ij} since the coefficients $I^{k\ell}_{ij}$ are zero on and outside
the boundaries of S^{2h}_{ij}.

Comparing equations (1.3.10) and (1.3.17) one concludes that
the coarse to fine mesh interpolation I^h_{2h} is the adjoint of the
residual weighting $^R I^{2h}_h$ since they have transposed coefficient
matrices.

The following result is obtained for uniform subdivisions
(Fig. 6a) with bilinear elements which corresponds to the uniform interpolation (1.3.12)

$$[I^{2h}_h \, R^h]_{ij} = R^h_{ij} + \frac{1}{2} (R^h_{i,j+1} + R^h_{i,j-1} + R^h_{i-1,j} + R^h_{i+1,j})$$

$$+ \frac{1}{4} (R^h_{i+1,j+1} + R^h_{i+1,j-1} + R^h_{i-1,j+1} + R^h_{i-1,j-1})$$

$$(1.3.19)$$

On an arbitrarily subdivided mesh (Fig. 6b) the situation is
different due to non overlapping integration domains for the
coarse and fine mesh.

However it is easy to show (Deconinck and Hirsch, 1981) that equation (1.3.17) remains a valuable approximation for (1.3.16) in the arbitrary mesh case, of course only when used with the arbitrary mesh interpolation coefficient $I_{ij}^{k\ell}$ already known from the non-uniform interpolation.

It remains equally valid on the Neumann boundaries of the physical domain where the summation extends over 6 fine meshpoints and 4 for boundary corners.

The same expression derived here was also obtained for orthogonal meshes by Nicolaides and Brandt based on the minimization approach. Brandt suggests that this "natural" choice is not always better than the residual injection which is simply given by:

$$[^R I_h^{2h} \ R^h]_{ij} = 4 \ R_{ij}^h \qquad (1.3.20)$$

in the uniform case.

On an arbitrary mesh residual injection could be constructed by supposing the function $g(\phi^h)$ constant over the coarse mesh residual integration domain leading to the following general expression:

$$[^R I_h^{2h} \ R^h]_{ij} = R_{ij}^h \ \frac{\int N_{ij}^{2h} \ dS}{\int N_{ij}^h \ dS} \qquad (1.3.21)$$

which reduces exactly to the form (1.3.20) in the uniform case.

Computational experience with (1.3.21) was highly unsatisfactory and showed that it is inapplicable, at least with the simple smoothing procedure used in the present investigations.

1.4. EXAMPLES OF COMPUTATIONAL RESULTS

In this section some of the computational results obtained with the finite element artificial compressibility methods will be discussed.

1.4.1. SLOR versus ADI

Both methods yield reliable results for moderate shock cases such as the non-lifting flow around the NACA 0012 airfoil at a free stream Mach number of .80 (see the pressure distribution, Fig. 7). The ADI method is faster than the line relaxation provided one can find an optimal or near optimal parameter sequence at reasonable cost. This is a difficult

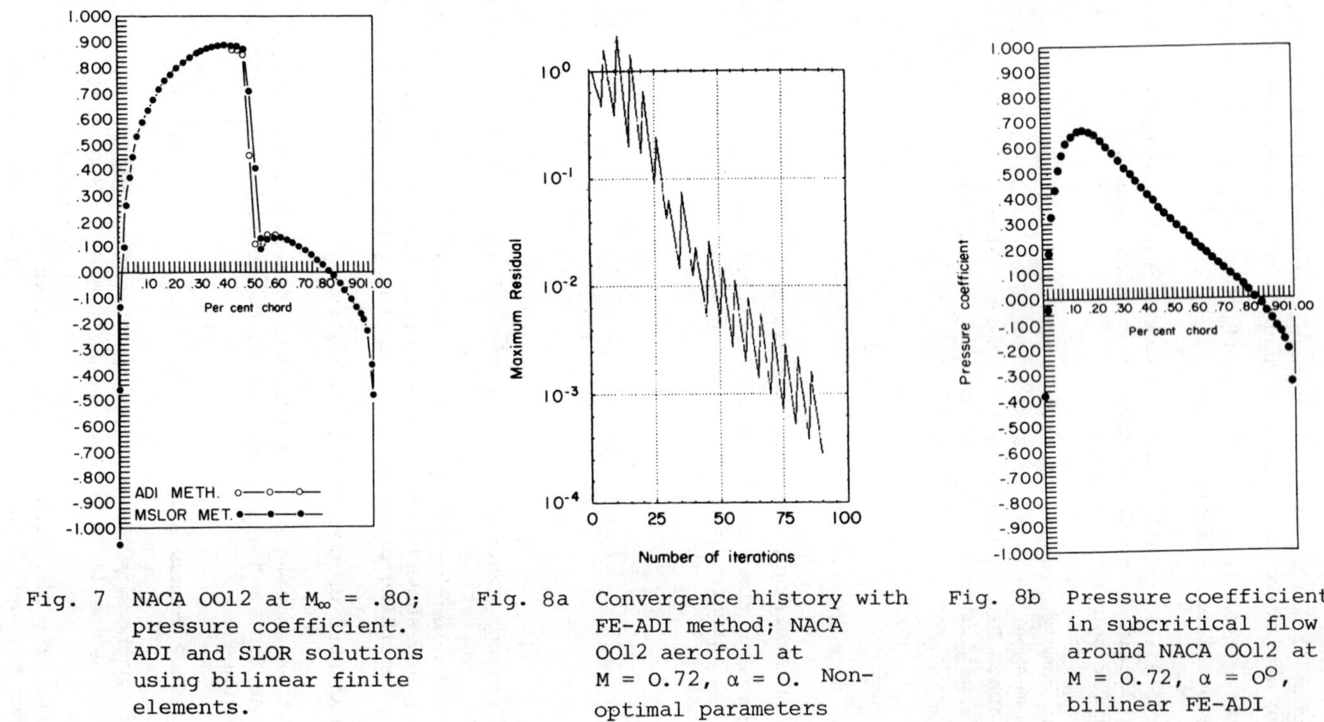

Fig. 7 NACA 0012 at M_∞ = .80; pressure coefficient. ADI and SLOR solutions using bilinear finite elements.

Fig. 8a Convergence history with FE–ADI method; NACA 0012 aerofoil at M = 0.72, α = 0. Non-optimal parameters

Fig. 8b Pressure coefficient in subcritical flow around NACA 0012 at M = 0.72, α = 0°, bilinear FE–ADI

task for grids with high and strongly varying meshwidth aspect
ratios.

Nevertheless reasonable convergence is obtained for entirely
subsonic flow for ADI with non-optimized parameters as is shown
in Fig. 8 for a freestream Mach number of .72.

It is clear that part of the efficiency of the AF1 scheme is
lost for transonic flows due to the hyperbolic character of the
equation in supersonic regions. This problem has been overcome
to some extent in F.D. applications by the introduction of the
AF2 scheme (Holst and Ballhaus, 1978) which has better convergence properties in supersonic regions due to the presence of
additional $\phi_{xt}^{\leftarrow}$ and $\phi_{yt}^{\leftarrow}$ terms in the iteration operator.

No attempts have been undertaken until now to implement a
F.E. version of this AF2 scheme.

1.4.2. Bilinear versus Biquadratic elements

A comparative study between both elements has been made
(Deconinck and Hirsch, 1980) leading to the following conclusions.

Both elements yield similar results for the Mach number or
pressure distributions except for the stagnation point region
where the zero velocity is better approximated by the quadratic
higher order element: M = .05 versus M = .22 in the leading
edge meshpoint for the NACA 0012 flow at M_∞ = .80.

This could be an important advantage for biquadratic elements
in cases where the stagnation region strongly influences the
overall flow behaviour as is the case for turbine cascades with
thick leading edge blades.

On the other hand the computational effort is lower for the
quadratic elements provided that one can use a constant density
approximation for each quadratic element which is an acceptable
assumption in the bilinear case.

For quadratic elements each composed of 4 bilinear elements
this assumption seems not appropriate especially in shock regions
where density variations are extremely large. With the more
appropriate 4 point Gauss quadrature the computational effort
for the residual evaluation with quadratic elements will be equal
to the cost for the residual computed with bilinear elements
and constant density assumption. Of course due to the higher
accuracy the number of meshpoints could be lower in the biquadratic approximation except in shock regions where a fine mesh is
needed in any case to be able to capture the correct shock
position.

Another important consideration is the algorithmic and programming complexity which is much higher for the quadratic
elements and which makes its use in a multigrid framework more
difficult.

1.4.3. Cascade computations with SLOR and grid refinement

Most of our routine applications are with SLOR and grid refining or nested iteration which means that a converged solution
on a coarse grid is used as initial approximation for the calculation on the next finer grid until the final solution is obtained on the finest mesh.

Body fitted cascade geometry grids are obtained by solving
an elliptic system of 2 equations for the curvilinear coordinates
ξ and η followed by a clustering on the ξ = constant lines towards
the blade boundaries. In Figs. 9 and 10 the finest mesh is
shown for the VKI gas turbine cascade and DCA 9.5° camber compressor cascade for which some results will be given.

For the turbine cascade at an inlet Mach number of .281 and
outlet Mach number of .975 the Mach number distribution is compared with experimental results (O and X) in Fig. 11 and shows
good agreement.

For the compressor cascade a first set of comparisons is
given at an inlet Mach number of 1.05 and different outlet Mach
number going from M = .80 to M = .86 (Fig. 12).

For all cases a shock is present detached from the blade
leading edge. However it is seen that with decreasing the back
pressure a second normal shock is built with increasing intensity
and slightly moving downstream. This double shock structure has
been observed in experiments and clearly corresponds to the
physical reality.

The same qualitative result is obtained with an inlet Mach
number of 1.125 and two different outlet Mach numbers of .734
and .800. Again the flow pattern changes from a one shock
situation to a double shock situation in which the second shock
is normal and the first shock oblique (Fig. 13).

1.4.4. Computations with Multigrid

To be acceptable for routine applications it is necessary
to accelerate the convergence speed of the previous calculations.
The results obtained with multigrid on two test cases, the non-
lifting NACA 0012 and a channel flow are very promising with
this respect.

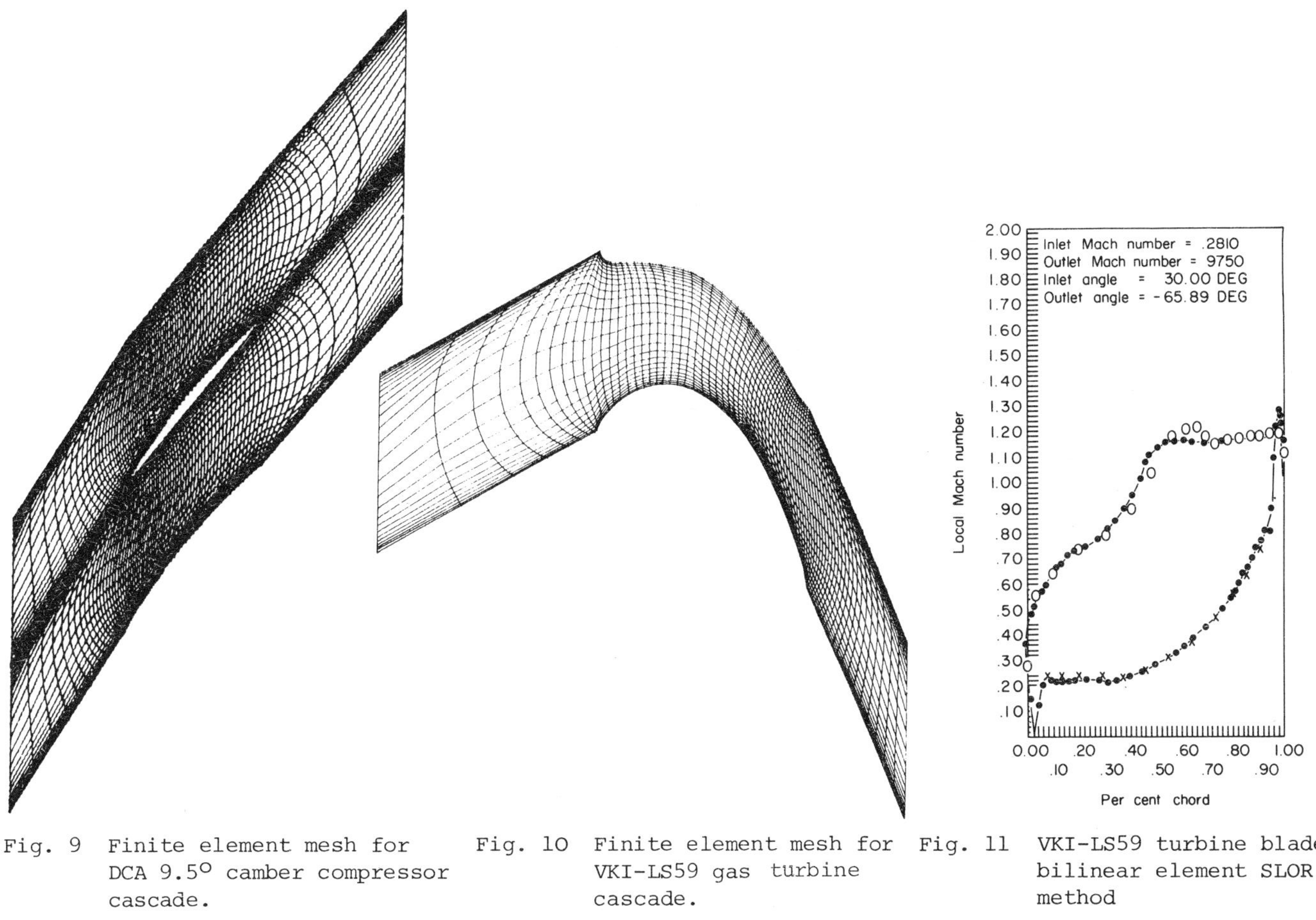

Fig. 9   Finite element mesh for DCA 9.5° camber compressor cascade.

Fig. 10   Finite element mesh for VKI-LS59 gas turbine cascade.

Fig. 11   VKI-LS59 turbine blade; bilinear element SLOR method

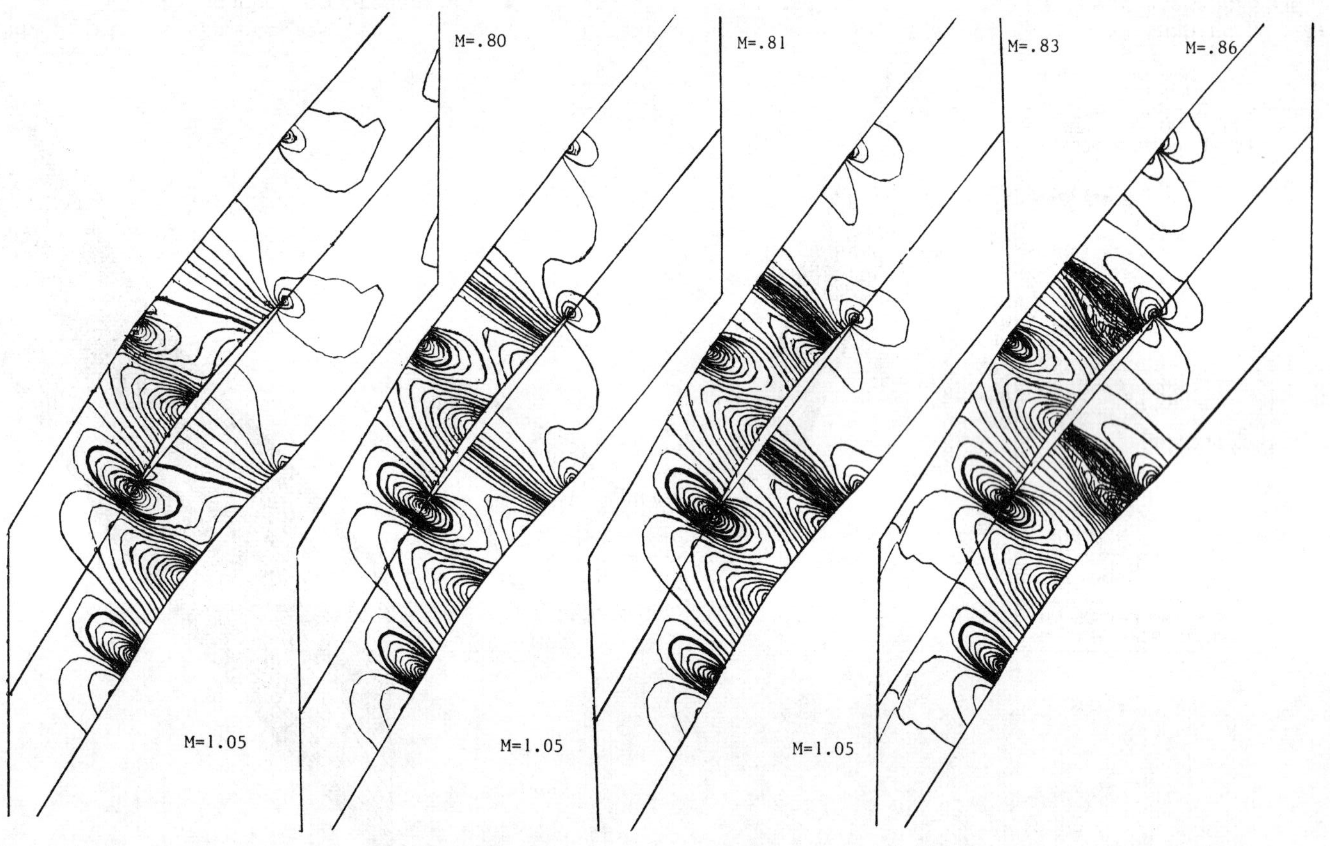

Fig. 12 For legend see opposite

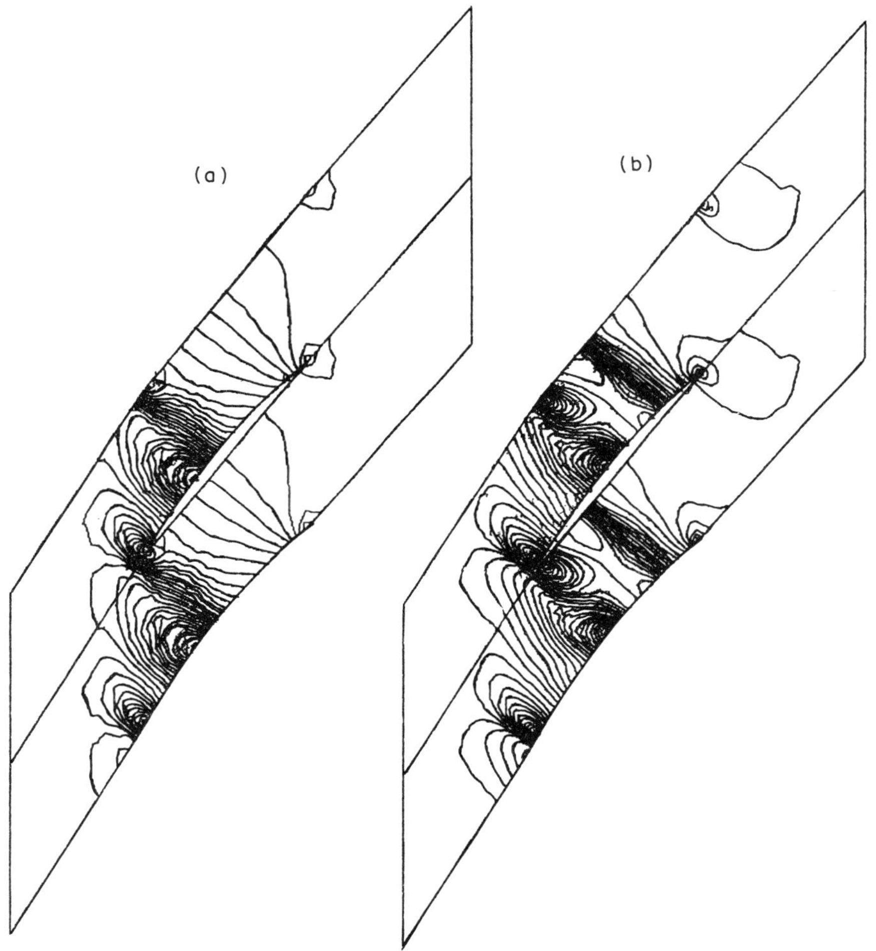

Fig. 13   DCA 9.5° camber compressor cascade with inlet Mach number
1.125.   Outlet Mach numbers 0.734 and 0.800.   Contours
as in Fig. 12 (a) M_{in} = 1.125 M_{out} = .734

(b) M_{in} = 1.125 M_{out} = 0.800

Fig. 12   DCA 9.5° camber compressor cascade with inlet Mach number
1.05.   Isomachlines for different outlet Mach numbers
from .80 to .86.   The heavy contours are at M = 1.0, and
the lighter contours at intervals of 0.025

In Fig. 14 the convergence history is compared for the NACA
OO12 case at M = .80 showing an average residual reduction per
fine grid relaxation sweep of .90 with the multigrid method
using four grids going to .967 with two grids. The simple grid
refining has a convergence rate which tends to 1 after a very
fast reduction in the initial iterations corresponding to a fast
elimination of the high frequence errors in the approximation.
The pressure distribution with multigrid and four grids is
fully converged within plottable accuracy after 13 multigrid
cycles, a computational effort equivalent to 3 fine grid relaxation sweeps.

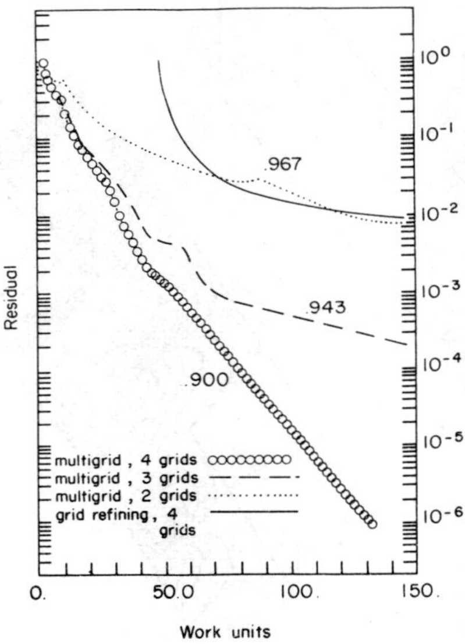

Fig. 14   Convergence history for NACA OO12 aerofoil; convergence
          rate is indicated. (M_∞ = O.80)

Figure 15 shows the convergence history at M = .85 where the
convergence rate is slowed down to .957. However, a small modification of the definition of the switching function following
Habashi and Hafez (1981) restores the rate to .92.

Equation (1.2.2) is replaced by

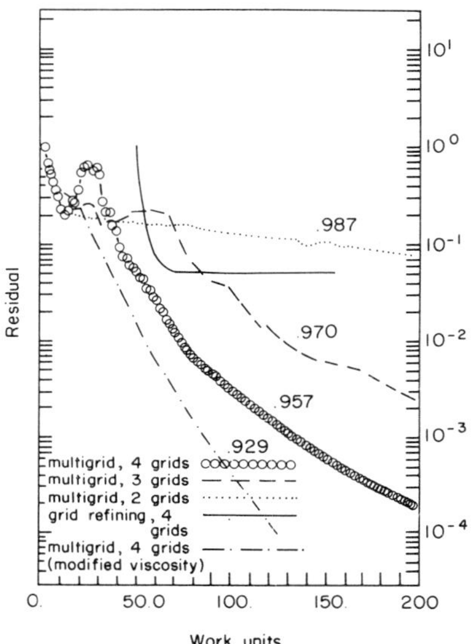

Fig. 15   Convergence history for NACA 0012 aerofoil at M = 0.85;
          convergence rate is indicated.

$$\nu_i = \max\left(0,\ 1 - \frac{1}{M_i^2},\ 1 - \frac{1}{M_{i-1}^2}\right)$$

This modification slightly changes the solution in the shock-
points. In Fig. 16 the converged pressure distribution obtained
after 20 multigrid cycles is compared with other conservative
methods.

The final multigrid result is a choked flow in a channel with
a circular arc bump at the lower wall. The choking Mach number
of .849 at the outlet was determined by successively increasing
the outlet Mach number until choking occurred. The isomach
lines plot is given in Fig. 18. The pressure distribution along
the bump and corresponding upper wall is compared with the choked
potential solution of Veuillot and Viviand obtained at M = .850.
For this difficult case it was not possible to obtain a converged
solution with the simple grid refining technique (Fig. 17).

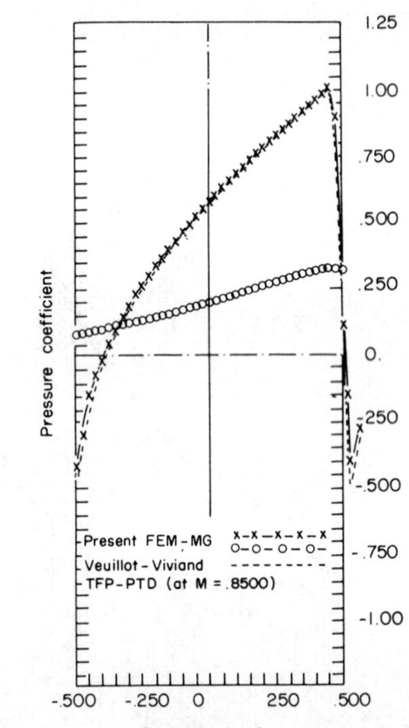

Fig. 16 NACA 0012 aerofoil at M = 0.850; comparison with other conservative methods

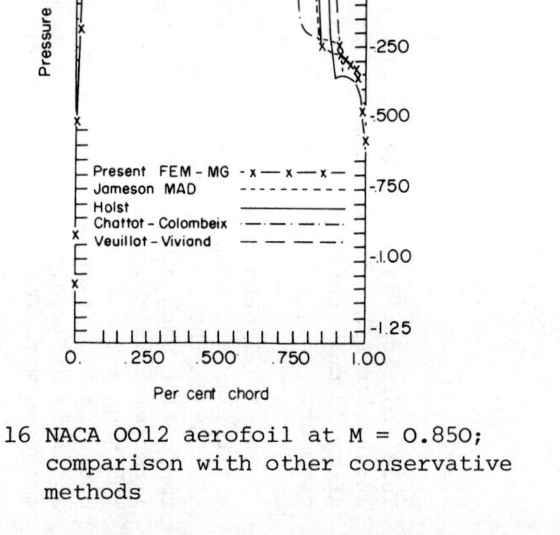

Fig. 17 Channel flow with circular arc bump at M = 0.849; comparison with other conservative methods

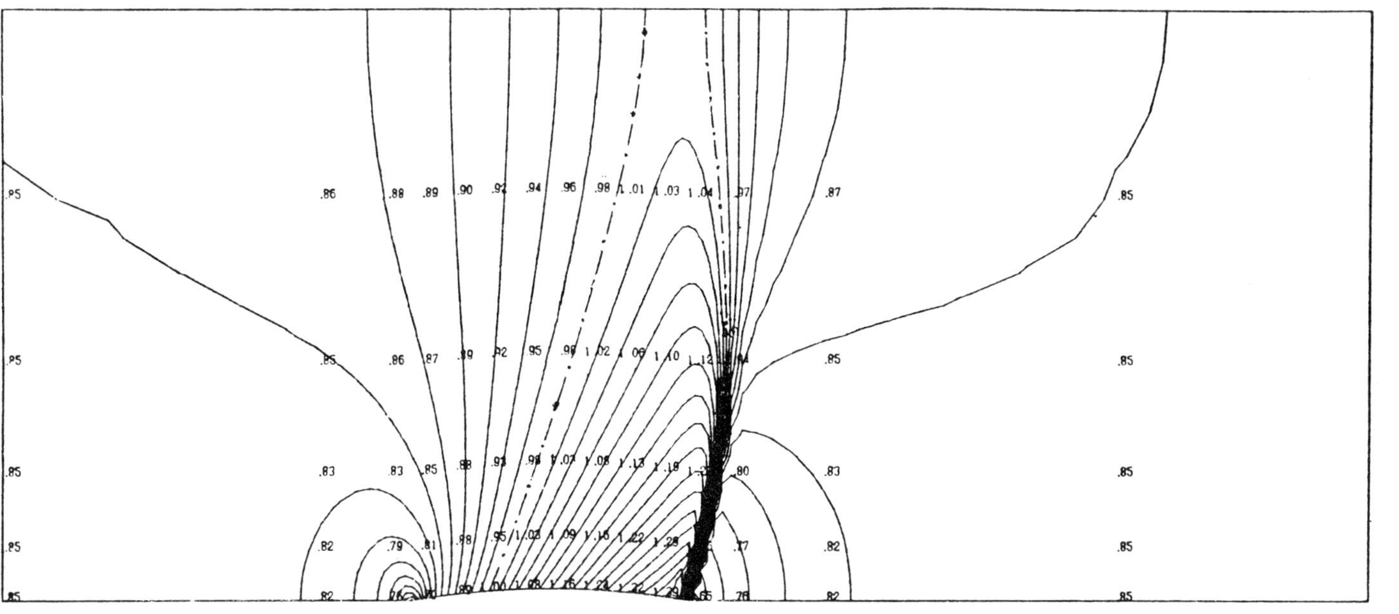

Fig. 18 Channel flow with circular arc bump at M = .8490 (choked). Contours are isomach lines, as in Fig. 12

1.5. CONCLUSION

The artificial density concept seems to be the most promising way to deal with the stability and non-uniqueness problems in the numerical solution of the transonic potential equation with finite element methods.

However the correct balancing of the amount of viscosity needed for stability reasons without too much disturbing the solution in the shock region remains a topic for further investigations.

The introduction of multigrid methods reduces the problem of finding a very efficient iterative solution method to the problem of finding an efficient smoothing algorithm which appears to be easier since classical relaxation methods such as SLOR already have excellent smoothing properties leading to very efficient methods when used in the multigrid approach.

2. EULER EQUATIONS

2.1 GALERKIN FINITE ELEMENT METHOD

The general formulation of Finite Element discretization in space can be applied, through a Galerkin approach to the system of the first order Euler equations.

The two-dimensional form of the Euler equations can be written as follows:

$$\frac{\partial U}{\partial t} + \nabla H = 0 \tag{2.1}$$

where:

$$U = \begin{Bmatrix} \rho \\ \rho u \\ \rho v \\ \rho E \end{Bmatrix} \qquad H_x = F = \begin{Bmatrix} \rho u \\ \rho u^2 + p \\ \rho uv \\ \rho u(E + \frac{p}{\rho}) \end{Bmatrix} \qquad H_y = G = \begin{Bmatrix} \rho v \\ \rho uv \\ \rho v^2 + p \\ \rho v(E + \frac{p}{\rho}) \end{Bmatrix} \tag{2.2}$$

The method of weighted residuals applied to (2.1) with weight functions W, leads to

$$\int_S \frac{\partial U}{\partial t} \cdot W \, ds + \int_S \vec{\nabla} \vec{H} \cdot W \, ds = 0 \tag{2.3}$$

and a Finite Element Galerkin approximation with shape functions N_i and weight functions N_j leads to the system of differential equations

$$M_{ij} \frac{\partial U_j}{\partial t} + \vec{H}_j \vec{K}_{ij} = O \qquad (2.4)$$

where:

$$M_{ij} = \int_S N_j W_i \, dS \qquad \vec{K}_{ij} = \int_S \vec{\nabla} N_j \cdot W_i \, dS \qquad (2.5a)$$

$$U = \sum_j U_j N_j \qquad W_i = N_i \qquad (2.5b)$$

The mass matrix M has non-zero elements at all nodes connected geometrically with node j and the system (2.4) is of implicit nature. Therefore inversion of M will generally be a time consuming task.

For instance, the one-dimensional equation

$$\frac{\partial U}{\partial t} + \frac{\partial F}{\partial x} = O \qquad (2.6)$$

with linear shape functions, leads to the system

$$\frac{U'_{i+1} + 4 U'_i + U'_{i-1}}{6} + \frac{\Delta t}{\Delta x} \frac{F_{i+1} - F_{i-1}}{2} = O \qquad (2.7)$$

The scheme obtained with a forward time difference is unstable while the backward time differencing gives an unconditionally stable scheme as is easily seen by a linearized Von Neumann stability analysis.

One obtains, with

$$w_i^{n+1} = U_i^{n+1} - U_i^n$$

$$w_{i+1}^{n+1} (1 + \alpha_{i+1}) + 4 w_i^{n+1} + (1 - \alpha_{i-1}) w_{i-1}^{n+1} = - \frac{3\Delta t}{\Delta x} (F_{i+1}^n - F_{i-1}^n)$$

$$(2.8)$$

where:

$$\alpha_i = \frac{3\Delta t}{\Delta x} \left(\frac{\partial F}{\partial U}\right)_i^n$$

A Taylor expansion shows that this scheme has a second order dissipation, leading to a strongly smeared shock (see Fig. 19), although it is not strictly dissipative in the sense of Kreiss and therefore strong non-linear post-shock oscillations will appear, e.g. for Burgers equation.

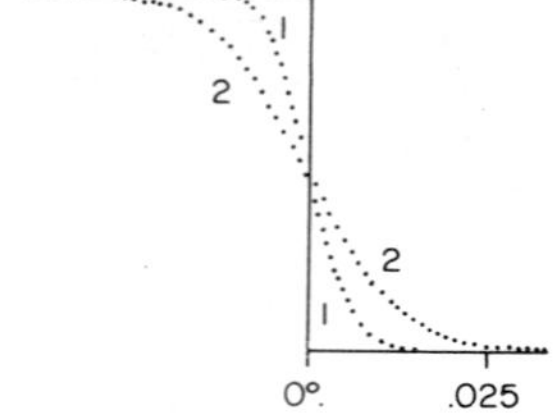

Fig. 19 $\dfrac{\partial u}{\partial t} + \dfrac{\partial u}{\partial x} = 0$ $\dfrac{\Delta t}{\Delta x} = 1$ 1: after 20 time steps,
2: after 100 time steps, -: exact solution, FD in time
(backward), FE (Galerkin) in space.

2.2. AN "ORTHOGONAL" FINITE ELEMENT METHOD (OFE)

It would be desirable to avoid the large mass matrix M,
obtained after assembly, while keeping the advantages of the FE
approach, in particular with regard to the arbitrary geometries
and computational meshes which can be handled with a FEM as
easily as a regular mesh system.

In order to obtain a diagonal mass matrix within the weighted
residual approach equation (2.5), one could define continuous
weight functions W_i orthogonal to the shape functions N_k in the
whole domain. This corresponds to the conjugate shape functions,
introduced by Oden (1972), which are non-zero over the whole
domain while each shape function N_j has only non-zero values
within few elements. Moreover the definition of these conjugate
shape functions requires the inversion of the full, assembled
mass matrix, leading to conjugate functions which are strongly
dependent on the geometry, in the sense that the addition of
a single element will change the values of these functions every
where. Hence, one of the main advantages of FE is lost in the
sense that one does not work anymore with locally defined
functions.

Therefore if one is willing to sacrifice the inter-element continuity of these conjugate functions and define locally orthogonal shape functions N_j^* through

$$\int_{S_e} N_i \, N_j^* \, dS_e = 0 \quad \text{for } i \neq j \tag{2.9}$$

where the integration is performed on a single element the mass
matrix would become diagonal on each element and remain so after
assembly.

Within one element, the orthogonal shape function N_j^* is defined through

$$\int_{S_e} N_i \, N_j^* \, dS = \mu^{(j)} \, \delta_{ij} \qquad (2.10)$$

Since the shape functions N_i form a complete set in an element one has

$$N_j^* \, (x,y) = C_{jk}^{(e)} \, N_k \, (x,y) \qquad (2.11)$$

where the C_{jk} are constants. Introducing (2.11) in (2.10) one obtains

$$C_{jk}^{(e)} \, M_{ki}^{(e)} = \mu^{(j)} \, \delta_{ij} \qquad (2.12)$$

and the matrix C is given by the inverse of the local mass matrix $M^{(e)}$

$$[C^{(e)}] = [\mu][M^{(e)}]^{-1} \qquad (2.13)$$

where $[\mu]$ is the diagonal matrix of the normalization coefficients.

The dimensions of these square matrices are equal to the number of nodes in each element and therefore inversion of $M^{(e)}$ is an easy task.

2.2.1. Linear Elements in One Dimension

Referring to Fig. 20, the linear shape functions for the element between nodes (i - 1) and i, element (1), are

$$N_{i-1}^{(1)} = \frac{x_i - x}{h_{i-1}} = 1 - \xi \qquad N_i^{(1)} = \frac{x - x_{i-1}}{h_{i-1}} = \xi \qquad (2.14)$$

where ξ is the local coordinate and the following orthogonal shape functions are obtained (Fig. 21)

$$N_{i-1}^{(1)*} = \mu_{i-1} \frac{2}{h_{i-1}^2} \cdot [-3(x-x_i)-h_{i-1}] = \mu_{i-1} \frac{2}{h_{i-1}} (-3\xi + 2) \qquad (2.15)$$

$$N_i^{(1)*} = \mu_i \frac{2}{h_{i-1}^2} [3(x-x_i)+2h_{i-1}] = \mu_i \frac{2}{h_{i-1}} (3\xi - 1) \qquad (2.16)$$

Generally, for an arbitrary mesh these functions will not be continuous at the inter-element boundaries. However in the linear case, choosing

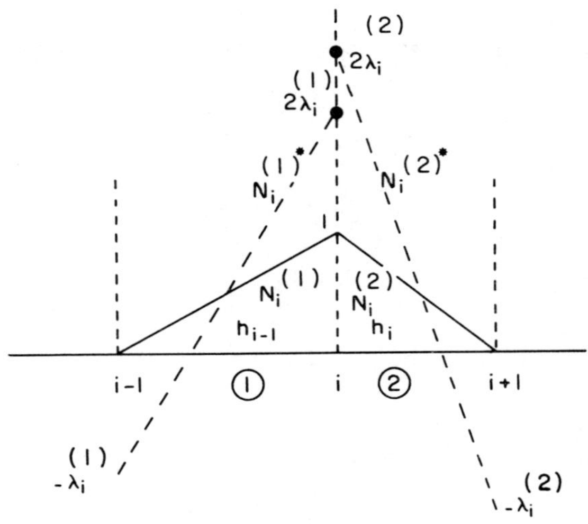

Fig. 20   Linear orthogonal shape functions

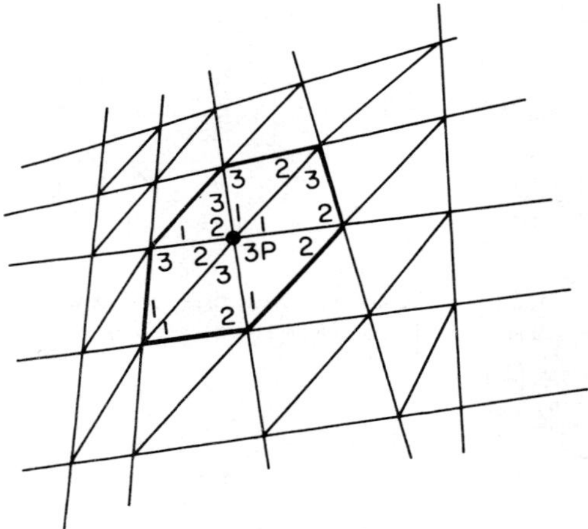

Fig. 21   Linear triangular elements

$$\mu_{i-1} = \mu_i = h_{i-1}$$

will lead to continuity of the orthogonal shape functions.

Applying the weighted residual method with the orthogonal shape functions to the linear equation (2.6), one obtains after assembling the two equations for node i

$$\frac{dU_i}{dt} + \frac{F_{i+1} - F_{i-1}}{h_{i-1} + h_i} = O \qquad \text{for } \mu_i^{(1)} = h_{i-1} \text{ and } \mu_i^{(2)} = h_i$$

$$(2.17)$$

$$\frac{dU_i}{dt} + \frac{F_i - F_{i-1}}{2h_{i-1}} + \frac{F_{i+1} - F_i}{2h_i} = O \qquad \text{for } \mu_i^{(1)} = \mu_i^{(2)} = 1 \qquad (2.18)$$

These are finite difference formulas for an arbitrary mesh spacing which are automatically generated through the OFE method.

2.2.2. Linear Elements in Two Dimensions

Considering a two-dimensional domain with linear triangular elements (Fig. 21) the shape functions are

$$N_i = \frac{1}{2\Delta} (a_i + b_i x + c_i y) \qquad i = 1,2,3 \qquad (2.19)$$

where a_i, b_i, c_i are coordinate dependent and Δ is the area of the triangle. The orthogonal shape functions become

$$N_1^* = \frac{3\mu_1}{\Delta} (3N_1 - N_2 - N_3)$$

$$N_2^* = \frac{3\mu_2}{\Delta} (-N_1 + 3N_2 - N_3) \qquad (2.20)$$

$$N_3^* = \frac{3\mu_3}{\Delta} (-N_1 - N_2 + 3N_3)$$

and are discontinuous at the interelement boundaries except for

$$\mu_1 = \mu_2 = \mu_3 = \Delta \qquad (2.21)$$

An interesting observation is that an application of the OFE method with the normalization (2.21) to the Euler equations (2.1), leads after straightforward calculation to the equations

$$S \frac{dU_{i,j}}{dt} + \sum_{k=1}^{6} \bar{F}_k \, \Delta y_k - \sum_{k=1}^{6} \bar{G}_k \, \Delta x_k = 0 \qquad (2.22)$$

where the summation on k extends to the 6 sides of the hexagonal
domain of Fig. 21. The bars on F and G indicate the arithmetic
average of the values at the two nodes of each side and Δx_k,

Δy_k are the differences in the y and x coordinates of these

nodes, S is the surface of the hexagon.

Equation (2.22) then clearly corresponds to the Finite Volume
discretization of the integral Euler equations applied to the
hexagonal domain S of Fig. 21

$$\int_S \frac{dU}{dt} \, dS + \oint_C (F \, dy - G \, dx) = 0 \qquad (2.23)$$

A similar analysis for quadrilateral elements (Fig. 22) and
normalization

$$\mu_i = S^{(e)}$$

where $S^{(e)}$ is the area of the element, shows that for a geometry
with parallel lines, the OFE method leads to the discretization
of the Finite Volume approach equation (2.23), but applied to
the volume indicated in heavy lines in Fig. 22* . However this
is not the case for an arbitrary mesh.

In this case, application of the OFE method leads to expressions for the spatial derivatives of equation (2.1) which contain the contribution from all eight nodes surrounding P. The
coefficients of the extreme corner nodes, like (i+1, j+1) vanish
when the mesh lines become parallel.

One obtains generally, an equation of the form

---

*In this case, the orthogonal shape functions are given by

$$N_j^* = \mu_j (1 + 3\xi\xi_j)(1 + 3\eta\eta_j)$$

in local coordinates (ξ,η) where ξ_j,η_j take the values ± 1. For
vertical i lines and arbitrary j lines, one obtains

$$N_j^* = \mu_j (1 + \alpha_j \xi\xi_j)(1 + 3\eta\eta_j)$$

where α_j is coordinate dependent.

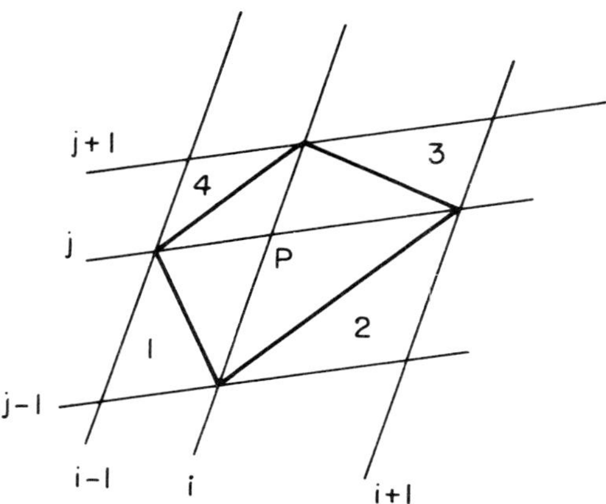

Fig. 22   Bilinear quadrilateral elements

$$\frac{dU_i}{dt} + \sum_{k=1}^{9} (S_k \, F_k + T_k \, G_k) = O \qquad (2.24)$$

where the summation is over all nodes around P (P included) and
the coefficients S_k and T_k are only dependent on the geometry.

For a regular orthogonal mesh, a central difference scheme
in space is obtained. Various finite difference schemes in
time can be adopted for equation (2.24) and a modified Lax scheme
(Couston, et al. 1975) called the damping surface technique is
used for the example shown in the following section.

2.2.3. Example Calculation

In order to illustrate the method, a transonic nozzle has
been tested with arbitrary geometry for the computational mesh.
The scheme proposed by Couston, et al. (1975), has been adopted,
although other schemes can be used, for instance by applying the
OFE-method to the time-discretized equation analyzed by Beam
and Warming (1976) in the space factored form. This allows an
automatic generation of numerical schemes for arbitrary geometries.

The damping surface technique applied to equation (2.24) leads
to

$$U_{i,j}^{n+1} = \overline{U^n}_{i,j} + \sum_{k=1}^{n_{i,j}} (S_k \, F_k + T_k \, G_k) - \alpha(U_{i,j}^n - U_{i,j}^L) \qquad (2.25)$$

where $U_{i,j}^n$ is the average of U on all nodes surrounding (i,j)
which contribute to the sum on k ($n_{i,j}$ nodes). The variables
at time indicated by L correspond to values which are updated
every L time iterations. The coefficient α is close to 1.
Figure 23 shows the Mach number distribution for the nozzle.

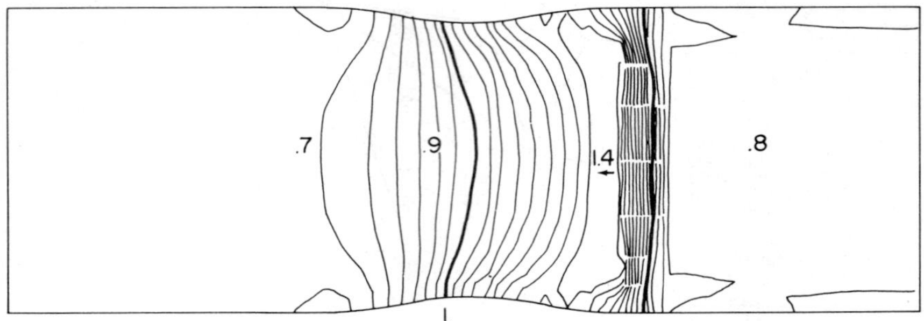

Fig. 23   Transonic nozzle calculation using Euler equations
          inlet Mach number = 0.7 and outlet Mach number 0.8;
          Isomach lines at intervals of 0.05. Heavy contours
          are M = 1.0.

REFERENCES

Beam, R. and Warming R. (1976). 'An implicit finite difference
algorithm for hyperbolic systems in conservation law form'.
J. Comp. Phys. **22**, 87–110.

Brandt, A. (1977).'Multi-level adaptive solutions to boundary
value problems', *Math. Comp.* **31**, 333–390.

Bristeau, M. O. (1978).'Application of a finite element method
to transonic flow problems using an optimal control approach'
Von Karman Institute, VKI-LS 1978-4.

Chan, S.T.K. and Bradshears, M.R. (1975). 'Finite element
analysis of transonic flow by the method of weighted residuals'
AIAA paper 7579.

Couston, M., McDonald, P.W. and Smolderen, J. (1975). 'The
damping surface technique for time-dependent solution to fluid
dynamics' *VIK-TN* 109.

Deconinck, H. and Hirsch, Ch. (1979a). 'A finite element method
solving the full potential equation with boundary layer interaction in transonic cascade flow. *AIAA* paper 79-0132.

Deconinck, H. and Hirsch, Ch. (1979b). 'Finite element methods for transonic flow calculations In "Proc. III GAMM Conf. on Num. Meth. in Fluid Mech.", Köln, 1979. (Notes on Numerical Fluid Mechanics, Vol. 2, Vieweg 1980).

Deconinck, H. and Hirsch, Ch. (1979c). 'Transonic flow computations with finite elements' In "Num. Meth. for the Computation of inviscid Transonic Flows with Shock Waves". Proceedings of the GAMM workshop Sept. 18—19, Stockholm, Sweden, 1979. A. Rizzi, H. Vivand (eds.) (Notes on Numerical Fluid Mechanics, Vol. 3, Vieweg 1981).

Deconinck, H. and Hirsch, Ch. (1980). 'Transonic Flow Calculations with Higher Order Finite Elements" Proc. of 7th Int. Conf. Num. Meth. Fl. Dyn., June 23—27, 1980, Stanford USA (Springer 1981).

Deconinck, H. and Hirsch, Ch. (1981). "A Multigrid Method for the Transonic Full Potential Equationdiscretized with Finite Elements on an Arbitrary Body fitted Mesh" NASA Symp. on Multigrid Methods, Oct. 1981, Ames Research Center, Moffett Field, CA.

Dendy, J. and Fairweather, G. (1975). 'Alternating direction Galerkin methods for parabolic and hyperbolic problems on rectangular polygons' *SIAM, Journal of Num. Anal.* **12**.

Douglas, J. and Dupont, T. (1975). "Alternating Direction Galerkin Methods on Rectangles" Proc. Symp. on Numerical Solution of Partial Differential Equations II SYNSPADE II, 1975, pp. 133—124.

Eberle, A. (1977). "Eine Methode Finiter Elements zur Berechnung der Transsonischen Potential Strömung um Profile" MBB Bericht UEE 1352(O).

Glowinski, R., Periaux, J. and Pironneau, O. (1976) "Transonic Flow Simulation by the Finite Element Method via Optimal Control" Second Int. Symp. on Finite Elements in Fluid Flow, Rapallo.

Habashi, W. G. and Hafez, M. M. (1982). 'Finite element method for transonic cascade flows'*AIAA Journal.*

Hafez, M. and Murman, E. M. (1978). "Recent Developments in Finite Element Analysis for Transonic Airfoils" Proc. of ATAR, Nasa Langley, March 1978.

Hafez, M., Murman, E.M. and South, J.C. (1978). 'Artificial compressibility methods for numerical solution of transonic full potential equation'.*AIAA* paper 78-1148.

Hafez, M.M. and South, J.C. (1979). 'Vectorization of Relaxation
Methods for Solving Transonic Full Potential Equation' Flow
Research Company.

Hirsch, Ch. and Warzee, G. (1979). "An orthogonal Finite Element
Method for Transonic Flow Calculations" Proc. 6th Int. Conf. on
Numerical Methods in Fluid Dynamics, Springer Verlag.

Holst, T.L. (1978). 'An implicit algorithm for the conservative
transonic full potential equation using an arbitrary mesh'
AIAA paper 78-1113, 1978 also *AIAA journal* **17**, no. 10, 1979.

Holst, T.L. and Ballhaus, W.F. (1978). "Conservative Implicit
Schemes for the Full Potential Equation Applied to Transonic
Flows" NASA TM 78469.

Jameson, A. (1975). "Numerical Computation of Transonic Flows
with Shock Waves" Symp. Transsonicum II, Springer Verlag.

Jameson, A. (1979). 'Acceleration of transonic potential flow
calculations on arbitrary meshes by the multiple grid method.
AIAA paper 79-1458.

Marchuk, G.I. (1975). "Methods of Numerical Mathematics" Springer
Verlag.

Oden, J.T. (1972). "Finite Elements of Non Linear Continua"
McGraw Hill.

Periaux, J. (1979). "Résolution de quelques problèmes non
linéaires en aérodynamique par des méthodes d'éléments finis
et de moindres carrés fonctionnels" Ph.D. Thesis, Univ. Pierre
et Marie Curie, Paris VI.

Polak, E. (1971). "Computational Methods in Optimization"
Academic Press.

South, J.C. and Brandt, A. (1977). 'Application of multi-level
grid method to transonic flow calculations' In "Transonic Flow
Problems in Turbomachinery" (eds. Adamson and Platzer) Hemi-
sphere, Washington.

AN IMPROVED TIME MARCHING METHOD FOR TURBOMACHINERY
FLOW CALCULATION

J.D. Denton

*(Whittle Laboratory,
Cambridge University Engineering Department, U.K.)*

SUMMARY

Time marching solutions of the Euler equations are now very
widely used for the calculation of flow through turbomachinery
blade rows. All methods suffer from the disadvantages of shock
smearing, lack of entropy conservation and comparatively long
run times. A new method is described which reduces all these
problems.

The method is based on the author's opposed difference scheme
but this is applied to a new type of grid consisting of quadrilateral elements which do not overlap and have nodes only at
their corners. The use of a non-overlapping grid reduces finite
differencing errors and gives complete freedom to vary the size
of the elements. Both these factors help to improve entropy
conservation. Considerable savings in run time (by a factor
of about 3) are obtained by using a simple multigrid method
whereby the solution is advanced simultaneously on a coarse and
on a fine grid. The resulting method is simpler, faster and
more accurate than its predecessor.

NOMENCLATURE

$\underline{A}$ Area vector of face of element

C Sonic velocity

CFP   Correction factor on pressure

CFRO   Correction factor on density

C_p Specific heat capacity at constant pressure

C_v Specific heat capacity at constant volume

E Specific internal energy

f Downwinding factor for pressure or density

H Specific stagnation enthalpy

I Streamwise grid point number

$\Delta\ell$ Length of element controlling stability

P   Static pressure

R   Gas Constant = $C_p - C_v$

RF   Relaxation factor

T   Static temperature

Δt Time step

$\underline{V}$ Velocity vector

ΔV Volume of element

α Distribution function for density

ρ Static density

SUBSCRIPTS

x $\left.\begin{array}{c} \\ \\ \\ \end{array}\right\}$

y   In coordinate directions

z

o   Stagnation conditions

*   At sonic condition

INTRODUCTION

Time dependent solutions of the Euler equations are now
widely used for the analysis of the flow through turbomachinery
blade rows. Their main attraction is the ability to compute
mixed subsonic-supersonic flows with automatic capturing of
shock waves. Solutions of the potential flow equation have
also recently been extended to compute transonic shocked flow
(e.g. Farrell and Adamczyk (1981)). Although these can be computationally much more efficient than solutions of the Euler
equations the limitation to potential flow rules them out for
applications where strong shock waves can occur. Solving the
Euler equations is also the most common way of computing fully
3D flow in turbomachinery, even for subsonic flow, since it is
not generally possible to assume irrotational flow.

The equations may be solved in either finite difference or
finite volume form. In the former (e.g. Veuillot (1976),
Gopalakrishnam and Bozzola (1971)) it is usual to transform the
computational domain into a uniform rectangular grid and to
express the derivatives of the flow variables in terms of values
at the nodes of this grid. Specialised numerical techniques
(e.g. McCormack or Lax-Wendroff schemes) are needed to ensure
stability of the integration of the equations through time until
a steady state is reached. In the finite volume form of the

method (e.g. McDonald (1971), Denton (1975)) the equations are
regarded as equations for the conservation of mass, energy and
momentum applied to a set of inter-locking control volumes
formed by a grid in the physical plane. When solved in this
way it is easier to ensure conservation of mass and momentum
than in the differential approach but similar numerical schemes
are necessary to ensure stability.

The argument as to whether finite difference or finite volume
schemes are preferable is not resolved and both types are still
used. The author's (not unbiased) view is that the finite
volume approach is superior because of its simplicity and its
automatic conservation of mass and momentum and also because
of the better physical understanding of the flow development
which is obtained from working in a physical grid. The latter
is an important consideration for design engineers who are not
usually specialists in numerical analysis.

The author's opposed-difference scheme for solving the Euler
equations in finite volume form has been wide used since its
publication (Denton (1975)). The basic philosophy of this
method is to take a very simple and fast first order scheme
and progressively add on a second or higher order correction as
the calculation converges. The resulting method appears to have
advantages of speed and simplicity over alternative second order
schemes and it is also extremely 'robust'. Because of these
factors it is the only method which has been widely used for
3D solutions of the Euler equations through blade rows (e.g.
Denton and Singh (1979), Kopper (1981), Sarathy (1981), Barber
(1981), Singh (1981)). The same basic algorithm has been used
for flow in the meridional plane by Spurr (1980), for unsteady
flow by Mitchell (1980) and for wet steam flow by Bakhtar *et al.*
(1980). As a result a great deal of user experience has been
accumulated (e.g. Bryce and Litchfield (1976)). The general
experience is that satisfactory accuracy can be obtained for
most turbine blades although Singh (1978) shows that in some
cases the inviscid solution is improved by the iterative
addition of a boundary layer calculation. For compressor blades,
however, Calvert and Herbert (1980) show that satisfactory
accuracy cannot usually be achieved without the addition of
boundary layer displacement to the inviscid calculation.

As a result of this experience several defects of the basic
scheme have come to light. They arise mainly from the use of
high order correction factors which, although ideal for smoothly
varying flows, can cause problems at points of discontinuity such
as stagnation points and shock waves. The method also becomes
unstable when the streamwise component of velocity is negative
(i.e. for backflow) and so is unable to deal accurately with
the leading edge flow on blades where the stagnation point
lies on the pressure surface.

It is well known that for steady inviscid flow along a streamline the streamwise momentum equation, the energy conservation
equation and the entropy conservation equation are not independent. Any two of these equations together imply the third
equation. Hence if two of the equations are satisfied in finite
difference form the third equation will also only be satisfied
in finite difference form with inevitably some numerical error.
When solving the Euler equations it is usual to solve the
momentum equations and the energy equation and hence entropy
will not be conserved exactly. Similarly if the momentum
equation is solved and isentropic flow assumed (e.g. McDonald
(1971)) then the stagnation enthalpy will not be conserved.
Since both stagnation enthalpy and entropy are associated with
a particular element of mass, errors in either will be convected
downstream from their source and will influence the whole of the
downstream flow. Finite differencing errors are particularly
likely to occur in the very rapid changes in flow around a
leading edge and these will then influence the flow on the whole
blade surface. Hence it is essential that accurate differencing
schemes and sufficient grid points are used around the leading
edge. At highly loaded leading edges it was found that the
author's scheme could produce changes of entropy (stagnation
pressure) which had an adverse effect on the blade surface
velocities particularly on the pressure surface. This is thought
to be a (unpublicised) characteristic of all time marching
methods.

The method described in this paper has been developed to try
to minimise these difficulties whilst retaining the advantages
of speed and simplicity of the original scheme.

EQUATIONS

The 2D Euler equations may be writen as conservation equations
for a control volume ΔV over a time step Δt to give

$$\text{Continuity} \qquad \Delta \rho = \Sigma_n (\rho \underline{V}.d\underline{A}) \; \Delta t/\Delta V \qquad (1)$$

$$\text{x Momentum} \qquad \Delta(\rho V_x) = \Sigma_n (P \; dA_x + \rho V_x \underline{V}.d\underline{A}) \; \Delta t/\Delta V \qquad (2)$$

$$\text{y Momentum} \qquad \Delta(\rho V_y) = \Sigma_n (P \; dA_y + \rho V_y \underline{V}.d\underline{A}) \; \Delta t/\Delta V \qquad (3)$$

$$\text{Energy} \qquad \Delta(\rho E) = \Sigma_n (\rho H \; \underline{V}.d\underline{A}) \; \Delta t/\Delta V \qquad (4)$$

where $d\underline{A}$ is a vector representing the area of the face of the
element in the direction of the inwards normal to the face and
the summations are over the n faces of the element. These
equations must be solved in conjunction with the perfect gas
relationships.

$$H = C_p T + \tfrac{1}{2} V^2$$

$$E = C_v T + \tfrac{1}{2} V^2 \qquad\qquad (5)$$

$$P = \rho RT$$

In 2D flow it is usual to replace the energy equation by
the assumption H = constant. This assumption is not correct
for an unsteady flow so the true time dependence of the solution
is lost. However, in a steady adiabatic flow with H constant at
inlet H is everywhere constant and so a correct steady state
solution can be obtained without the need to solve the energy
equation. A similar condition of constant rothalpy $(H - \Omega r V_\theta)$

can be used in quasi-3D flow through rotating blade rows but in
fully 3D flow it is generally necessary to solve the energy
equation.

GRID

The finite volume elements used for the new scheme are formed
by the same pitchwise lines and quasi-streamlines as were used
for the original method (Fig. 1.) However, instead of having a
node at the centre of each element nodes are now located at
each of the 4 corners. The fluxes of mass momentum and energy
through each face are then found using averages of the flow
properties stored at the ends of that face. These fluxes may
then be used in the RHS of equations 1—3 to obtain the changes
in ρ, ρV_x, ρV_y for the element in time Δt.

The question now arises as to how these changes should be
distributed between the 4 corners of the element. It is
important to realise that this distribution chiefly affects the
stability and time dependence of the method rather than the
steady solution. A steady solution is obtained when each
node of the grid receives from its surrounding elements changes
which sum to zero. It is found that this usually happens
because the changes within each element are individually small.
That is to say, the sum of the fluxes of each conserved variable
over the faces of each element will be small. Hence, the conservation equations will be closely satisfied for each element,
irrespective of how the changes were distributed. Moreover,
it is easy to show, because of the periodic and solid surface
boundary conditions that global conservation is exact. The
manner of distribution must therefore be chosen to satisfy
stability considerations rather than accuracy. The latter is
only limited by the accuracy with which the flux across a face
can be estimated from an average of the flow properties at its
ends.

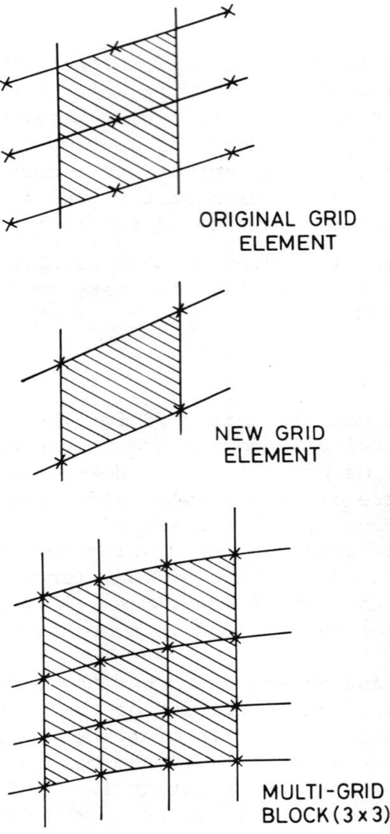

Fig. 1 Grid Systems

DIFFERENCING SCHEMES

A variety of stable distribution schemes have been discovered
for this grid. The exact analogy of the opposed-difference
scheme is to send the changes of all flow quantities to the two
downstream corners of the element and then to let the average
pressure calculated on the downstream face act on the upstream
face when solving the axial momentum equation (Eqn. 2). Like

the basic opposed-difference scheme this method is only of
1st order accuracy unless it is corrected, using a lagged correction factor to correct the downwinded pressure to a value
close to the true one, i.e. the pressure, $P_{A,I}$, acting on face
I (Fig. 1) is taken as

$$P_{A,I} = P_{I+1} + CFP_I \tag{6}$$

where after every time step (or every few steps)

$$CFP_{I,\,NEW} = (1 - RF)\, CFP_{I,\,OLD} + RF(P_I - P_{I+1}) \tag{7}$$

RF is a relaxation factor whose value is typically 0.05. In
the steady state equations 6 and 7 become

$$P_{A,I} = P_I \tag{8}$$

This scheme will be referred to as scheme A, it has exactly
the same stability mechanism as the original opposed difference
scheme, Denton (1975), but is simpler because correction factors
are only needed for pressure and because the correction factors
are not based upon interpolation but on the difference between
pressures stored at two nodes.

A second scheme (scheme B) was discovered as the result of
trying to eliminate the use of correction factors completely.
In scheme A, as in the opposed difference scheme, only pressure
moves upwind, however, at low Mach numbers pressure is very
closely tied to density so a scheme whereby the changes of
density were sent to the upstream corners of the element was
tried, the changes in ρV_x and ρV_y still being sent to the downstream corners. This scheme proved stable, without any correction factors or damping, at low Mach numbers but instability
was found to develop at Mach numbers around unity and above.
Despite this limitation scheme B was found to have one important
advantage, it remained stable for negative values of streamwise
velocity and permitted solutions with a stagnation point on the
pressure surface and reverse flow around the leading edge.
No theoretical explanation of this tolerance of reverse flow has
been found but a physical explanation must be related to the
fact that mass can now be transported upstream by the upwinding
of density. By using an upwinded density to obtain velocity
from ρV it was possible to stabilise this scheme for all cases
where the axial Mach number was subsonic (hence covering most
turbomachinery applications) but this loss of generality removes
one of the main attractions of time marching methods.

A third scheme (scheme C) was discovered whilst trying to combine the advantages of schemes A and B. As in scheme A changes in all variables are sent to the downstream corners of the element but the pressure at any point is now calculated from the density at the next downstream point plus a correction factor i.e.

$$P_I = (\rho_{I+1} + CFRO_I)\ R\ T_I \qquad\qquad (9)$$

where after every time step (or few steps)

$$CFRO_{I_{NEW}} = (1 - RF)\ .\ CFRO_{I_{OLD}} + RF\ (\rho_I - \rho_{I+1})$$

so that in the steady state

$$P_I = \rho_I R\ T_I\ . \qquad\qquad (10)$$

This scheme was found to have good shock capturing properties. It was stable for all Mach numbers (with an appropriate time step) but would not permit reversed flow.

Further attempts to develop a single scheme which would permit both reverse flow and supersonic flow were not successful so a linear combination of schemes B and C was adopted for the final program. The density change obtained for each element is distributed between upstream and downstream corners of the element according to

$$\Delta\rho_U = \alpha\ \Delta\rho$$

$$\Delta\rho_D = (1 - \alpha)\ \Delta\rho \qquad\qquad (11)$$

where α is a function of Mach number such that $\alpha \to 1$ as $M \to O$ and $\alpha \to O$ as $M \to \infty$. The precise form of $\alpha(M)$ is not critical as regards either the stability or the steady state solution. A variation which is discontinuous at $M = 1$ i.e.

$$\alpha = 1 \text{ for } M < 1$$

$$\alpha = O \text{ for } M > 1$$

gives very good shock fitting in 1D flow but the discontinuity caused problems in 2D flow. Because temperature is a variable already calculated in the program α is more conveniently expressed as a function of T than of M and a linear variation of the form

$$\alpha = 0.5 \left(1 + \frac{T-T^*}{T_O - T^*}\right) \tag{12}$$

$$\alpha \geqslant 0$$

was chosen. This expression decreases from 1.0 at stagnation, through 0.5 at a source point, to zero at some supersonic Mach number, above which we choose $\alpha = 0$.

Having chosen the distribution of density change in this way the pressure is calculated as an average, weighted with respect to α, of the pressures from schemes B and C

$$\text{i.e. } P_I = RT_I(\alpha\rho_I + (1 - \alpha)(\rho_{I+1} + CFRO_I)) \tag{13}$$

The distribution function α does not affect the steady state solution significantly and is merely a device to ensure stability of the scheme at both high and low Mach numbers and with reverse flow. As such it is similar but not exactly analogous to the rotated difference schemes used for potential flow calculations. The difference arises because the latter introduce numerical damping which the distribution functon does not.

With the pressure calculated in this way all fluxes and pressures on the faces of the elements are, in the steady state, obtained from central difference formulae using values stored at the corners of the elements. As a result the solution is second order in space for smoothly varying flows but does not have enough numerical damping to capture shocks without overshoots and undershoots being produced. It is known that numerical schemes need an artificial viscosity term proportional to the square of the first derivative of the flow properties in order to model the natural dissipative processes in a shock wave. Hence such a term was explicitly introduced into the equations in the form of an artificial pressure proportional to the square of the density gradient in the streamwise direction. Equation 13 is thereby modified to become

$$P_I = RT_I(\alpha\rho_I + (1 - \alpha)(\rho_{I+1} + CFRO_I)$$

$$+ (\rho_{I+1} - \rho_I)(\rho_{I+1} - \rho_{I-1})/\rho_I) \tag{14}$$

The last term is negligible for smoothly varying flows.

The scheme has been described so far as if it were applied to a 1D flow. In practice a grid such as that shown in Fig. 2 is used for a blade-blade calculation. The flow variables at each node are then updated by half the change calculated for the element below and half that from the one above the node, with the distribution function α being evaluated as an average

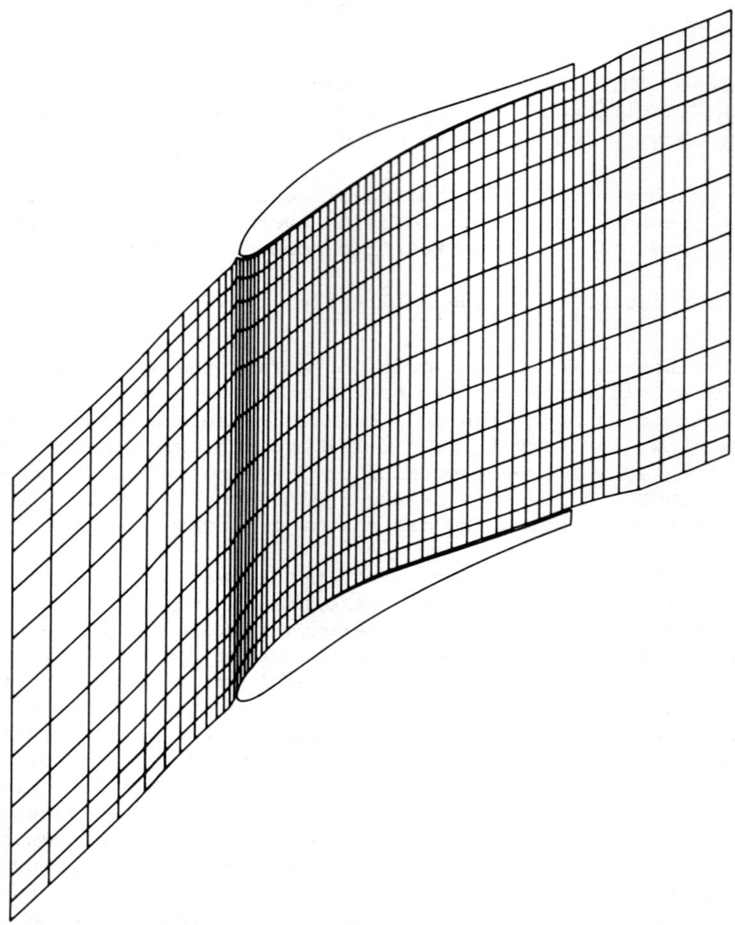

Fig. 2.   Computational Mesh

value for each element. As shown in Fig. 2 the scheme allows
complete flexibility in choosing the spacing of the elements in
both the pitchwise and axial directions and as such is an improvement on the original method where non-uniform spacing, although
possible, caused significant complication. It is also apparent
that when a steady state is reached the conservation equations
are satisfied for each individual element. In the original
scheme the elements overlapped in the pitchwise direction and so,
for the same number of grid points, were of twice the pitchwise
extent. Hence finite differencing in the pitchwise direction is
considerably more accurate on the new grid.

Cusps, which were used at the leading and trailing edges of a blade in the original method, are not strictly necessary with the new grid. However, a cusped trailing edge is felt to be a better approximation to the real viscous flow than is a blunt trailing edge and so is usually used. At a leading edge a large number of grid points are necessary to resolve the flow accurately and a cusp is a useful means of minimising the number used when details of the leading edge flow are not required.

STABILITY

As with all explicit time marching methods the theoretical maximum stable time step is determined by the CFL condition i.e.

$$\Delta t < \frac{\Delta \ell}{C+V}$$

where $\Delta \ell$ is usually the streamwise distance between the upstream and downstream faces of an element.

In practice the stability is found to depend more on the axial Mach number than on the absolute one and the less restrictive condition

$$\Delta t < \frac{\Delta x}{C+V_x}$$

can usually be used. For grids which are very closely spaced in the pitchwise direction the spacing perpendicular to the streamwise lines may become the limit on stability.

As pointed out by Denton and Singh (1979) it is not necessary to take the same physical time step for each element or even for each equation to obtain the correct steady state solution. Equations 1—4 show that as long as the conservation equations are satisfied for every element the solution is independent of the magnitude of Δt. Hence the maximum stable time step can be chosen for each individual element to obtain the fastest convergence to the steady state. This spatial variation of time step permits typically about 30% reduction in computer time but means that the transients of the calculation have no physical significance.

This ability to use variable time steps leads to a very powerful but simple means of controlling stability. If the local time step is made inversely proportional to the local rate of change of the property then incipient instability will immediately (i.e. for the current property in the current time loop) reduce the time step, and hence the actual change produced in the element concerned. This technique is analogous to negative feedback and means that local instabilities do not grow and cause failure of the calculation. As a result the timestep does not need to be reduced below the value which is stable for the converged solution in order to cope with large amplitude initial transients. Typically a 25% reduction in the number of time steps to convergence can be obtained in this way.

BOUNDARY CONDITIONS

   The conditions applied at the inlet and outlet flow boundaries are the same as in most other time marching methods. At the outflow boundary the static pressure is specified and held constant which is physically correct as long as the axial Mach number is subsonic. At the inflow boundary the relative stagnation temperature and stagnation pressure are specified together with either the relative flow direction or the relative whirl velocity (V_y). The latter condition must be used if the relative inflow Mach number is supersonic and the calculation will then automatically satisfy the unique incidence condition.

   The periodicity condition on the bounding streamwise lines upstream and downstream of the blade row is easily satisfied by first treating points on these lines as interior points and then equating values at corresponding points on the two boundaries. This periodicity applied immediately downstream of the trailing edge is found to be sufficient to very closely satisfy the Kutta condition at the trailing edge and no explicit Kutta condition need be applied.

   The boundary conditions on the blade surfaces are the most difficult to satisfy accurately and usually a 1st order boundary condition (e.g. assuming the second derivative in the y direction to be zero on the surface) has to be applied. At first sight it would appear that with the new grid no special condition, other than zero flow through the surface, is needed. The boundary nodes can be updated by the changes calculated for the elements adjacent to the boundary and the only difference between them and internal points is that the latter receive changes from two adjacent elements and the former from only one. However, this treatment ignores the fact that if there are n streamwise lines within the blade row then there are only n-1 elements. Hence the number of equations is less than the number of unknowns and this leads to a condition whereby the sum of the changes between the initial guess and final solution on even numbered nodes is related to the sum of the changes on odd numbered nodes. The final solution is therefore not independent of the initial guess. As a simple illustration of this effect consider a case with only 3 nodes across the pitch. For each

variable the change of the centre node must equal the average
of the changes on the two boundary nodes. Hence if the initial
guess was a linear variation across the pitch the final solution
must also be linear. With more quasi-streamlines the dependence
on the initial guess becomes much weaker but remains an undesirable constraint upon the solution. The solution may be
made unique by applying a boundary condition on one surface,
e.g. by obtaining all flow properties on the pressure surface by
extrapolation (not necessarily linear) from the interior points.
However, a better method is to remove the dependence on the
initial guess by a slight smoothing of the flow properties across
the pitch. By providing a link between changes on odd and even
points the smoothing relaxes the condition that changes on them
must be related and replaces it by a condition that the pitchwise
variation must be in some sense smooth. The mathematical implications of the smoothing have not yet been understood but the
treatment works well and is very simple to apply. In practice
the smoothing factor is dropped to near zero as convergence is
approached.

A further refinement is possible for points on the blade surfaces when the upstream flow can be assumed isentropic. As
mentioned previously the streamwise momentum equation can then
be replaced by the equations for conservation of enthalpy and
entropy. Hence if the flow direction is known, as it is on the
blade surfaces, and if the flow from the inlet to some point on
the blade surface can be assumed reversible and adiabatic (i.e.
H and S constant) the density computed at that point can be used
to obtain the flow velocity via

$$T = T_0 (\rho/\rho_0)^{\gamma-1}$$

$$V = \sqrt{2\ C_p (T_0 - T)} \tag{15}$$

Hence the surface velocity can be found without using the streamwise momentum equation and so without its inevitable finite differencing errors. The assumptions used on the blade surface then
become exactly the same as those used in potential flow methods,
ie., only the continuity equation is solved in finite volume form
and constant enthalpy and entropy are assumed. This treatment
is particularly valuable around a heavily loaded but shock free
leading edge where finite differencing errors are large and the
associated entropy changes can influence the whole of the downstream flow. It may be applied over the whole surface if the
blade is known to be shock free or to have only weak shocks.

MULTIGRID ANALYSIS

The method described so far is marginally faster than the
original method (1.2×10^{-4} sec/point/timestep cf 1.3×10^{-4}
secs) because of the simpler boundary conditions and fewer correction factors and the rate of convergence is similar. Very
large savings in computer time have been obtained for potential
flow methods using the so-called multigrid method (e.g. Jameson
(1979)). The basic philosophy of this approach is to allow information regarding the overall flow pattern to propagate rapidly
on a coarse grid whilst fine detail is resolved on one or more
finer grids. Changes in potential on the coarse grid are interpolated into the finer grid. This philosophy is applicable in
principle to solutions of the Euler equations but to the author's
knowledge has not been previously used.

A coarse grid may be considered as being formed by combining
a group of elements into a block (Fig. 1) of say 3×3 elements.
This block may then be treated as a single large element and
the conservation equations applied to it exactly as for basic
elements. However, the CFL condition applied to the block shows
that much larger stable timesteps can be taken for the block,
e.g. for a 3×3 block 3 times the timestep can be used. The
changes of flow properties for the block can be found either
from summing the fluxes around its faces or, more easily, by
summing the changes already calculated for the elements within
it. Each element within the block is then updated by its own
individual change plus the (much larger) change for the block.
Since the changes involve velocity and density it is not necessary to interpolate the block changes onto the fine grid as it
is when dealing with the velocity potential which has to be subsequently differentiated. It is also apparent that the stability
mechanism of the fine grid (i.e. opposed differencing) carries
over unchanged to the coarse grid so no special techniques are
needed to ensure convergence.

The multigrid method described above is very easy to program
and involves little extra computer work per time step. With
two levels of grid, block sizes of 3×3 elements appear about
optimal and hence permit timesteps which are effectively 3 times
as large as for the fine grid alone. Figure 3 compares the rate
of convergence of the multigrid and single grid methods for the
test case of Fig. 6. Mass balance is a suitable measure of convergence since it is found that the momentum balances converge
at about the same rate. The convergence of the multigrid method
is typically about 4 times faster than a single grid and uses
about one-third the amount of computer time.

No difficulties have been encountered in using the multigrid
method on a large number of test cases. In general the maximum

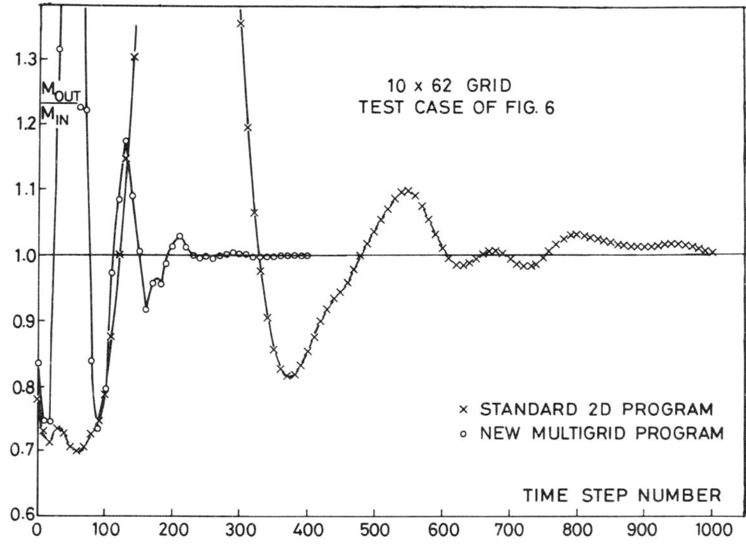

Fig. 3. Convergence

stable timestep is slightly less than for a single grid and it
is preferable to have an integral number of blocks across the
pitch (e.g. 9 elements can be formed into three three-sided
blocks). The method is equally applicable to the original grid
and has been implemented on this in three dimensional calculations. It should be possible to obtain further savings in computer time by using more than two levels of grid but preliminary
attempts to use three levels have so far given no improvement.
In principle the multigrid approach should be applicable to
other numerical schemes for solving the Euler equations the main
requirement being that the stability mechanism for the fine grid
carries over into the coarse grid.

EXAMPLES OF APPLICATION

The method has been programmed for two dimensional and quasi-
three dimensional blade to blade flow, the later permitting
changes of stream tube thickness and radius. Because of the
lack of exact solutions and of experimental data for Q3D flow
all the examples given here are for 2D flow.

Figure 4 shows computed Mach number distributions in a 1D
nozzle designed to produce a linear variation of Mach number
with distance for isentropic flow. Solutions are shown for 3
different back pressures all of which produce shocked flow. In
all cases the shock position and exit Mach number (hence

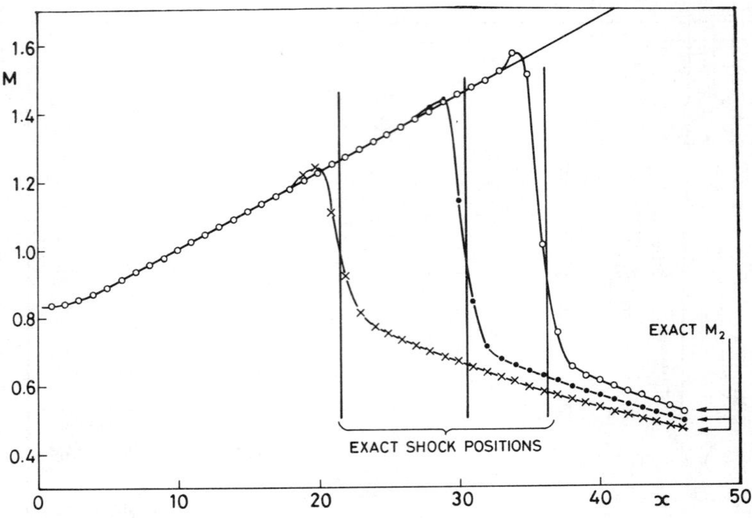

Fig. 4. Shocks in a one-dimensional nozzle

stagnation pressure loss) are predicted almost exactly. The
shock is smeared over 3—4 grid points in all cases with negligible overshoot or undershoot. The ability to capture a wide
range of shock strengths with equal smearing indicates that the
correct form of artificial viscosity (Eqn. 14) has been used.

To illustrate the shock capturing ability of the method for
oblique shocks Fig. 5 shows the solution for a cascade of wedges
with inlet Mach number 2.0 and completely supersonic flow. The
leading edge shock should be exactly cancelled at the upstream
corner giving a uniform flow between the two parallel surfaces
and an expansion off the downstream corner. The computed
results show a sharp leading edge shock which is more highly
smeared on reflection and probably as a result of this does
not meet the adjacent blade exactly at the corner. Consequently
cancellation is not complete and a weak expansion penetrates into
what should be the region of uniform flow. The periodicity
condition is applied downstream of the trailing edge and an
interesting result is the prediction of two oblique shock waves,
typical of those formed at the trailing edge of a turbine blade
with supersonic exit flow. A 13 × 64 point grid was used for
this example and the CPU time was 50 secs on an IBM 370–165.

Figure 6 shows a comparison with experimental measurements on
a transonic turbine blade tested at VKI, results from which are
given in Sieverding (1976). The results are compared at exit
Mach numbers of 1.05 and 1.42. With this type of blade the trailing edge shock from one blade reflects off the suction surface

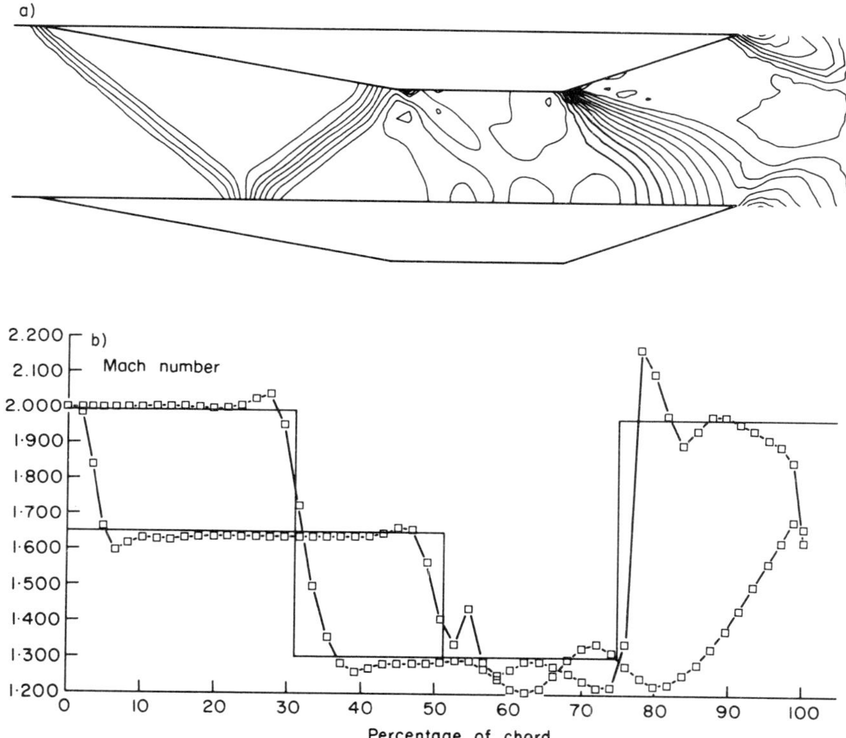

Fig. 5.   Supersonic flow through a cascade of wedges

of the adjacent blade and accurate results cannot be obtained
unless this shock system is modelled correctly. Despite using
a comparatively coarse grid a trailing edge shock system was
discernible in the solution although, as shown by Fig. 6, this
was smeared by the time it reached the suction surface. Better
results have recently been obtained by modelling the base flow
region behind the trailing edge. Figure 7 compares the computed
and measured exit angles from this blade over a range of exit
Mach numbers. The trend of the computed results is correct and
the discrepancy of order 2° is in the direction expected from
viscous effects. This solution took 48 secs CPU time for a
10 × 62 point grid.

Figure 8 compares the computed and exact design solutions for
the Bauer, Garabedian and Korn (1975) shock free supercritical
compressor blade (Fig. 2). This blade has a highly loaded
leading edge which initially caused significant loss of stagnation pressure on the surface streamlines. This was prevented

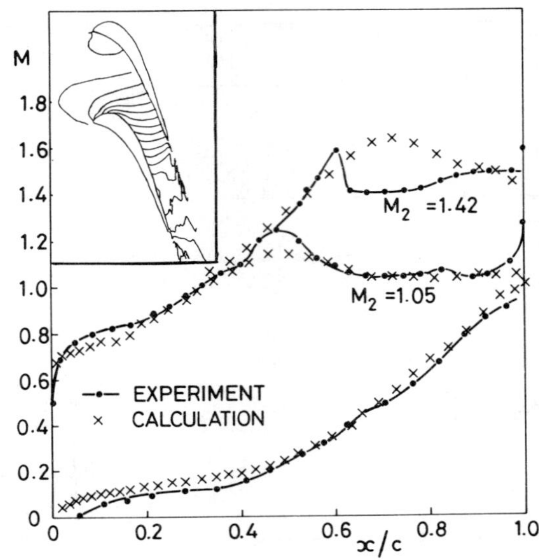

Fig. 6.   VKI rotor blade-surface Mach numbers

by use of the isentropic surface boundary condition which lead
to the good agreement shown in the Figure.   For this case 65
secs CPU time were needed on a 13 × 60 point grid.

Perhaps the most important practical application of time

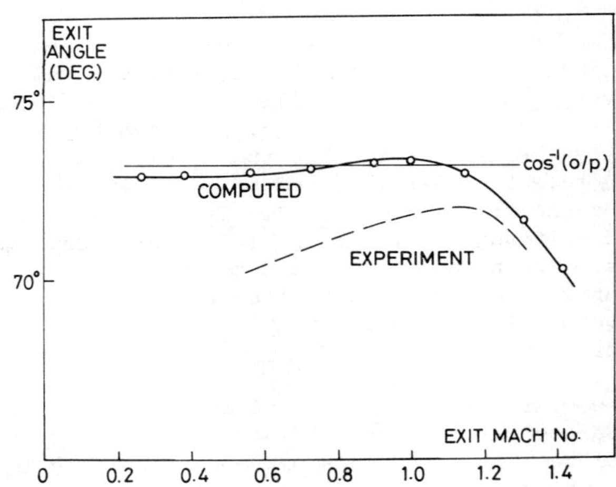

Fig. 7.   Exit flow angle from VKI rotor blade

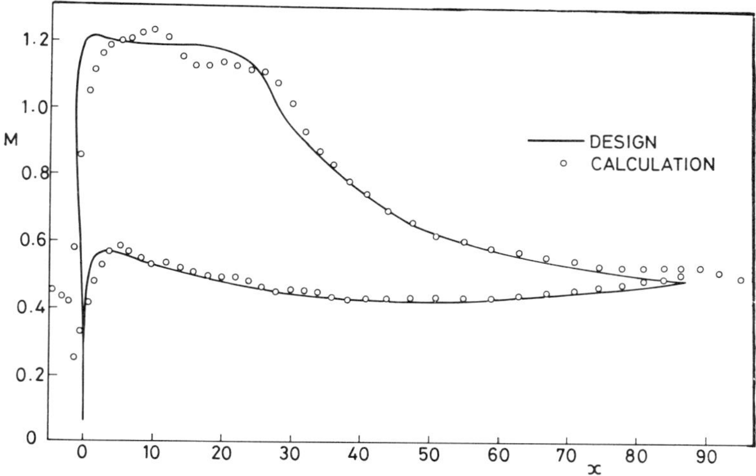

Fig. 8. Supercritical compressor cascade.

marching methods is the calculation of flow through supersonic
fan blades where the strong shocks present make all other methods
inapplicable. The flow through such blades is strongly influenced
by viscous effects, particularly downstream of the shock, so
comparison of inviscid calculations with test data is not realistic. Figure 9 illustrates the ability of the method to compute
such flows for a typical fan blade with an inlet Mach number of
1.4 and a strong shock attached at the leading edge. The shock
is much cleaner than would be obtained from the original method
despite the use of a fairly coarse 10 × 50 point grid. CPU time
for this solution was 29 secs.

DISCUSSION AND CONCLUSIONS

   Many more test cases than those presented here have been calculated with the new method and there is no doubt that the method
is faster and more accurate than its predecessor. The improved
accuracy comes about mainly because interpolation is no longer
needed to evaluate the correction factors. Such interpolations
can increase accuracy for a smooth flow but experience is that
they cause problems at shock waves and at leading edges. The
isentropic surface boundary condition is an important aid to
improving accuracy at highly loaded leading edges. The only
non-second order term in the new method is the numerical damping
which is under the direct control of the user and which has
negligible effect except at shock waves. Hence it seems likely
that accuracy can only be further improved by using more complex higher order schemes or by using more grid points.

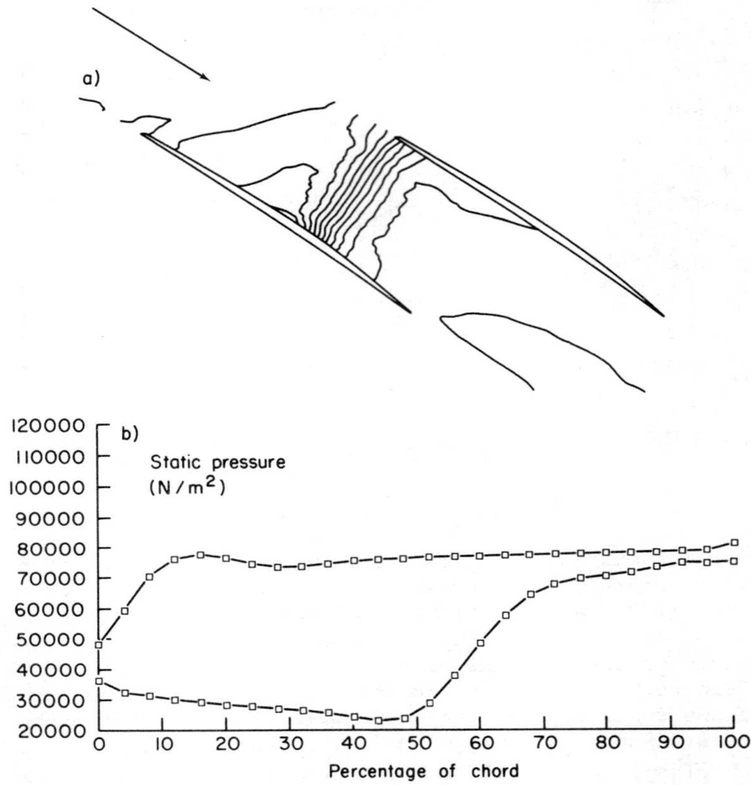

Fig. 9.   Transonic fan blade.   (a) Mach number contours (b) Surface static pressure distribition.

For 2D calculations the number of grid points is unlikely to
be limited by computer storage and the main limitation is likely
to be the consequent increase in CPU time. Use of the multigrid
method allows more grid points to be used for the same CPU
time (about twice as many) and so permits substantially greater
accuracy for the same cost. With the present method a large
modern computer could obtain solutions for a grid of 5000 or
more points in few minutes CPU time. Such a large number of
points should permit very detailed resolution of the flows
around highly loaded leading edges and around supersonic trailing
edges which are the main problem areas with previous calculations.
It is considered that the multi-grid method has potential to provide further increases in speed by the use of more levels of grid
and this deserves further investigation.

The new grid geometry has recently been extended to 3D flows.
However, difficulty was found in solving the energy equation (as

is necessary in 3D) using schemes B and C and so it proved necessary to use scheme A, which does not permit reverse flow, for 3D calculations.

ACKNOWLEDGEMENTS

The original idea of using a grid with nodes only at the corners of the elements was obtained from Dr R. Ni of Pratt and Whitney Aircraft.

The wedge case test of Fig. 5 was provided by Brown Boveri & Co., Baden, Switzerland.

REFERENCES

Bakhtar, F., and Tochai, M. (1980) 'An Investigation of 2D Flows of Nucleating and Wet Steam by the Time-Marching Method', *International Journal Heat and Fluid Flow*, 2.1.

Barber, T.J. (1981) 'Analysis of Shearing Internal Flows', AIAA Paper 81-0005.

Bauer, F., Garabedian, P., Korn, D., and Jameson, A. (1975) 'Supercritical Wing Sections II', Lecture Notes in Economics and Mathematical Systems, Vol. 108, Springer Verlag.

Bryce, J.D., and Litchfield, M., (1976) 'Experience of the Denton Blade-Blade Time Marching Programs', NGTE Note 1050.

Calvert, W.J., and Herbert, M.V. (1980) 'An Inviscid-Viscous Interaction Method to Predict the Blade-Blade Performance of Axial Compressors', NGTE Memo 80012.

Denton, J.D. (1975) 'A Time Marching Method for Two and Three Dimensional Blade to Blade Flow', ARC R. & M. 3775.

Denton, J.D., and Singh, U.K. (1979) 'Time Marching Methods for Turbomachinery Flow Calculation', VKI Lecture Series on Transonic Flows for Turbomachinery.

Farrell, C., and Adamczyk, J. (1981) 'Full Potential Solution of Transonic Quasi-3D Flow Through a Cascade using Artificial Compressibility', ASME paper 81-GT-70.

Gopalakrishnan, S., Bozzola, R. (1971) 'A Numerical Technique for Calculation of Transonic Flows in Turbomachinery', ASME paper 71-GT-42.

Jameson, A. (1979) 'Acceleration of Transonic Potential Flow Calculations on Arbitrary Meshes by the Multiple Grid Method',

AIAA paper 79-1458.

Kopper, F., Milano, R., and Vanco., M. (1981) 'An Experimental
Investigation of Endwall Profiling in a Turbine Vane Cascade',
AIAA Journal. 19, 8.

McDonald, P.W. (1971) 'The Computation of Transonic Flow Through
Two-Dimensional Gas Turbine Cascades', ASME paper 71-GT-89.

Mitchell, N. (1980) 'A Time Marching Method for Unsteady 2D
Flow in a Blade Passage', *International Journal Heat and Fluid
Flow*, 2.4.

Sarathy, K.P. (1981) 'Computation of Three-Dimensional Flow
Fields Through Rotationg Blade Rows and Comparison with Experiment', ASMI paper 81-GT-121.

Sieverding, C. (1976) 'Base Pressure in Supersonic Flow', VKI
Lecture Series on Transonic Flows in Turbomachinery.

Singh, U.K. (1978) 'Computation of Transonic Flow in Cascade
with Shock and Boundary Layer Interaction', *Proceedings 1st
Int. Conf. Num. Meth.* in Laminar and Turbulent Flow, Swansea.

Singh, U.K. (1981) 'A computation and comparison with Measurements of Transonic Flow in an Axial Compressor Stage with Shock
and Boundary Layer Interaction', ASME paper 81-Gr/GT-5.

Spurr, A. (1980) 'The Prediction of 3D Transonic Flow in Turbomachinery Using a Combined Throughflow and Blade-to-Blade Time
Marching Method'. *International Journal Heat and Fluid Flow*.

Veuillot, J.P. (1976) 'Calculation of Quasi-3D flow in a Turbo-
machine Blade Row', ASME paper 76-GT-56.

NUMERICAL MODELLING OF SHOCKWAVES AND OTHER DISCONTINUITIES

P. L. Roe

*(Fluid Mechanics Division,
Royal Aircraft Establishment, Bedford, UK)*

ABSTRACT

Numerical implications of the distinction between equilibrium
and evolutionary problems are considered. The simple example of
scalar advection shows that for evolutionary problems accuracy
is gained by giving the algorithm a bias towards the direction
of propagation. Methods are described which apply appropriate
bias to complex problems where many interacting effects may be
present. A computer program is described which applies one of
these methods to solve the Euler equations, and an application
to transonic flow over a plane aerofoil is given.

1. INTRODUCTION: EVOLUTIONARY AND EQUILIBRIUM PROBLEMS

In one of his Essays, the sixteenth-century philosopher
Francis Bacon wrote that "as in nature things move violently to
their place, and calmly in their place, so virtue in ambition is
violent, and in authority settled and calm." To make his poli-
tical analogy, he drew strikingly on the distinction between
what we would nowadays describe in technical language as
problems of evolution, and problems of equilibrium. The two types
are exemplified most simply by Laplace's equation

$$\phi_{xx} + \phi_{yy} = 0 \qquad (1.1)$$

representing the calm movement of things in their place, and by
the wave equation

$$\phi_{xx} - \phi_{yy} = 0 \qquad (1.2)$$

representing the violent movement of things to their place.

For Laplace's equation, and other equations which share its

elliptic nature, we know that solutions will be smooth (infinitely differentiable) unless we actually impose a singularity as a
boundary condition and even then the solution elsewhere will
remain smooth. Also we know that the solution at any point
depends on all the boundary conditions. For the wave equation,
by contrast, singularities will propagate away from the boundaries along characteristics, and even worse, will appear spon-
taneously away from the boundaries, if non-linear terms are added
to the equation. Moreover, the solution at any point depends only
on part of the data, and is independent of the rest.

From the beginning, numerical analysts have of course been
aware that these distinctions must be reflected somehow in the
numerical methods. The great pioneer L. F. Richardson wrote of
equilibrium problems as "jury problems" which had to be decided
by the unanimous consent of all participants. Evolution problems
he proposed to solve by what he called "marching" methods,
setting out from that part of the boundary containing the relevant data and proceeding step by step into the interior. However,
the full numerical implications of the distinction are by no
means clear, even today. A point which the early workers failed
to grasp can be illustrated by quoting Richardson again: "Finite
differences have, in themselves, but little importance to the
student of matter and ether. They are here regarded simply as a
makeshift for infinitesimals; and the understanding is always
that we will eventually make the differences so small that the
errors due to their finite size will be less than the errors of
experiment or practical working, and may therefore be dis-
regarded" (Richardson, 1910). Clearly, Richardson thought of the
physical problem as being modelled by a mathematical problem
expressible in the language of applied calculus. In turn the
mathematical problem was to be modelled by an arithmetical
problem, and it was rather casually taken for granted that the
arithmetical solution would converge to the analytic solution.

The deficiencies of this viewpoint were rather slow to be
appreciated, chiefly because the early workers concentrated on
equilibrium problems. The smoothness of the solutions helped to
ensure convergence, and the central differencing schemes, which
were adopted with a view to accuracy, were also physically
appropriate because they allowed information to propagate equally
in all directions. One might say that the calm and settled nature
of things moving in their place was well represented, both by the
mathematical model and also by its arithmetical replacement.

Matters are quite otherwise, however, with the violent motion
of things to their place. Difficulties arise from many causes.
In their famous paper, Courant, Friedrichs and Lewy (1928)
showed that explicit marching methods cannot advance more quickly
than a certain limiting rate. The cause of the instability is

that the numerical process does not have access to all the relevant boundary conditions. The converse of this leads to what we
may call the paradoxical form of the Courant-Friedrichs-Lewy
(CFL) condition; the fact that irrelevant information is necessary to ensure stability. Similarly, we can only achieve a high
degree of formal accuracy by using large amounts of data, some
of which is irrelevant, and whose use can only be justified if
the data is very smooth. Indeed, there is a theorem due to
Godunov (1959), to the effect that formal accuracy which is
second-order or better can only be obtained if we are prepared
to tolerate spurious oscillations in regions where the solution
is not smooth. Such oscillations, aptly known as 'wiggles', seem
to be the inevitable consequence of applying naïve numerical
methods to problems of violent evolution. The past decade has
seen great effort devoted to understanding and alleviating the
problem. A full bibliography would certainly list hundreds of
publications, ranging from fundamental mathematics to *ad hoc*
programming devices for specific applications.

This article is not a comprehensive review, but concentrates
on a particular line of approach which seems to be favoured by an
increasing number of workers, many of whom have aeronautical
applications in mind. I have attempted to extract a common philosophy from a variety of sources, whilst also highlighting significant differences of detail. The core of the philosophy is that
the wave motions which characterise evolutionary problems are
essentially directional; 'information', in some sense, is carried
in various directions by various mechanisms. If only one mechanism is involved, the best numerical methods appear to have a
corresponding bias. We illustrate this in Section 2, where we
discuss the seemingly simple problem of linear advection from a
numerical point of view.

If more than one mechanism is present, and no single mechanism
is clearly dominant, the first task seems to be to disentangle
the flow of information, so that each component can be suitably
directed. This amounts to analysing the different physical effects
which may be present. Several ways of doing this, for an evolution problem in one space dimension, are discussed in Sections 3
and 4. In Section 5 we show that this directional analysis allows
us to achieve high accuracy without introducing wiggles. In
Section 6 we describe how these concepts have been put to use
within the Royal Aircraft Establishment, to create a program which
computes two-dimensional inviscid solutions of the Euler equations, with cleanly and accurately captured shockwaves.

2.   SCALAR ADVECTION AND UPWIND BIAS

We will discuss the problem

$$u_t + au_x = 0 \tag{2.1}$$

where u is a scalar unknown, and a is a constant. It is quite
remarkable that so much remains to be said, and indeed discovered,
about this simple equation. An analyst would dismiss it with the
general solution, thus. If

$$u(x,0) = f(x) \qquad (2.2)$$

then
$$u(x,t) = f(x-at) \qquad (2.3)$$

which clearly satisfies (2.1) if f is any differentiable function.
If f is not differentiable, then (2.3) is said to be a weak solution of (2.1). The reader may consult Lax (1954) for a full discussion of weak solutions to more general equations.

Interesting questions arise if we insist on solving (2.1) the
hard way (numerically on a discrete grid) say at points $(x_i,t_n)=$
$(x_0 + i\Delta x, t_0 + n\Delta t)$, obtaining approximate values of
$u(x_i,t_n) = u_i^n$. A survey of this topic has been given by Roe
(1981c). Of course the numerical problem is, because of (2.3),
equivalent to simple interpolation. We can write

$$u(x_i,t_n) = u(x_i - a\Delta t, t_{n-1}) \qquad (2.4)$$

where $x_i - a\Delta t$ will not in general coincide with a data point.
Any interpolation formula of the form

$$u(x_i - a\Delta t, t_{n-1}) = \sum_{k=-r}^{k=s} c_k u(x_{i+k},t_{n-1}) \qquad (2.5)$$

where $\{c_k\}$ are a set of weighting coefficients will solve our
problem. There will be some truncation error, of course, defined
most simply as the difference between the numerical solution and
the exact solution. The general form of the truncation error
comes out as

$$E = \Delta x^p \cdot g(\nu) \frac{\partial^p u}{\partial x^p} \qquad (2.6)$$

where p is the formal order of accuracy of the process, g is some
function of the Courant number $a\Delta x/\Delta t$, a function which depends
on the particular algorithm, and the derivative term is to be
evaluated at some unknown point in some smooth interpolation of
the data.

A classical numerical analyst might observe, as did
Richardson in the passage quoted earlier, that the error can be
made as small as we please merely by choosing p large enough, or
Δx small enough. As a refinement we may choose the algorithm
which minimises some measure of the function g. These observations are usually adequate for the treatment of equilibrium

problems, but for the non-smooth solutions which arise in evolutionary problems, the derivative term becomes large, and the formal accuracy is made meaningless. Under these circumstances it seems preferable to concentrate on *qualitative* features of the numerical solution, and it is at this point that the subject begins to drift away from the secure harbour of mathematics. One can readily propose what seem to be useful criteria, formulate these mathematically, and begin proving theorems. The problem is to know whether the most sensible criteria have been chosen. This kind of work is therefore characterized by a good deal of numerical experiment, and by spirited public debate.

An account of what has actually been proved can be organised around a recent result due to Iserles (1981). We assume that a is positive, so that information is being convected from left to right. In equation (2.5), the summation is over the range $[-r,s]$, so that in the numerical solution r points act in what one might intuitively suppose to be a correct way, and s points act to send information in the contrary direction. Now, by considering the degrees of freedom involved, it is obvious that the order of accuracy (p) of the algorithm must satisfy

$$p \leqslant r + s . \tag{2.7}$$

Iserles has shown that if the algorithm is to be *stable for small time steps,* then additionally

$$p \leqslant 2 \min (r, s+1) \tag{2.8}$$

	r=0	1	2	3	4
4	0	2	4	6	8
3	0	2	4	6	7
2	0	2	4	5	6
1	0	2	3	4	4
s=0	0	1	2	2	2

Fig. 1   Maximum order of accuracy (p) obtainable for linear advection problem using r upwind data points and s downwind data points.

Fig. 1 shows the maximum accuracy achievable by an algorithm with given r, s when these inequalities are combined. The effect of taking stability into account is to reduce the accuracy below the simple estimate (2.7) for all algorithms except those lying on the three marked diagonals, i.e. those for which r = s, s + 1, or s + 2. This confirms our expectation that the flow of numerical information should be biassed in the proper direction, but also shows that our options for implementing that strategy are

quite restricted. It is sometimes argued that the flow of information must be wholly in the true direction, that is to say we
must choose s = 0. If this should be necessary, it would imply
$p \leqslant 2$, a surprisingly severe restriction. (This special case of
Iserles' result was obtained by Engquist and Osher (1980)).

Further insight is obtained by considering more informative
measures of the error. For example, one can consider separately
the phase errors and amplitude errors for each Fourier component
of the solution. It then emerges that there is a fundamental distinction between algorithms for which p is odd or even. A proof
is given in Roe (1981c). When p is odd the leading error terms
in both phase and amplitude are of order θ^{p+1}, where θ is the
Fourier angle. However, when p is even the errors in phase and
amplitude are respectively of order θ^p, θ^{p+2}. Graphs which display
these errors, as functions of θ, for various Courant numbers, and
for algorithms having r = s and r = s+1, are given by Davies
(1980). Davies observes that the sequence of algorithms for which
r = s+1 and p is odd appear to converge (to the analytical solution, as p increases) much faster than the algorithms having
r = s and p even.

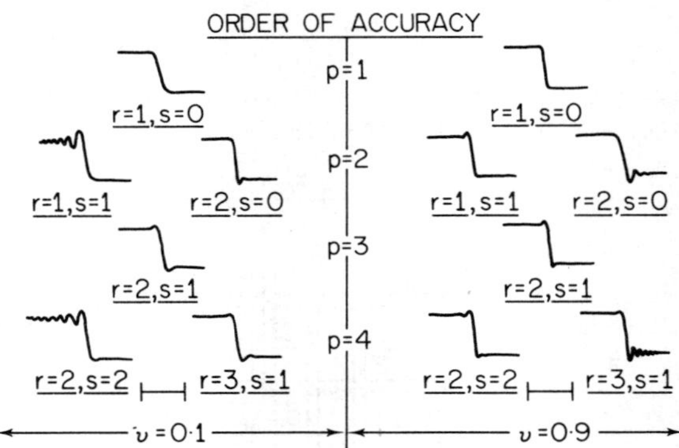

Fig. 2   Numerical results for the linear advection of step data

The results of numerical experiments using step function data
(so that the problem somewhat resembles a shockwave problem) are
shown in Fig. 2. The width of the initial discontinuity spreads
like $t^{1/(p+1)}$. (Harten (1977)).   For the choice of ordinate
employed in Fig. 2 a sufficient quantity of numerical data eventually forms a continuous curve. Results are shown for the first-
order algorithm (r = 1, s = 0), for the second-order algorithms
(r = 1, s = 1) and (r = 2, s = 0), for the third-order algorithm
(r = 2, s = 1) and the fourth-order algorithms (r = 2, s = 2)
and (r = 3, s = 1). For the even-order methods, the consequences
of having imbalanced phase and amplitude errors are apparent. The
step function data implies the existence of waves at all frequen-

cies. The higher frequencies propagate at the wrong speed, but
are only lightly damped. This combination of errors is most
unfortunate, because it leads to the spread of wiggles which
contaminate the solution over a wide area.

From these studies, the algorithms with $r = s+1$ begin to emerge
as a privileged class. The important special case $r = 1$, $s = 0$,
is the simplest example of upwind differencing; when some authors
use the term this is the only algorithm they have in mind. It was
shown by Godunov (1959) to be the algorithm with least truncation
error among all algorithms of the form (2.5) which *preserve monotonicity*. That is, if $u(i\Delta x)$ is a monotonically increasing
(decreasing) function of i at some time level n, it remains so
for all subsequent n. We shall refer to this important result as
Godunov's Theorem. Note the implication: there are no monotonicity preserving algorithms of the form (2.5) for which $p > 1$.

The third-order algorithm $r = 2$, $s = 1$ has been studied by
several authors, e.g. Warming, Kutler, and Lomax (1973). It was
enthusiastically recommended by Leonard (1979), on heuristic
grounds, and as a result of extensive application to hydraulic
modelling.

The outcome of this section can be summarised by saying that
if we want to solve (2.1) numerically by formulae of the type
(2.5), then it appears to be a good thing to bias the algorithm
by choosing $r > s$, but that the restriction to $s = 0$ seems
excessively severe. There seems to be a suggestion that $r = s + 1$
is optimal.

Let us suppose, then, that a *prima facie* case has been made
for acknowledging in the numerical treatment of an evolution
problem the direction in which information travels. This approach
is widely referred to as 'upwind differencing', and we will use
this term from now on, despite some misgivings about its approp-
riateness. (For example, the mechanism which transfers the information may well be a wave rather than a convection current). We
now proceed to apply this concept to systems of equations.

3. UPWIND DIFFERENCING WITHOUT CONSERVATION

The earliest, and perhaps in principle still the best, example
of upwind differencing is the classical method of characteristics.
Information, in the form of Riemann invariants, propagates along
the characteristic paths in a way which clearly is a correct
reflection of the true physical situation. Despite this, the
method of characteristics has not found wide favour for automatic
computation. The programming logic is tricky, if many shockwaves
are present, although Hoskin and Lambourne (1970) present some
useful ideas.

   The basic idea of the method of characteristic has, however,
been incorporated into a number of so-called "hybrid" schemes,
most of which descend from Courant, Isaacson and Rees (1952).
The 'CIR' method is illustrated in Fig. 3. We assume that data is
available on a line t = 0 and the solution is required at t = Δt.
Since we know the slopes of characteristics at A we can estimate

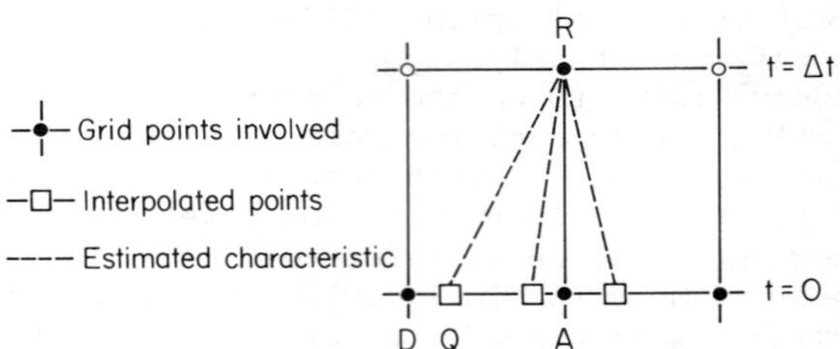

Fig. 3   The Courant, Isaacson, and Rees, hybrid method of
                         characteristics

where the characteristics through R would intersect t = 0, and
find conditions at Q, for example, by linear interpolation in
the interval AD which contains Q. We then employ the method of
characteristics on this interpolated data. It is clear that we
are in effect applying our upwind differencing scheme (r = 0,
s = 1) to the characteristic form of the equations. At Salford
University, Walkden (1966, 1972, 1978) has made extensive use of
such methods to compute flows of aeronautical interest. He uses
a multidimensional version (1978) to compute steady, supersonic,
three-dimensional flow. It is worth remarking that any multi-
dimensional method based on the theory of characteristics must
involve interpolation amongst the data, so that all such methods
are, in effect, hybrid. See, for example, Pandolfi and Zanetti
(1980).

   By using more sophisticated interpolation methods, but still
retaining some upwind bias, hybrid methods can be constructed
which are better than first order accurate. Recent publications
along these lines include Carver (1980), Moretti (1979), and
Chakravarthy et al. (1980). It is a strong attraction of such
methods that because of their physically realistic basis they
are able to incorporate boundary conditions in a simple and satisfying way, as described by Moretti and de Neef (1980). When these
hybrid methods are applied to flows involving shock-waves, they
tend to produce results which feature only relatively mild wiggles,
but the location of the shockwave transition may sometimes be
imprecise, because the conservation laws have not been strictly
enforced. A combination of hybrid characteristic differencing
with shock fitting would avoid this difficulty, and is undoubtedly one of the most attractive options open at the present time.
There is, however, another possible way forward, and since this

is the one in which I have worked personally, it is the one which
I shall develop in more detail. It consists of combining the concepts of hybrid differencing with the strict enforcement of conservation.

4. UPWIND DIFFERENCING WITH CONSERVATION

4.1 *A definition of conservation*

We begin this section by defining a conservative differencing
scheme. This definition is a little different from the usual
definition, due to Lax (1954), although it can easily be shown
equivalent to it for evenly spaced grids in one space dimension.
The advantage of the present definition is that it generalises
readily to irregular, multidimensional grids. Our exposition will
be in terms of an irregular one-dimensional mesh. Recall that a
one-dimensional conservation law is of the form,

$$\frac{\partial u}{\partial t} + \frac{\partial F}{\partial x} = 0 \qquad\qquad (4.1.1)$$

where u is often said to be a "conserved quantity". It is, however,
more correct and more helpful, to call it a density of a conserved quantity. When we write the Euler equations in one, two or
three dimensions we deal with the densities (per unit length,
area, or volume) of mass, momentum and energy. The quantity F has
the dimensions of a density times a velocity, that is, quantity
per unit time. It is called the flux function, and represents the
flow (past a point, over a line, or through a surface) of the
conserved quantities. In addition it represents those gradient
terms, such as pressure, which redistribute momentum and energy.
Now suppose that we have data for a one-dimensional problem as
sketched below

$$
\begin{array}{ccccc}
F_{i-2} & F_{i-1} & F_i & F_{i+1} & F_{i+2} \\
\hline
x_{i-2} & x_{i-1} & x_i & x_{i+1} & x_{i+2}
\end{array}
\quad .
$$

Consider the quantity

$$\Delta m_{i+\frac{1}{2}} = -\Delta t \Delta F_{i+\frac{1}{2}} = \Delta t (F_i - F_{i+1})$$

where Δt is a small time interval. Dimensionally this is a density
times a length; it is the actual mass (or momentum, or energy)
which would accumulate inside the interval (i, i+1) in time Δt.
We define our computing scheme to be conservative, if this mass
increment, having been subdivided in any arbitrary way, is then
distributed over the various nodes of the grid without any residue remaining unaccounted for. When we have done this for all
intervals the total allocation of mass increments will be

$\Delta t [\ldots + (F_{i-1} - F_i) + (F_i - F_{i+1}) + (F_{i+1} - F_{i+2}) + \ldots]$ which is, as it should be, zero except for the fluxes through the boundaries. Therefore the system has gained or lost precisely the correct amount of mass. We finally recover the densities at each node by dividing the total mass which now resides there by some measure S_i of the space surrounding that node. The sum of these measures should be the total length (area, volume) of the computational domain. This completes one time step of the process, and clearly the total mass present in the system $(\Sigma m_i = \Sigma S_i u_i)$ will only be affected by events on the boundaries.

4.2 *Directing the fluxes*

In the above definition of conservation we allowed ourselves to distribute each mass increment in a quite arbitrary way. We shall now try to distribute it in a physically realistic way. To do this, we need to ascribe to each observed flux difference $\Delta F_{i+\frac{1}{2}}$ a physical explanation. This is possible if we collect together all the flux differences (mass, momentum, energy) which we observe in a given interval, and form a vector-valued flux difference $\Delta \underline{F}_{i+\frac{1}{2}}$. Each of the physical effects which we might suppose to be present (waves of various kinds) will give rise to a distinctive type of vector (see Appendix). Analysing the part-icular vector which we have observed then tells us which physical effects are actually present. Basically we have to do an eigen-vector analysis, complicated by non-linear effects.

A very active area of current research is concerned with the details of this analysis, taking account of such considerations as accuracy, speed, and reliability. The various proposals have been put forward with rather different explanations, but they can, in fact, almost all be represented by rather similar wave diagrams, and it is by this means that we present them now. In this way we impose some unity on the proposed methods and make their properties easy to compare. However, to keep the wave dia-grams simple and easily visualised, we restrict ourselves to a problem with only two dependent variables. The simplified Euler equations derived with $\gamma = 1$ in the Appendix involve only density and velocity. The set of states (ρ, v) which can be connected to a given state (ρ_1, v_1) by a simple wave is the set which satisfies

$$\rho = \rho_1 \exp (\pm (v - v_1)/a) \qquad (4.2.1)$$

and the set of states which can be connected to (ρ_1, v_1) by a shockwave is the set which satisfies

$$\pm (\rho \rho_1)^{\frac{1}{2}} (v - v_1) = a (\rho - \rho_1) \qquad (4.2.2)$$

and for which ρ is greater than ρ_1.

We will draw wave diagrams which trace these loci in a space
$(\rho v, \rho)$, i.e. in u-space. For a general account of the way wave
diagrams can be used to solve interaction problems, see Courant
and Friedrichs (1948).

4.2.1 Gudunov's Method   Gudunov (1959) supposed that if the
fluid were in a state $\underline{u}_L$ at the left-hand end of some interval
(x_i, x_{i+1}), and in a state $\underline{u}_R$ at the right-hand end, then the
whole of the change took place discontinuously at the mid-point
of that interval. Thus he was led to consider the classical
Riemann problem (see Lax (1972)), i.e. the interaction of two
adjacent constant states. The graphical solution of the Riemann
problem is achieved by drawing the C^--characteristic through
state $\underline{u}_L$, and the C^+-characteristic through state $\underline{u}_R$, to find
their intersection F. (See Fig. 4a). If either LF or RF turns out
to represent a compression wave, as does RF in Fig. 4a, it is
deleted and replaced with the shock polar originating at L or R,
giving a new position for F. Even in the simple case $\gamma = 1$, the
algebraic solution must be done by iteration if we have one shock
and one expansion wave.

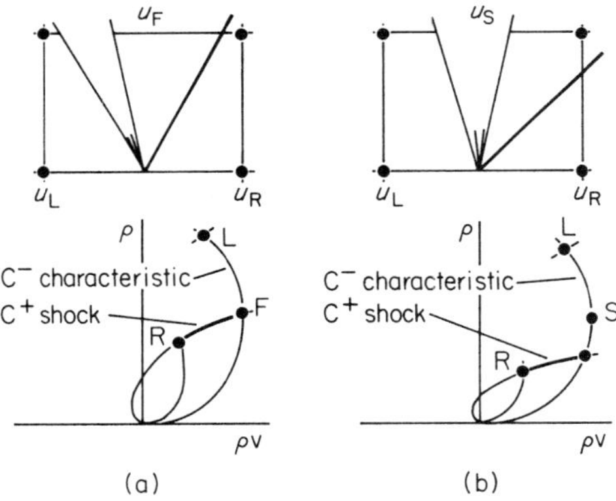

Fig. 4   Directing the fluxes according to Godunov's method.
Case (a) is a C^- expansion and a C^+ shock. Case (b)
is the same, but the expansion wave spans the time
axis, and therefore contains a sonic point. The upper
diagrams show part of the computational (x,t) space,
and the lower diagrams show the wave interactions in
u-space.

If it turns out that both waves are left-running, then the
whole of the flux difference $(\underline{F}(\underline{u}_L) - \underline{F}(\underline{u}_R))$ is assigned to x_i
(or to x_{i+1} if both waves are right-running). It is then multi-
plied by $\Delta t/\Delta x$ and used to increment $\underline{u}_i$ (resp u_{i+1}). If one wave
runs in each direction, then $\underline{F}(\underline{u}_L) - \underline{F}(\underline{u}_F)$ is assigned to x_i,
and $\underline{F}(\underline{u}_F) - \underline{F}(\underline{u}_R)$ to x_{i+1}. The last, but important, special case

is where one wave runs partly in each direction, i.e. it is an
expansion fan which spans the time axis (see Fig. 4b). The time
axis is therefore a sonic line, represented by a sonic point $\underline{u}_S$
in the wave diagram. We assign $\underline{F}(\underline{u}_L) - \underline{F}(\underline{u}_S)$ to x_i, and $\underline{F}(\underline{u}_S) - \underline{F}(\underline{u}_R)$ to x_{i+1}.

4.2.2 Osher's Method   Osher and Solomon (1982) report a scheme
which bears a rather curious relationship to Gudunov's method.
Instead of solving the Riemann problem, which asks how the states
$\underline{u}_L$, $\underline{u}_R$ will interact in the future, they ask instead what past
history of the flow could have brought about a discontinuous
junction between $\underline{u}_L$ and $\underline{u}_R$. We might call this the backward
Riemann problem, although the question is not, as it stands, well
posed. For example, if $\underline{u}_L$, $\underline{u}_R$ are separated by a shockwave the
explanation could be that the shockwave has always been there.
Alternatively, it might have been just formed by a coalescing
compression wave (together with, in the case of the full 3 x 3
Euler equations, a contact discontinuity carrying the necessary
entropy jump). Osher and Solomon cleverly exploit this ambiguity
by always choosing to construct their solution by means of wave
paths, and never shock paths. It then turns out not only that
the solution to this "restricted backward Riemann problem" is
unique, but that its solution for a polytropic gas emerges in
closed form, so eliminating the iterations needed by Gudunov's
method. Since the interest this time is in analysing the hypothet-
ical history of the flow, rather than at once attempting to
predict its future, the wave diagram has to be drawn rather
differently. The path to be followed away from L is the C^+ rather
than the C^- characteristic. We seek the intersection of this path
with the C^- characteristic through R. In Fig. 5a, we carry out
this construction for the same data as we used in Fig. 4a to illustrate Godunov's method. The subarcs of the path from L to R are
again interpreted as instructions to direct the fluxes, but now
$\underline{F}(\underline{u}_L) - \underline{F}(\underline{u}_P)$ is assigned to x_{i+1}, and $\underline{F}(\underline{u}_P) - \underline{F}(\underline{u}_R)$ to x_i. It is
easy to see that if $\underline{u}_L$, $\underline{u}_R$ are not widely separated, Osher's
method and Godunov's method are equivalent, because then LPRF will
be a parallelogram.

The interesting distinctions arise as this parallelogram
becomes increasingly distorted. An important special case is that
where $\underline{u}_L$, $\underline{u}_R$ can be connected by a stationary shock. Osher's
method does not, of course, connect them in this way. The path
which it actually follows is shown in Fig. 5b. It consists of a
short arc of the C^+ characteristic LP and a long arc of C^- characteristic PSR. The mapping of this wave diagram onto physical
space is shown in the top part of Fig. 5b. As an interpretation of
past events this is perfectly reasonable, but for future times
the physical space becomes folded and cannot be given any realistic interpretation. Nevertheless, it does provide a perfectly
feasible device for directing the fluxes. The arcs LP, PS are both

right-running waves, so that $\underline{F}(\underline{u}_L) - \underline{F}(\underline{u}_P) + \underline{F}(\underline{u}_P) - \underline{F}(\underline{u}_S)$ is assigned to x_{i+1}, and the left-running wave SR assigns $\underline{F}(\underline{u}_S) - \underline{F}(\underline{u}_R)$ to x_i. So, unlike Godunov's scheme, Osher's scheme does not accept the data $(\underline{u}_L, \ldots \underline{u}_L, \underline{u}_R, \ldots \underline{u}_R)$ as a stationary finite difference solution. It does, however, as shown by Osher and Solomon, accept the solution $(\underline{u}_L, \ldots \underline{u}_L, \underline{u}_\ell, \underline{u}_r, \underline{u}_R, \ldots \underline{u}_R)$ where $\underline{u}_r$ is an arbitrary state on the arc RS and $\underline{u}_\ell$ lies on SP and obeys

$$\underline{F}(\underline{u}_\ell) + \underline{F}(\underline{u}_r) = \underline{F}(\underline{u}_L) + \underline{F}(\underline{u}_S)$$

4.2.3. Roe's Method Roe (1981a,b), takes a different path from $\underline{u}_L$ to $\underline{u}_R$. It is not necessarily a path which has a direct physical interpretation, as in Godunov's scheme, or even an indirect interpretation, as in Osher's scheme, although when $\underline{u}_L$, $\underline{u}_R$ are close, the path is once again two sides of the same parallelogram. In the general case, the path has linear elements (see Fig. 6) which are eigenvectors of a matrix $\tilde{A}$ which depends on $\underline{u}_L$ and $\underline{u}_R$ in such a way as to represent average properties of the mapping $\underline{u} \rightarrow \underline{F}$ in the region between $\underline{u}_L$ and $\underline{u}_R$. The matrix is not uniquely specified, but is required to possess the following four properties, collectively described as Property U by Roe (1981a).

i   It is a linear mapping from $\underline{u}$-space to $\underline{F}$-space

ii   if $\underline{u}_L \rightarrow \underline{u}_R \rightarrow \underline{u}$, $\tilde{A}(\underline{u}_L, \underline{u}_R) \rightarrow A(\underline{u})$;

iii $\underline{F}(\underline{u}_L) - \underline{F}(\underline{u}_R) = \tilde{A}(\underline{u}_L, \underline{u}_R) \times (\underline{u}_L - \underline{u}_R)$;

iv   the eigenvectors of $\tilde{A}$ are independent.

Consider the case where $\underline{u}_L$, $\underline{u}_R$ are connected by an arbitrarily moving shock. Then (see Appendix)

$$\underline{F}(\underline{u}_L) - \underline{F}(\underline{u}_R) = S(\underline{u}_L - \underline{u}_R)$$

for some scalar S equal to the shock speed. By property (iii) S is an eigenvalue of $\tilde{A}$, and by property (iv) the whole of $(\underline{u}_L - \underline{u}_R)$ projects onto the corresponding eigenvector. In this special case we recover Godunov's method. The representation of a stationary shock by Roe's method can be shown to tend at large time towards a situation with one (exceptionally no) state intermediate between $\underline{u}_L$ and $\underline{u}_R$.

The practical value of the method turns in part on the ability to construct, for any given set of conservation laws, a matrix $\tilde{A}$, having the specified properties, in an analytical form. In Roe (1981b) a technique is developed which gives simple results for the Euler equations. If $\gamma = 1$, $\tilde{A}$ reduces to

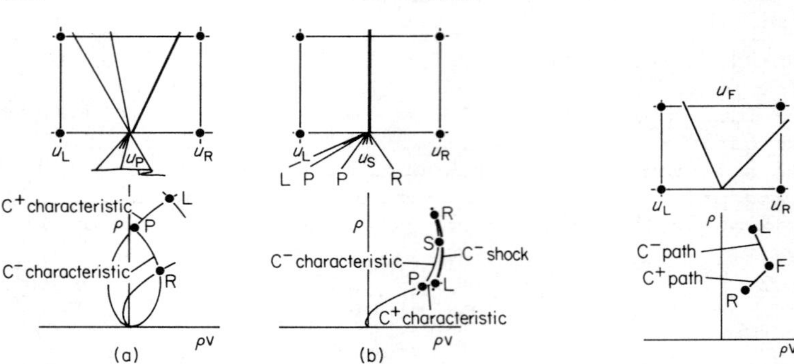

Fig. 5 Directing the fluxes according to Osher's method. Case (a) is the same data as Fig. 4(a). Case (b) is a stationary C^- shock.

Fig. 6 Direction the fluxes according to Roe's method. The data is as in Figs 4(a) and 5(a)

$$\tilde{A} = \begin{bmatrix} 0 & 1 \\ \frac{a^2}{2} - \tilde{v}^2 & 2\tilde{v} \end{bmatrix} \qquad\qquad (4.2.3.1)$$

where

$$\tilde{v} = \frac{\rho_L^{\frac{1}{2}} V_L + \rho_R^{\frac{1}{2}} V_R}{\rho_L^{\frac{1}{2}} + \rho_R^{\frac{1}{2}}} \qquad\qquad (4.2.3.2)$$

The theoretical drawback to Roe's method is that expansion fans are represented only by their average velocity, so that their finite extent is not recognised. This can lead to errors near a sonic line.

4.2.4 The Homogeneous method The following observation was apparently first made by members of the Computational Fluid Dynamics Branch at NASA Ames. The Euler equations are homogeneous, in the sense that $\underline{F}(k\underline{u}) = k\underline{F}(\underline{u})$ for any scalar k. From Euler's theorem on homogeneous functions (did Euler himself notice this connection between two of his claims to fame?) it follows that $\underline{F} = A(\underline{u})$ where $A = \partial \underline{F}/\partial \underline{u}$. Several neat and surprising results follow from this (Warming and Beam (1977), Steger and Warming (1981)), but it is not yet really clear what their practical application is. The property can be used, for example, to direct flux differences in the following way.

Let T be a transformation which diagonalises A, i.e.

$$A = TDT^{-1}$$

for some diagonal matrix D whose elements will be the eigenvalues of A. Let $D = D^+ + D^-$, where D^+ contains all the positive elements, and D^- all the negative elements. Consider $A^+ = TD^+T^{-1}$

and $A^- = TD^-T^{-1}$. Clearly $d\underline{F} = (A^+ + A^-)d\underline{u} = (d\underline{F})^+ + (d\underline{F})^-$ represents a decomposition into right- and left-travelling waves, of an observed *flux difference*. It is tempting to argue that $\underline{F} = (A^+ + A^-)\underline{u} = \underline{F}^+ + \underline{F}^-$ must represent a similar decomposition of an observed *flux*. If we accept this, and follow through the indicated algebra for the Euler equations with $\gamma = 1$, then, if the eigenvalues have opposite signs,

$$\underline{F}^- = \frac{\rho(v-a)}{2}\begin{pmatrix}1\\v-a\end{pmatrix}, \quad \underline{F}^+ = \frac{\rho(v+a)}{2}\begin{pmatrix}1\\v+a\end{pmatrix}.$$

These vectors are parallel to the characteristic paths at the point considered. (If both eigenvalues are positive, then $\underline{F}^- = 0$, $\underline{F}^+ = \underline{F}$, and *vice versa*). The increment assigned to x_i will be $\underline{F}^-_L - \underline{F}^-_R$ and the increment assigned to x_{i+1} will be $\underline{F}^+_L - \underline{F}^+_R$. Steger and Warming (1981) discuss several alternative flux-splitting techniques, which are generalisations of the above.

4.3 *Some Comparisons*

Of the methods for directing fluxes set out above, Godunov's is certainly the one which contains the most careful modelling of the physics. It recognises shocks and simple waves, and passes them wholly in the true direction, whenever this is unambiguously defined. If an expansion fan should propagate partly in each direction, this too is recognised. The other methods are all equivalent to Godunov's method for sufficiently weak waves. Osher's method is perhaps a little artificial in its treatment of strong shock-waves. The mapping of the wave diagram (5b) onto physical (x,t) space is multivalued, and this causes part of the flux difference to be misdirected. Roe's method is exact for pure shock discontinuities (and therefore third-order accurate for simple waves), but does not give a true account of fans which should spread both ways and may in some circumstances lead to rarefaction shocks. (Work is in hand to remedy this defect, by providing an artificial dispersion if one eigenvector represents an expansion across the sonic line). Further comparisons between these three methods have been made by Van Leer (1981) and also by Harten, Lax and van Leer (1982). The homogeneous method leads to the simplest algebra, but at present it is unclear what special properties it may possess.

5 ACHIEVING HIGHER-ORDER ACCURACY

5.1 *General Considerations*

All four methods described in the last section are, as they stand, merely first-order accurate. If they were to be applied to a system of linear equations (for which, of course, the character-

istics have constant slope, in x-t space or in $\underline{u}$-space) then all
four would become identical, not merely to each other, but also
to the CIR hybrid method. This is, as remarked earlier, the
(r = 1, s = 0) algorithm for linear advection, applied to the
characteristic form of the equations. Now, for aeronautical
applications, first-order methods do not usually provide either
the accuracy needed to predict forces, or the resolution needed
to treat interference and interaction problems.

Naïve attempts to secure higher order accuracy come to grief
against Godunov's (1959) theorem concerning the inevitable production of wiggles by linear algorithms of accuracy p > 1. For
specific applications, especially where there is a single trouble-
some feature, such as the shockwave above a transonic aerofoil,
fairly simple measures may be adequate to combat the wiggles. The
use of upwind differences for *all* terms in supersonic regions of
the flow, with central differences elsewhere, and a judicious
touch of artificial viscosity near the shockwave, is a formula
that has been applied in papers too numerous to mention. A useful
refinement is to devise mixed differencing schemes, based on
geometrical interpretations of the CFL condition (Albone (1974),
Jameson (1974), Albone and Hall (1980)). These latter developments
have, as yet, only been worked out for the equations of potential
flow or transonic small-perturbation theory.

In this section we follow up some of the ideas put forward in
Section 4; in particular we develop further the notion that the
key to success in complex situations will lie in a detailed analysis of the physical processes involved. We will also describe,
however, some alternative approaches which attempt to relate the
algorithm to the data without resort to such analysis.

5.2 Van Leer's Approach

Godunov's method assumes, when analysing the data, that $\underline{u}(x)$
is piecewise *constant* between the mid-points of intervals. Van
Leer (1979) developed a similar analysis, based on the assumption
that $\underline{u}(x)$ is piecewise *linear between* mid-points and may be *discontinuous at* mid-points. He solves the resulting wave interaction
problems by an approximate characteristic theory. If he were to do
this consistently everywhere in the flow, the resulting second-
order algorithm would not escape Godunov's Theorem, and would
create overshoots. In fact the solution is monitored to see
whether state variables within any given cell exceed certain
limits obtained by considering the neighbour cells. Whenever this
happens, the assumed initial distribution within each cell is
revised to make it more nearly constant.

The full details of the method are complicated, and no attempt
will be made to describe them here. The method has been programmed

to solve one- and two-dimensional problems in unsteady gas
dynamics. The results obtained have much sharper resolution of
discontinuities than the original Godunov scheme, and are virtually free from spurious oscillations. A simpler implementation
of the same approach is proposed by van Leer (1981).

5.3 Roe's Approach

Roe (1981a,c) achieves higher-order accuracy by a different
technique. Each flux difference $\Delta \underline{F}_{i+\frac{1}{2}}$ is analysed as in Section
4.2 to produce increments of $\underline{u}$ associated with each wave system.
The refinement is that these increments will not now be added
merely at x_i for left-running waves or x_{i+1} for right-running
waves, but will be broken up into finer components, and spread
over a wider range of targets, according to rules which depend on
a Courant number

$$\nu_{i+\frac{1}{2},j} = a_{i+\frac{1}{2},j} \frac{\Delta t}{\Delta x} \qquad (5.3.1)$$

where $a_{i+\frac{1}{2},j}$ is the average speed of the jth wave between x_i and
x_{i+1}. If Roe's analysis (Section 4.2.3) of the flux differences
is used, $a_{i+\frac{1}{2},j}$ is the jth eigenvalue of $\tilde{A}_{i+\frac{1}{2}}$, but the other
analyses could be used, and would probably give very similar
results. Various rules for spreading the fluxes are possible, and
there is a correspondence between these rules and the advection
algorithms discussed in Section 2 (Roe (1981c)). For example,
the rule which assigns each increment to x_i, x_{i+1} in proportions
$\frac{1}{2}(1-\nu)$, $\frac{1}{2}(1+\nu)$ corresponds to the second order centred scheme
$r = s = 1$. To the upwind second order scheme $r = 2$, $s = 0$, there
corresponds the following rule. If ν is positive, assign the
increment to x_{i+1}, x_{i+2} in proportions $\frac{1}{2}(3-\nu)$, $\frac{1}{2}(\nu-1)$; if ν is
negative, assign it to x_i, x_{i-1} in proportions $\frac{1}{2}(3+\nu)$, $-\frac{1}{2}(1+\nu)$.

As with van Leer's approach, an element of feedback (from the
data to the algorithm) is required to circumvent Godunov's
Theorem. Roe (1981c) discusses various possibilities, one of
which is given below, in a formulation due to Baines (1980).
Essentially it is a tidier version of a technique devised by Roe
(1979).

i   Compute increments $\Delta \underline{u}_{i+\frac{1}{2},j}$ due to the jth wave in the
cell $(i, i+1)$.

ii   Compare each of these with the increment which arose
in the cell immediately upstream, i.e. with $\Delta \underline{u}_{i+\frac{1}{2}-k,j}$
where $k = \text{sign}(a_{i+\frac{1}{2},j})$. Of these two vectors, let
the one which contains the density increment of least
absolute magnitude be called $\Delta \underline{u}^*_{i+\frac{1}{2},j}$.

iii Distribute $\Delta \underline{u}^*_{i+\frac{1}{2},j}$ to $(i, i+1)$ in proportions $\frac{1}{2}(1-\nu)$,

$\frac{1}{2}(1+\nu)$.

iv Send $(\Delta u_{i+\frac{1}{2},j} - \Delta u^{*}_{i+\frac{1}{2},j})$ wholly to the downstream
    node of the cell. This step may not, of
    course, be required.

Some properties of this scheme can be proved very easily; for
example it clearly meets the conditions for conservation. If the
data is such that we find ourselves consistently choosing the
local increment at stage (ii) we are using the centred scheme
$r = s = 1$. If, on the other hand, we find ourselves consistently
choosing the upwind increment, then it can be shown that we are
using $r = 2$, $s = 0$. Sweby (1981) has investigated other properties
of the scheme. His most important conclusion is that when the
scheme is specialised to linear advection, monotone data remains
monotone for all time, whatever sequence of choices we are led
to make at stage (ii). Excellent results have been obtained on a
wide variety of one-dimensional test problems.

5.4 *Flux-corrected transport and other approaches*

Flux-corrected transport is a technique originated by Boris
and Book (1973) which in practice stands perhaps mid-way in
sophistication between the artificial viscosity methods and the
upwind differencing methods. It has recently been redefined by
Zalesak (1979), however, in very general terms. We quote his
description below with small changes of notation.

We suppose that $\underline{u}$ is to be advanced in time by some finite-
difference scheme of the form

$$u^{n+1}_{\underline{i}} - u^{n}_{\underline{i}} = - \left(\frac{\Delta t}{\Delta x}\right) \left[F^{n}_{\underline{i}+\frac{1}{2}} - F^{n}_{\underline{i}-\frac{1}{2}}\right] \qquad (5.4.1)$$

where $F^{n}_{\underline{i}+\frac{1}{2}}$ is some estimate of the average flux in (x_i, x_{i+1})
arrived at by any means whatsoever from all or part of the data
at time level n. In fact, any algorithm consistent with the conservation law can be put into the form (5.4.1), and one way to
implement the algorithm is to begin by computing the implied
values of $F^{n}_{\underline{i}+\frac{1}{2}}$, for all i, and then to apply (5.4.1). Zalesaks'
scheme is this:

i   Compute $F^{L}_{\underline{i}+\frac{1}{2}}$, corresponding to a low-order algorithm
     known not to create wiggles.

ii   Compute $F^{H}_{\underline{i}+\frac{1}{2}}$, corresponding to a more ambitious high
     order scheme.

iii Define $\underline{A}_{\underline{i}+\frac{1}{2}} = F^{H}_{\underline{i}+\frac{1}{2}} - F^{L}_{\underline{i}+\frac{1}{2}}$ (the "anti diffusive flux").

iv   Compute the "safe" solution

$$u_{-i}^{s} = u_{-i}^{n} - \left(\frac{\Delta t}{\Delta x}\right)\left[F_{-i+\frac{1}{2}}^{L} - F_{-i-\frac{1}{2}}^{L}\right].$$

v   "Limit" each anti-diffusive flux to a "corrected" value

$$A_{-i+\frac{1}{2}}^{C} = C_{i+\frac{1}{2}} A_{-i+\frac{1}{2}} \qquad 0 \leqslant C_{i+\frac{1}{2}} \leqslant 1$$

vi   Compute the "flux limited solution"

$$u_{-i}^{n+1} = u_{-i}^{s} - \left(\frac{\Delta t}{\Delta x}\right)\left[A_{-i+\frac{1}{2}}^{C} - A_{-i-\frac{1}{2}}^{C}\right].$$

There are, evidently, three ingredients in this recipe: a low-order algorithm, a high-order algorithm, and some strategy which enables the limiting factors $C_{i+\frac{1}{2}}$ to be chosen. The usual strategy is to insist that the solution u^{n+1} contains no extreme values not found in $\underline{u}^{n}$ or $\underline{u}^{s}$. Subject to this constraint $C_{i+\frac{1}{2}}$ is taken as large as possible. Zalesak discusses this point in detail. It is also important that the safe algorithm $\underline{F}^{L}$ be reasonably accurate, and the accurate algorithm $\underline{F}^{H}$ be not too prone to wiggles. Of course both $\underline{F}^{L}$ and $\underline{F}^{H}$ could be constructed on upwind principles, but this is not the usual practice in FCT codes. Since FCT obliges us to compute the solution twice over and do some elaborate comparisons between the two, it is obviously desirable to keep each solution fairly simple.

Zalesak's formulation of FCT leaves so many options open to choice that it may take some time before the best combinations are discovered. However, a recent paper by Book *et al.* (1980) contains impressively realistic calculations of the flow due to shock diffraction by a wedge.

Various other techniques were reviewed by Sod (1978), and compared on the basis of a shock tube calculation. It should be noted, however, that the 'upwind' algorithm which he tested was a simplified method which only accounted for the particle velocity and took no account of wave speeds. A novel approach is described by Knorr and Mond (1980) who attempt to reconstruct discontinuities from distinctive patterns of wiggles. If the solution contains waves travelling in each direction, then the 'clipping' stage of their algorithm involves scanning the solution from left to right and then back again.

6.   THE SELLS AEROFOIL PROGRAM

In this section we describe briefly how some of the ideas mentioned above have been put to an aeronautical application, the

inviscid flow past a two-dimensional aerofoil. A more complete
description is available in Sells (1980). Inevitably, the application to a real technical problem involves, at present, a number
of empirical extensions. The listing of these serves also to
highlight the areas on which future research must concentrate.

The most fundamental extension is from one space dimension to
two. This is done by adapting the techniques of finite-volume
analysis (e.g. Rizzi (1975)). The computing space is partitioned
into quadrilaterals, which are thought of as containing the conserved quantities. The amount contained in a typical cell is
$A_{jk}\underline{u}_{jk}$, where $A_{j,k}$ = area of $(j,k)^{th}$ cell, $\underline{u}$ = $(\rho,\rho u,\rho v,e)^{T}$.
From $\underline{u}$ we compute flux vectors $\underline{F}$ = $(\rho u,p+\rho u^2,\rho uv,u(e+p))^{T}$,
$\underline{G}$ = $(\rho v,\rho uv,p+\rho v^2, v(e+p))^{T}$, and then compute the flow between
the $(j,k)^{th}$ and $(j, k+1)^{th}$ cells, for example, (see Fig. 7) as

$$(\underline{F}_{j,k+1} - \underline{F}_{j,k})\ (y_R - y_S) - (\underline{G}_{j,k+1} - \underline{G}_{j,k})(x_R - x_S).$$

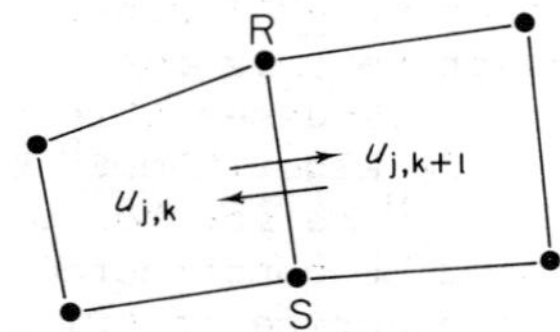

Fig. 7   Interaction of two states occupying adjacent finite
         volumes in Sells' method.

This flux is analysed into Mach waves, contact discontinuities,
or slip discontinuities moving normal to RS. In the version which
Sells describes, a wave is used entirely to increment the cell
into which it moves, giving an algorithm analogous to the first-
order schemes of Section 4.2. Second-order accuracy is achieved
by adding a corrector step, and monotonicity is ensured by the
shifting device described in Roe (1979). Since then, Sells has
modified his program along the lines described in Section 5.3.
This has the considerable advantage that the wave analysis now
has to be performed only once per cell interface per time step.
The results are scarcely altered by this modification.

At the outermost edges of the outermost cells, fluxes are
computed between these cells and imaginary cells containing
fluid at free stream conditions. Any outgoing waves indicated by
this analysis are simply discarded. In general this process provides a very satisfactory 'transparent boundary condition'. Sells
(1980) discusses refinements appropriate to supersonic flow and
to lifting flows. Surface boundary conditions are dealt with by
including 'image cells' just inside the solid boundary, and set-
ting the pressure within these according to $\partial p/\partial n = \rho(u^2+v^2)/R$.

The basic mode of operation of the program is that of explicit
time-marching, but in our applications we are usually interested
only in the final steady state. Convergence towards this is very
slow, partly because it is found that accuracy near the stagnation point requires the local use of very small cells, and hence,
by the CFL condition, global use of a very small time step. To
overcome this, the program is run initially in a 'pseudo time
dependent' mode, where larger cells are allowed to advance more
rapidly. The program reverts to real time toward the end of the
run. As originally described by Sells, the program made use of
the differential form of the energy equation and allowed enthalpy
to vary as the solution developed. His later versions assume
constant enthalpy, which simplifies and shortens the wave analysis, and has a negligible effect on the asymptotic steady state.
Nevertheless, the program remains rather time-consuming, and
currently some of our work aims to speed it up, perhaps by the
kind of approximate factorisation implicit method described by
Baker in this volume, perhaps by the pseudo-time explicit methods
reviewed by Viviand (1980).

In its present form the program is too slow for routine use
as a design tool. Its main use has been to calibrate, and where
necessary tune, predictions based on less exact models of the
flow. To this extent the program is being used, with due caution,
as a substitute for wind-tunnel testing. The article by J. E.
Green in these proceedings describes, briefly, such an application.

7 CONCLUDING REMARKS

In order to compute the behaviour of 'things moving violently
to their place' many more methods have been devised than I have
had the space to discuss. I have barely mentioned shock fitting
methods, for example, and amongst my other sins of omission are
Lagrangean (particle-tracking) methods, pseudo-Boltzmann methods,
spectral methods, finite element methods, random-choice methods
and many more. The subject itself, indeed, is 'moving to its
place', and one can scarcely do more than report recent events
from one's own area. The upwind differencing approach is, however,
rapidly gaining adherents, as attested by the very recent dates
on many of the references. My own allegiance to it arises from its
strongly physical motivation, and the rather clean mathematical
structure which seems to be emerging out of the earlier empiricism. Both these aspects indicate that rewarding new developments
may still lie ahead.

The foundations of upwind differencing are the methods for
one-dimensional initial-value problems described in Sections 4
and 5. This is a very active area of current research, which
would already be in quite a satisfactory state if one-dimensional

initial-value problems were what we actually wished to solve.
What we really need from this phase, however, is the simplest
method which also provides a secure base on which to build
further. When the most promising method is established it should
be possible to find the numerical boundary conditions which are
best suited for use with it.

The Sells aerofoil program has shown that two-dimensional
problems can be tackled by a form of operator splitting in which
the waves are supposed to travel normal to the mesh lines. As
long as this supposition is roughly true the results are very
good, but shockwaves which lie obliquely across the mesh are not
so cleanly captured. (At least, this has been our experience at
the RAE; preliminary results reported by Osher and Solomon (1981)
appear not very sensitive to shock orientation). It would be
more in the spirit of the enterprise to develop a method which
could 'recognise' oblique waves and deal with them appropriately.
Work on this aspect of the problem has only just begun.

The other main requirement, as we have already noted in Section 6, is to find ways of speeding up the solution. Essentially
this would be by taking larger steps in time (or pseudo-time)
than the explicit marching methods permit. There is, however,
little point in doing this if the problem is one in which we wish
to make accurate observations of strong transient effects.
Although implicit methods may be stable, they are notoriously
inaccurate if the discontinuities move by more than about one
cell width in the course of one time step. Implicit methods come
into their own at that stage where the violence of the initial
motion is giving way to the calm of the converged state. The
difficulty is that the final equilibrium may be marked by stationary discontinuities, which are, so to speak, the geological
evidence of a violent history. An inappropriate treatment at this
stage can bring such 'fault lines' unwelcomely back to life. It
is notable that the great success of approximate factorisation
methods depends upon distinguishing waves which move in various
directions, so that we never attempt to send signals across a
stationary shockwave.

Preliminary results by Pike (1980), using an implicit method
based on the analysis of Section 4.2.3, have shown that it is
possible to speed up considerably the convergence of a one-
dimensional nozzle problem featuring a stationary shock. His
methods are unconditionally stable for weak disturbances, but as
the conditions of the problem are varied to make the final
strength of the shock greater, the stable rate of advance becomes
less. However, it seems that the time required can always be
reduced by a factor of at least 4.0. More work along these lines
is still needed.

To attempt a broad generalisation in conclusion, it would seem that as Richardson's "student of matter and ether" attempts to make his mathematical description of the physical world more exact, so the seemingly insignificant details of the arithmetic come increasingly into prominence. The ultimate task is to find rules which combine effectiveness with simplicity. The extent to which that can be done will largely determine the role played by computation in the aerodynamics of the future.

APPENDIX: THE BASIC THEORY OF CONSERVATION LAWS,
    ILLUSTRATED BY A SIMPLE CASE OF THE EULER
    EQUATIONS

At various points in the text we have felt the need to supplement the discussion of certain ideas by showing how they would be applied to a particular set of equations. The choice of a good illustrative set is not easy. The full Euler equations can be distractingly complex, but if we simplify too much we may omit essential structure. J. Pike at RAE Bedford has investigated various alternatives. His suggestion is to set $\gamma = 1$ in the unsteady Euler equations. The energy equation then becomes a statement that the speed of sound a is constant along particle paths. If we now agree only to consider initial data in which a is constant, then a will remain constant for all time, and in the momentum equation we can set $p_x = a^2 \rho_x$.

We are left with the simple system

$$\text{(Continuity)} \qquad \rho_t + (\rho v)_x = 0 \qquad\qquad \text{(A-1)}$$

$$\text{(Momentum)} \qquad (\rho v)_t + (\rho v^2 + \rho a^2)_x = 0 \quad . \qquad \text{(A-2)}$$

Pike has used these equations in order to test various numerical schemes. For this purpose they form a useful stepping-stone between the full Euler equations and the commonly used inviscid Burgers equation $v_t + (\tfrac{1}{2}v^2)_x = 0$, to which they reduce as $a \rightarrow 0$.

We develop below the basic analytic results for these equations. This also serves to introduce informally the concepts and notation of conservation laws. For a more general and rigorous presentation the reader may consult Jeffrey (1976). To achieve a compact notation we introduce two vectors.

$$\underline{u} = \begin{pmatrix} u_1 \\ u_2 \end{pmatrix} = \begin{pmatrix} \rho \\ \rho v \end{pmatrix} , \quad \underline{F} = \begin{pmatrix} F_1 \\ F_2 \end{pmatrix} = \begin{pmatrix} \rho v \\ \rho (v^2 + a^2) \end{pmatrix} \qquad \text{(A-3)}$$

and then write the problem as

$$\underline{u}_t + \underline{F}_x = 0 \quad . \qquad\qquad \text{(A-4)}$$

This is said to be in *conservation form* because it is the form in which physical conservation laws are naturally expressed, as a result of applying them to a small control volume. It is also said to be in *divergence form,* because it can be expressed even more concisely as

$$\text{div } (\underline{u}, \underline{F}) = 0 \tag{A-5}$$

If the control volume is not small, but is an arbitrary closed contour in the (x,t) plane, we can obtain the *integral form*

$$\oint \underline{u} \ dx - \underline{F} \ dt = 0 \tag{A-6}$$

and by specialising the control volume to a narrow strip lying between $t = t_o$, $t = t_o + dt$, and extending from x_L to x_R, we obtain the *semi-integral form*

$$\frac{\partial}{\partial t} \int_{x_L}^{x_R} \underline{u} \ dx + [\underline{F}(x_R) - \underline{F}(x_L)] = 0 \ . \tag{A-7}$$

Finally, we present the *matrix form,* which in our example would be

$$\begin{pmatrix} u_1 \\ u_2 \end{pmatrix}_t + \begin{bmatrix} 0 & 1 \\ a^2 - u_2^2/u_1^2 & 2u_2/u_1 \end{bmatrix} \begin{pmatrix} u_1 \\ u_2 \end{pmatrix}_x = 0 \tag{A-8}$$

or concisely,

$$\underline{u}_t + A \ \underline{u}_x = 0 \tag{A-9}$$

in which the *Jacobian matrix* A has its $(i,j)^{\text{th}}$ element equal to $\partial F_i/\partial u_j$. A crucial role in the theory is played by the eigenvalues and eigenvectors of A. If the eigenvalues are all real, the problem is said to be *hyperbolic,* and we will consider only this case.

In our example the right eigenvectors are

$$\underline{e}_1 = \begin{pmatrix} 1 \\ v-a \end{pmatrix}, \ \underline{e}_2 = \begin{pmatrix} 1 \\ v+a \end{pmatrix} \tag{A-10}$$

and the eigenvalues are

$$\lambda_1 = v-a, \ \lambda_2 = v+a \ . \tag{A-11}$$

If $\underline{u}_x$ happens to be proportional to the kth eigenvector of A, then so is $\underline{u}_t$. In fact we will have, if λ_k is the kth eigenvalue,

$$\underline{u}_t + \lambda_k \ \underline{u}_x = 0 \tag{A-12},$$

so that we have a special solution in which the values of $\underline{u}$ are convected with *wavespeed* λ_k along *characteristic lines* $dx/dt = \lambda_k$. The special data which admits such a *simple wave* solution satisfies

$$d\underline{u} \quad \alpha \quad \underline{e}_k \qquad \text{(A-13)}$$

which defines a direction in $\underline{u}$-space. This direction is constant for linear problems (A is then a constant matrix) but variable for non-linear problems. Starting from a given state $\underline{u}_o$, which in the general case can be thought of as a point in a phase space of n dimensions, (A-13) yields a curve in that space, and any point on that curve (which we shall call a *wave path*) can be reached from $\underline{u}_o$ by means of a simple wave in physical space. In our example (A-13) yields

$$\frac{d\rho}{d(\rho v)} = \frac{1}{v \pm a}$$

which integrates to give

$$\log \left(\frac{\rho}{\rho_o}\right) \pm \frac{(v-v_o)}{a} = 0 \qquad \text{(A-14)}$$

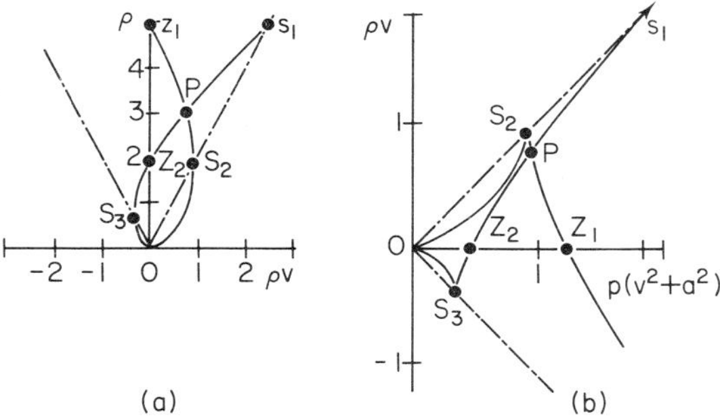

Fig. A-1.   Characteristic paths in (a) $\underline{u}$-space, (b) $\underline{F}$-space. The curve OS_2PZ_1 is a C^- characteristic; $OS_3Z_2PS_1$ is a C^+ characteristic.

These wave paths can be plotted in any convenient space. In Fig. (A-1) we show the two wave paths through a particular point P, traced out on the left in $\underline{u}$-space, and on the right in $\underline{F}$-space. Units of measurement have been chosen so that $a = 0.5$, which makes the sonic line $\rho v = \frac{1}{2}\rho$ in $\underline{u}$-space, or $\rho v = \rho(v^2 + a^2)$ in $\underline{F}$-space. At corresponding points of the paths, the curve in $\underline{u}$-space is parallel to the curve in $\underline{F}$-space (because $d\underline{u}$ is an eigenvector of the mapping $d\underline{u} \to d\underline{F}$). The reciprocal of that slope is, in our example, the wave speed in (x,t) space, but this result is not true in general. Note the typical feature of most non-linear

problems, that the u-space maps onto a folded surface in F-space.
The fold lies along the sonic line, where the wavespeed is
either 0 or 2a. By drawing more wave paths in u-space we obtain
Fig. A-2a. These *wave diagrams* (discussed very fully in Courant

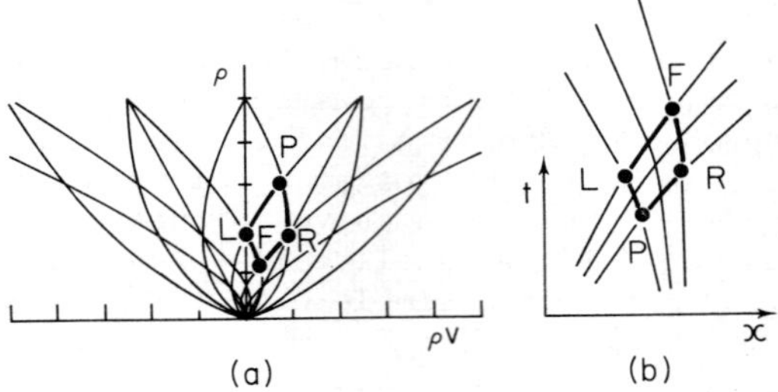

Fig. A-2.   Four uniform regions separated by simple waves
            which pass through an interaction.

and Friedrichs (1948)) are the basis of the analysis in Section
4.2. They allow the solution to many interaction problems by the
following typical process. Consider four regions of uniform flow
in x-t space separated by intersecting waves as in Fig. 4-2b. We
label these regions LRPF (left, right, past and future). Since
these regions are separated by simple waves, they must be connected by wave paths as indicated in Fig. A-2a. Given any two of
the states, we can construct the others, taking due care with
regard to the order in which the paths are traced. For instance,
if $\underline{u}_L$, $\underline{u}_R$ are given, and $\underline{u}_F$ is required, we need the intersection
of the C^- path through $\underline{u}_L$ and the C^+ path through $\underline{u}_R$.

Note that as we move along a characteristic in (x,t) space,
any changes of state must be due to the action of the other wave
family. If we have smooth data of a more general kind than (A-13),
we can obtain a local solution by expressing $\underline{u}_x$ in terms of the
eigenvectors as basis, that is

$$\underline{u}_x = \sum \alpha_k \underline{e}_k \tag{A-15}$$

whence

$$\underline{u}_t = -\sum \alpha_k \lambda_k \underline{e}_k \ . \tag{A-16}$$

Along a characteristic line $dx/dt = \lambda_k$ we have

$$d\underline{u} = dt \, (\underline{u}_t + \lambda_k \underline{u}_x)$$

from which the contribution due to $\underline{e}_k$ vanishes, so that $d\underline{u}$ must
lie in the subspace spanned by the other eigenvectors. If those
eigenvectors are used to form (n-1) columns of an n x n matrix,

M_k, with $d\underline{u}$ as the remaining column, then,

$$\det (M_k) = 0 \qquad\qquad (A-17)$$

which is the *characteristic equation* which holds along the kth
family of characteristic lines. It is a slightly confusing coin-
cidence that in the special case n = 2 Eqn. (A-17) leads to an
equation, satisfied along one family of characteristics, which is
identical to the wave path (Eqn. (A-13)), for the *other* family of
characteristics. In general, however, a wave path is a curve in
n-dimensional $\underline{u}$-space, but a characteristic equation is an (n - 1)
dimensional manifold. In our example, the characteristic equations
are merely (A-14) with the '±' sign reversed.

The equations which hold across a discontinuity are analogous
to the wave paths, but since $\underline{u}(x)$ is no longer differentiable
they must be derived from the integral form of the equations. Let
the discontinuity D move with speed S and let it separate two
constant states $\underline{u}_1$, $\underline{u}_2$. Apply (A-6) to the control volume shown
below in Fig. A-3a.

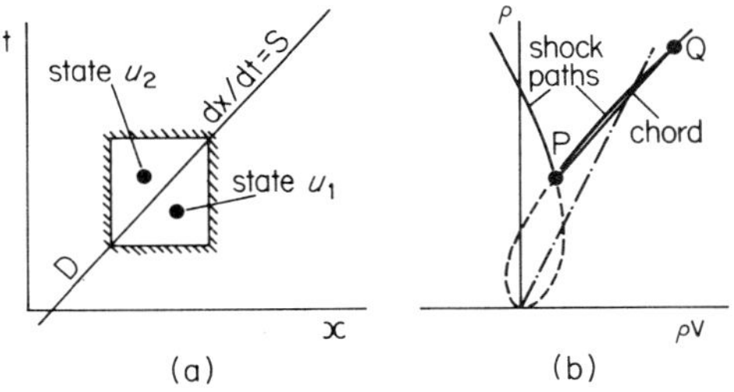

Fig. A-3.(a) The control volume for analysing a shock
              transition.

   A-3(b)   The corresponding diagram in $\underline{u}$-space.
            P = $\underline{u}_1$, Q = $\underline{u}_2$.

We obtain

$$S(\underline{u}_1 - \underline{u}_2) = \underline{F}_1 - \underline{F}_2 \qquad\qquad (A-18)$$

so that two states $\underline{u}_1$, $\underline{u}_2$ can be connected by a *shockwave* if
(A-18) is true for some scalar S.   In our example

$$S(\rho_1 - \rho_2) = (\rho_1 v_1 - \rho_2 v_2) \qquad\qquad (A-19)$$

$$S(\rho_1 v_1 - \rho_2 v_2) = (\rho_1 v_1^2 + \rho_1 a^2 - \rho_2 v_2^2 - \rho_2 a^2) \quad (A-20)$$

After S has been eliminated, a little algebra yields

$$\rho_1 \rho_2 (v_1 - v_2)^2 = a^2 (\rho_1 - \rho_2)^2 \qquad\qquad (A-21)$$

as the equation of a *shock path,* where we may think of $\underline{u}_1$ as
given, and $\underline{u}_2$ as the set of states which may be reached from it
via a shockwave. If a shock path and a wave path originate at
the same point, then at that point they have the same first and
second derivatives.

   The two shock paths through a particular point P in $\underline{u}$-space
are shown in Fig. (A-3b). For any other point Q on the path the
reciprocal slope of the chord PQ gives (see Eqn. (A-19)) the
speed of the shockwave. The parts of the shock paths which lie
below P have been drawn dotted, to indicate that they are not
physically realistic. Our reasons for discarding them have nothing to do with any thermodynamic definition of entropy because
in the limit $\gamma = 1$ there is no entropy change across shockwaves.
However, there are other reasons, just as fundamental. If, in
(x,t) space, we draw the physical pattern of characteristics
corresponding to a transition on one of the dotted paths, we
find that one family of characteristics appears to branch out of
the shockwave (Fig. A-4a) rather than converging into it (Fig.
A-4b).

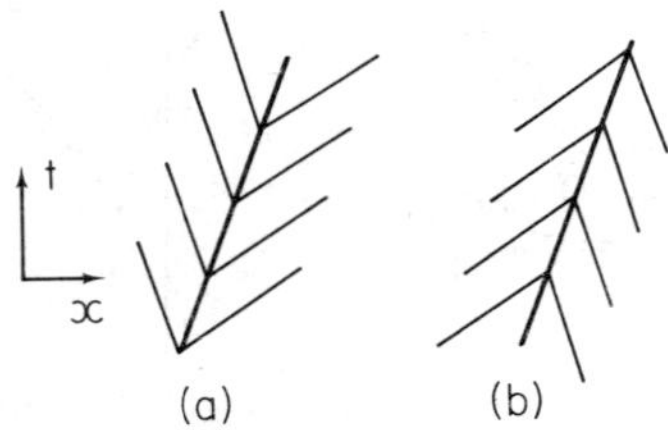

Fig. A-4.   Shocks and characteristics. In Fig. A-4a the
           characteristics belonging to the same wave
           family as the shock diverge from it, which is
           physically inadmissible. In Fig. A-4b they
           converge into it, which is physically correct.

It can be shown (Jeffrey (1976)) that only the solution with converging characteristics is stable against small perturbations.
This is called the *evolutionary condition* for discriminating
between physically acceptable or unacceptable shockwaves. See
also Oleinik (1963), Lighthill (1978). The discriminations made
by the evolutionary condition can be shown under a wide range of
circumstances to be equivalent to those made on the basis of
entropy, or those made by accepting only such solutions as are
valid limits of a slightly viscous solution. All these discriminations reflect different aspects of *irreversibility,* and are
collectively called *entropy conditions*. Recent discussions
include Lax (1971) and Harten and Lax (1981).

It is at the present time a much debated issue how to incorporate entropy conditions into numerical methods. Common practice
is to incorporate artificial terms representing a localised viscosity, but there are obvious difficulties in combining this
accurately with a small, but significant, physical viscosity.
Fundamental observations have been made by Harten, Hyman and Lax
(1976), and by LeRoux (1977).

Discontinuities are also possible in which the characteristics,
although of variable slope, do not intersect. Such discontinuities do not arise in our simple example, but the particle paths,
across which density (also temperature and entropy) may jump
whilst pressure and velocity remain constant, are examples in
the full Euler equations. Such discontinuities are said to be
degenerate, semilinear, or not *genuinely nonlinear*. The numerical
difficulties to which they give rise are very similar to those
occurring in the linear advection problem of Section 2.

REFERENCES

Albone, C.M. (1974). A finite-difference scheme for computing
supercritical flow in arbitrary coordinate systems. RAE Technical
Report TR 74090.

Albone, C.M. and Hall, M.G. (1980). A scheme for the improved
capture of shockwaves in potential flow calculations.RAE Technical Report TR 80126.

Baines, M.J. (1980). Private communication. Dept. of Mathematics,
University of Reading.

Book, D., Boris, J., Oran, E., Picone, M., Zalesak, S., and Kuhl,
A. (1980). Simulation of complex shock reflections from wedges
in inert and reactive gaseous mixtures.Naval Research Laboratory
Memorandum Report 4333.

Carver, M.B. (1980). Pseudo characteristic method of lines solution of the conservation equations. *J. Comp. Phys.*, **35**, 1, pp.
57-76.

Chakravarthy, S.R., Anderson, D.A., and Salas, M.D. (1980). The
split-coefficient matrix method for hyperbolic systems of gas
dynamic equations. AIAI Paper 80-0268, 18th Aerospace Sciences
Meeting.

Courant, R., and Friedrichs, K.O. (1948). Supersonic flow and
shockwaves. Wiley, Interscience, New York.

Courant, R., Friedrichs, K.O., and Lewy, H. (1928). Über die
partiellen differenzengleichungen der mathematischen physik.

Math. Ann., **100**, p.32.

Courant, R., Isaacson, E., and Rees, M. (1952). On the solution of nonlinear hyperbolic differential equations by finite differences. *Comm. Pure and Appl. Math.* **5**, p.243.

Davies, H.C. (1980). A pseudo-upstream differencing scheme for advection. *J. Comp. Phys.* **37**, pp.280-286.

Engquist, B., and Osher, S. (1980). One-sided difference schemes and transonic flow. *Proc. Nat. Acad. Sci. U.S.A.*

Godunov, S.K. (1959). A difference method for the numerical calculation of discontinuous solutions of hydrodynamic equations. *Mat. Sbornik* **47**, 3, pp.271-306. Translated as JPRS 7225 by US Dept. of Commerce, November 1960.

Harten, A. (1977). The artificial compression method for computation of shocks and contact discontinuities. I. Single conservation laws. *Comm. Pure and Appl. Maths.* <u>XXX</u> pp.611-638.

Harten, A., Hyman, J.M. and Lax, P.D. (1976). On finite difference approximations and entropy conditions for shocks. *Comm. Pure and Appl. Maths.* <u>XXIX</u>, p.297.

Harten, A., and Lax, P.D. (1981). A random choice finite difference scheme for hyperbolic conservation laws. SIAM. *Jnl. Num. Anal.* **18**, p.289.

Harten, A., Lax, P.D., and van Leer, B. (1982). On upstream differencing and Godunov-type schemes for hyperbolic conservation laws. ICASE Report 82-5.

Hoskin, N.E., and Lambourn, B.D. (1971). The computation of general problems in one-dimensional unsteady flow by the method of characteristics. Proc. 2nd Int. Conf. on Numer. Meth. in Fl. Dyn., Springer.

Iserles, A. (1981). Order stars and a saturation theorem for first-order hyperbolics. Univ. of Cambridge DAMTP NA3/81.

Jameson, A. (1974). Iterative solution of transonic flows over aerofoils and wings, including flows at Mach 1. *Comm. Pure and Appl. Math.*, **27**, pp.283-309.

Jeffrey, A. (1976). Quasilinear hyperbolic systems and waves. Research notes in mathematics No.5. Pitman.

Knorr, G., and Mond, M. (1980). The representation of shock-like solutions in an Eulerian mesh. *J. Comp. Phys.* **38** pp.185-211.

Lax, P.D. (1954). Weak solutions of nonlinear hyperbolic equations and their numerical computation. *J. Comp. Phys.* **7** pp 159-193.

Lax, P.D. (1971). Shock waves and entropy, in Zarantanello, E.H. (ed). Contributions to nonlinear functional analysis. Academic Press.

Lax, P.D. (1972). Hyperbolic systems of conservation laws and the mathematical theory of shock waves. SIAM Regional Conference Series, Lectures in Applied Mathematics 11.

Leonard, B.P. (1979). A stable and accurate convective modelling procedure. *Computer Methods in Appl. Mech. and Eng.*, **19**, pp 59-98.

Le Roux, A.Y. (1977). A numerical conception of entropy for quasi-linear equations. *Math. Comp.* **31** pp 848-872.

Lighthill, M.J. (1978). Waves in fluids. C.U.P.

Moretti, G. (1979). The λ-scheme. *Computers and Fluids*, **7**, pp. 191-206.

Moretti, G., and de Neef, T. (1980). The "post-correction" technique for the fitting of shocks and other boundaries, in Hirschel, E.H. (ed). Proc. 3rd GAMM Conference on Numer. Meth. in Fl. Mech. Vierveg.

Oleinik, O.A. (1963). *Amer. Math. Soc. Trans.* Sec. 2, **33** p.285.

Osher, S. and Solomon, F. (1982). Upwind difference schemes for hyperbolic systems of conservation laws. *Math. Comp.* (to appear).

Pandolfi, M., and Zannetti, L. (1981). A physical approach to solve numerically complicated hyperbolic flow problems. Proc. 7th Int. Conf. on Numer. Meth. in Fl. Dyn. Springer.

Pike, J. (1980). Private communication, RAE Bedford.

Richardson, L.F. (1910). The approximate arithmetical solution by finite differences of physical problems involving differential equations, with an application to the stresses in a masonry dam. *Phil. Trans. Roy. Soc.* A **210** pp 307-357.

Rizzi, A.W. (1976). Transonic solutions of the Euler equations by the finite volume method, in Oswatitsch, K., and Rues, D. (eds.) Symposium Transsonicum II, Springer.

Roe, P.L. (1979). An improved version of MacCormack's shock-capturing algorithm. RAE Technical Report TR 79041.

Roe, P.L. (1981a). The use of the Riemann problem in finite-difference schemes. Proc. 7th Int. Conf. on Numer. Meth. in Fl. Dyn., Springer.

Roe, P.L. (1981b). Approximate Riemann solvers, parameter vectors, and difference schemes, *J. Comp. Phys.*, **43**, p.357.

Roe, P.L. (1981c). Numerical algorithms for the linear wave equation. RAE Technical Report TR 81047.

Sells, C.C.L. (1980). Solution of the Euler equations for transonic flow past a lifting aerofoil. RAE Technical Report TR 80065.

Steger, J.L. (1978). Coefficient matrices for implicit finite-difference solution of the inviscid fluid conservation law equations. *Comp. Meth. in Appl. Mech. and Eng.*, **13**, pp 175-188.

Steger, J.L., and Warming, R.F. (1981). Flux vector splitting of the inviscid gasdynamic equations with application to difference schemes. *J. Comp. Phys.* **40**, p.263.

Sweby, P.K. (1980). Private communication. Dept. of Mathematics, University of Reading.

van Leer, B. (1979). Towards the ultimate conservative differencing scheme. V. A second-order sequel to Godunov's method. *J. Comp. Phys.* **32**, pp 101-136.

van Leer, B. (1981). On the relationship between the upwind-differencing schemes of Godunov, Engquist-Osher, and Roe. ICASE Report 81-11.

Viviand, H. (1981). Pseudo-unsteady methods for transonic flow computations. Proc. 7th Int. Conf. on Numer. Meth. in Fl. Dyn., Springer.

Walkden, F., and Sellars, J.E. (1966). A pseudo-viscous method for the calculation of two-dimensional supersonic flow fields. *Aero. Q.* **17**, pp 285-301.

Walkden, F., and Caine, P. (1972). Application of a pseudo-viscous method to the calculation of the steady supersonic flow past a waisted body. *Inst. J. Numer. Meth. in Eng.* **5** pp 151-162.

Walkden, F., Caine, P., and Laws, G.T. (1978). A locally two-dimensional method for calculating three-dimensional supersonic flows. *J. Comp. Phys.* **27**, p.103.

Warming, R.F., Beam, R.M. (1977). On the construction and applic-

ation of implicit factored schemes for conservation laws. SIAM/-
AMS Symoposium on Comp. Fl. Dyn., New York.

Warming, R.F., Kutler, P., and Lomax, H. (1973). Second- and
third-order noncentred difference schemes for nonlinear hyperbolic equations. AIAA Jul. **11**, pp 196-204.

Zalesak, S. (1979). Fully multidimensional flux-corrected transport algorithms for fluids. *J. Comp. Phys.* **31**, p.335.

A NEW FINITE-VOLUME METHOD FOR THE EULER EQUATIONS
WITH APPLICATIONS TO TRANSONIC FLOWS*

A. Lerat** and J. Sidès

(Office National d'Etudes et de Recherches Aerospatiales,
29 Avenue de la Division Leclerc, 92320 Chatillon, France)
**(Ecole Nationale Supérieure d'Arts et Métiers, 75640 Paris,
Cédex 13 and Laboratoire de Mécanique Théorique,
Université Paris VI. Consultant at ONERA)

ABSTRACT

A finite volume method is presented for the solution of the
Euler equations. The method is based on a new explicit scheme
of second-order accuracy which has been selected in an extended
class of S_β^α schemes for a hyperbolic system of conservation laws
in two space variables. The method is applied to the calculation
of transonic flows over the NACA 0012 airfoil ($M_\infty = 0.85$,
$\alpha = 0^O$ and $M_\infty = 0.80$, $\alpha = 1^O25$) and over the RAE 2822 supercritical airfoil ($M_\infty = 0.75$, $\alpha = 3^O$). The increase in accuracy
over a classical method clearly appears in the numerical results:
an improved pressure jump across shock waves, a weakening of
spurious oscillations and a reduction of the entropy error in
the expansion waves upstream of the shocks. The calculation of
an unsteady transonic flow over an oscillating airfoil is also
presented.

1. INTRODUCTION

A great number of calculation methods have been proposed for
the simulation of unsteady transonic flows, but only a few of
them concern the solution of the complete Euler equations (Beam
and Warming, 1976; Chyu and Schiff, 1981; Chyu et al., 1981;
Laval, 1976; Lerat and Sidès, 1977; Lerat and Sidès , 1979;
Magnus and Yoshihara, 1975; Magnus and Yoshihara, 1976; Steger,
1978.). In Lerat and Sidès (1979), the present authors used
a finite-volume method based on the explicit MacCormack scheme
without time-splitting, to compute transonic flows over oscillating airfoils. An improvement of this method is presented here,

---

*Work partly supported by Direction de Recherches Etudes et
 Techniques, of French Ministry of Defence.

based on the introduction of a new scheme, deduced from a general
study of predictor-corrector schemes for hyperbolic systems in
two space variables (Lerat, 1981, ch. 3). This study is a continuation of the work (Lerat and Peyret, 1975) for a one-dimensional
system, in which the equivalent system approach was used to explain the origin of spurious oscillations in the numerical solutions and then to select an "optimal scheme".

In Section 2, we present the main results of the study of
predictor-corrector schemes on a cartesian mesh. In Section 3,
we formulate the optimal scheme on a moving curvilinear mesh, as
a finite-volume method. Finally, in Section 4, the new method
is applied to the calculation of steady and unsteady transonic
flows and the numerical results are compared to those obtained
with the MacCormack scheme.

2.   PREDICTOR-CORRECTOR SCHEMES ON A CARTESIAN MESH (Lerat, 1981)

2.1 *Nine-point schemes in conservation form*

We consider the system of conservation laws:

$$w_t + f(w)_x + g(w)_y = 0 \tag{1}$$

where w is an unknown m-component vector and f, g are given
vector functions of w. This system is supposed to be hyperbolic,
that is the matrix

$$P_\mu(w) = \mu^x A(w) + \mu^y B(w) \tag{2}$$

has only real eigenvalues and can be uniformly diagonalized. In
(2), μ^x and μ^y denote scalars such that

$$(\mu^x)^2 + (\mu^y)^2 = 1$$

and

$$A(w) = \frac{df}{dw}(w), \qquad B(w) = \frac{dg}{dw}(w) .$$

We study nine-point schemes of the form:

$$w_{i,j}^{n+1} = H(w_{i-1,j-1}^n,\ w_{i-1,j}^n,\ w_{i-1,j+1}^n,\ w_{i,j-1}^n,\ w_{i,j}^n,$$

$$w_{i,j+1}^n,\ w_{i+1,j-1}^n,\ w_{i+1,j}^n,\ w_{i+1,j+1}^n) \tag{3}$$

where $w_{i,j}^n$ is the numerical solution at $x = i\Delta x$, $y = j\Delta y$ and $t = n\Delta t$, Δx and Δy are the space increments and Δt is the time step.

By extending the Lax and Wendroff definition (Lax and Wendroff, 1960) to the case of two space variables, we shall say that scheme (3) is "in conservation form" if it can be written as:

$$w_{i,j}^{n+1} = w_{i,j}^n - \sigma^x (h_{i+\frac{1}{2},j}^x - h_{i-\frac{1}{2},j}^x) - \sigma^y (h_{i,j+\frac{1}{2}}^y - h_{i,j-\frac{1}{2}}^y) \qquad (4a)$$

where

$$\sigma^x = \Delta t/\Delta x, \qquad \sigma^y = \Delta t/\Delta y$$

$$h_{i+\frac{1}{2},j}^x = h^x(w_{i,j-1}^n, w_{i,j}^n, w_{i,j+1}^n, w_{i+1,j-1}^n, w_{i+1,j}^n, w_{i+1,j+1}^n)$$
$$h_{i,j+\frac{1}{2}}^y = h^y(w_{i-1,j}^n, w_{i,j}^n, w_{i+1,j}^n, w_{i-1,j+1}^n, w_{i,j+1}^n, w_{i+1,j+1}^n), \qquad (4b)$$

The numerical fluxes h^x and h^y are functions of six **variables** which must only meet the consistency requirements

$$h^x(u,u,u,u,u,u) = f(u), \qquad h^y(u,u,u,u,u,u) = g(u), \qquad (4c)$$

for all u in $\mathbb{R}^m$.

Similarly as in (Lax and Wendroff, 1960), it can be shown that if a solution of the scheme (4) converges boundedly almost everywhere to a function w^*, then w^* is a weak solution of the system (1). In other words, the scheme (4) is able to capture shock waves or contact discontinuities.

Furthermore, the scheme (4) approximates the system (1) with a second-order accuracy if and only if:

$$(\bar{h}_{u_{-1}}^x + \bar{h}_{u_0}^x + \bar{h}_{u_1}^x - \bar{h}_{v_{-1}}^x - \bar{h}_{v_0}^x - \bar{h}_{v_1}^x)(u) = \sigma^x A^2(u)$$

$$(\bar{h}_{u_{-1}}^x - \bar{h}_{u_1}^x + \bar{h}_{v_{-1}}^x - \bar{h}_{v_1}^x)(u) = \frac{1}{2} \sigma^y A(u)\, B(u)$$

$$(\bar{h}_{u_{-1}}^y - \bar{h}_{u_1}^y + \bar{h}_{v_{-1}}^y - \bar{h}_{v_1}^y)(u) = \frac{1}{2} \sigma^x B(u)\, A(u) \qquad (5)$$

cont......

$$(\bar{h}^y_{u_{-1}} + \bar{h}^y_{u_0} + \bar{h}^y_{u_1} - \bar{h}^y_{v_{-1}} - \bar{h}^y_{v_0} - \bar{h}^y_{v_1})(u) = \sigma^y \, B^2(u) \qquad (5)$$

for all u in $\mathbb{R}^m$. The subscripts u_{-1}, u_0, u_1, v_{-1}, v_0 and v_1 denote respectively partial differentiation with respect to the first, second, . . . , sixth variable of h^x or h^y, and a bar over a function indicates a restriction to the set $u_{-1} = u_0 = u_1 = v_{-1} = v_0 = v_1$. For instance

$$\bar{h}^x_{u_{-1}}(u) = \frac{\partial h^x}{\partial u_{-1}}(u,u,u,u,u,u).$$

2.2 *One-predictor schemes*

To avoid computations of the matrices $A(w)$ and $B(w)$, we consider nine-point schemes of predictor-corrector type, that is schemes in the form:

$$\tilde{w}_{i,j} = \sum_{\substack{k=0,1 \\ \ell=0,1}} (a_{k,\ell} w^n_{i+k,j+\ell} + b^f_{k,\ell} f^n_{i+k,j+\ell} + b^g_{k,\ell} g^n_{i+k,j+\ell}), \qquad (6a)$$

$$w^{n+1}_{i,j} = \sum_{\substack{k=-1,0,1 \\ \ell=-1,0,1}} (c_{k,\ell} \, w^n_{i+k,j+\ell} + d^f_{k,\ell} f^n_{i+k,j+\ell} + d^g_{k,\ell} g^n_{i+k,j+\ell})$$

$$\qquad\qquad (6b)$$

$$+ \sum_{\substack{k=-1,0 \\ \ell=-1,0}} (e^f_{k,\ell} \tilde{f}_{i+k,j+\ell} + e^g_{k,\ell} \tilde{g}_{i+k,j+\ell})$$

where

$$f^n_{i,j} = f(w^n_{i,j}), \qquad \tilde{f}_{i,j} = f(\tilde{w}_{i,j}) \text{ and the same for } g.$$

Schemes (6) involve 47 scalar coefficients. By requiring that these schemes be in conservation form and satisfy conditions (5), we can obtain the class $\mathcal{C}$ of predictor-corrector schemes of second-order accuracy for system (1). The class $\mathcal{C}$ depends on 17 parameters. Here, we shall consider only the subclass of symmetric schemes. A more general study including six and seven point schemes can be found in (Lerat, 1981).

The nine-point scheme (3) is said to be "symmetric" if it is invariant when:

a) one exchanges $w^n_{i+1,j+k}$ and $w^n_{i-1,j+k}$, $k = -1,0,1$ while replacing f(w) by -f(w).

b) one exchanges $w^n_{i+k,j+1}$ and $w^n_{i+k,j-1}$, $k = -1,0,1$ while replacing g(w) by -g(w).

Such a scheme is suitable for the computation of flows around airfoils.

The subclass of symmetric schemes depends on four parameters α, γ, δ^x and δ^y ($\alpha \neq 0$). It can be written as:

$$\tilde{w}_{i,j} = \frac{1}{4}(w^n_{i,j} + w^n_{i+1,j} + w^n_{i,j+1} + w^n_{i+1,j+1})$$

$$- \frac{\alpha\sigma^x}{2}(f^n_{i+1,j} - f^n_{i,j} + f^n_{i+1,j+1} - f^n_{i,j+1}) \tag{7a}$$

$$- \frac{\alpha\sigma^y}{2}(g^n_{i,j+1} - g^n_{i,j} + g^n_{i+1,j+1} - g^n_{i+1,j})$$

$$w^{n+1}_{i,j} = w^n_{i,j} - \gamma \left[w^n_{i+1,j+1} - 2w^n_{i+1,j} + w^n_{i+1,j-1} - 2(w^n_{i,j+1} - 2w^n_{i,j} + w^n_{i,j-1}) \right.$$

$$\left. + w^n_{i-1,j+1} - 2w^n_{i-1,j} + w^n_{i-1,j-1} \right]$$

$$- \frac{\alpha^x}{4\alpha} \left[(2\alpha-1-4\alpha\delta^x)(f^n_{i+1,j} - f^n_{i-1,j}) + 2\alpha\delta^x(f^n_{i+1,j+1} - f^n_{i-1,j+1}) \right. \tag{7b}$$

$$\left. + f^n_{i+1,j-1} - f^n_{i-1,j-1}) + \tilde{f}_{i,j} - \tilde{f}_{i-1j} + \tilde{f}_{i,j-1} - \tilde{f}_{i-1,j-1} \right]$$

$$- \frac{\sigma^y}{4\alpha} \left[(2\alpha-1-4\alpha\delta^y)(g^n_{i,j+1} - g^n_{i,j-1}) + 2\alpha\delta^y(g^n_{i+1,j+1} - g^n_{i+1,j-1}) \right.$$

$$\left. + g^n_{i-1,j+1} - g^n_{i-1,j-1}) + \tilde{g}_{i,j} - \tilde{g}_{i,j-1} + \tilde{g}_{i-1,j} - \tilde{g}_{i-1,j-1} \right]$$

The predictor $\tilde{w}_{i,j}$ is a first-order approximation at $x = (i+\frac{1}{2})\Delta x$, $y = (j+\frac{1}{2})\Delta y$ and $t = (n+\alpha)\Delta t$ (see Fig. 1). The schemes of Burstein (1967a, 1967b) are obtained by setting $\alpha = \frac{1}{2}$ or $\alpha = 1$ and $\gamma = \delta^x = \delta^y = 0$, but the original scheme of Richtmyer (1962) (see Richtmyer and Morton 1967), does not belong to the class C since it needs a staggered mesh. The subclass (7) also contains the two-parameter family of schemes proposed by Turkel

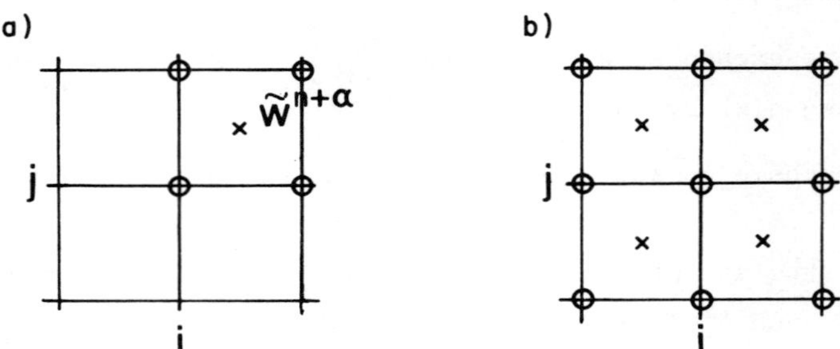

Fig. 1 Symmetric predictor-corrector schemes. a) predictor
        b) corrector.

(1974). This family corresponds to $\delta^x = \delta^y = 0$. Let us note
that the term with coefficient γ introduces an artificial damping if $\gamma > 0$.

However, schemes (7) are not quite satisfactory. Indeed, by
setting

$$\gamma = 0 \text{ and } \delta^x = \delta^y = \frac{1}{4} - \frac{1}{8\alpha}$$

we can obtain linear stability (in L^2) under the **usual** Courant-
Friedrichs-Lewy condition, but this choice leads to a large
dispersive error. On the contrary, if we set

$$\gamma = 0 \text{ and } \delta^x = \delta^y = - \frac{1}{8\alpha}$$

the dispersive error is reduced, but this yields linear instability.

2.3 Two-predictor schemes

To overcome this difficulty, we shall consider more general
schemes. First, we notice that numerical fluxes h^x and h^y
relative to a predictor-corrector scheme are coupled since they
both make use of the same predictor. Nevertheless, numerical
fluxes which meet the consistency requirement (4c) and the
accuracy condition (5) can be independently defined in the x
and y direction. In order to obtain uncoupled numerical fluxes,
we consider schemes involving one predictor in each space direction.
Such a scheme seems to have been first introduced by Thommen
(1966) for Navier-Stokes equations as a variant of the Richtmyer
scheme allowing a more compact space-differencing. Two-predictor
schemes for system (1) are in the following form:

$$\widetilde{w}^{x}_{i,j} = \sum_{\substack{k=0,1 \\ \ell=-1,0,1}} (a^{x}_{k,\ell} w^{n}_{i+k,j+\ell} + b^{f,x}_{k,\ell} f^{n}_{i+k,j+\ell} + b^{g,x}_{k,\ell} g^{n}_{i+k,j+\ell}) , \qquad (8a)$$

$$\widetilde{w}^{y}_{i,j} = \sum_{\substack{k=-1,0,1 \\ \ell=0,1}} (a^{y}_{k,\ell} w^{n}_{i+k,j+\ell} + b^{f,y}_{k,\ell} f^{n}_{i+k,j+\ell} + b^{g,y}_{k,\ell} g^{n}_{i+k,j+\ell}) , \qquad (8b)$$

$$w^{n+1}_{i,j} = \sum_{\substack{k=-1,0,1 \\ \ell=-1,0,1}} (c_{k,\ell} w^{n}_{i+k,j+\ell} + d^{f}_{k,\ell} f^{n}_{i+k,j+\ell} + d^{g}_{k,\ell} g^{n}_{i+k,j+\ell}) \qquad (8c)$$

$$+ \sum_{k=-1,0} e^{f}_{k,\ell} \widetilde{f}^{x}_{k,\ell} f^{x}_{i+k,j+\ell} + \sum_{\ell=-1,0} e^{g}_{k,\ell} \widetilde{g}^{y}_{i+k,j+\ell} ,$$

where
$$\widetilde{f}^{x}_{i,j} = f(\widetilde{w}^{x}_{i,j}) \quad \text{and} \quad \widetilde{g}^{y}_{i,j} = g(\widetilde{w}^{y}_{i,j}) .$$

We note that the predictors (8a) and (8b) involve more points than the predictor (6a) (6 instead of 4) and that the two predictor-scheme (8) introduces more scalar coefficients than the one-predictor scheme (6) (67 instead of 47).

In the general class of two-predictor schemes of second-order accuracy, the following subclass has the most interesting properties:

$$\widetilde{w}^{x}_{i,j} = (1-\beta^{x}) w^{n}_{i,j} + \beta^{x} w^{n}_{i+1,j} - \alpha^{x} \sigma^{x} (f^{n}_{i+1,j} - f^{n}_{i,j}) \qquad (9a)$$

$$- \alpha^{x} \frac{\sigma^{y}}{4} (g^{n}_{i+1,j+1} + g^{n}_{i,j+1} - g^{n}_{i+1,j-1} - g^{n}_{i,j-1})$$

$$\widetilde{w}^{y}_{i,j} = (1-\beta^{y}) w^{n}_{i,j} + \beta^{y} w_{i,j+1} - \alpha^{y} \sigma^{y} (g^{n}_{i,j+1} - g^{n}_{i,j}) \qquad (9b)$$

$$- \alpha^{y} \frac{\sigma^{x}}{4} (f^{n}_{i+1,j+1} + f^{n}_{i+1,j} - f^{k}_{i-1,j+1} - f^{n}_{i-1,j})$$

$$w^{n+1}_{i,j} = w^{n}_{i,j} - \frac{\sigma^{x}}{2\alpha^{x}} \Big[(\alpha^{x} - \beta^{x}) f^{n}_{i+1,j} + (2\beta^{x} - 1) f^{n}_{i,j} + (1 - \alpha^{x} - \beta^{x}) f^{n}_{i-1,j}$$

$$+ \widetilde{f}^{x}_{i,j} - \widetilde{f}^{x}_{i-1,j} \Big]$$

$$- \frac{\sigma^{y}}{2\alpha^{y}} \Big[(\alpha^{y} - \beta^{y}) g^{n}_{i,j+1} + (2\beta^{y} - 1) g^{n}_{i,j} + (1 - \alpha^{y} - \beta^{y}) g^{n}_{i,j-1}$$

$$+ \widetilde{g}^{y}_{i,j} - \widetilde{g}^{y}_{i,j-1} \Big]$$

where α^X, β^X, α^Y and β^Y are four parameters ($\alpha^X \neq 0, \alpha^Y \neq 0$).
The predictors $\widetilde{w}^X_{i,j}$ and $\widetilde{w}^Y_{i,j}$ are first-order approximations at

$$x = (i+\beta^X)\Delta x, \quad y = j\Delta y, \quad t = (n+\alpha^X)\Delta t$$

and $x = i.\Delta x$, $y = (j+\beta^Y)\Delta y$, $t = (n+\alpha^Y)\Delta t$, respectively.

The subclass (9) appears to be a straightforward extension of the schemes S^α_β in one-space variable (Lerat and Peyret, 1973) that is:

$$\widetilde{w}_i = (1-\beta) w^n_i + \beta w^n_{i+1} - \alpha\sigma (f^n_{i+1} - f^n_i) \tag{10a}$$

$$w^{n+1}_i = w^n_i - \frac{\sigma}{2\alpha}[(\alpha-\beta) f^n_{i+1} + (2\beta-1) f^n_i + (1-\alpha-\beta) f^n_{i-1} + \widetilde{f}_i - \widetilde{f}_{i-1}] \tag{10b}$$

where $\sigma = \Delta t/\Delta x$.

Schemes (9) can be written in the form (4) with the numerical fluxes:

$$h^X_{i+\frac{1}{2},j} = \frac{1}{2\alpha^X}[(\alpha^X-\beta^X) f(w^n_{i+1,j}) + (\alpha^X+\beta^X-1) f(w^n_{i,j}) + f(\widetilde{w}^X_{i,j})]$$

$$\tag{11}$$

$$h^Y_{i,j+\frac{1}{2}} = \frac{1}{2\alpha^Y}[(\alpha^Y-\beta^Y) g(w^n_{i,j+1}) + (\alpha^Y+\beta^Y-1) g(w^n_{i,j}) + g(\widetilde{w}^Y_{i,j})]$$

They are symmetric if and only if: $\beta^X = \beta^Y = \frac{1}{2}$.

The scheme of Thommen (1966) belongs to the subclass (9) and corresponds to:

$$\alpha^X = \alpha^Y = \beta^X = \beta^Y = \frac{1}{2}.$$

This scheme has been used by Singleton (1968) and Magnus and Yoshihara (1975) for the Euler equations. Palumbo and Rubin (1972) have proposed another two-predictor scheme for the Navier-Stokes equations corresponding to:

$$\alpha^X = \alpha^Y = 1 \quad \text{and} \quad \beta^X = \beta^Y = \frac{1}{2}.$$

It is worth noting that one can easily generalize the schemes (9) to the case of three space variables by introducing a third predictor (see Lerat, 1981). To a certain extent, multi-predictor schemes share some of the advantages of time-splitting methods.

A real time-splitting approach using the S^α_β schemes has been considered by Laval (1981).

For the schemes (9), we have derived the equivalent system up to third-order, that is the system approximated by the schemes with a third-order accuracy. This system can be written in the divergence form:

$$w_t + f(w)_x + g(w)_y = -\frac{\Delta t^2}{6} w_{ttt} - \frac{\Delta x^2}{6} f(w)_{xxx} - \frac{\Delta y^2}{6} g(w)_{yyy}$$

$$+ \frac{\Delta x^2}{4\alpha^x} \{f''(w) [(1-\beta^x) w_x - \alpha^x \sigma^x w_t,\ \beta^x w_x + \alpha^x \sigma^x w_t]\}_x \qquad (12)$$

$$+ \frac{\Delta y^2}{4\alpha^y} \{g''(w) [(1-\beta^y) w_y - \alpha^y \sigma^y w_t,\ \beta^y w_y + \alpha^y \sigma^y w_t]\}_y$$

where $f''(w)$ and $g''(w)$ are symmetric bilinear operators which denote the second derivatives of f and g at point w.

We remark that the equivalent system (12) has no dispersive terms with derivatives $g(w)_{xxy}$ or $f(w)_{xyy}$ contrary to the case of schemes involving only one predictor. This is due to the fact that in the first predictor (9a), the derivative $g(w)_y$ is approximated by a centred four-point formula rather than by $(g^n_{i,j+1} - g^n_{i,j})/\Delta y$, and likewise in the second predictor (9b) with the derivative $f(w)_x$. A similar centred formula cannot be used at the first step of a one-predictor scheme (6), because for such a scheme, it is not compatible with the requirement of a nine-point scheme.

When the matrices $A(w)$ and $B(w)$ are supposed to be constant, all the schemes (9) reduce to:

$$w^{n+1}_{i,j} = w^n_{i,j} - \frac{1}{2}\sigma^x A (w^n_{i+1,j} - w^n_{i-1,j}) - \frac{1}{2}\sigma^y B (w^n_{i,j+1} - w^n_{i,j-1})$$

$$+ \frac{1}{2}(\sigma^x A)^2 (w^n_{i+1,j} - 2w^n_{i,j} + w^n_{i-1,j}) + \frac{1}{2}(\sigma^y B)^2 (w^n_{i,j+1} - 2w^n_{i,j} + w^n_{i,j-1})$$

$$+ \frac{1}{8} \sigma^x \sigma^y (AB+BA) (w^n_{i+1,j+1} - w^n_{i+1,j-1} - w^n_{i-1,j+1} + w^n_{i-1,j-1}). \qquad (13)$$

Scheme (13) is nothing but the linear version of the original one-step scheme proposed by Lax and Wendroff (1964). In the case where A and B are symmetrical, Lax and Wendroff (1964) have shown that a sufficient condition for stability of (13) is:

$$\sigma^x \rho(A) \le \frac{1}{\sqrt{8}} \qquad\qquad \sigma^y \rho(B) \le \frac{1}{\sqrt{8}}$$

where ρ denotes the spectral radius. In gas dynamics, numerical experiment indicates that one can use a time-increment greater than the limit allowed by the above inequalities. By performing a numerical study of the amplification matrix relative to scheme (13) with $\Delta y = \Delta x$, Burstein (1967a) has found the following stability condition:

$$\sigma(|\vec{V}| + a) \le 0.5406 \tag{14}$$

where $\vec{V}$ is the fluid velocity, a is the sound speed and $\sigma = \sigma^x = \sigma^y$.

Let us note that schemes (9) are dissipative in the sense of Kreiss (1964), inside their strict stability domain, except for some special states w such as stagnation or sonic states in gas dynamics. For the shortest wavelengths along the diagonal directions of the mesh ($\lambda^x = \pm 2\Delta x$, $\lambda^y = \pm 2\Delta y$), the amplification matrix becomes:

$$G = I - 2\left[(\sigma^x A)^2 + (\sigma^y B)^2\right]$$

where I is the identity matrix. On the contrary, for these shortest wavelengths, we get $G = I$ for the Richtmyer scheme or the one-predictor schemes (7) with $\gamma = 0$ (no artificial damping), so that these schemes are not dissipative in the sense of Kreiss.

2.4 Optimal scheme

Now, we want to choose the parameters α^x, α^y, β^x and β^y of the two-predictor schemes (9). These parameters are related to the nonlinear properties of the schemes since they are not present in the linear version (13). In the one-dimensional case, an optimal S_β^α scheme has been found by analysing the equivalent system up to third-order (Lerat and Peyret, 1975). The optimal scheme is dissipative at second-order in compression waves and shocks and its dissipation is minimal among the schemes S_β^α. It corresponds to $\alpha = 1 + \sqrt{5}/2$ and $\beta = \frac{1}{2}$.

In the equivalent system (12), the terms multiplied by Δx^2 are exactly those obtained for the schemes S_β^α applied to the system $w_t + f(w)_x = 0$ with $\alpha = \alpha^x$ and $\beta = \beta^x$. In the same way, the terms multiplied by Δy^2 are exactly those obtained for the schemes S_β^α applied to the system $w_t + g(w)_y = 0$ with $\alpha = \alpha^y$ and $\beta = \beta^y$. So, we can obtain an optimal scheme in the x and y directions by setting:

$$\alpha^x = \alpha^y = 1 + \frac{\sqrt{5}}{2}, \qquad \beta^x = \beta^y = \frac{1}{2}$$

This scheme happens to be symmetric (see Fig. 2).

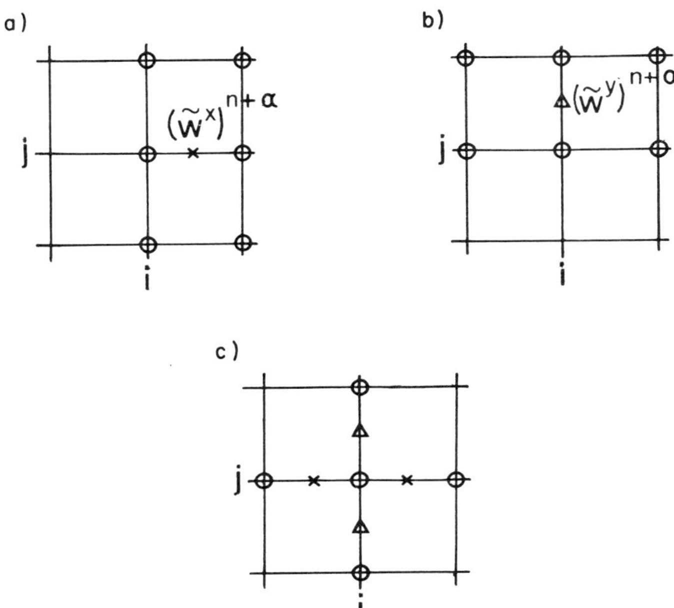

Fig. 2 Optimal scheme ($\alpha = 1 + \sqrt{5}/2$). a) first-predictor, b) second-predictor, c) corrector.

3. FINITE-VOLUME FORMULATION ON A CURVILINEAR MESH

Cartesian meshes are seldom used in practical computations. The necessity to close up the mesh points in the regions where the gradients of the solution are large and also the need to correctly treat the boundary conditions often prescribe the use of a curvilinear mesh well suited to the problem. In this section, we shall formulate the optimal scheme on a curvilinear mesh. In order to solve problems in domains with moving boundaries, we shall consider curvilinear meshes in motion. Since we look for weak solutions, we have to preserve the conservative property in a rigorous way. To reach this aim, we use the finite-volume approach, first introduced in (MacCormack and

Paullay, 1972; Rizzi and Inouye, 1973 and MacCormack et al.,
1976) for the MacCormack scheme.

Systems of conservation laws are originally stated in integral
form.   In a bounded domain $\Omega(t)$ with boundary $\Gamma(t)$ in motion
with respect to an absolute cartesian frame $(O;x,y)$, system (1)
can be expressed as:

$$\iint_{\Omega(t)} w\,dx\,dy \Big|_{t_1}^{t_2} + \int_{t_1}^{t_2} dt \int_{\Gamma(t)} h(w,\vec{\nu},\vec{s})\,d\Gamma = 0 \qquad (16a)$$

where

$$h(w,\vec{\nu},\vec{s}) = [f(w)-s^x w]\nu^x + [g(w)-s^y w]\nu^y,$$

ν^x and ν^y are the components of the unit outward normal $\vec{\nu}$ to
$\Gamma(t)$, s^x and s^y, are the components of the displacement velocity
$\vec{s}$ of the points of $\Gamma(t)$ (see Fig. 3) and (t_1,t_2) is a time interval.

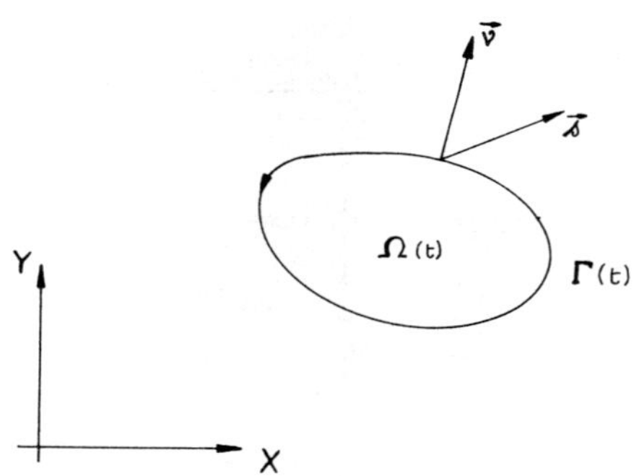

Fig. 3   General domain $\Omega(t)$.

For the computation of the flux, we note that

$$h(w,\vec{\nu},\vec{s})\,d\Gamma = f(w)\,dy - g(w)\,dx - w\,ds , \quad \text{on } \Gamma(t) \qquad (16b)$$

with

$$ds = s^x dy - s^y dx.$$

Let us now divide the time axis into intervals (t^n,t^{n+1}) and
the space domain into a number of cells $\Omega_{i,j} = \Omega_{i,j}(t)$ delimited
by a moving curvilinear mesh and denote by $\Gamma_{i,j} = \Gamma_{i,j}(t)$ the
boundary of $\Omega_{i,j}$ (see Fig. 4).

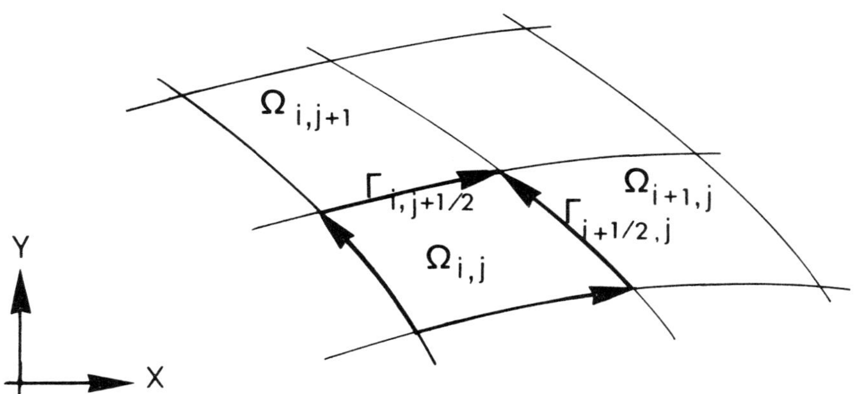

Fig. 4   Typical mesh cell.

We define the mean value of w in the cell $\Omega_{i,j}$ at time t^n:

$$\bar{w}^n_{i,j} = \frac{1}{S^n_{i,j}} \iint_{\Omega_{i,j}} w\,dx\,dy \Big|_{t=t^n}$$

where $S^n_{i,j}$ is the area of $\Omega_{i,j}(t^n)$.

We also define the mean values during the time interval (t^n, t^{n+1}) of the fluxes across $\Gamma_{i+\frac{1}{2},j} = \Gamma_{i,j} \cap \Gamma_{i+1,j}$ and across $\Gamma_{i,j+\frac{1}{2}} = \Gamma_{i,j} \cap \Gamma_{i,j+1}$, that is:

$$\bar{H}^{n+\frac{1}{2}}_{i+\frac{1}{2},j} = \frac{1}{\Delta t} \int_{t^n}^{t^{n+1}} dt \int_{\Gamma_{i+\frac{1}{2},j}} h(w,\vec{\nu},\vec{s})\,d\Gamma$$

$$\bar{H}^{n+\frac{1}{2}}_{i,j+\frac{1}{2}} = \frac{1}{\Delta t} \int_{t^n}^{t^{n+1}} dt \int_{\Gamma_{i,j+\frac{1}{2}}} h(w,\vec{\nu},\vec{s})\,d\Gamma$$

where $\Delta t = t^{n+1} - t^n$, $\Gamma_{i+\frac{1}{2},j}$ and $\Gamma_{i,j+\frac{1}{2}}$ are respectively oriented in the direction of increasing j and of increasing i.

With these notations, the *exact system* (16) can be rewritten as:

$$S_{i,j}^{n+1}\bar{w}_{i,j}^{n+1} = S_{i,j}^{n}\bar{w}_{i,j}^{n} - \Delta t(\bar{H}_{i+\frac{1}{2},j}^{n+\frac{1}{2}}-\bar{H}_{i-\frac{1}{2},j}^{n+\frac{1}{2}}-\bar{H}_{i,j+\frac{1}{2}}^{n+\frac{1}{2}}+\bar{H}_{i,j-\frac{1}{2}}^{n+\frac{1}{2}}) \qquad (17)$$

Let $w_{i,j}^{n}$ be an approximation for $\bar{w}_{i,j}^{n}$, given by an explicit numerical method. We shall say that the method is "in conservation form" if it can be written as:

$$S_{i,j}^{n+1}w_{i,j}^{n+1} = S_{i,j}^{n}w_{i,j}^{n} - \Delta t(H_{i+\frac{1}{2},j}^{n+\frac{1}{2}}-H_{i-\frac{1}{2},j}^{n+\frac{1}{2}}-H_{i,j+\frac{1}{2}}^{n+\frac{1}{2}}+H_{i,j-\frac{1}{2}}^{n+\frac{1}{2}}) \qquad (18)$$

where the numerical fluxes $H_{i+\frac{1}{2},j}^{n+\frac{1}{2}}$ and $H_{i,j+\frac{1}{2}}^{n+\frac{1}{2}}$ are consistent approximations of the exact expressions of the fluxes $\bar{H}_{i+\frac{1}{2},j}^{n+\frac{1}{2}}$ and $\bar{H}_{i,j+\frac{1}{2}}^{n+\frac{1}{2}}$, respectively. These approximations involve vectors $w_{i+k,j+\ell}^{n}$ at time t^{n} in cells $\Omega_{i+k,j+\ell}$ close to $\Gamma_{i+\frac{1}{2},j}$ or $\Gamma_{i,j+\frac{1}{2}}$. For a method based on a nine-point scheme, the numerical flux $H_{i+\frac{1}{2},j}^{n+\frac{1}{2}}$ (resp. $H_{i,j+\frac{1}{2}}^{n+\frac{1}{2}}$) may only depend on values relative to the six cells $\Omega_{i+k,j+\ell}$ for $k = 0,1$ and $\ell = -1,0,1$ (resp. $k = -1,0,1$ and $\ell = 0,1$).

The main outcome of this definition is the following one: if we use (18) to strike a balance of Sw on a domain made of mesh cells, then the interior numerical fluxes exactly cancel out two by two, whatever the order of accuracy of (18) may be. When applied on a fixed cartesian mesh, the above definition is nothing else but the one given in the previous section.

Let us now present a finite-volume method based on the optimal scheme (9), (15). We shall consider the sides $\Gamma_{i+k,j+\ell}$ ($k =\pm \frac{1}{2}$, $\ell = 0$ or $k = 0$, $\ell = \pm\frac{1}{2}$) of a cell $\Omega_{i,j}$ as vectors and denote by $(\Delta x,\Delta y)_{i+k,j+\ell}$ the components of $\Gamma_{i+k,j+\ell}$ and by $(s^{x},z^{y})_{i+k,j+\ell}$ the components of the mean velocity of $\Gamma_{i+k,j+\ell}$. For the computation of the predictors, we define (moving) cells $\Omega_{i+\frac{1}{2},j}^{1}$ with sides $\Gamma_{i,j}^{1}$, $\Gamma_{i+1,j}^{1}$, $\Gamma_{i+\frac{1}{2},j-\frac{1}{2}}^{1}$, $\Gamma_{i+\frac{1}{2},j+\frac{1}{2}}^{1}$ and areas:

$$S_{i+\frac{1}{2},j} = \frac{1}{2}(S_{i,j}+S_{i+1,j})$$

and likewise cells $\Omega_{i,j+\frac{1}{2}}^{2}$ with sides $\Gamma_{i,j}^{2}$, $\Gamma_{i,j+1}^{2}$, $\Gamma_{i-\frac{1}{2},j+\frac{1}{2}}^{2}$, $\Gamma_{i+\frac{1}{2},j+\frac{1}{2}}^{2}$ and areas:

$$S_{i,j+\frac{1}{2}} = \frac{1}{2}(S_{i,j} + S_{i,j+1}),$$

as shown on Fig. 5. Finally, we denote by $(\Delta x^p, \Delta y^p)_{i+k,j+\ell}$ the components of a side $\Gamma^p_{i+k,j+\ell}$ and by $[(s^x)^p, (s^y)^p]_{i+k,j+\ell}$ the components of its mean velocity, for $p = 1,2$.

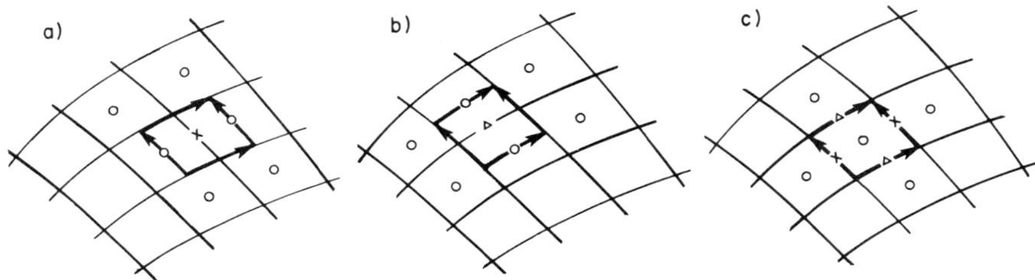

Fig. 5.   Finite-volume method based on the optimal scheme. a) first-predictor in $\Omega^1_{i+\frac{1}{2},j}$, b) second-predictor in $\Omega^2_{i,j+\frac{1}{2}}$, c) corrector in $\Omega_{i,j}$.

a) <u>First-predictor</u>: Mean values $\widetilde{w}^1$ in cells $\Omega^1_{i+\frac{1}{2},j}$ at time $t^n + \alpha\Delta t\,(\alpha = 1+\sqrt{5}/2)$ are defined by:

$$(S\widetilde{w}^1)^{n+\alpha}_{i+\frac{1}{2},j} = \frac{1}{2}\left[(Sw)^n_{i,j} + (Sw)^n_{i+1,j}\right]$$

$$-\alpha\Delta t\left[(H^1)^n_{i+1,j} - (H^1)^n_{i,j} - (H^1)^n_{i+\frac{1}{2},j+\frac{1}{2}} + (H^1)^n_{i+\frac{1}{2},j-\frac{1}{2}}\right]$$

(19a)

b) <u>Second-predictor</u>: Mean values $\widetilde{w}^2$ in cells $\Omega^2_{i,j+\frac{1}{2}}$ at time $t^n + \alpha\Delta t$ are defined by:

$$(S\widetilde{w}^2)^{n+\alpha}_{i,j+\frac{1}{2}} = \frac{1}{2}\left[(Sw)^n_{i,j} + (Sw)^n_{i,j+1}\right]$$

$$-\alpha\Delta t\left[(H^2)^n_{i+\frac{1}{2},j+\frac{1}{2}} - (H^2)^n_{i-\frac{1}{2},j+\frac{1}{2}} - (H^2)^n_{i,j+1} + (H^2)^n_{i,j}\right]$$

(19b)

In the expressions of the predictors:

$$(H^p)^n_{i+k,j+\ell} = (f\,\Delta y^p - g\Delta x^p - w\Delta s^p)^n_{i+k,j+\ell}, \qquad p = 1,2$$

where

$$\Delta s^p = (s^x)^p\Delta y^p - (s^y)^p\Delta x^p$$

$$w^n_{i+\frac{1}{2},j+\frac{1}{2}} = \frac{1}{4}(w^n_{i,j}+w^n_{i+1,j}+w^n_{i,j+1}+w^n_{i+1,j+1})$$

$$f^n_{i+\frac{1}{2},j+\frac{1}{2}} = \frac{1}{4}(f^n_{i,j}+f^n_{i+1,j}+f^n_{i,j+1}+f^n_{i+1,j+1})$$

$$f^n_{i,j} = f(w^n_{i,j})$$

and the same for g.

c) <u>Corrector</u>: Mean values $w^{n+1}_{i,j}$ in cells $\Omega_{i,j}$ at time t^{n+1} are obtained from the general expression (18) of a method in conservation form, with the following definition of the numerical fluxes:

$$H^{n+\frac{1}{2}}_{i+\frac{1}{2},j} = \frac{2\alpha-1}{2\alpha}(f\Delta y-g\Delta x-w\Delta s)^n_{i+\frac{1}{2},j} + \frac{1}{2\alpha}(\tilde{f}^1\Delta y-\tilde{g}^1\Delta x-\tilde{w}^1\Delta s)^{n+\alpha}_{i+\frac{1}{2},j}$$

$$H^{n+\frac{1}{2}}_{i,j+\frac{1}{2}} = \frac{2\alpha-1}{2\alpha}(f\Delta y-g\Delta x-w\Delta s)^n_{i,j+\frac{1}{2}} + \frac{1}{2\alpha}(\tilde{f}^2\Delta y-\tilde{g}^2\Delta x-\tilde{w}^2\Delta s)^{n+\alpha}_{i,j+\frac{1}{2}} \qquad (19c)$$

where

$$\Delta s = s^x\Delta y-s^y\Delta x$$

$$w^n_{i+\frac{1}{2},j} = \frac{1}{2}(w^n_{i,j}+w^n_{i+1,j}) , \qquad w^n_{i,j+\frac{1}{2}} = \frac{1}{2}(w^n_{i,j}+w^n_{i,j+1})$$

$$f^n_{i+\frac{1}{2},j} = \frac{1}{2}(f^n_{i,j}+f^n_{i+1,j}) , \qquad f^n_{i,j+\frac{1}{2}} = \frac{1}{2}(f^n_{i,j}+f^n_{i,j+1})$$

$$(\tilde{f}^1)^{n+\alpha}_{i+\frac{1}{2},j} = f[(\tilde{w}^1)^{n+\alpha}_{i+\frac{1}{2},j}] , \qquad (\tilde{f}^2)^{n+\alpha}_{i,j+\frac{1}{2}} = f[(\tilde{w}^2)^{n+\alpha}_{i,j+\frac{1}{2}}]$$

and the same for g.

4. APPLICATION TO THE CALCULATION OF TRANSONIC FLOWS

4.1 Numerical treatment of transonic flow problems

Unsteady flows of an inviscid compressible fluid are governed by the Euler equations which express the conservation of mass, momentum and energy, and form a hyperbolic system. In the case of a two-dimensional flow, the Euler equations can be expressed in the integral form (16) where w, f(w) and g(w) are the four-component vectors:

$$w = \begin{bmatrix} \rho \\ \rho u \\ \rho v \\ \rho E \end{bmatrix}, \quad f(w) = \begin{bmatrix} \rho u \\ \rho u^2 + p \\ \rho uv \\ (\rho E + p) u \end{bmatrix}, \quad g(w) = \begin{bmatrix} \rho v \\ \rho uv \\ \rho v^2 + p \\ (\rho E + p) v \end{bmatrix}$$

where ρ is the density, u and v are the components of the fluid velocity $\vec{V}$ in the frame $(O; x, y)$, $E = e + (u^2 + v^2)/2$ and e is the specific internal energy. The pressure p is related to ρ and e by an equation of state. In the calculations, we have taken

$$p = (\gamma - 1)\rho e, \quad \text{with} \quad \gamma = 1.4.$$

When the unsteady finite-volume method is applied to the calculation of transonic flow, two kinds of simplification can be performed.

a) The first kind of simplification concerns a steady flow over an airfoil with uniform freestream. In this case, we use a fixed mesh and we replace the unsteady energy equation by the condition of constant total enthalpy:

$$\frac{\gamma}{\gamma - 1} \frac{p}{\rho} + \frac{1}{2}(u^2 + v^2) = H_\infty$$

where H_∞ is the freestream total enthalpy.

To reduce further the computing time, we use the maximal local time-step $\Delta t = \Delta t_{i,j}$ allowed by the stability condition for each cell $\Omega_{i,j}$. We obtain a "pseudo-unsteady method" which is only consistent when the steady-state has been reached. In a more general way, various types of pseudo-unsteady methods have been developed by Viviand and Veuillot (Viviand and Veuillot, 1978 and Viviand, 1980).

b) The second kind of simplification occurs for the unsteady flow over an airfoil in rigid body motion. In this case, we choose a mesh moving with the airfoil and rewrite the integral system (16) by using space coordinates in a relative cartesian frame (A, ξ, η) with origin A at the leading edge and ξ-axis directed toward the trailing edge. If we define the airfoil motion by the trajectory of the leading edge: $x_A = x_A(t)$, $y_A = y_A(t)$ and by the angle $\theta = \theta(t)$ from the x-axis to the ξ-axis (see Fig. 6), the integral system (16) becomes:

$$\iint_\Omega wd\xi d\eta \Big|_{t_1}^{t_2} + \int_{t_1}^{t_2} dt \int_\Gamma f_r(w)d\eta - g_r(w)d\xi = 0 \qquad (20)$$

where the domain ω is fixed in the relative frame (A,ξ,η) and

$$f_r(w) = \begin{bmatrix} \rho u_r \\ \rho uu_r + p\cos\theta \\ \rho vu_r + p\sin\theta \\ \rho Eu_r + pu' \end{bmatrix}, \quad g_r(w) = \begin{bmatrix} \rho v_r \\ \rho uv_r - p\sin\theta \\ \rho vv_r + p\cos\theta \\ \rho Ev_r + pv' \end{bmatrix}$$

with

$$u' = u\cos\theta + v\sin\theta \quad , \quad v' = -u\sin\theta + v\cos\theta$$

$$u_r = (u-\dot{x}_A)\cos\theta + (v-\dot{y}_A)\sin\theta + \eta\dot\theta \quad , \quad v_r = -(u-\dot{x}_A)\sin\theta + (v-\dot{y}_A)\cos\theta - \xi\dot\theta$$

The dots denote time-differentiation. In the transport terms, u_r and v_r are the components of the relative velocity in the relative frame. In the energy equations, u' and v' are the components of the absolute velocity in the relative frame.

   With such an approach, we can calculate the unsteady flow as if the mesh were fixed.

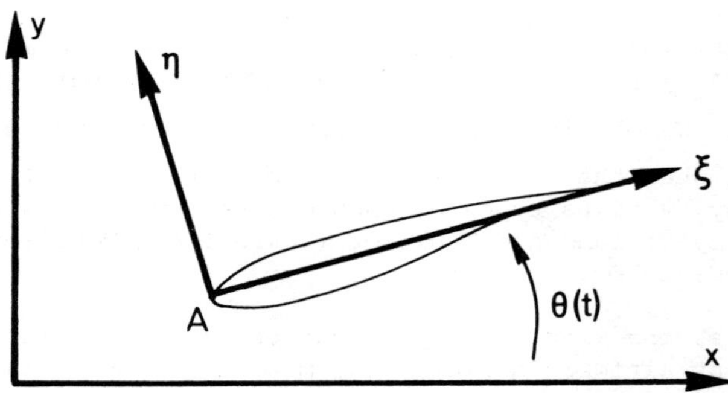

Fig. 6   Airfoil in rigid body motion.

   We now return to the general situation described by equations (16) and consider the treatment of the boundary condition on the airfoil, that is the slip condition:

$$(u-s^x)\nu^x + (v-s^y)\nu^y = 0.$$

In the integral system (16), this yields:

$$h(w,\vec{v},\vec{s})\,d\Gamma = p \cdot \begin{bmatrix} 0 \\ dy \\ -dx \\ ds \end{bmatrix}, \qquad \text{on the airfoil} \qquad (21)$$

where $ds = s^x dy - s^y dx$. Thus to calculate the flux across the airfoil, we need only the pressure on the airfoil.

For the computation of the cells $\Omega_{i,1}$ located along the airfoil, the finite-volume method is slightly modified as follows. The boundary of $\Omega_{i,1}$ is composed of the four sides $\Gamma_{i+\frac{1}{2},1}$, $\Gamma_{i-\frac{1}{2},1}$, $\Gamma_{i,3/2}$ and $\Gamma_{i,\frac{1}{2}}$ (see Fig. 7b). Across the side $\Gamma_{i,\frac{1}{2}}$ lying on the airfoil, the flux is obtained from (21) with a pressure p deduced from the parabolic extrapolation of the interior pressure at time $t^{n+\frac{1}{2}}$. Across the side $\Gamma_{i,3/2}$, the flux is computed as any interior flux. However, the flux across $\Gamma_{i+\frac{1}{2},1}$ (or $\Gamma_{i-\frac{1}{2},1}$) must be modified since, in the first predictor (19a), the flux $(H^1)^n_{i+\frac{1}{2},\frac{1}{2}}$ across the airfoil cannot be defined by the current formula (compare Figs. 5a and 7a). This flux is calculated by taking into account the slip condition, that is by using again (21) and a parabolic extrapolation for p at time t^n.

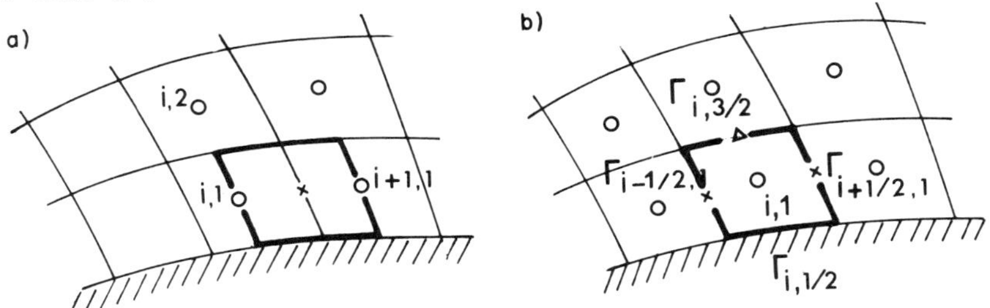

Fig. 7   Computation of the cells along the airfoil.   a) first-
         predictor, b) corrector.

At the trailing edge, we ensure an exact conservation of all fluxes, as sketched on Fig. 8. The Kutta-Joukowsky condition happens to be automatically satisfied in practice, probably due to the dissipative properties of the numerical method.

Across a subsonic inflow boundary, the flux is computed from the data of total enthalpy, entropy and velocity direction, and

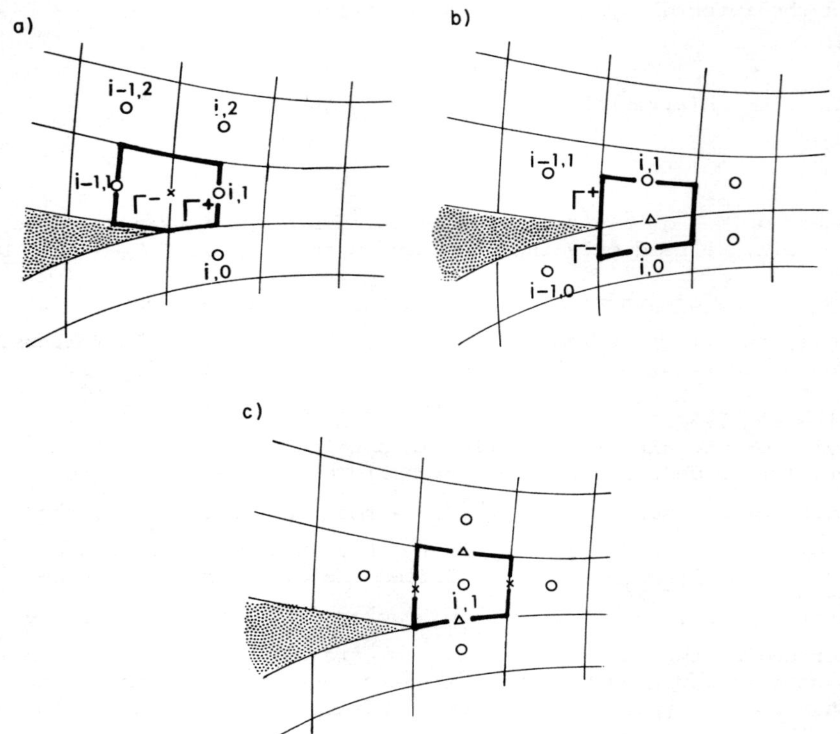

Fig. 8   Computation of the trailing-edge cells. a) first-predictor
(slip condition on Γ^-, flux across Γ^+ from $\Omega_{i,0}$ and $\Omega_{i,1}$)
b) second predictor (flux across Γ^- from $\Omega_{i-1,0}$ and
$\Omega_{i,0}$, flux across Γ^+ from $\Omega_{i-1,1}$ and $\Omega_{i,1}$) c) corrector.

an extrapolation for pressure. Across a subsonic outflow boundary, the flux is computed from the pressure data and extrapolations of the other quantities.

4.2 Steady transonic flows

   a) Symmetrical transonic flow over the NACA 0012 airfoil

   As a first application, the method is applied to the calculation of the flow over the NACA 0012 airfoil at zero incidence
for a freestream Mach number $M_\infty = 0.85$. Figure 9 shows the computational mesh obtained from the parabolic transformation of
Jameson (1974). The total number of mesh cells is 188 × 24 with
128 cells along the airfoil. However, owing to the symmetry of

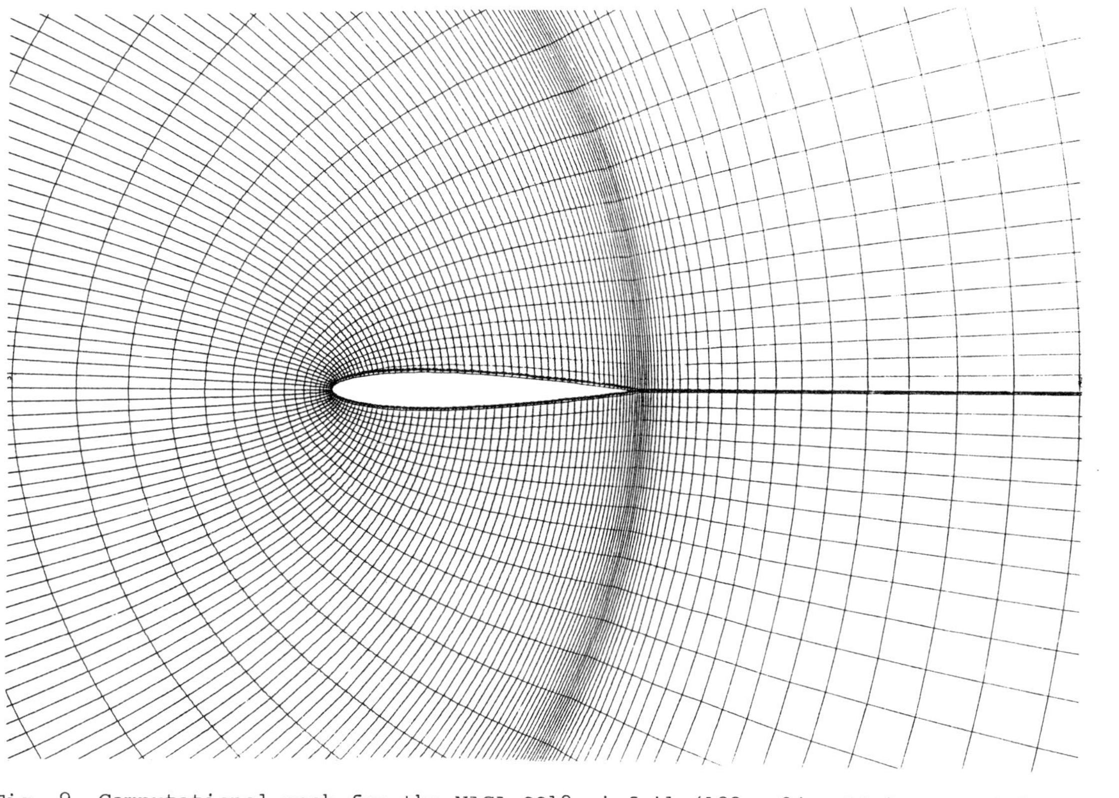

Fig. 9 Computational mesh for the NACA 0012 airfoil (188 × 24 cells). Partial view

the flow, the computational domain has been restricted to a half
plane. A comparison is made with a finite-volume method based on
the classical MacCormack scheme (without time-splitting). The
two calculations have been performed in the same conditions: same
mesh, same finite-volume approach, same treatment of boundary
conditions. In both methods, the artificial viscosity term
described in Lerat and Sidès (1979) has been added in the momentum equations. This correction maintains the conservation and
also the second-order accuracy. The dimensionless coefficient
χ has been taken equal to 0.8 in both cases. In transonic problems, artifical viscosity is necessary to avoid marginal stability
on sonic lines, when using a nine-point scheme of second-order
accuracy.

Iso-Mach lines obtained from the MacCormack scheme and the
optimal scheme are presented on Fig. 10. The corresponding
pressure distributions on the airfoil are shown on Fig. 11.
One observes that the optimal scheme reduces the spurious oscillations in the shock wave and gives a better resolution of the
trailing edge region. It is also interesting to compare the
entropy distribution on the airfoil (Fig. 12). Inside the
numerical shock profile, the entropy is not a monotonic increasing function: it reaches a maximal value, similarly as in the
shock structure of the Navier-Stokes equations (see for example
Zel'Dovich and Raizer (1967), vol. 2, p. 474). In the expansion
wave upstream of the shock, the quantity $\Sigma = (p/p_\infty)/(\rho/\rho_\infty)^\gamma - 1$
should be equal to zero, so that Σ is a measure of the entropy
error. Upstream of the shock, Σ is found to be lower than 1.4%
with the MacCormack scheme and lower than 0.65% with the optimal
scheme. This reduction of entropy error can be explained by the
theoretical properties of the schemes. Indeed, the optimal
scheme is more dissipative than the MacCormack scheme in a shock
wave, but it is less dissipative in an expansion wave.

b) Transonic flow over a supercritical airfoil

We now consider the more difficult problem of the computation
of a transonic flow over a supercritical airfoil: the RAE 2822
airfoil at Mach $M_\infty = 0.75$ with an angle of attack $\alpha = 3^\circ$. In
order to capture the suction peak, the computational mesh (Fig.
13)* is more refined around the leading edge than in the previous
problem. The total number of cells is now 224×29 with 152
cells along the airfoil.

Figure 14 shows the pressure distribution on the airfoil given
by two methods applied in the same conditions, as previously (here

---

*Mesh generation devised by J.C. Leballeur at ONERA.

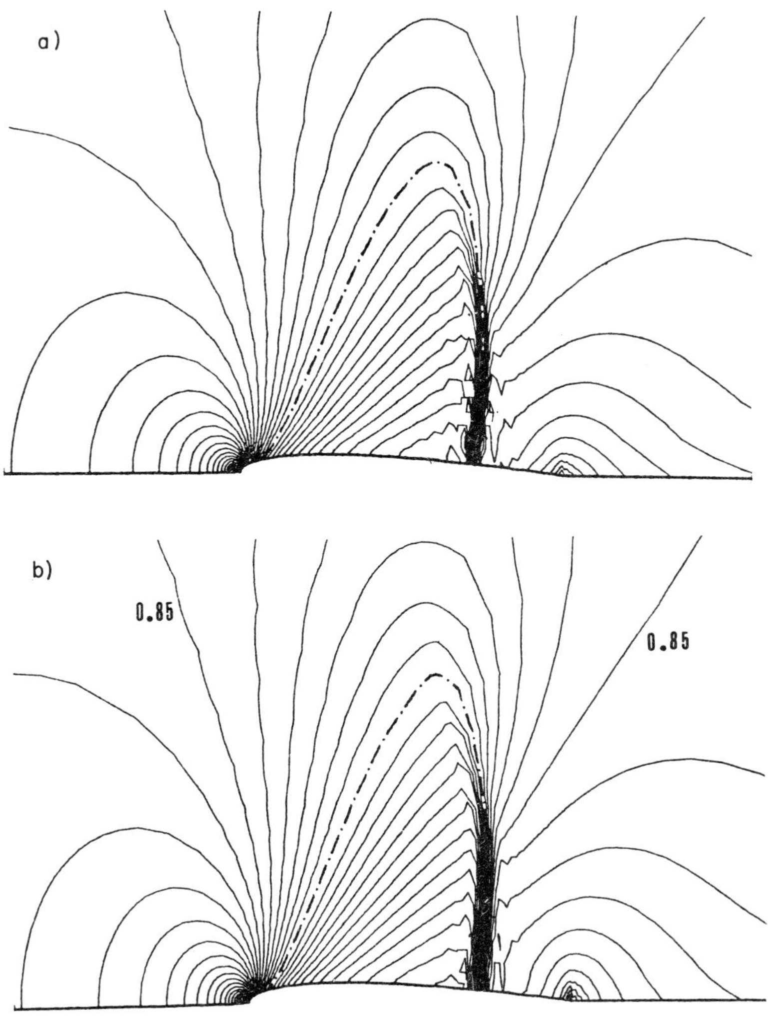

Fig. 10   Iso-Mach lines over the NACA 0012 airfoil at M_∞ = 0.85
          and $\alpha = 0°$ (Increment: ΔM = 0.025).   a) MacCormack
          scheme, b) optimal scheme.

χ = 0.75).   On the pressure curves, we have indicated the pressure level just downstream of the shock deduced from the data of
the pressure just upstream of the shock and the total freestream
pressure, by using the exact Rankine—Hugoniot relations. The
use of the optimal scheme provides a correct pressure jump,
a weakening of the spurious oscillations, and also a better

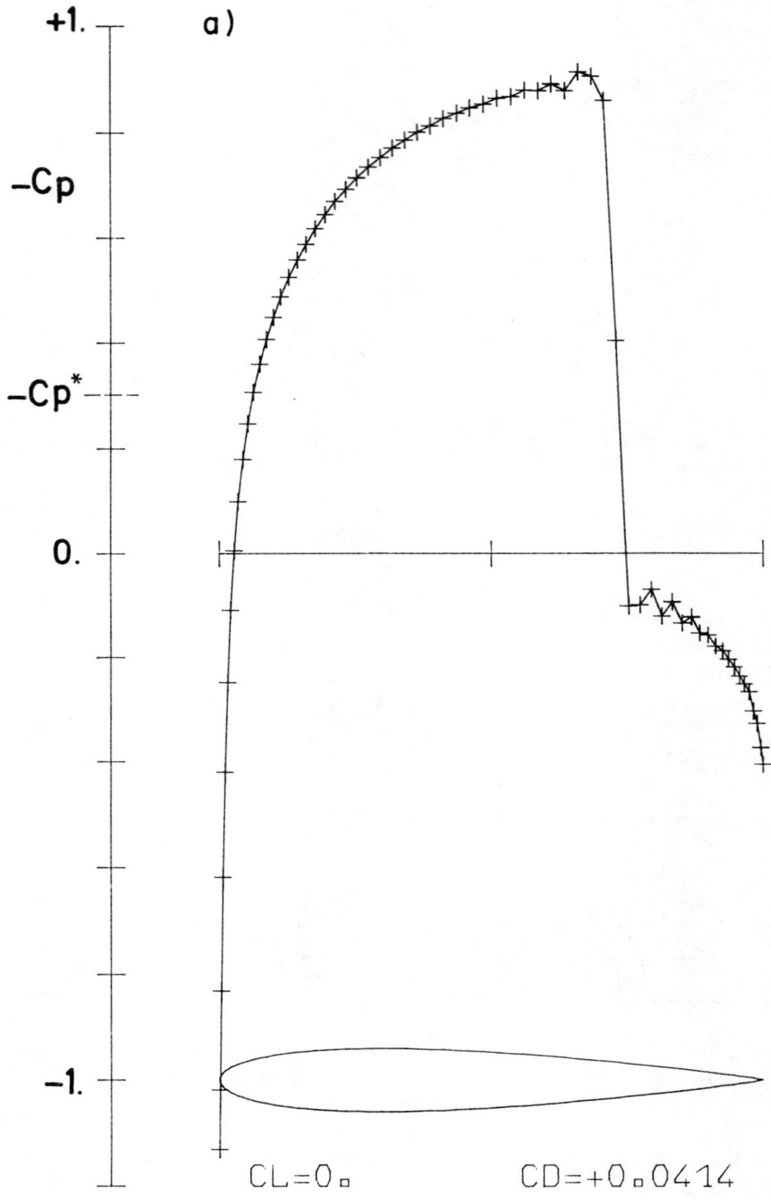

Fig. 11   Pressure distribution on the NACA 0012 airfoil at
$M_\infty = 0.85$ and $\alpha = 0^\circ$.   a) MacCormack scheme

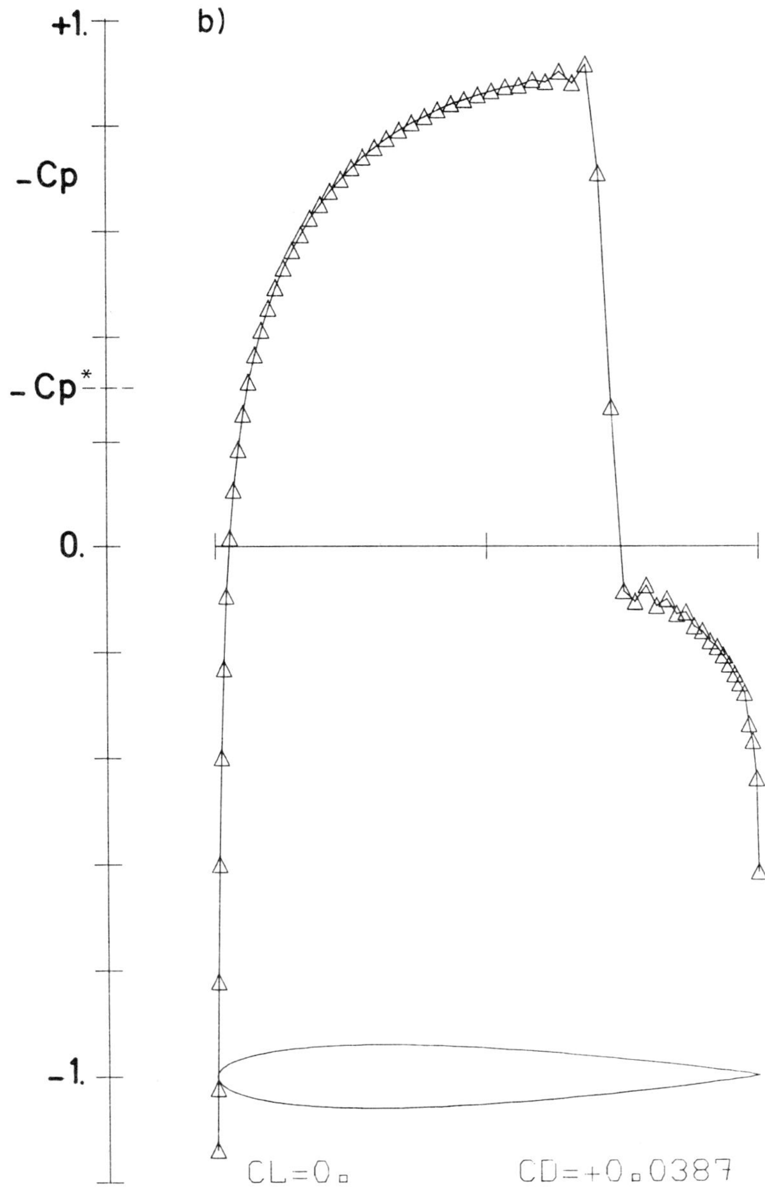

Fig. 11 b) Optimal scheme

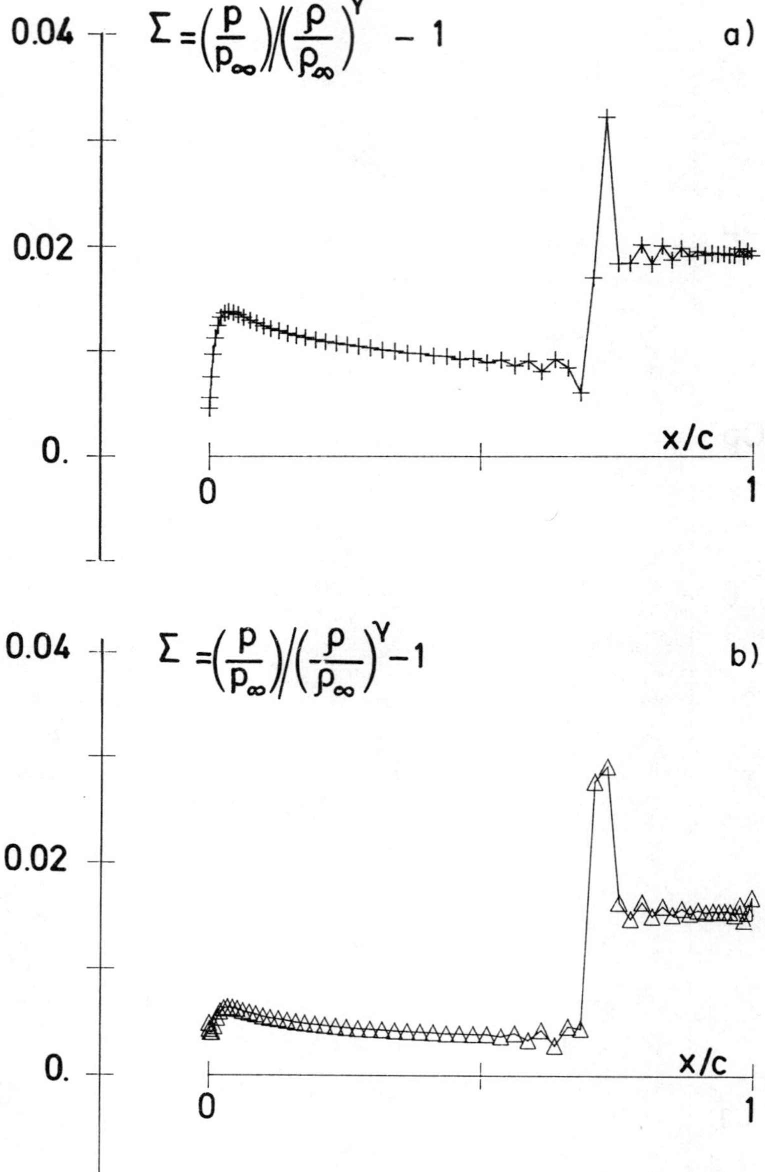

Fig. 12   Entropy distribution on the NACA 0012 airfoil at $M_\infty = 0.85$ and $\alpha = 0^\circ$.   a) MacCormack scheme, b) optimal scheme.

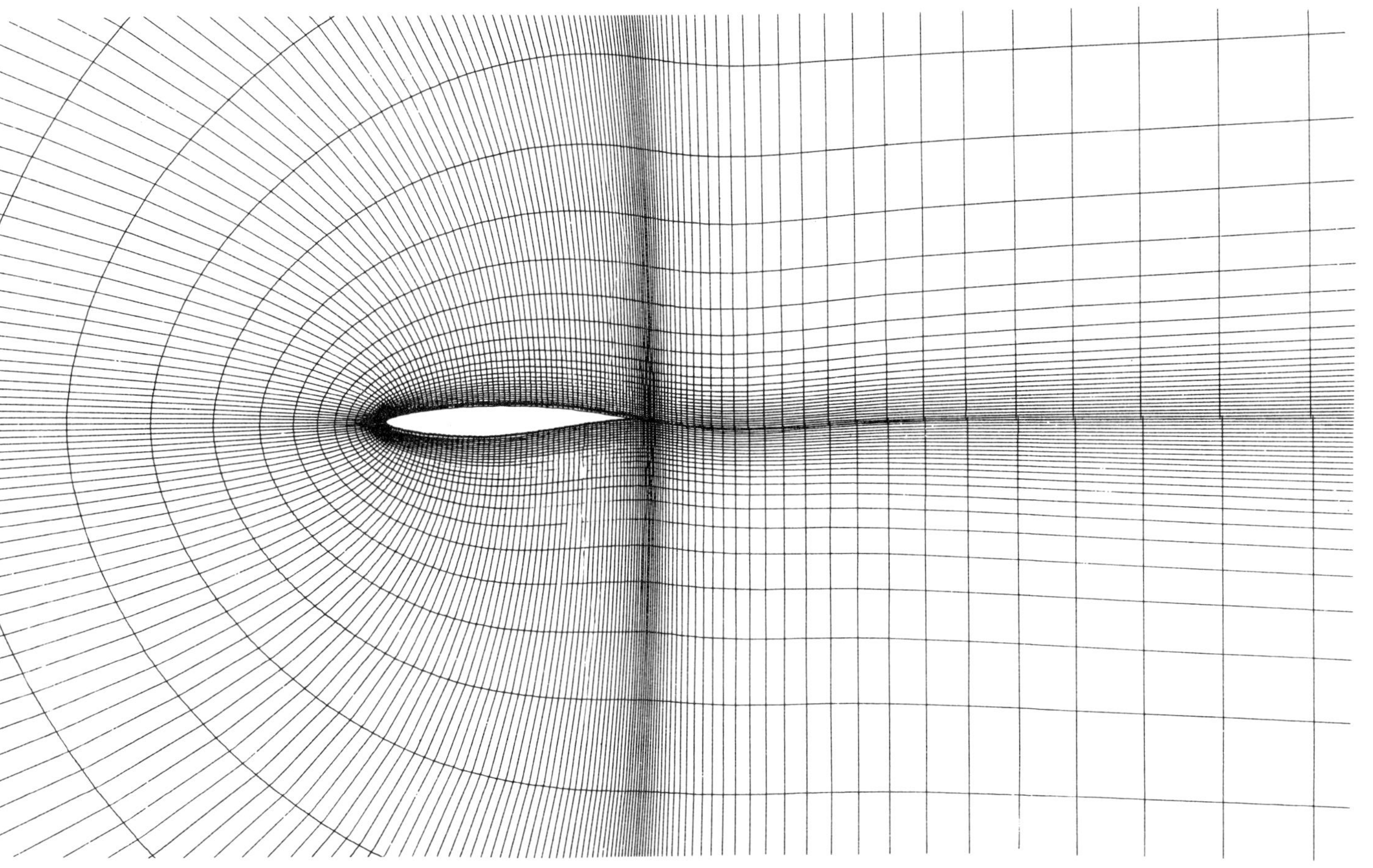

Fig. 13 Computational mesh for the RAE 2822 airfoil (224 × 29 cells). Partial view

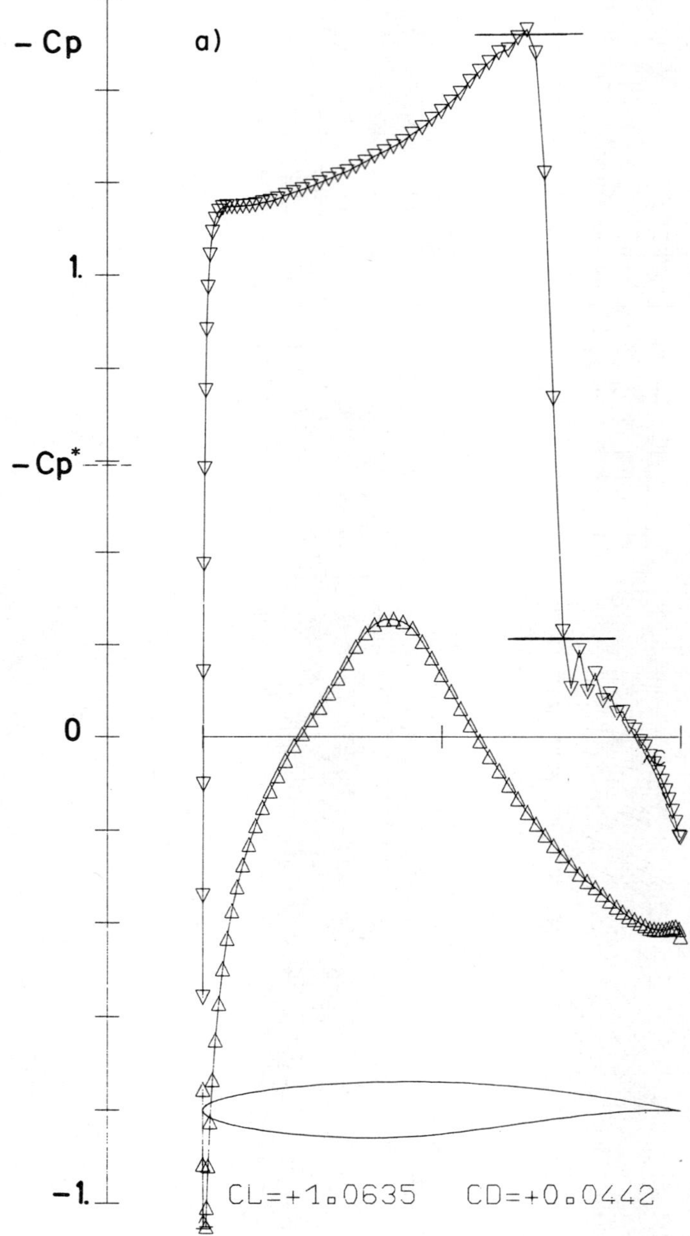

Fig. 14   Pressure distribution on the RAE 2822 airfoil at $M_\infty =$ 0.75 and $\alpha = 3^\circ$.   a) MacCormack scheme.

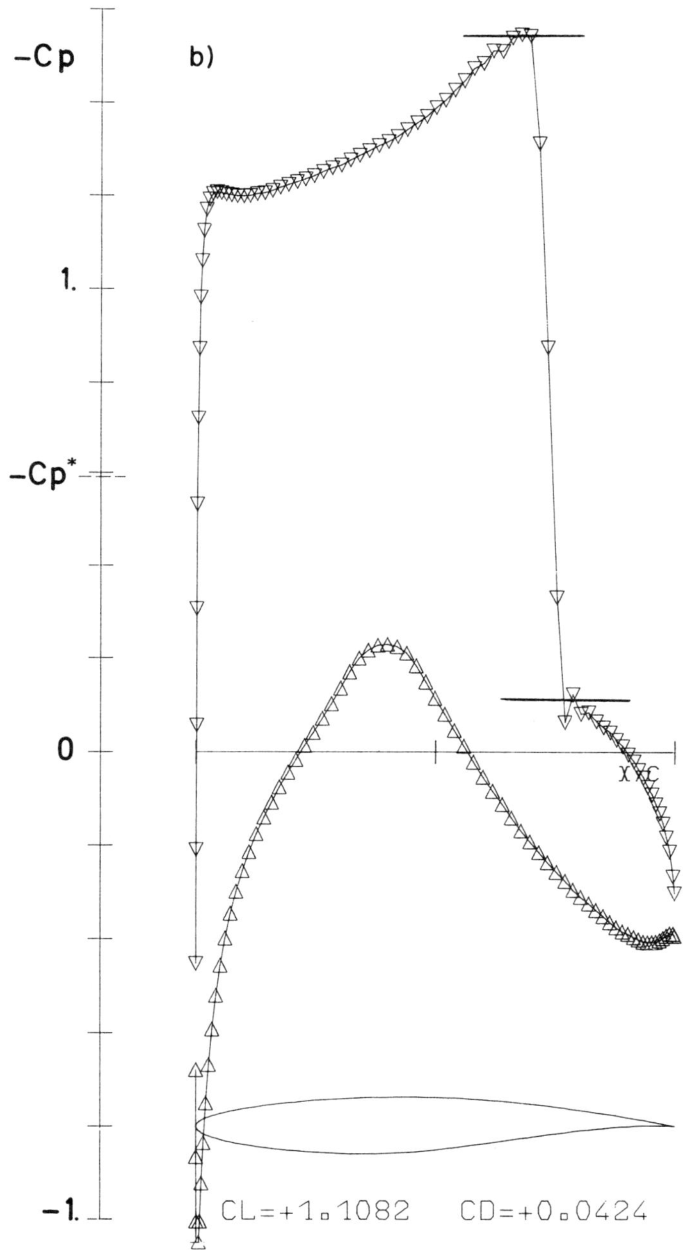

Fig. 14 b) optimal scheme

    LERAT and SIDÈS

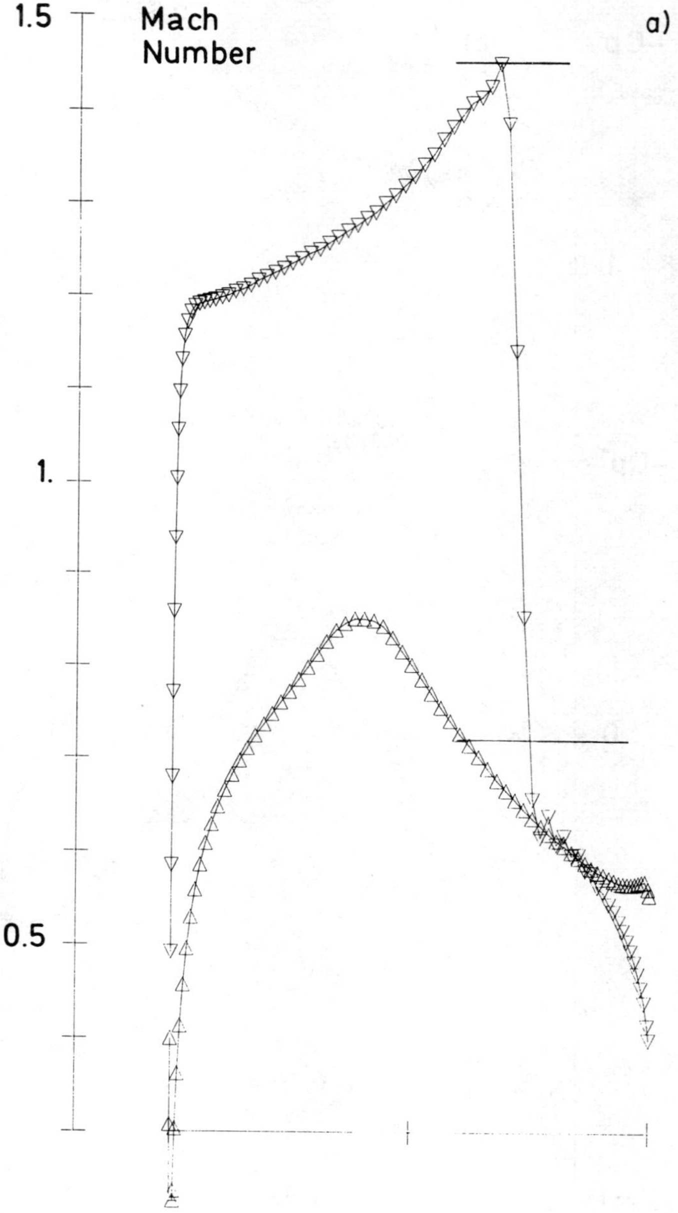

Fig. 15   Mach number distribution on the RAE 2822 airfoil at $M_\infty = 0.75$ and $\alpha = 3^\circ$.   a) MacCormack scheme

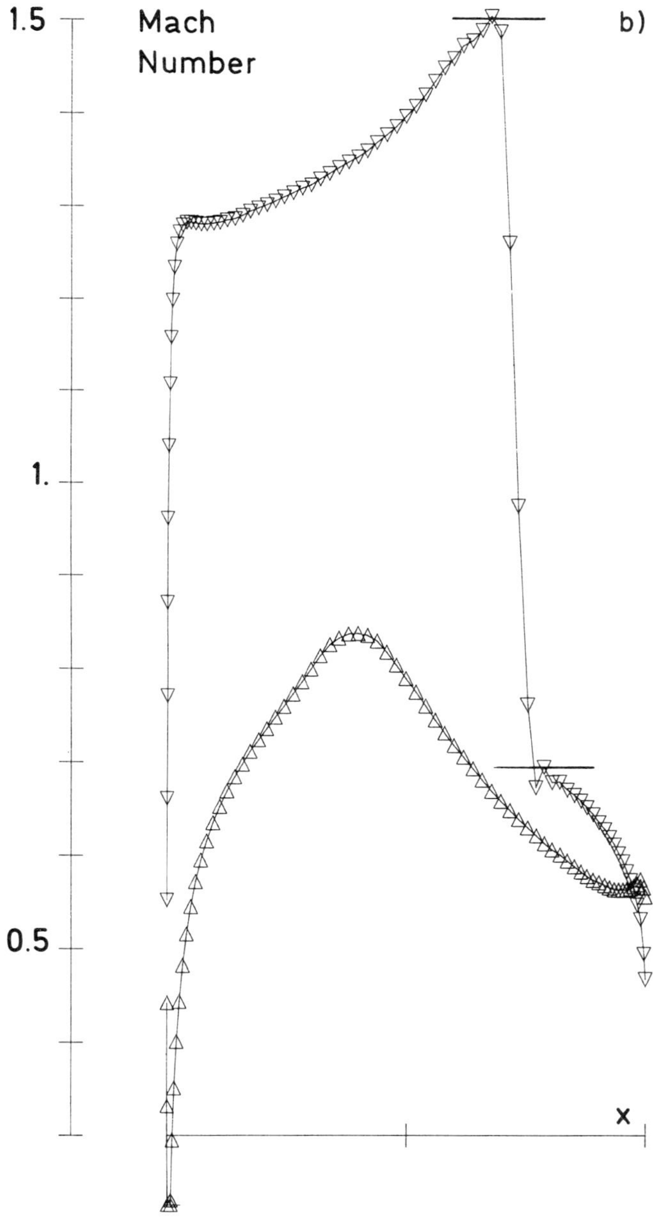

Fig. 15 b) optimal scheme.

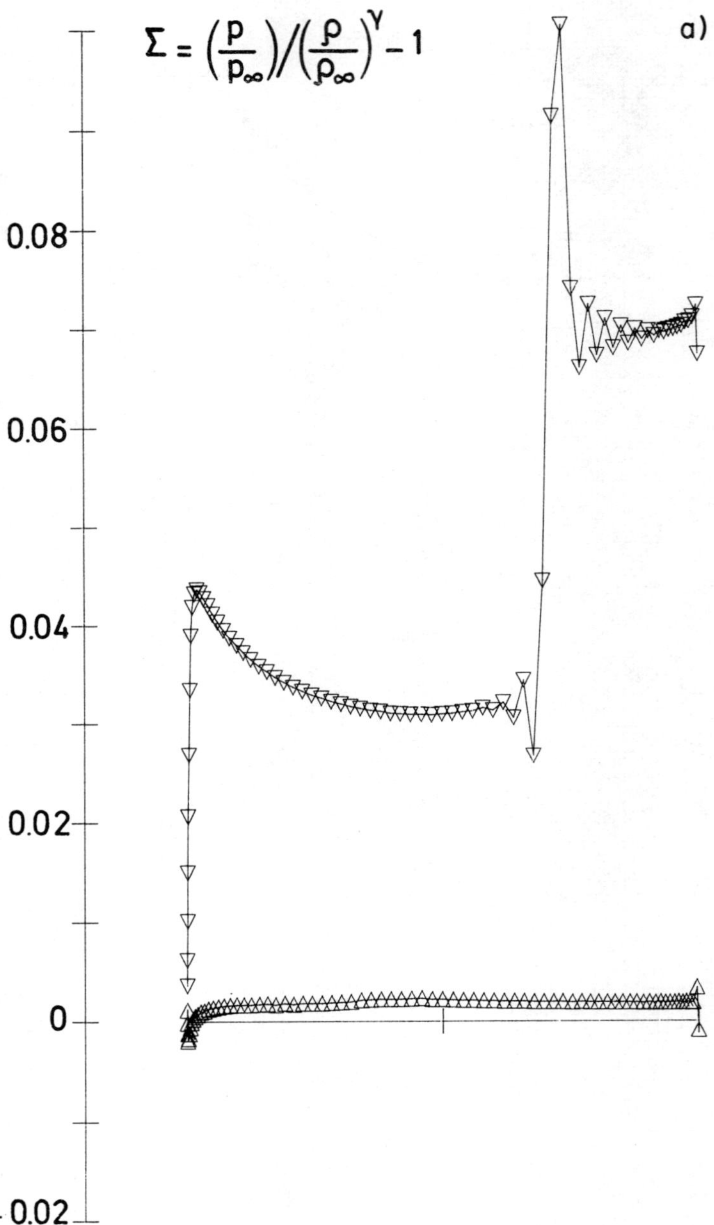

Fig. 16   Entropy distribution on the RAE 2822 airfoil at M_∞ = 0.75 and $\alpha = 3^{\circ}$.   a) MacCormack scheme.

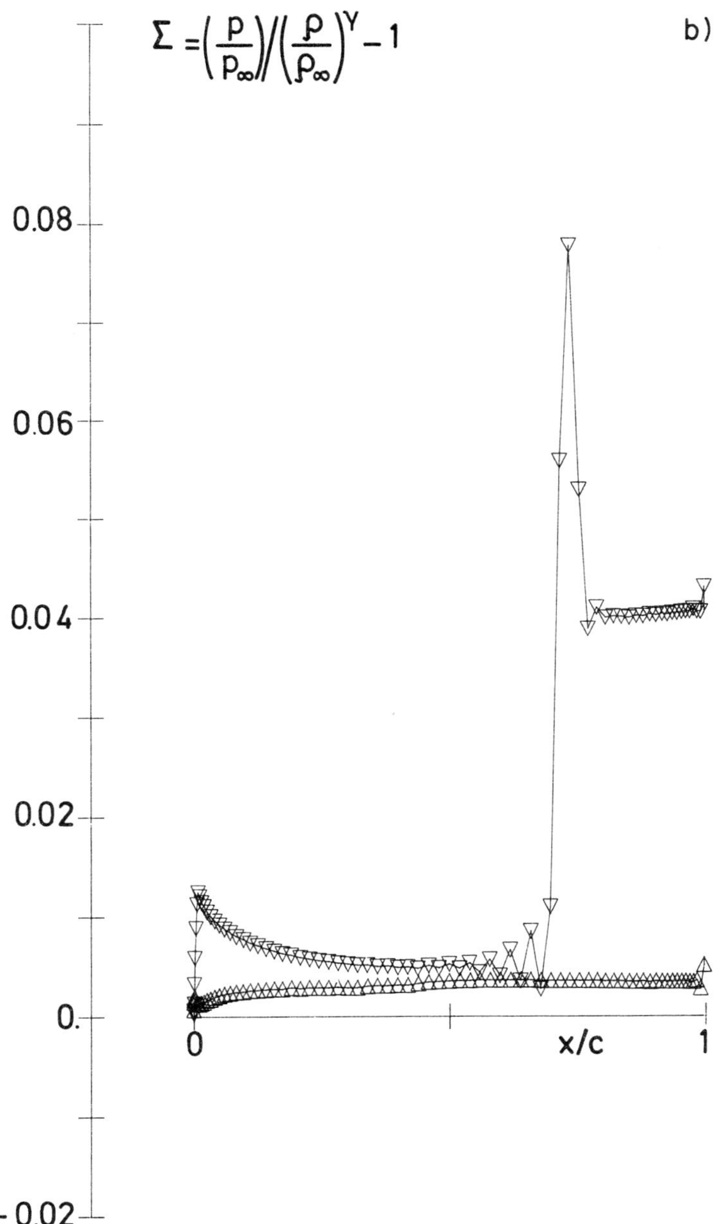

Fig. 16 b) optimal scheme

representation of the suction peak. A similar comparison is
presented in Fig. 15 for the Mach number on the airfoil. It
reveals a large discrepancy between the two solutions on the
upper surface upstream of the shock, with an erroneous jump in
Mach number for the method based on the MacCormack scheme. The
reason for this difference between the two numerical results
can be found by looking at the entropy distributions on the airfoil (Fig. 16). In this difficult problem, we see that the Mac-
Cormack scheme creates a much larger entropy error in the expansion wave upstream of the shock. Thus, it can be pointed out
that numerical accuracy cannot be appreciated only from a plot
of pressure distribution on the airfoil.

   c) Transonic flow over the NACA 0012 airfoil in a fine mesh

We discuss here an attempt to obtain a very accurate solution
of a classical test case, that is the flow over the NACA 0012
airfoil at Mach 0.8 for an angle of attack equal to $1^{o}25$. To
reach this aim, we have applied the finite-volume method based
on the optimal scheme in a fine mesh. This mesh is deduced
from Jameson's transformation as the one on Fig. 9, but it is
composed of 298×34 cells with 200 cells along the airfoil.
The second-order artificial viscosity coefficient is $\chi = 1$.

Inspection of the iso-Mach lines (Fig. 17) confirms that the
method is able to properly capture the upper and lower shocks
and also the slip line running from the trailing edge of the
airfoil. The isentropic lines around the airfoil are shown
on Fig. 18. They are equally spaced with a Σ-increment equal to
10^{-3}. It can be seen that the lower shock is nearly isentropic.
The slip line and the entropy wake downstream of the shocks
clearly appear on the plot. Outside this region, one checks
that the flow is homentropic. In the entropy wake, the isentropic lines also represents the streamlines. Far downstream,
these lines are parallel to the freestream direction.

The pressure distribution on the airfoil is shown on Fig. 19.
The pressure jump computed across the upper shock is in agreement with the Rankine—Hugoniot relations. Unfortunately, this
is not true for the lower shock. The small expansion behind
this shock corresponds to the Von Zierep singularity (1958),
which occurs on a curved wall. This singularity also appears
on the Mach number contours (see Fig. 17). In a coarse mesh,
such a singularity cannot be seen. In the present mesh, it
seems that the singularity is not sufficiently resolved. A
local mesh refinement should probably improve the calculation.

Finally, the entropy distribution on the airfoil is shown
on Fig. 20. Upstream of the shocks, the entropy error is very
low ($\Sigma < 0.2\%$). However, we observe that the entropy jump across

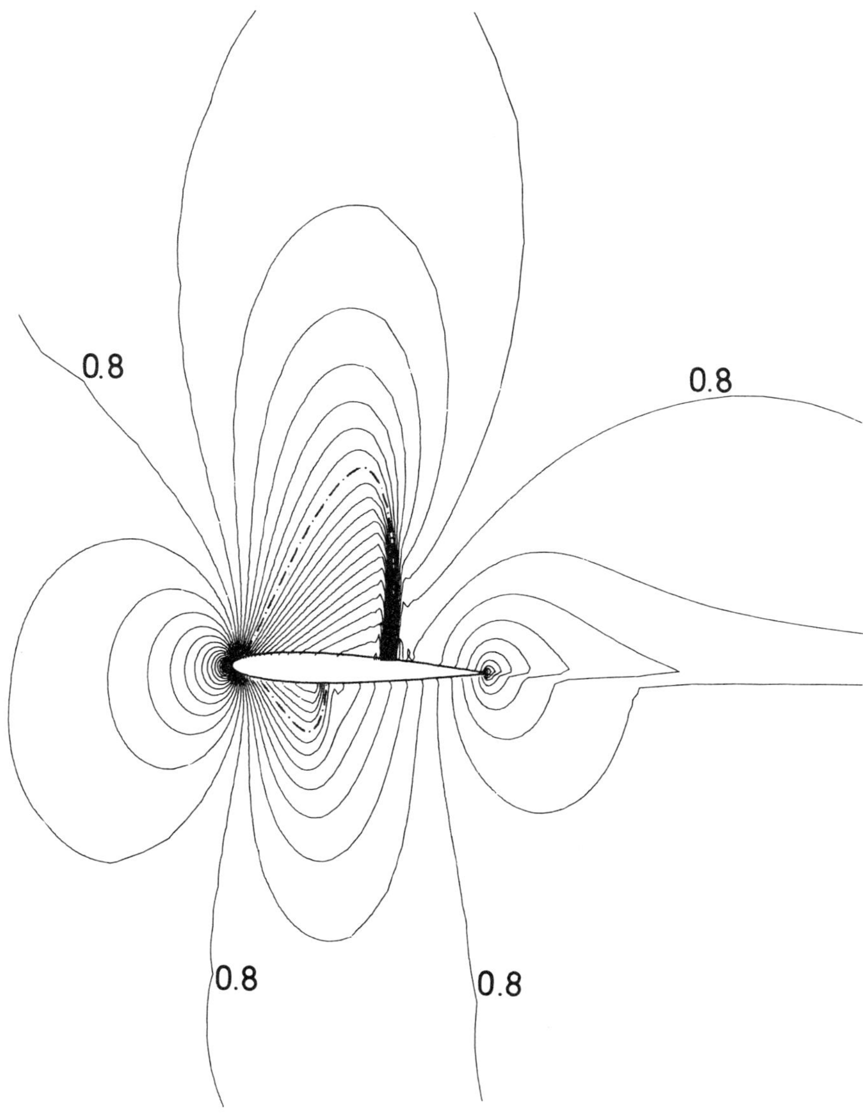

Fig. 17   Iso-Mach lines over the NACA 0012 airfoil at $M_\infty = 0.80$
and $\alpha = 1^\circ 25$ (increment: $M = 0.025$). Optimal scheme
in a fine mesh.

the lower shock is of the order of the upwind entropy error in
the present mesh. As previously, we find a maximal value for
entropy in the middle of the numerical upper-shock structure.

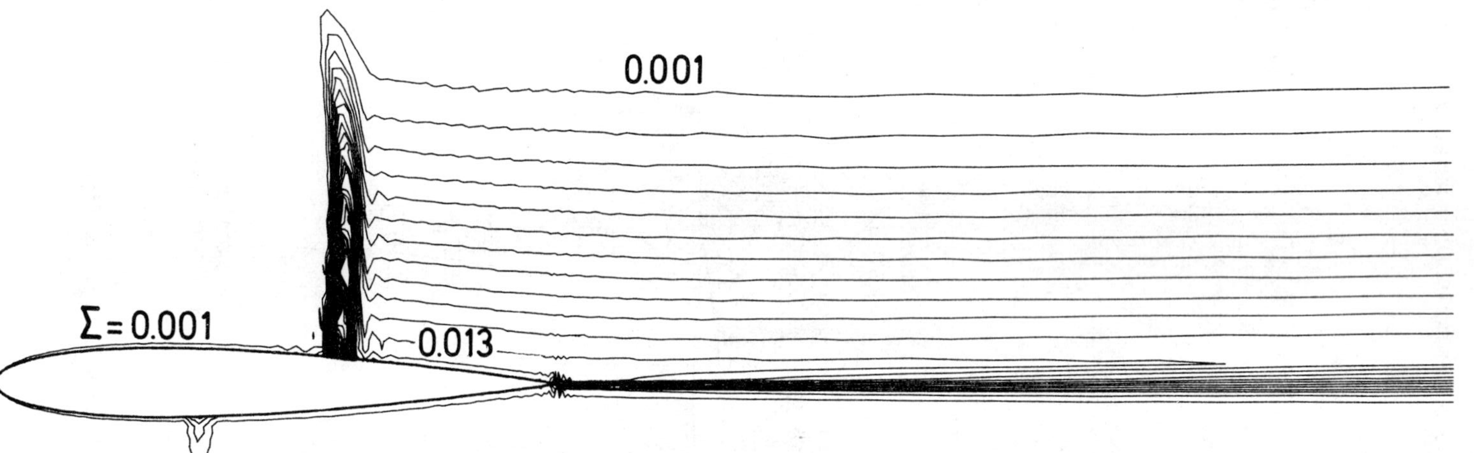

Fig. 18   Isentropic lines over the NACA 0012 airfoil at $M_\infty = 0.80$ and $\alpha = 1°25$ ($\Sigma = (p/p_\infty)/(\rho/\rho_\infty)^\gamma - 1$; increment $\Delta\Sigma = 0.001$).   Optimal scheme in a fine mesh.

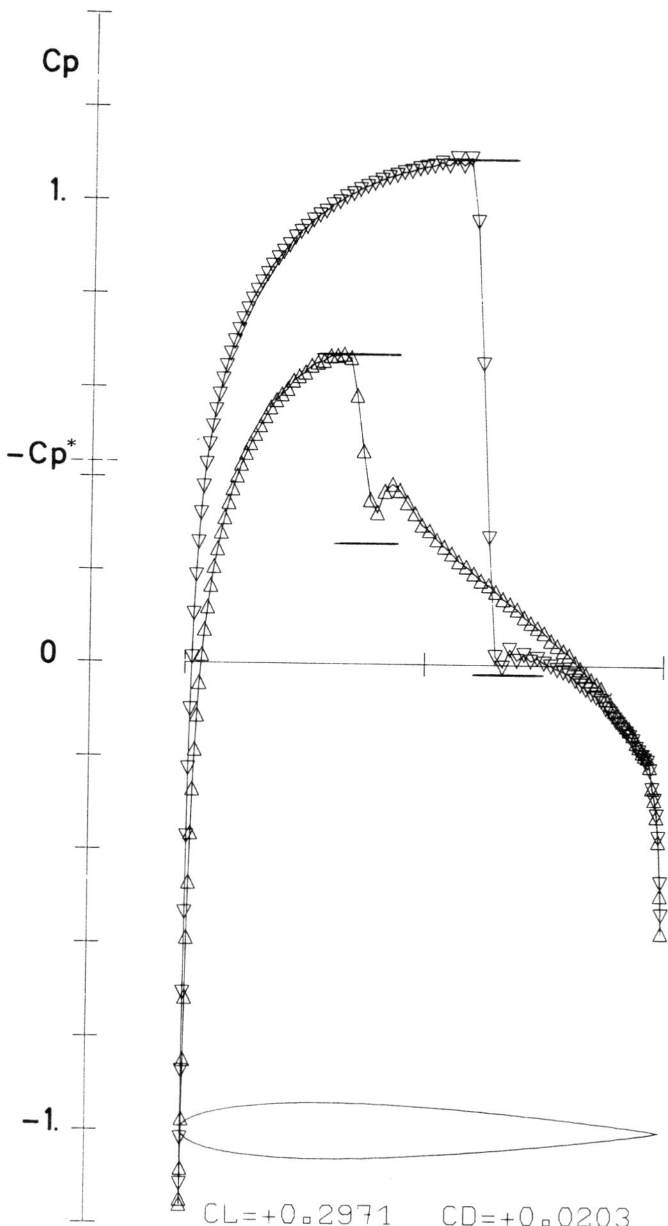

Fig. 19 Pressure distribution on the NACA 0012 airfoil at M_∞ = 0.80 and α = 1°25. Optimal scheme in a fine mesh.

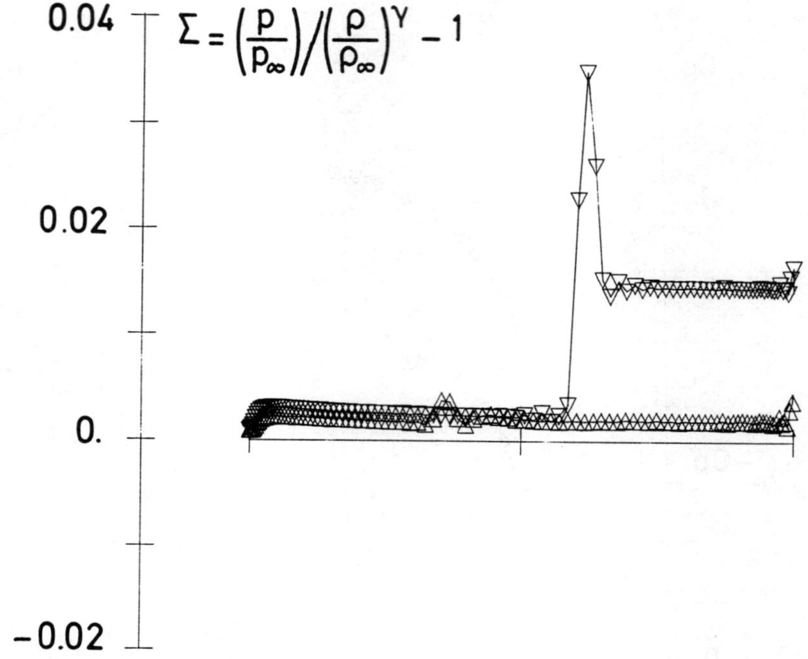

Fig. 20   Entropy distribution on the NACA 0012 airfoil at M_∞ = 0.80 and α = 1°25.   Optimal scheme in a fine mesh.

4.3   Unsteady transonic flow

The ability of the new method to treat a real unsteady problem is now investigated. We consider the unsteady flow over the NACA 0012 airfoil oscillating in translation with a great amplitude at a low frequency. The freestream Mach number is M_∞ = 0.536 and there is no incidence, so that the flow is symmetric. The Mach number relative to the airfoil $M_{r,\infty}$ varies with time according to the law:

$$M_{r,\infty} = M_\infty + M_0 \sin kt$$

where M_0 = 0.327 and the reduced frequency k = 0.185. Such a motion simulates the flow over a section of helicoptor rotor blade near the tip of the blade.   It is a severe test for a numerical method because of the important displacement of the shock waves occurring in transonic regime. In order to limit the computing time, we consider here the transient evolution from a steady flow during the first half-cycle of the airfoil motion. During this half-cycle (forward blade motion), the relative Mach number goes from 0.536 up to 0.863 and down again

to O.536. During the second half cycle (backward blade motion),
the flow will remain subsonic. The calculation makes use of the
optimal scheme on the computational mesh of the first application (Fig. 9), with X = 1.5. No difficulty has been encountered
in the unsteady calculation.

Figure 21 shows the pressure coefficient on the airfoil,
defined with the freestream velocity:

$$C_p = \frac{p-p_\infty}{\frac{1}{2}\gamma p_\infty M_\infty^2} \quad .$$

The pressure coefficient is given at several times of the
evolution, that is kt = 30°, 60°, 90°, 120° and 150°. One can
see that a shock wave forms, moves downstream and then it quickly
moves upstream. The importance of the unsteady effects appears
through a comparison of the solutions for two values of kt
corresponding to the same relative Mach number. For example,
the pressure distributions for kt = 60° and kt = 120° are quite
different. Although $M_{r,\infty}$ is equal to O.819 in both cases, only
the second distribution involves a shock wave. Isobar contours
are plotted on Fig. 22, for kt = 60° and kt = 120°.

5. CONCLUSION

A second-order accurate scheme for hyperbolic systems in
several space variables has been selected in a class of schemes
involving one predictor per space variable, by using the equivalent system approach. The scheme has been formulated on a
moving curvilinear mesh in a strictly conservative way and its
application to inviscid transonic flows with shock waves has led
to appreciable improvements in accuracy. We hope that the presented numerical results can contribute to the determination of
reference solutions of the Euler equations in Aeronautical
Fluid Dynamics.

REFERENCES

Beam, R. and Warming, R.F. (1976) An implicit finite-difference
algorithm for hyperbolic systems in conservation law form. *J.
Comput. Phys.* **22**, 87—110.

Burstein, S.Z. (1967a) High order accurate difference methods
in hydrodynamics. *in* Nonlinear Partial Differential Equations
(ed. Ames, W.F.) Academic Press, p. 279—290.

Burstein, S.Z. (1967b) Finite-difference calculations for hydrodynamic flows containing discontinuities. *J. Comput. Phys.* **2**,
198-222.

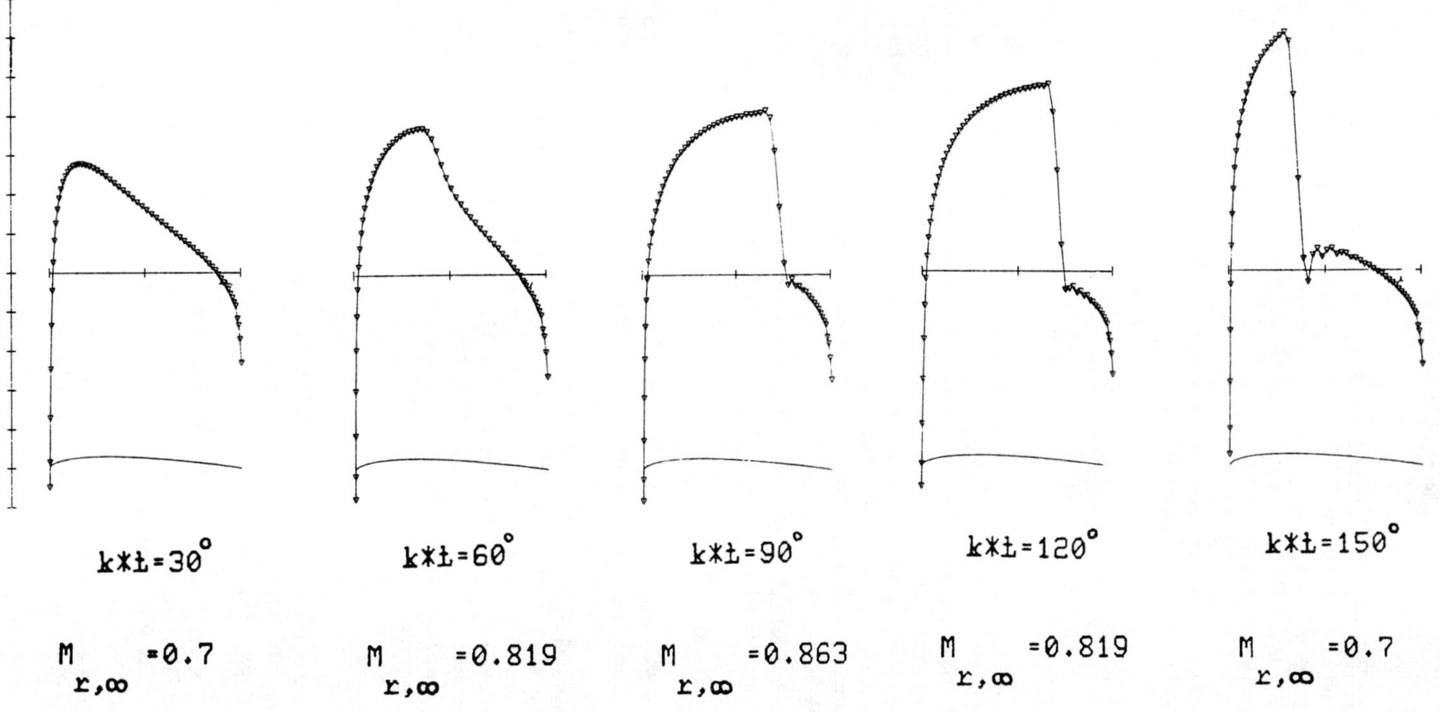

Fig. 21 Evolution of the pressure distribution on the NACA 0012 airfoil oscillating in translation. Optimal scheme.

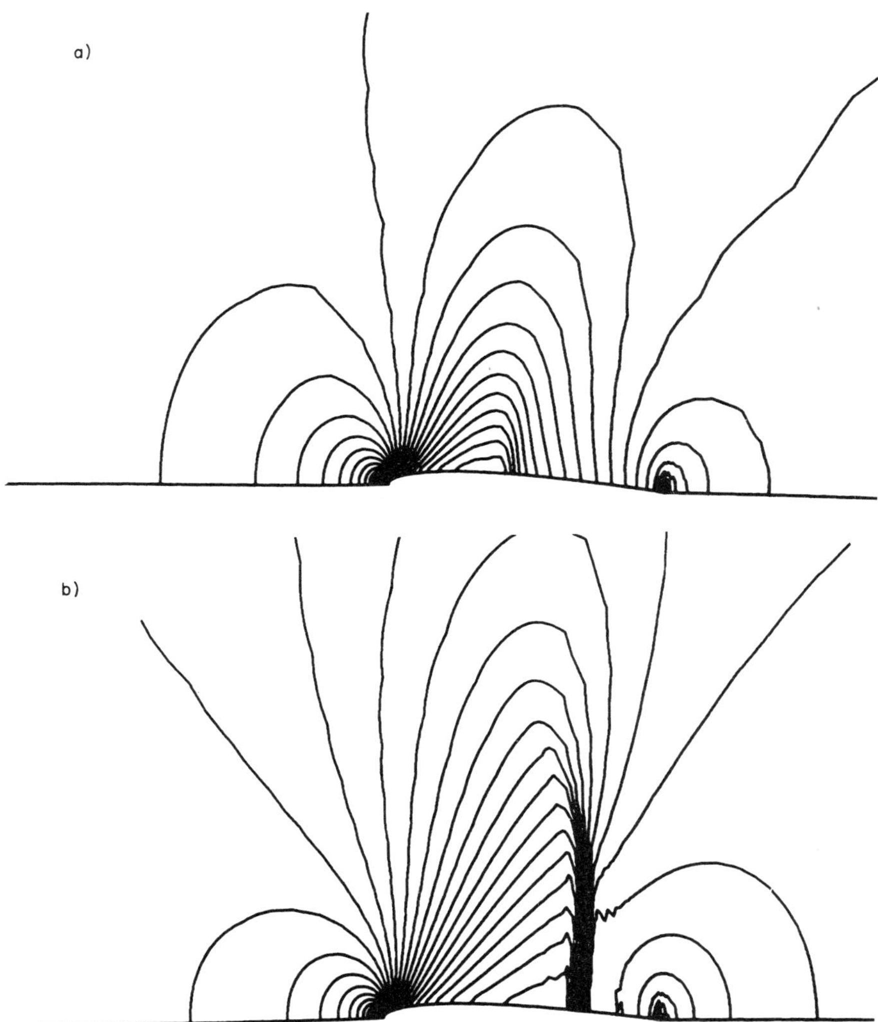

Fig. 22   Isobar lines over the NACA 0012 airfoil oscillating in
          translation. Optimal scheme. a) kt = 60°. b) kt =
          120°.

Chyu, W.J., Davis, S.S. and Chang, K.S. (1981)   Calculation of
unsteady transonic flow over an airfoil. *AIAA Journal* **19**,
pp.684,690.

Chyu, W.J. and Schiff, L.B. (1981) Nonlinear aerodynamic modelling
of flap oscillations in transonic flow - A numerical validation.
AIAA Paper 81-0073.

Jameson, A. (1974) Three dimensional flows around airfoils with
shocks. Lecture Notes in Computer Science, **11**, p. 185—212.

Kreiss, H.O. (1964) On difference approximations of the dissipative type for hyperbolic differential equations. *Comm. Pure
Appl. Math.* **17**, 335—353.

Laval, P. (1976) Calcul de l'écoulement instationnaire autour
d'un profil oscillant par une méthode à pas fractionnaires.
Arch. Mech. **28**, 855—880.

Laval, P. (1981) Schémas explicites de désintégration du second
ordre pour la résolution des problèmes hyperboliques non linéaires:
théorie et applications aux écoulements transsoniques. Thèse de
doctorat d'état. Université Paris VI

Lax, P.D. and Wendroff, B. (1960) Systems of conservation laws.
Comm. Pure Appl. Math. **13**, 217—237.

Lax, P.D. and Wendroff, B. (1964) Difference schemes for hyperbolic equations with high order of accuracy. *Comm. Pure Appl.
Math.* **17**, 381—398.

Lerat, A. (1981) Sur le calcul des solutions faibles des
systèmes hyperboliques de lois de conservation à l'aide de schémas
aux différences. Publ. ONERA no. 1981-1.

Lerat, A. and Peyret, R. (1973) Sur le choix de schemas aux
differences du second ordre fournissant des profils de choc
sans oscillation. *Comptes Rendus Acad. Sc. Paris*, t. 277 A,
p. 363—366.

Lerat, A. and Peyret, R. (1975). Propriétés dispersives et dis-
sipatives d'une classe de schémas aux différences pour les
systèmes hyperboliques non linéaires. Rech. Aerosp. no. 1975—2,
p. 61—79. English translation: European Space Agency, TT-205.

Lerat, A. and Sidès, J. (1977) Calcul numérique d'écoulements
transsoniques instationnaires. AGARD Conf. Proceed., **226**,
P. 15.1—15.10. English translation available as TP ONERA
1977-19E.

Lerat, A. and Sides, J. (1979) Numerical simulation of unsteady
transonic flows using the Euler equations in integral form.
Israel J. Techn. **17**, 302—310.

MacCormack, R.W. and Paullay, A.J. (1972) Computational efficiency achieved by time-splitting of finite-difference operators. AIAA Paper no. 72—154.

MacCormack, R.W., Rizzi, A.W. and Inouye, M. (1976) 'Steady supersonic flowfields with embedded subsonic regions' *in* Computational Methods and Problems in Aeronautical Fluid Dynamics (ed. Hewitt, B.L. *et al.*) Academic Press, p. 424—447.

Magnus, R. and Yoshihara, H. (1975) Unsteady transonic flows over an airfoil. *AIAA J.* **13**, 1622—1628.

Magnus, R. and Yoshihara, H. (1976) The transonic oscillating flap. AIAA Paper 76—327.

Palumbo, D.J. and Rubin, D.L. (1972) Solution of the two-dimensional unsteady compressible Navier-Stokes equations using a second-order accurate numerical scheme. *J. Comput. Phys.* **9**, 466—495.

Richtmyer, R.D. (1962) A survey of difference methods for non steady gas dynamics. *Nat. Center Atm. Res. Techn.* Note 63—2, Boulder

Richtmyer, R.D. and Morton, K.W. (1967) Difference methods for initial value problems, Interscience Publ.

Rizzi, A.W. and Inouye, M. (1973) A time-split finite-volume technique for three-dimensional blunt-body flow. *AIAA J.* **11**, 1478—1485.

Singleton, R.E. (1968) Lax-Wendroff difference scheme applied to the transonic airfoil problem. AGARD Conf. Proceed., **35**, p. 2.1—2.9.

Steger, J.L. (1978) Implicit finite-difference simulation of flow about arbitrary two dimensional geometries. *AIAA J.* **16**, 679—686.

Thommen, H.U. (1966) Numerical integration of the Navier—Stokes equations. *Z.A.M.P.* **17**, 369—384.

Turkel, E. (1974) Phase error and stability of second-order methods for hyperbolic problems. (I). *J. Comput. Phys.* **15**, 226—250.

Viviand, H. (1981) Pseudo-unsteady methods for transonic flow computations. Proceedings of 7th Int. Conf. Num. Methods in Fluid Dynamics, Standford (1980). Lecture Notes in Physics, **141**, p. 44—54.

Viviand, H. and Veuillot, J.P. (1978) Methodes pseudo-instation-
naires pour le calcul d'écoulements transsoniques. Publ. ONERA
no. 1978—4.

Von Zierep, J. (1958) Der senkrechte Verdichtungsstoss am
Gekrummten Profil. *Z.A.M.P.* **9b**, 764-776.

Zel'dovich, Ya.B. and Raizer, Yu.P. (1967) Physics of shock
waves and high-temperature hydrodynamic phenomena. Academic
Press.

TRANSONIC AEROFOIL CALCULATIONS USING THE EULER EQUATIONS

A. Jameson

Princeton University, U.S.A.[*]

I am going to tell you about some of the transonic aerofoil calculations I have been doing recently based on the Euler equations. During the past year I have been collaborating with Wolfgang Schmidt of Dornier in an effort to find a reasonably fast method for solving these equations for engineering applications. The calculations I shall describe have mostly been performed on an 'O'-grid, for a non-lifting case. They use a potential flow as a starting solution, although a direct solution starting with uniform flow is also possible.

I have actually been trying to do the Euler equations on and off for several years, but only extensively in the last year. The history of the particular scheme I am going to tell you about began last summer with a visit to Dornier where I spent several weeks. There was an Euler code there developed by Schmidt and Rizzi, and results from that code were presented at the Stockholm workshop a couple of years ago (Rizzi and Viviand, 1981). It was a finite-volume code based on a time-split version of the MacCormack scheme, so that it took separate account of the gradients in each coordinate direction. I took that code as a starting point, and converted it to a scheme which uses central differencing in space and a three-stage time-stepping process which is stable for Courant numbers up to 2.0. It seemed to work quite well. Dornier have gone on to produce an engineering version of it which handles lifting cases and incorporates boundary layers. Rizzi has now implemented the new differencing scheme in his three-dimensional wing calculations. I have also been cooperating with Eli Turkel in Tel Aviv, and have followed his advice on how to treat the outer boundaries.

For the past nine months I have been trying to get the scheme to work to my satisfaction. The problem with both the earlier Rizzi-Schmidt code and the first version that I developed from it was that they converged quite quickly down to residuals of about 10^{-2} and then tended to oscillate. After a very painful nine

---

[*]Prepared by the editor from a transcript of the oral presentation.

months I seem to have made it converge right down to 10^{-8} in
about 2000 time steps. For engineering purposes you could probably stop after 200-300 steps but I do feel that to be really
useful an Euler code should have the ability to converge as far
as possible, in fact down to the kind of residual we are accus-
tomed to in potential flow calculations. For example, imagine we
are trying to study the behaviour of a supposedly shock-free
design, like a Korn aerofoil, around its design conditions. We
want to see that shock-free solution at the design Mach number,
and when we go up or down 0.001 in Mach number we want to see if
a shockwave forms, and if so how much drag it produces. I don't
think you can resolve those questions with a code in which the
far-field Mach number wanders around between 0.74 and 0.76.

I have already said that the code I am working with uses the
finite volume formulation. The other general point I want to make
is that I work with a four-equation model. That is to say, I
solve the four conservation laws for mass, momentum and energy.
I could discard the energy equation and replace it by a condition
of constant enthalpy, because I know that condition must hold in
the steady state, but I prefer to let the enthalpy vary, because
I believe that I can add a useful damping term proportional to
the difference between the local value of enthalpy and its free
stream value. However, I will say more about that later.

The Euler equations we want to solve are, in conservation form

$$\frac{\partial w}{\partial t} + \frac{\partial f}{\partial x} + \frac{\partial g}{\partial y} = 0 \tag{1}$$

where

$$w = \begin{pmatrix} \rho \\ \rho u \\ \rho v \\ \rho E \end{pmatrix} \quad ; \quad f = \begin{pmatrix} \rho \\ \rho u^2 + p \\ \rho u v \\ \rho u H \end{pmatrix} \quad ; \quad g = \begin{pmatrix} \rho v \\ \rho u v \\ \rho v^2 + p \\ \rho v H \end{pmatrix} \tag{2}$$

in which E = total energy, H = total enthalpy, and pressure p is
given by

$$p = (\gamma - 1)\rho (E - \tfrac{1}{2}(u^2 + v^2)) \ . \tag{3}$$

We can transform these to an arbitrary set of coordinates
(X,Y) for which the transformation derivatives are contained in
the matrix

$$H = \begin{pmatrix} x_X & x_Y \\ \\ y_X & y_Y \end{pmatrix} \ . \tag{4}$$

The result is

$$\frac{\partial W}{\partial t} + \frac{\partial F}{\partial X} + \frac{\partial G}{\partial Y} = 0 \tag{5}$$

where

$$\left. \begin{aligned}
W &= \det(H)w = hw \\
F &= y_Y f - x_Y g \\
G &= x_X g - y_X f
\end{aligned} \right\} \tag{6}$$

We shall identify the lines X,Y = constant with the boundaries of our finite volumes. Note that h = det(H) is the volume in (x,y) space of a cell having unit volume in (X,Y) space. If we define the contravariant velocities

$$\begin{pmatrix} U \\ V \end{pmatrix} = H^{-1} \begin{pmatrix} u \\ v \end{pmatrix} \tag{7}$$

(which are simply the velocity components normal to the cell sides) we can write

$$w = \begin{pmatrix} \rho h \\ \rho uh \\ \rho vh \\ \rho Eh \end{pmatrix} \; ; \quad F = \begin{pmatrix} U\rho h \\ U\rho uh + y_Y p \\ U\rho vg - x_Y p \\ U\rho Hh \end{pmatrix} \; ; \quad G = \begin{pmatrix} V\rho h \\ V\rho uh - y_X p \\ V\rho vh + x_X p \\ V\rho Hh \end{pmatrix} \tag{8}$$

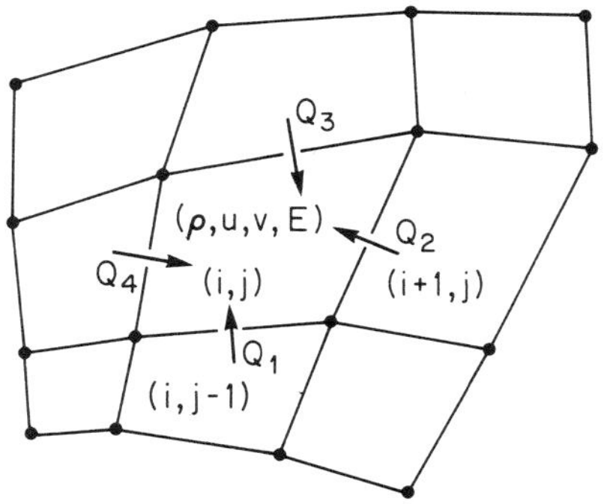

Fig. 1. A typical finite-volume cell

The discretisation scheme which I have made out of this uses a finite volume form with cell-centred quantities (Fig. 1), so that (ρ, u, v, E) are supposed to represent values at some central point in each cell. You calculate a "flux velocity" across each cell face, according to

$$Q_k = \Delta y_k u_k - \Delta x_k v_k \tag{9}$$

where k = 1, 2, 3, 4, and u_2, for example, is the average velocity for the two cells (i,j) and $(i + 1,j)$.

The equation which gives me, let us say, the local rate of change of x-momentum is

$$\frac{\partial}{\partial t} (\rho uh) + \sum_{k=1}^{k=4} (Q_k (\rho u)_k + \Delta y_k p_k) \tag{10}$$

where, as before, $(\rho u)_k$ and p_k are the appropriate average values.

So that is how I do the flux balancing. If you follow it through you will see that it amounts to carrying out a central differencing operation on the quantities F and G. As a digression I might say that in the fifty or so Euler codes I have written by now a lot of alternatives have been tried, and some of them may yet be made to work. I still haven't found out whether it is really better to store things at the centres of cells or at the corners, but the present scheme uses the cell centres, and in effect does central differencing on them. However, it is well known that central differencing is unstable unless you add some dissipation terms, so what I actually solve is

$$\frac{\partial}{\partial t} (hw) + Qw - Dw = 0 \tag{11}$$

where Q is my flux balancing operator (not to be confused with the Q_k defined in equation (9)) and D is some dissipative operator. A point that I want to emphasize is that I have kept the time-stepping scheme quite independent from the spatial differencing, so that my final steady state should be one in which $Qw-Dw = 0$. With other differencing schemes this does not always happen; for example with a MacCormack scheme the 'steady state' may depend on Δt.

The dissipation operator which I have found to work is actually a combination of second and fourth differences. These are done separately in the X and Y directions so we will just consider the one-dimensional case and show how it applies to the density equation. I smooth the density according to

$$D\rho = D_X \rho + D_Y \rho$$

where

$$D_X \rho = d_{i+\frac{1}{2},j} - d_{i-\frac{1}{2},j} \tag{12}$$

and

$$d_{i+\frac{1}{2},j} = \frac{h_{i+\frac{1}{2},j}}{\Delta t^*_{i+\frac{1}{2},j}} \left[\epsilon^{(2)}_{i+\frac{1}{2},j} (\rho_{i+1,j} - \rho_{i,j}) \right.$$

$$\left. - \epsilon^{(4)}_{i+\frac{1}{2},j} (\rho_{i+2,j} - 3\rho_{i+1,j} + 3\rho_{i,j} - \rho_{i-1,j}) \right]$$

$$(13)$$

where Δt^* is the time step which would bring about a local Courant number of unity. It varies from cell to cell, and I have computed its average value on the interface between cells (i,j), $(i+1.j)$. Likewise $h_{i+\frac{1}{2},j}$ is an average of two cell volumes. Dimensionally, eqn (13) is a rate of change of mass (since the ϵ are numerical constants) which is as it should be for use in eqn (11). The factor outside the square brackets in eqn (13) is of order (Δx), and the terms inside would be respectively of order (Δx), $(\Delta x)^3$ if $\epsilon^{(2)}$, $\epsilon^{(4)}$ were each of order unity. In fact I am going to make $\epsilon^{(2)}$ of order $(\Delta x)^2$, so the whole expression will be of order $(\Delta x)^4$, it adds to the spatial differencing operator Q a relative error which is merely third order. You can easily see that the added terms are actually a mixture of second and fourth differences.

The way that these ϵ coefficients are chosen is to adapt them to the local flow gradients. For the second-order coefficient I take

$$\epsilon^{(2)}_{i+\frac{1}{2},j} = \max (V_{i+1,j}, V_{i,j}) \qquad (14)$$

where

$$V_{i,j} = K^{(2)} \frac{|p_{i+1,j} - 2p_{i,j} + p_{i-1,j}|}{|p_{i+1,j}| + 2|p_{i,j}| + |p_{i-1,j}|} \qquad (15)$$

In eqn (15), $K^{(2)}$ is a constant, and I am not sure whether the normalisation factor in the denominator is needed. However, the outcome is that $\epsilon^{(2)}$ is positive, of order $(\Delta x)^2$, and proportional to the second difference of pressure. It does seem to do a nice job suppressing the oscillations around shocks.

However, what I found was that the second-order dissipation was not enough by itself; the solution would still go down to moderately small residuals and then oscillate. The oscillations were most noticeable in the small cells near the trailing edge,

presumably because a given increment in mass looks like a big
increment in density if you have a small cell, and it was density
that I was monitoring. The total excursion in density would be
about ± 1% and there might be a limit cycle of around 130 time
steps. It turned out that those fluctuations could be eliminated by adding the fourth difference term, but then overshoots
returned near the shockwave; the obvious answer was to switch
off that term near the shock, which was quite easy because that
was where $\epsilon^{(2)}$ was being turned on. In fact, I set

$$\epsilon^{(4)}_{i+\frac{1}{2},j} = \max \left\{ 0, \left(K^{(4)} - \epsilon^{(2)}_{i+\frac{1}{2},j} \right) \right\} \tag{16}$$

where $K^{(4)}$ is a second constant, and that does the trick. What
it comes down to is that the second difference is usually multiplied by a factor of order $(\Delta x)^2$, rising to order unity near a
shock; the fourth difference is multiplied by a factor which is
usually of order unity, but which falls to zero through a shock.

Now we go on to how the time-stepping is done. I began by
devising a three-step scheme which can be written as follows.
We call the complete space operator, flux balance and dissipation, P, so that

$$Pw = Qw - Dw \tag{17}$$

and then we go from time level n to time level (n+1) by means of

$$\left. \begin{aligned}
w^{(0)} &= w^n \\
w^{(1)} &= w^{(0)} - \Delta t\, Pw^{(0)} \\
w^{(2)} &= w^{(0)} - \frac{\Delta t}{2} Pw^{(0)} - \frac{\Delta t}{2} Pw^{(1)} \\
w^{(3)} &= w^{(0)} - \frac{\Delta t}{2} Pw^{(0)} - \frac{\Delta t}{2} Pw^{(2)} \\
w^{n+1} &= w^{(3)}
\end{aligned} \right\} \tag{18}$$

That was the scheme I used at Dornier last summer, and at the
time I thought it was rather neat, although it turned out not to
be new, having been proposed by Gary in 1964. If you do the
usual stability analysis you find that it is stable for Courant
numbers less than 2.0, and you get that rate of advance for the
price of evaluating the spatial operator three times. For
comparison, the MacCormack scheme gives you a Courant number of
1.0 for two evaluations.

Then I realised that (18) belongs to the class of Runge-Kutta
methods, and that a lot is known about the properties of such

methods, especially in the context of ordinary differential
equations, such as

$$\frac{\partial w}{\partial t} + Aw = 0 \tag{19}$$

The most useful concept seems to be that of the stability
region. If we apply some specified algorithm to eqn (19) it
turns out that there are values of $A\Delta t$, with A in the complex
plane, for which the algorithm will be stable, and that region
of the complex plane is the stability region. There is a book
by Stetter (1973), in which plots of these regions are given
for a variety of schemes.

The way that we can apply this concept to partial differential equations such as

$$\frac{\partial w}{\partial t} + \frac{A \partial w}{\partial x} = 0 \tag{20}$$

is just to take the Fourier transform of w, say $\hat{w}$ and to realise
that the central difference operator multiplies $\hat{w}$ by a factor
which is purely imaginary, but that the dissipative terms add a
purely real component to that factor.

In the plotted stability regions that I am going to show you
(Fig. 2) the imaginary part is merely what we call Courant
number, and the real part is a measure of the added dissipation.
You can now see that the first order scheme is not stable any-
where on the real axis but if we add enough dissipation (for
example by using one sided differences) we can get ourselves
inside the circle. In fact the second-order two-step scheme is
also unstable everywhere on the imaginary axis, but the third-
order, three-step scheme does include a segment of the imaginary
axis. The three step scheme which I showed you just now (eqn(18))
goes up to a Courant number of 2.0, and that is more efficient
than MacCormack or Lax-Wendroff, in terms of time advanced
for each evaluation of the spatial operator P. In fact we can do
even better by going to the standard fourth-order scheme thus

$$
\left.
\begin{aligned}
w^{(0)} &= w^n \\
w^{(1)} &= w^{(0)} - \frac{\Delta t}{2} Pw^{(0)} \\
w^{(2)} &= w^{(0)} - \frac{\Delta t}{2} Pw^{(1)} \\
w^{(3)} &= w^{(0)} - \Delta t\, Pw^{(2)} \\
w^{(4)} &= w^{(0)} - \frac{\Delta t}{6} Pw^{(0)} - \frac{\Delta t}{3} Pw^{(1)} - \frac{\Delta t}{3} Pw^{(2)} - \frac{\Delta t}{6} Pw^{(3)}
\end{aligned}
\right\} \tag{21}
$$

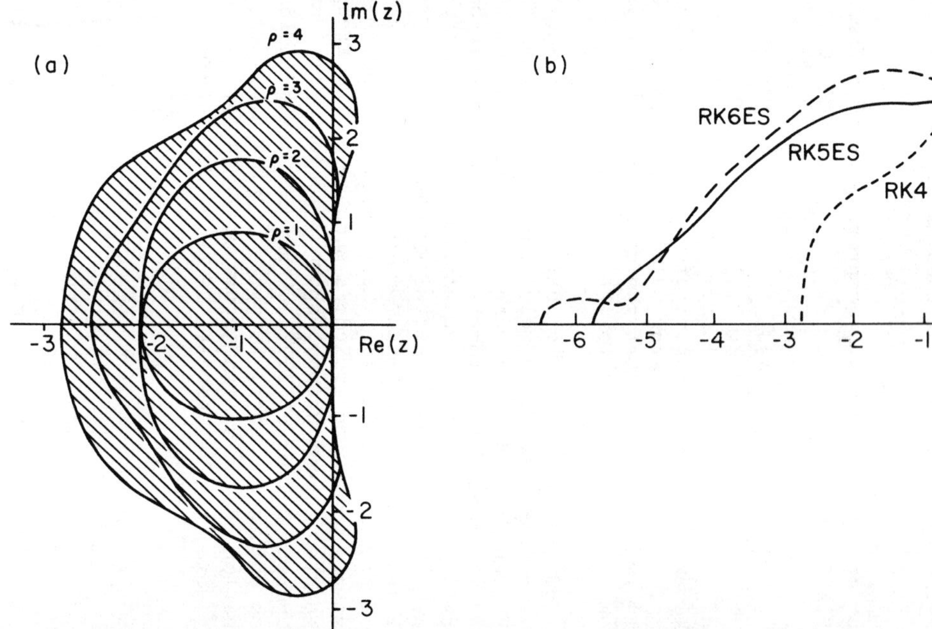

Fig. 2. Stability regions for various Runge-Kutta time-stepping schemes: (a) standard schemes of order 1,2,3,4 (Stetter, 1973), (b) non-standard schemes of order 4,5,6 (Lawson, 1966)

and this gives us a Courant number of $2\sqrt{2} = 2.8$ for four evaluations of the operator. Also I have got quite a lot of room to add dissipative terms if I need to, near the shocks. That is shown by the shaded region to the left of the axis.

A very interesting possibility is that when you add extra stages to a Runge-Kutta scheme you can do so either with the intention of gaining more speed or of gaining more accuracy. And although the standard four-step method is fourth-order accurate in time there is not very much point in that when you are only second-order accurate in space. So perhaps by backing off in accuracy you could gain even more speed. Lawson (1966) has done some work on these lines, and he has a six-stage scheme which is stable out to Courant numbers of 3.50, with a huge amount of room in the real direction, and I think there is scope for a lot of investigation here, to find the most efficient possibilities.

Something else that I have been thinking about a lot recently is the point I mentioned earlier about how to make use of the fact that we expect to have constant enthalpy in the steady state. That is obvious when you look at the equations (2) and notice that when the time derivatives vanish the fourth equation is just H times the first equation. We can try to exploit that in two different ways. We could assert that H will always be H_∞ and then solve the non-physical system given by the first three equations (see the chapter by Lerat, for example). The alternative is something for which I do not yet have a real mathematical justification, but for which I can offer analogies based on experience in potential flow calculations.

Suppose we set out to solve the potential equation by means of line relaxation. Then it turns out that the relaxation operator effectively adds something like a φ_t term, turning the unsteady potential equation into a kind of telegraph equation which has damped waves as its fundamental solutions. On the other hand, the true linear wave equation

$$\varphi_{tt} = \varphi_{xx} + \varphi_{yy} \tag{22}$$

is undamped, as you can see if you multiply by φ_t and integrate

$$\iint \varphi_t \, \varphi_{tt} \, dx \, dy = \iint \varphi_{xx} \, \varphi_t \, dx \, dy + \iint \varphi_{yy} \, \varphi_t \, dx \, dy \tag{23}$$

If you integrate by parts and neglect the boundary terms, or consider an infinite domain, then

$$\iint \varphi_t \, \varphi_{tt} \, dx \, dy + \iint \varphi_x \, \varphi_{xt} \, dx \, dy + \iint \varphi_y \, \varphi_{yt} \, dx \, dy = 0$$

and so

$$\frac{\partial P}{\partial t} = 0 \qquad (24)$$

where $\quad P = \tfrac{1}{2}\iint (\varphi_t^2 + \varphi_x^2 + \varphi_y^2)\ dx\ dy \qquad (25)$

so the quantity P, which is a kind of energy, does not decay.
The equation which mimics the relaxation process is

$$\varphi_{tt} + \alpha\varphi_t + \varphi_{xx} + \varphi_{yy} \qquad (26)$$

and then the same piece of algebra gives

$$\frac{\partial P}{\partial t} + \alpha\iint\varphi_t^2\ dx\ dy = 0 \qquad (27)$$

Therefore, if α is positive, the positive quantity P must decay,
until we reach a steady state with $\varphi_t = 0$ everywhere. The coefficient α is actually proportional to $(2/\omega - 1)$ where ω is the
relaxation factor, and since we all know that you have $\omega < 2$
for stability, that makes the whole argument fairly believable.

Now I would like to get that sort of damping into the Euler
equations if I possibly can, but unfortunately I no longer have
a potential. However, you may recall that the unsteady Bernoulli
equation you get from the potential equation tells you that
$\varphi_t + H = $ const, so maybe adding some terms proportional to
$(H - H_\infty)$ will put some damping into the Euler equations. I feel
that will be the case if you do it right, but I am not yet sure
of the details. What I am sure of is that you had better be
certain that H = constant really is a solution, not merely of
the differential equations, but of the difference equations you
use to solve them. And that is not easy to ensure in something like a MacCormack scheme, with mixtures of backward and
forward differences. But it is quite easy to ensure in the kind
of central difference scheme I have told you about, with this
proviso, that the dissipation operator in the energy equation is
applied to ρH rather than to ρE.

What I tried was to add a term proportional to $H - H_\infty$ in the
continuity equation, thus

$$\frac{\partial \rho}{\partial t} + \rho\nabla.q + \alpha\rho(H - H_\infty) = 0 \qquad (28)$$

and then I found that I needed to add corresponding terms to all
other three equations, so as to retain conservation. In the
first place I tried the extra terms out in a code which only had
the second-order dissipation and really was not converging very
well and I found that the enthalpy damping would make it converge. Subsequently I added the fourth-order dissipation, and

then the scheme would converge whether or not I included the
enthalpy damping. The enthalpy damping does produce quicker convergence but I still have to find the best combination, and to
decide whether the extra improvement is worthwhile.

For non-lifting flow past the NACA 0012 at M = 0.08 I can get
an adequate engineering solution within 600 time steps if I start
by creating the aerofoil impulsively in a uniform stream, or I
can bring that down to 300 if I start from the potential solution.
At higher Mach numbers with a stronger shock the potential solution is less useful. It is interesting that the method appears
to give very respectable shockwaves without the need to switch
differences, or to get involved with any complicated logic depending on the signs of eigenvalues. I have not attempted any
kind of multigrid solution yet because until now the basic code
was not coverging to my satisfaction. But now it does, and so
maybe this is the time to try and implement a multigrid version.
What needs to be observed, though, is that multigrid depends on
having some element of smoothing, and does not fit readily with
problems where the basic mechanism is wave propagation.

The last thing I want to tell you about is boundary conditions.
At the outer boundary what I do is to calculate the characteristic
variables, on the basis of a local linearisation. These turn out
to be $(p - c^2\rho)$, $(p + cu\rho)$, $(p - cu\rho)$ and v, where v is the tangential velocity component, and u is the normal velocity component
relative to the outer boundary. The first and last of these are
carried along the streamline, with velocity u, and the second and
third are carried along Mach waves with velocities (u + c), (u − c).
Now the theory says that if you are at a subsonic outflow boundary, the third quantity may be imposed, but the other three have
to be determined from the interior, whereas if you are at a subsonic inflow boundary the second must be determined from the
interior, but the other three may be imposed. And basically that
is what I do, but I convert all these conditions so that they
relate to the conservation variables I am actually using. Some
people say that you can get away with over-specifying the conditions at the outer boundary, simply fixing everything at its
free-stream value, but when I have tried that it has sometimes
gone badly wrong. With a more dissipative difference scheme it
seems less critical what you do.

To apply a boundary condition on the body is very easy for
steady flows if you look back to eqn (10). All you have to do is
to set Q_k = 0 on the interface concerned, or to give it whatever
value you want it to have. And you specify a surface pressure by
extrapolating from the interior according to $\partial p/\partial n = \rho q^2/R$, where
n is the surface normal, and R is the surface curvature.

Finally, I would like to show you a few of the results, beginning with the non-lifting flow past a NACA 0012 aerofoil at Mach number 0.8. The inner part of my computing mesh for this problem is shown in Fig. 3.

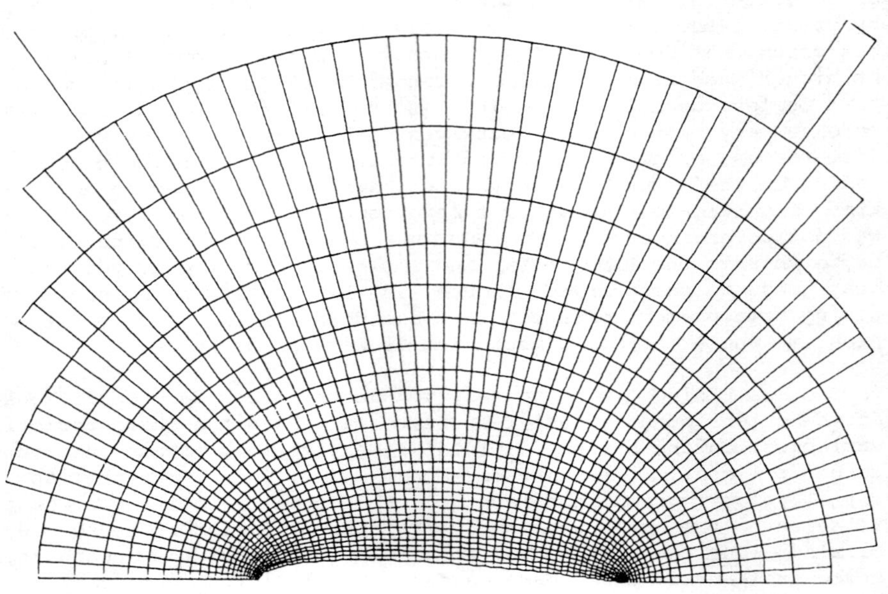

Fig. 3.   Part of the computing grid for a NACA 0012 Aerofoil. The total grid has 64 cells from front to rear and 32 cells radially

The development of the flow from impulsive starting conditions is shown in Fig. 4, where I have not used enthalpy damping, and in Fig. 5 where I have. You can see that in each case the plotted pressure distributions develop smoothly. However, if we measure the error as the r.m.s. value of $\Delta\rho/\Delta t$ within cells, then this quantity converges quite a lot faster using enthalpy damping. In Fig. 6 we have the corresponding information, for the rather harder case $M_\infty = 0.85$, with the enthalpy damping.

REFERENCES

Rizzi, A. and Viviand, H. (1981). Numerical methods for the computation of inviscid transonic flows with shock waves. Notes on Numerical Fluid Mechanics, vol 3, Vieweg.

Stetter, H. (1973). Analysis of discretisation error for ordinary differential equations. Tracts in Natural Philosophy, no 23 Springer.

Gary, J. (1964). On certain finite difference schemes for hyperbolic systems. *Math. Comp.* **18** pp 1 - 18.

Lawson, J.D. (1966). An order 5 Runge-Kutta process with extended region of stability. *SIAM. J. Num. Anal.* **3**, no. 4, pp 593 - 606.

EDITOR'S NOTE

Since Professor Jameson gave the above lecture, he has gone on to develop his scheme into a three-dimensional method. Essentially no new mathematical problems arise. The analysis of the flux balance is done exactly as above, with each cell receiving increments of mass, momentum and energy due to fluxes across six faces, computed from average values of pressure, normal velocities etc., on those faces. A smoothing operation is carried out consisting of three one-dimensional operations, each identical to that given in the lecture, and then the solution can be advanced in time by the same four-stage Runge-Kutta scheme. The problems which do arise are concerned with programming. Many details of the coding have been altered to remove the pressure on storage.

Professor Jameson has very kindly made available some results which this method has produced. They relate to the ONERA M6 wing under test conditions $M_\infty = 0.84$, $\alpha = 3.06°$. The computational mesh consists of 128 cells around each aerofoil section, 32 cells along each spanwise line, and 16 cells going outward from the wing. The results of the Euler code are shown as solid lines and results from a potential method on the same grid as dashed lines. Fig. E1 (a)-(d) shows pressure distributions along those chordwise lines which lie at 12.5, 37.5, 62.5 and 87.5 per cent of the semi-span. Fig. E2 shows the Euler results only, at 20 chordwise sections. Fig. E2 (a) shows results for the upper surface of the wing, and Fig. E2 (b) shows results for the lower surface.

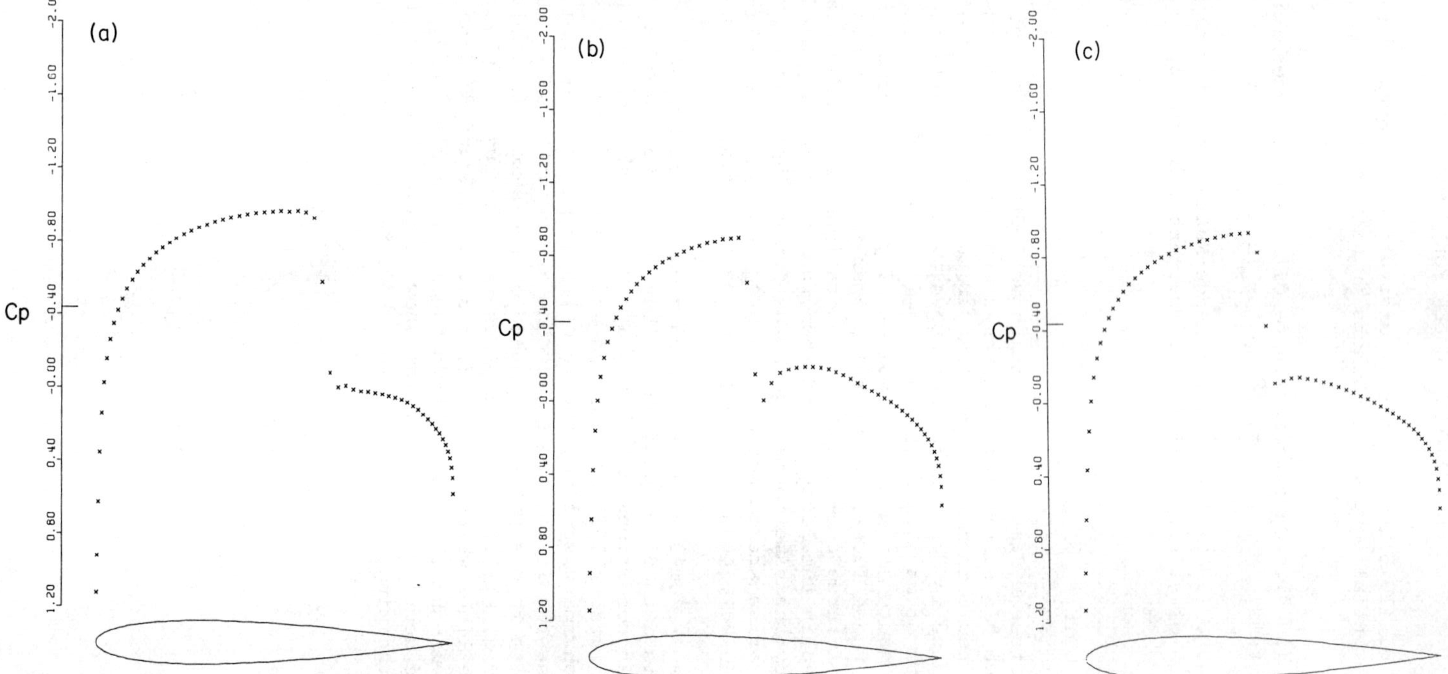

(a)
Cp
-2.00 -1.60 -1.20 -0.80 -0.40 -0.00 0.40 0.80 1.20
(b)
Cp
-2.00 -1.60 -1.20 -0.80 -0.40 -0.00 0.40 0.80 1.20
(c)
-2.00 -1.60 -1.20 -0.80 -0.40 -0.00 0.40 0.80 1.20
Cp

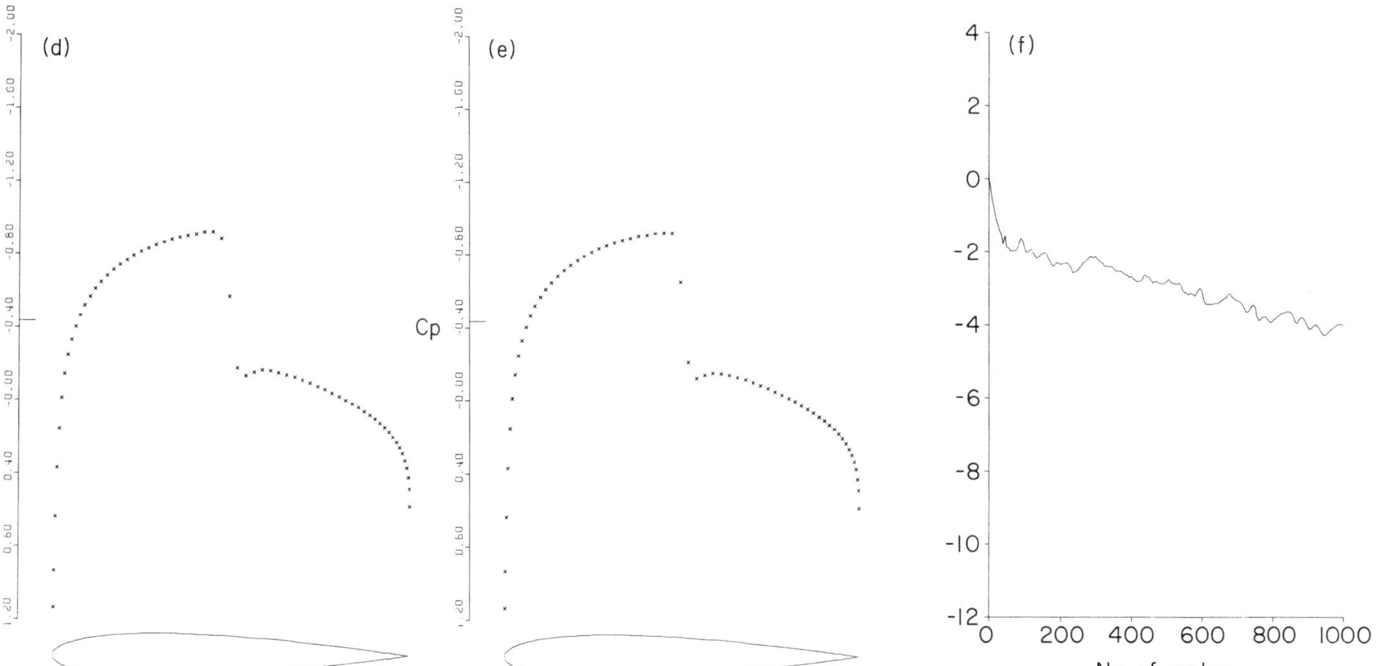

Fig. 4. Solution of the flow past a NACA 0012 aerofoil section at $M_\infty = 0.80$, $\alpha = 0°$; (a)-(e) show pressure distributions after 200, 400, 600, 800 and 1000 cycles of the Runge-Kutta process; (f) shows the logarithm of the r.m.s. error. This calculation does not use enthalpy damping.

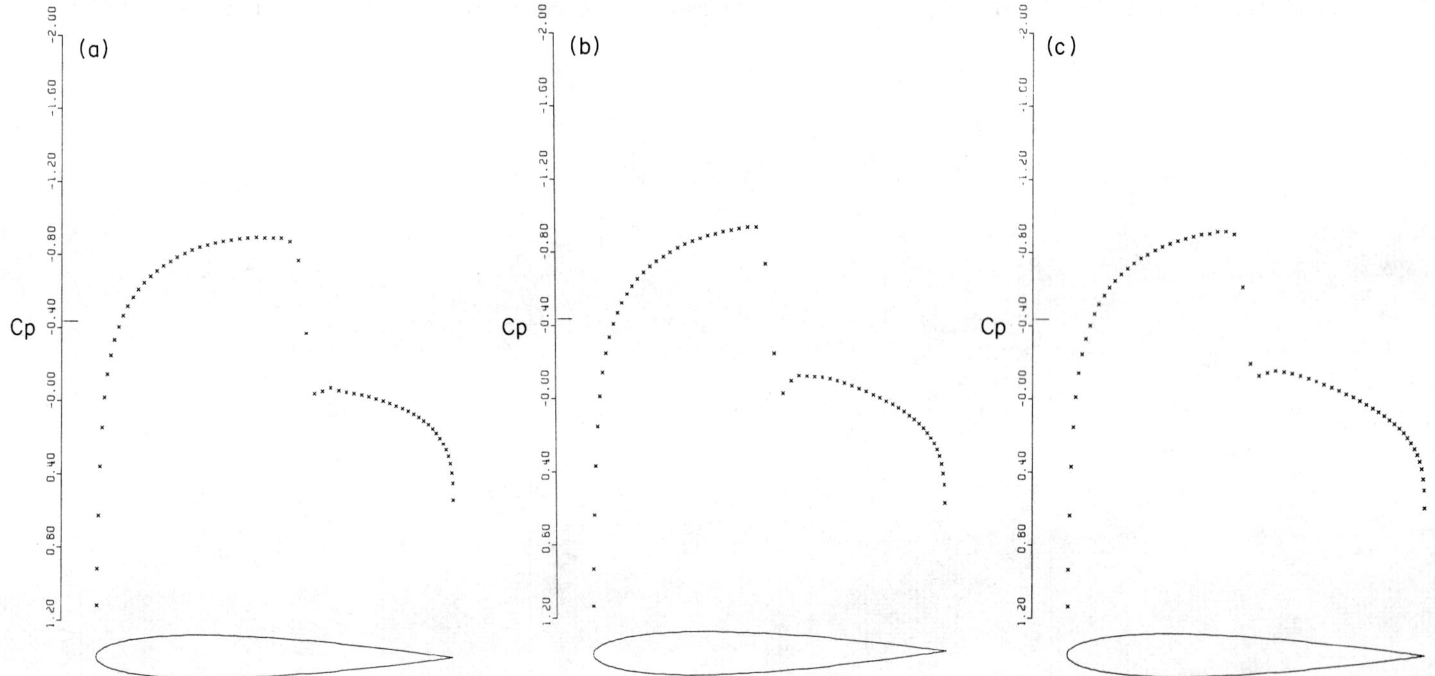

(a)
Cp
(b)
Cp
(c)
Cp

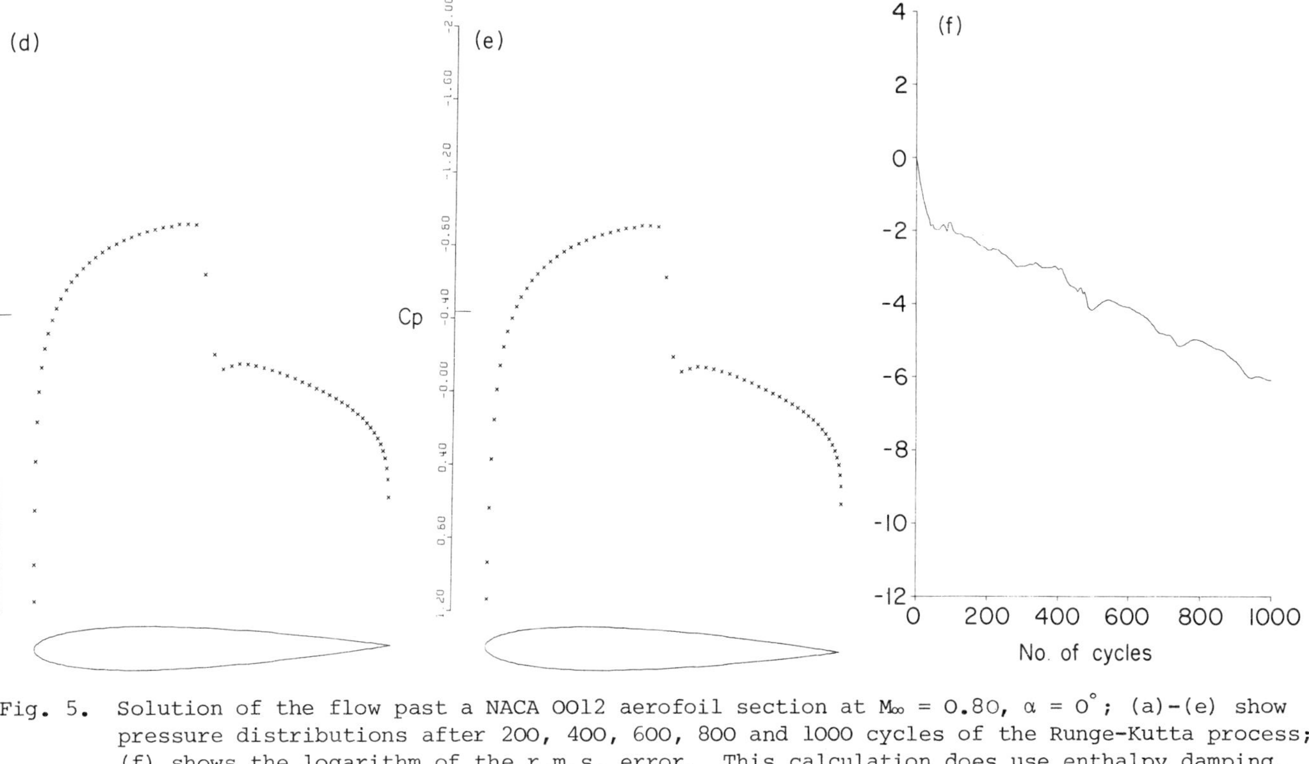

Fig. 5. Solution of the flow past a NACA 0012 aerofoil section at $M_\infty = 0.80$, $\alpha = 0°$; (a)-(e) show pressure distributions after 200, 400, 600, 800 and 1000 cycles of the Runge–Kutta process; (f) shows the logarithm of the r.m.s. error. This calculation does use enthalpy damping.

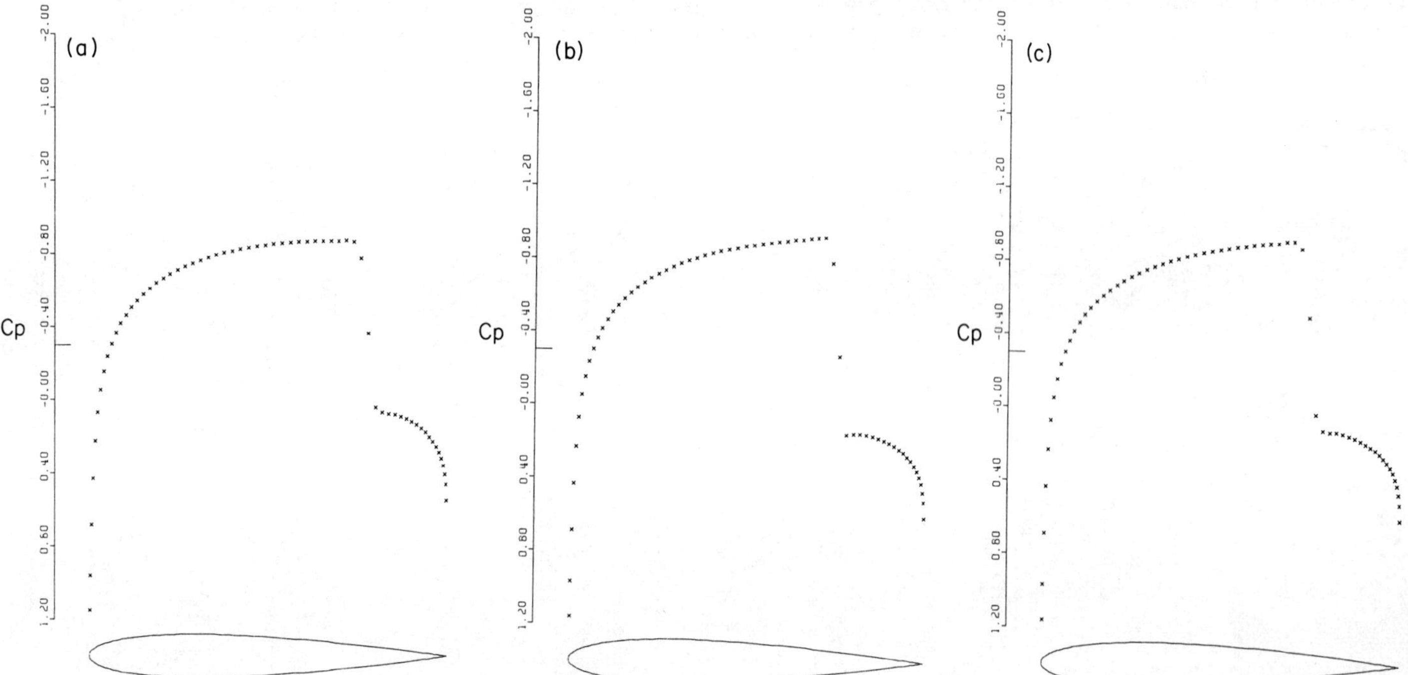

(a)
Cp
-2.00
-1.60
-1.20
-0.80
-0.40
-0.00
0.40
0.80
1.20
(b)
Cp
-2.00
-1.60
-1.20
-0.80
-0.40
-0.00
0.40
0.80
1.20
(c)
Cp
-2.00
-1.60
-1.20
-0.80
-0.40
-0.00
0.40
0.80
1.20

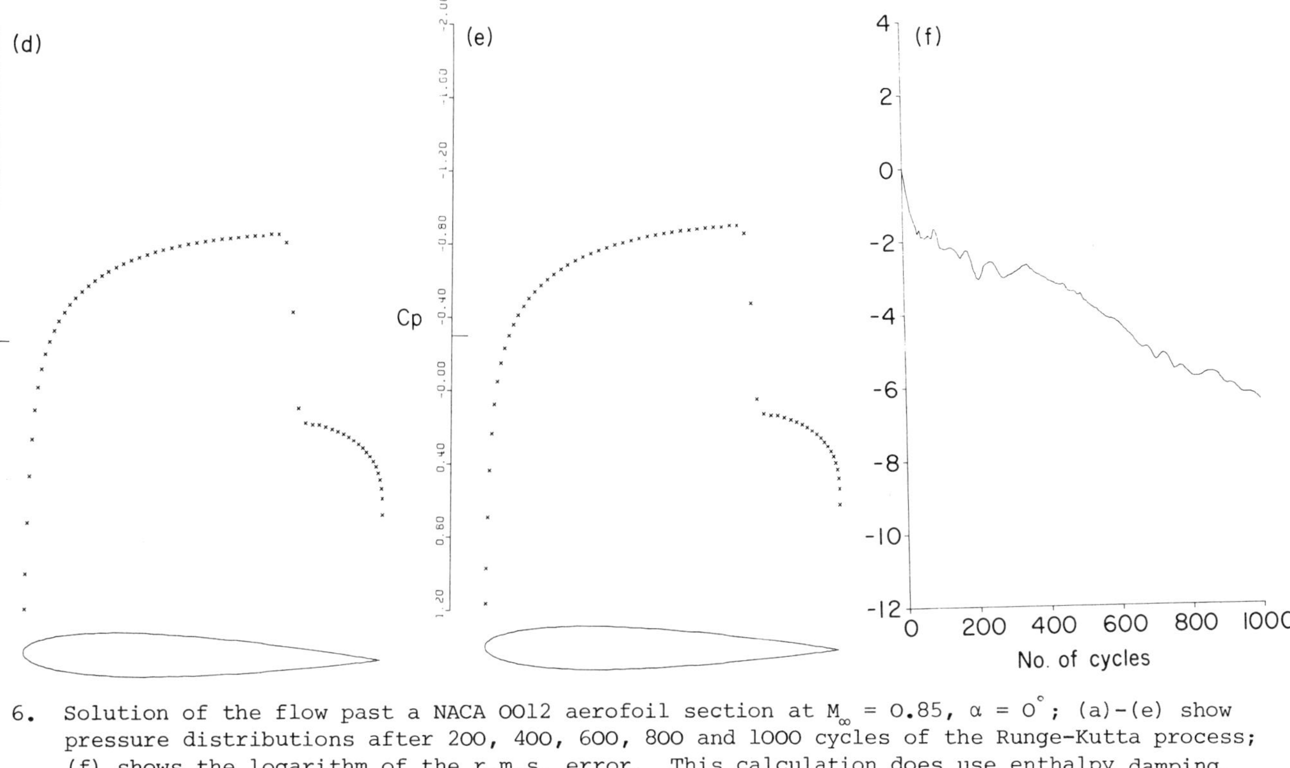

Fig. 6. Solution of the flow past a NACA 0012 aerofoil section at $M_\infty = 0.85$, $\alpha = 0°$; (a)-(e) show pressure distributions after 200, 400, 600, 800 and 1000 cycles of the Runge-Kutta process; (f) shows the logarithm of the r.m.s. error. This calculation does use enthalpy damping.

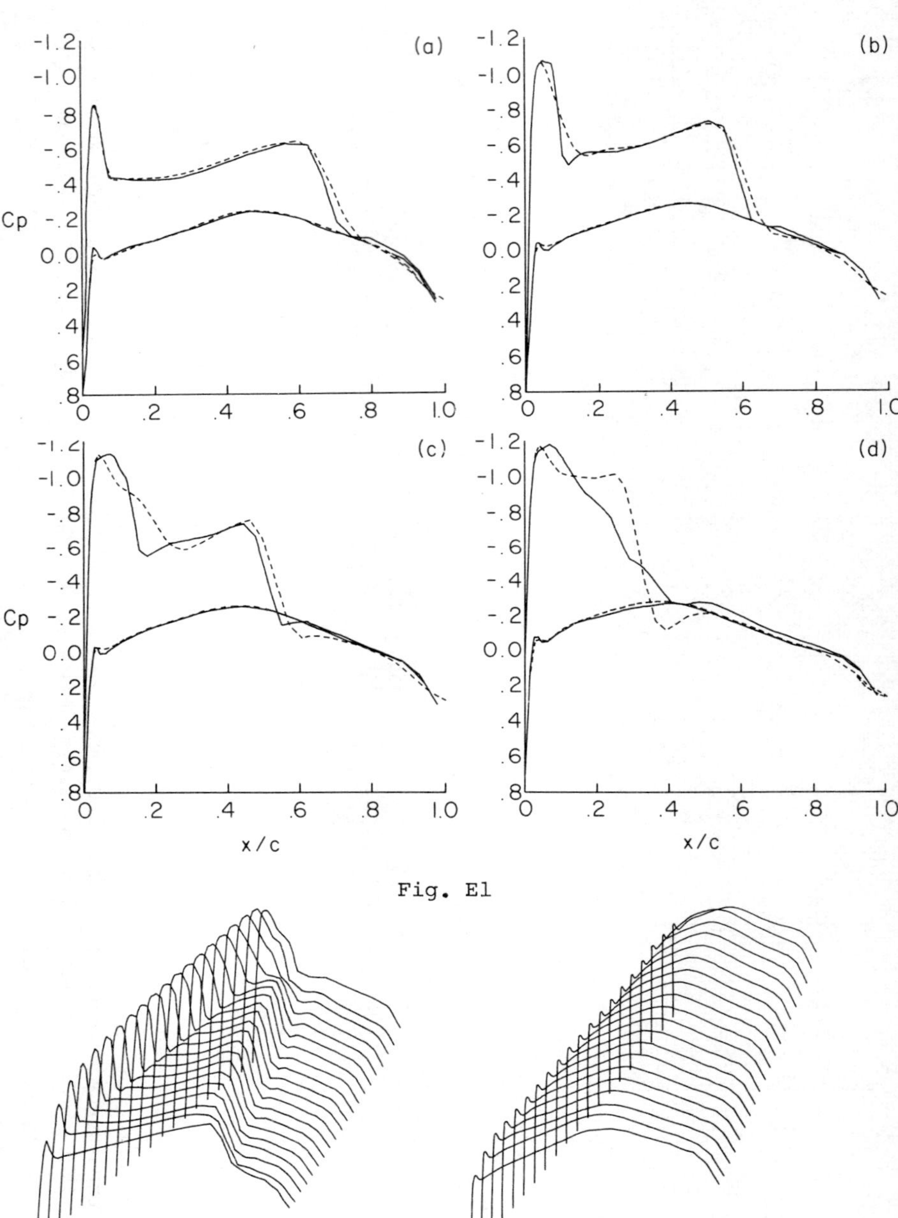

Fig. E1

Fig. E2

NUMERICAL SOLUTION OF THE TIME DEPENDENT
NAVIER-STOKES EQUATIONS FOR INCOMPRESSIBLE VISCOUS
FLUIDS BY FINITE ELEMENT AND ALTERNATING DIRECTION METHODS

R. Glowinski

(*Laboratoire d'Analyse Numérique,
Universite Pierre et Marie Curie, France*)

B. Mantel and J. Periaux

(*Avions Marcel Dassault/Breguet Aviation,
Saint-Cloud, France*)

1. INTRODUCTION

We describe in this paper some new methods for solving the
time dependent Navier-Stokes equations for incompressible viscous fluids; these methods combine finite elements for the space
discretization and alternating directions for the time discretization. The key idea is to use the splitting associated to the
alternating direction methods to decouple the two main difficulties of the original problem, namely nonlinearity and
incompressibility.

The methods which follow are a natural extension of those
described in Bristeau et al(1980a) and Bristeau et al (1980b)
since least square and conjugate gradient methods are still
used to treat the nonlinearity; however due to the decoupling
mentioned above the present methods are in fact more efficient
since they require less computer time and lead to more accurate
numerical results. They provide in particular quite efficient
methods for solving the steady Navier-Stokes equations.

The following paper is closely related to Glowinski (1982)
to which we refer for more details and also for a substantial
bibliography concerning the Navier-Stokes equations and their
numerical treatment.

2. FORMULATION OF THE TIME DEPENDENT NAVIER-STOKES EQUATIONS
FOR INCOMPRESSIBLE FLUIDS

Let us consider a Newtonian incompressible viscous fluid. If
Ω and Γ denote the region of the flow ($\Omega \subset \mathbb{R}^N$, N=2,3 in practice)
and its boundary, respectively, then this flow is governed by

the following <u>Navier-Stokes equations</u>

$$\frac{\partial \underset{\sim}{u}}{\partial t} - \nu \Delta \underset{\sim}{u} + (\underset{\sim}{u}.\underset{\sim}{\nabla})\underset{\sim}{u} + \nabla p = \underset{\sim}{f} \ \underline{in} \ \Omega, \tag{1}$$

$$\underset{\sim}{\nabla}.\underset{\sim}{u} = 0 \ \underline{in} \ \Omega \ (\text{incompressibility condition}). \tag{2}$$

In (1),(2)

(a) $\quad \underset{\sim}{\nabla} = \{\frac{\partial}{\partial x_i}\}_{i=1}^{N}$, $\Delta = \underset{\sim}{\nabla}^2 = \sum_{i=1}^{N} \frac{\partial^2}{\partial x_i^2}$,

(b) $\quad \underset{\sim}{u} = \{u_i\}_{i=1}^{N}$ is the <u>flow velocity</u>

(c) $\quad$ p is the <u>pressure</u>,

(d) $\quad \nu$ is the <u>viscosity</u> of the fluid ($\nu = 1/Re$
Re: Reynold's number),

(e) $\quad \underset{\sim}{f}$ is a <u>density of external forces</u>.

In (1), $(\underset{\sim}{u}.\underset{\sim}{\nabla})\underset{\sim}{u}$ is a <u>symbolic notation</u> for the nonlinear (vector) term

$$\left\{ \sum_{j=1}^{N} u_j \frac{\partial u_i}{\partial x_j} \right\}_{i=1}^{N} .$$

<u>Boundary conditions</u> have to be added; for example in the case of the airfoil B of Figure 1, we have (since the fluid is <u>viscous</u>) the following <u>adherence condition</u>

$$\underset{\sim}{u} = \underset{\sim}{0} \ \underline{on} \ \partial B = \Gamma_B; \tag{3}$$

typical conditions at infinity are

$$\underset{\sim}{u} = \underset{\sim}{u}_\infty \tag{4}$$

where $\underset{\sim}{u}_\infty$ is a <u>constant</u> vector (with regard to the space variables at least).

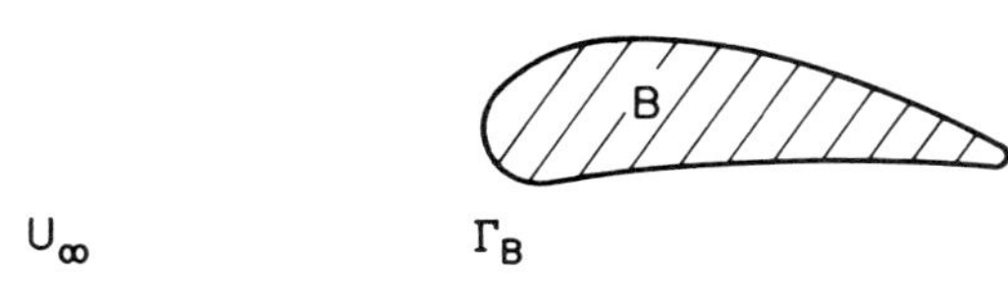

Figure 1

If Ω is a bounded region of $\mathbb{R}^N$ we may prescribe as boundary condition

$$\underset{\sim}{u} = \underset{\sim}{g} \text{ on } \Gamma \tag{5}$$

where (from the incompressibility of the fluid) the given function $\underset{\sim}{g}$ has to satisfy

$$\int_\Gamma \underset{\sim}{g} . \underset{\sim}{n} \ d\Gamma = 0, \tag{6}$$

where $\underset{\sim}{n}$ is the outward unit vector normal at Γ.

Finally for the time dependent problem (1), (2) an initial condition such as

$$\underset{\sim}{u}(x,0) = \underset{\sim}{u}_0(x) \text{ a.e. on } \Omega, \tag{7}$$

with $\underset{\sim}{u}_0$ given, is usually prescribed.

From the above equations we observe three difficulties (even for flows at low Reynold's numbers in bounded regions Ω) which are

(i)   The nonlinear term $(\underset{\sim}{u}.\underset{\sim}{\nabla})\underset{\sim}{u}$ in (1),

(ii)   The incompressibility condition (2),

(iii) The fact that the solutions of the Navier-Stokes equations are vector-valued functions of x,t, whose components are coupled by the nonlinear term $(\underset{\sim}{u}.\underset{\sim}{\nabla})\underset{\sim}{u}$ and by the incompressibility condition $\underset{\sim}{\nabla}.\underset{\sim}{u} = 0$.

Using convenient alternating direction methods for the time discretization of the Navier-Stokes equations we shall be able to decouple the difficulties due to the nonlinearity and to the incompressibility, respectively.

For simplicity we suppose from now on that Ω is <u>bounded</u> and that we have (5) as boundary condition (with $\underset{\sim}{g}$ satisfying (6) and possibly depending upon t).

3. TIME DISCRETIZATION BY ALTERNATING DIRECTION METHODS

Let Δt (> 0) be a <u>time discretization</u> step and θ a parameter such that $0<\theta<1$.

3.1 A first alternating direction method

We consider first the following alternating direction method (of Peaceman-Rachford type):

$$\underset{\sim}{u}^{0} = \underset{\sim}{u}_{0},\qquad\qquad(8)$$

<u>then for</u> $n \geq 0$ <u>compute</u> $\{\underset{\sim}{u}^{n+1/2}, p^{n+1/2}\}$ <u>and</u> $\underset{\sim}{u}^{n+1}$,

<u>from</u> $\underset{\sim}{u}^{n}$, <u>by solving</u>

$$\begin{cases} \dfrac{\underset{\sim}{u}^{n+1/2}-\underset{\sim}{u}^{n}}{(\Delta t/2)} - \theta\nu\Delta\underset{\sim}{u}^{n+1/2} + \nabla p^{n+1/2}=\underset{\sim}{f}^{n+1/2}+(1-\theta)\nu\Delta\underset{\sim}{u}^{n}-(\underset{\sim}{u}^{n}.\nabla)\underset{\sim}{u}^{n} \quad \underline{in}\ \Omega, \\[2em] \underset{\sim}{\nabla}.\underset{\sim}{u}^{n+1/2} = 0 \ \underline{in}\ \Omega, \\[2em] \underset{\sim}{u}^{n+1/2} = \underset{\sim}{g}^{n+1/2} \ \underline{on}\ \Gamma, \end{cases}\qquad(9)$$

<u>and</u>

$$\begin{cases} \dfrac{\underset{\sim}{u}^{n+1}-\underset{\sim}{u}^{n+1/2}}{(\Delta t/2)} - (1-\theta)\nu\Delta\underset{\sim}{u}^{n+1}+(\underset{\sim}{u}^{n+1}.\nabla)\underset{\sim}{u}^{n+1}=\underset{\sim}{f}^{n+1}+\theta\nu\Delta\underset{\sim}{u}^{n+1/2}-\nabla p^{n+1/2} \\[1em] \hspace{10cm} \underline{in}\ \Omega, \\[1em] \underset{\sim}{u}^{n+1} = \underset{\sim}{g}^{n+1} \ \underline{on}\ \Gamma, \end{cases}$$
$$(10)$$

<u>respectively.</u>

We use the notation $f^{j}(x) = f(x,j\Delta t)$, $g^{j}(x) = g(x,j\Delta t)$, and $u^{j}(x)$ is an approximation of $u(x,j\Delta t)$.

3.2 A second alternating direction method

We consider now the following alternating direction method (of Strang type):

$$\underset{\sim}{u}^{0} = \underset{\sim}{u}_{0} \ , \tag{11}$$

then for $n \geq 0$ and starting from $\underset{\sim}{u}^{n}$ we solve

$$\left\{ \begin{array}{l} \dfrac{\underset{\sim}{u}^{n+1/4} - \underset{\sim}{u}^{n}}{(\frac{\Delta t}{4})} - \theta \nu \Delta \underset{\sim}{u}^{n+1/4} + \nabla p^{n+1/4} = \underset{\sim}{f}^{n+1/4} + (1-\theta) \nu \Delta \underset{\sim}{u}^{n} - \underset{\sim}{u}^{n} . \nabla) \underset{\sim}{u}^{n} \\ \qquad\qquad\qquad\qquad\qquad\qquad \underline{\text{in}} \ \Omega, \\[2mm] \nabla . \underset{\sim}{u}^{n+1/4} = 0 \ \underline{\text{in}} \ \Omega, \\[2mm] \underset{\sim}{u}^{n+1/4} = \underset{\sim}{g}^{n+1/4} \ \underline{\text{on}} \ \Gamma, \end{array} \right. \tag{12}$$

$$\left\{ \begin{array}{l} \dfrac{\underset{\sim}{u}^{n+3/4} - \underset{\sim}{u}^{n+1/4}}{(\Delta t/2)} - (1-\theta) \nu \Delta \underset{\sim}{u}^{n+3/4} + (\underset{\sim}{u}^{n+3/4} . \nabla) \underset{\sim}{u}^{n+3/4} = \\[2mm] \underset{\sim}{f}^{n+3/4} + \theta \nu \Delta \underset{\sim}{u}^{n+1/4} - \nabla p^{n+1/4} \ \underline{\text{in}} \ \Omega, \\[2mm] \underset{\sim}{u}^{n+3/4} = \underset{\sim}{g}^{n+3/4} \ \underline{\text{on}} \ \Gamma, \end{array} \right. \tag{13}$$

$$\left\{ \begin{array}{l} \dfrac{\underset{\sim}{u}^{n+1} - \underset{\sim}{u}^{n+3/4}}{(\frac{\Delta t}{4})} - \theta \nu \Delta \underset{\sim}{u}^{n+1} + \nabla p^{n+1} = \underset{\sim}{f}^{n+1} + (1-\theta) \nu \Delta \underset{\sim}{u}^{n+3/4} \\ \qquad\qquad\qquad\qquad\qquad - (\underset{\sim}{u}^{n+3/4} . \nabla) \underset{\sim}{u}^{n+3/4} \ \underline{\text{in}} \ \Omega, \\[2mm] \nabla . \underset{\sim}{u}^{n+1} = 0 \ \underline{\text{in}} \ \Omega \\[2mm] \underset{\sim}{u}^{n+1} = \underset{\sim}{g}^{n+1} \ \underline{\text{on}} \ \Gamma. \end{array} \right. \tag{14}$$

3.3 Some comments and remarks concerning the alternating direction schemes (8) - (10) $\underline{and}$ (11) - (14).

Using the two alternating direction schemes described in Secs. 2.1, 2.2 we have been able to decouple <u>nonlinearity</u>

and _incompressibility_ in the Navier-Stokes equations (1), (2).
We shall describe in the following sections the specific treatment of the subproblems encountered at each step of (8)-(10)
and (11)-(14); we shall consider first the case where the subproblems are still continuous in space (since the formalism of
the continuous problems is much simpler), and then the discrete
case where a finite element method is used to approximate in
space the Navier-Stokes equations.

Scheme (8)-(10) has a _truncation error_ in $O(\Delta t)$; due to the
symmetrization process involved in it, scheme (11)-(14) has a
truncation error in $O(|\Delta t|^2)$.

We observe that $u^{n+1/2}$ and $u^{n+1/4}$, u^{n+1} are obtained from the
solution of _linear problems_ ((9) and (12), (14), respectively)
very close to the _steady Stokes_ problem. Despite its greater
complexity scheme (11) - (14) is almost as economical to use as
scheme (8) - (10); this is mainly due to the fact that the
"quasi" steady Stokes problems (9) and (12), (14) (actually
convenient finite element approximations of them) can be solved
by quite efficient solvers so that most of the computer time
used to solve a full alternating direction step ((9), (10) or
(12) - (14)) is in fact used to solve the nonlinear subproblem
((10) or (13)).

The good choice for θ is $\theta = 1/2$ (resp. $\theta = 1/3$) if one uses
scheme (8) - (10) (resp. (11) - (14)); with the above choices
for θ, many computer subprograms can be used for both the linear
and nonlinear subproblems, resulting therefore in quite substantial computer core memory savings.

Remark 3.1: A variant of scheme (8) - (10) is the following
(it corresponds to $\theta = 1$):

$$u^0 = u_0 \tag{15}$$

then for $n \geq 0$ and starting from u^n,

$$
\left\{
\begin{array}{l}
\dfrac{u^{n+1/2} - u^n}{(\Delta t/2)} - \nu\Delta u^{n+1} + \nabla p^{n+1/2} = f^{n+1} - (u^n \cdot \nabla)u^n \ \underline{in}\ \Omega, \\[2em]
\nabla \cdot u^{n+1/2} = 0\ \underline{in}\ \Omega, \\[2em]
u^{n+1/2} = g^{n+1/2}\ \underline{on}\ \Gamma,
\end{array}
\right.
\tag{16}
$$

$$\begin{cases} \dfrac{\underset{\sim}{u}^{n+1}-\underset{\sim}{u}^{n+1/2}}{(\Delta t/2)} + (\underset{\sim}{u}^{n+1/2}.\underset{\sim}{\nabla})\underset{\sim}{u}^{n+1} = \underset{\sim}{f}^{n+1} + \nu\Delta\underset{\sim}{u}^{n+1/2} \quad \underline{in}\ \Omega, \\[2em] \underset{\sim}{u}^{n+1} = \underset{\sim}{g}^{n+1} \quad \underline{on}\ \Gamma_{\underline{\ }}^{n+\frac{1}{2}}, \end{cases} \tag{17}$$

<u>where</u>

$$\Gamma_{\underline{\ }}^{n+\frac{1}{2}} = \{x \mid x\epsilon\Gamma,\ \underset{\sim}{g}^{n+\frac{1}{2}}(x).\underset{\sim}{n}(x) < 0\}.$$

Both subproblems (16) and (17) are <u>linear</u>; the first one is
also a "quasi" steady Stokes problem and the second which is
a <u>first order system</u> can be solved by a <u>method of character-
istics</u>.

A similar remark holds for scheme (11) - (14).

Such methods have been used by several authors, the space
discretization being done by finite element methods very close
to those described in Sec. 6 of this paper (see Benque et al,
1980 and Ibler, 1981 for a discussion of those characteristics-
finite element methods for solving Navier-Stokes equations).
In our opinion these characteristics-finite element methods are
still too dissipative and will not be discussed here any longer
(we are presently working at such schemes with very small
dissipation).

4. LEAST SQUARE-CONJUGATE GRADIENT SOLUTION OF THE NONLINEAR
SUBPROBLEMS

4.1 Classical and Variational Formulations. Synopsis.

At each full step of the alternating direction methods (8) -
(10) and (11) - (14) we have to solve a <u>nonlinear elliptic
system</u> of the following type

$$\begin{cases} \alpha\underset{\sim}{u} - \nu\Delta\underset{\sim}{u} + (\underset{\sim}{u}.\underset{\sim}{\nabla})\underset{\sim}{u} = \underset{\sim}{f}\ \underline{in}\ \Omega, \\[1em] \underset{\sim}{u} = \underset{\sim}{g}\ \underline{on}\ \Gamma, \end{cases} \tag{18}$$

where α and ν are two positive parameters and where f and g are
two <u>given</u> functions defined on Ω and Γ, respectively.

We shall not discuss here the existence and uniqueness of
solutions for problems (18).

We introduce now the following functional space of Sobolev's type (see, e.g., Adams, (1975), Necas, (1967) and Oden-Reddy, (1976) for information on Sobolev spaces):

$$H^1(\Omega) = \{\phi \mid \phi \in L^2(\Omega), \ \frac{\partial \phi}{\partial x_i} \in L^2(\Omega) \ \forall i = 1,\ldots N\}, \qquad (19)$$

$$H_O^1(\Omega) = \{\phi \mid \phi \in H^1(\Omega), \phi = 0 \ \underline{on} \ \Gamma\}, \qquad (20)$$

$$V_O = (H_O^1(\Omega))^N, \qquad (21)$$

$$V_g = \{\underset{\sim}{v} \mid \underset{\sim}{v} \in (H^1(\Omega))^N, \ \underset{\sim}{v} = \underset{\sim}{g} \ \underline{on} \ \Gamma\}; \qquad (22)$$

if $\underset{\sim}{g}$ is sufficiently smooth then V_g is $\underline{nonempty}$.

We shall use in the sequel the following notation

$$dx = dx_1 \ldots dx_N,$$

and if $\underset{\sim}{u} = \{\underset{\sim}{u}_i\}_{i=1}^N, \ \underset{\sim}{v} = \{\underset{\sim}{v}_i\}_{i=1}^N$

$$\begin{cases} \underset{\sim}{u}.\underset{\sim}{v} = \sum_{i=1}^N u_i \, v_i, \\[2em] \underset{\sim}{\nabla}\underset{\sim}{u}.\underset{\sim}{\nabla}\underset{\sim}{v} = \sum_{i=1}^N \underset{\sim}{\nabla}\underset{\sim}{u}_i.\underset{\sim}{\nabla}\underset{\sim}{v}_i = \sum_{i=1}^N \sum_{j=1}^N \frac{\partial u_i}{\partial x_j} \frac{\partial x_i}{\partial x_j} . \end{cases}$$

Using Green's formula we can prove that for sufficiently smooth functions $\underset{\sim}{u}$ and $\underset{\sim}{v}$ belonging to $(H^1(\Omega))^N$ and V_O, respectively, we have

$$-\int_\Omega \Delta\underset{\sim}{u}.\underset{\sim}{v} \ dx = \int_\Omega \underset{\sim}{\nabla}\underset{\sim}{u}.\underset{\sim}{\nabla}\underset{\sim}{v} \ dx. \qquad (23)$$

It can also be proved that if $\underset{\sim}{u} \in V_g$ is a solution of (18) it is also a solution of the $\underline{nonlinear \ variational \ problem}$

$$\begin{cases} \text{Find } \underset{\sim}{u} \in V_g \ \underline{such \ that} \\[1.5em] \alpha\int_\Omega \underset{\sim}{u}.\underset{\sim}{v} \ dx + \nu\int_\Omega \underset{\sim}{\nabla}\underset{\sim}{u}.\underset{\sim}{\nabla}\underset{\sim}{v} \ dx + \int_\Omega ((\underset{\sim}{u}.\underset{\sim}{\nabla})\underset{\sim}{u}).\underset{\sim}{v} \ dx = \int_\Omega \underset{\sim}{f}.\underset{\sim}{v} \ dx \ \forall \underset{\sim}{v} \in V_O, \end{cases}$$

$$\qquad (24)$$

and $\underline{conversely}$.

We observe that (18), (24) $\underline{\text{is not equivalent}}$ to a problem of the $\underline{\text{Calculus of Variations}}$ since there is no functional of $\underset{\sim}{v}$ with $(\underset{\sim}{v}.\nabla)\underset{\sim}{v}$ as differential; however using a convenient $\underline{\text{least}}$ $\underline{\text{square formulation}}$ we shall be able to solve (18), (24) by efficient methods from $\underline{\text{Nonlinear Programing}}$, like conjugate gradient, for example.

The $\underline{\text{finite element}}$ approximation of problem (18), (24) will be discussed in Sec. 6.

4.2. *Least square formulation of (18), (24)*

Let $\underset{\sim}{v} \in V_g$; from $\underset{\sim}{v}$ we define $\underset{\sim}{y}$ $(= \underset{\sim}{y}(\underset{\sim}{v})) \in V_0$ as the solution of

$$\begin{cases} \alpha \underset{\sim}{y} - \nu \Delta \underset{\sim}{y} = \alpha \underset{\sim}{v} - \nu \Delta \underset{\sim}{v} + (\underset{\sim}{v}.\nabla)\underset{\sim}{v} - \underset{\sim}{f} \ \underline{\text{in}} \ \Omega, \\[2em] \underset{\sim}{y} = 0 \ \underline{\text{on}} \ \Gamma. \end{cases} \tag{25}$$

We observe that $\underset{\sim}{y}$ is obtained from $\underset{\sim}{v}$ via the solution of N uncoupled linear Poisson problems (one for each component of $\underset{\sim}{y}$); using (23) it can be shown that problem (25) is actually $\underline{\text{equivalent}}$ to the $\underline{\text{linear variational problem}}$

$\underline{\text{Find } \underset{\sim}{y} \in V_0 \text{ such that } \forall z \in V_0 \text{ we have}}$

$$\begin{cases} \alpha \displaystyle\int_\Omega \underset{\sim}{y}.\underset{\sim}{z} \ dx + \nu \int_\Omega \nabla \underset{\sim}{y}.\nabla \underset{\sim}{z} \ dx = \alpha \int_\Omega \underset{\sim}{v}.\underset{\sim}{z} \ dx + \nu \int_\Omega \nabla v.\nabla \underset{\sim}{z} \ dx + \\[2em] + \displaystyle\int_\Omega ((\underset{\sim}{v}.\nabla).\underset{\sim}{v}).\underset{\sim}{z} \ dx - \int_\Omega \underset{\sim}{f}.\underset{\sim}{z} \ dx, \end{cases} \tag{26}$$

$\underline{\text{which has a unique solution.}}$

Suppose now that $\underset{\sim}{v}$ is a solution of the nonlinear problem (18), (24); the corresponding $\underset{\sim}{y}$ (obtained through the solution of (25), (26)) is clearly $\underset{\sim}{y} = \underset{\sim}{0}$. From these observations it is quite natural to introduce the following $\underline{\text{(nonlinear) least square}}$ formulation of problem (18), (24):

$$\begin{cases} \underline{\text{Find } \underset{\sim}{u} \in V_g \text{ such that}} \\[2em] J(\underset{\sim}{u}) \le J(\underset{\sim}{v}) \ \forall \underset{\sim}{v} \in V_g, \end{cases} \tag{27}$$

where $J: \ (H^1(\Omega))^N \to \mathbb{R}$ is that function of $\underset{\sim}{v}$ defined by

$$J(\underset{\sim}{v}) = \frac{1}{2} \int_\Omega \{\alpha|\underset{\sim}{y}|^2 + \nu|\underset{\sim\sim}{\nabla y}|^2\} \, dx, \tag{28}$$

with $\underset{\sim}{y}$ defined from $\underset{\sim}{v}$ by solving the linear problem (25), (26).

We observe that if u is solution of (18), (24) it is also a solution of (27) such that $J(u) = 0$; conversely if $\underset{\sim}{u}$ is a solution of (27) such that $J(\underset{\sim}{u}) = 0$ it is also a solution of (18), (24).

4.3 Conjugate gradient solution of the least square problem (27)

4.3.1 Description of the algorithm

We use the Polak-Ribière version (see Polak, (1971)), of the conjugate gradient method to solve the minimization problem (27); we have then (with $J'(\underset{\sim}{v})$ the differential of J at $\underset{\sim}{v}$)

<u>Step O: Initialization</u>

$$\underset{\sim}{u}^O \in V_g, \ \underline{\text{given}}, \tag{29}$$

<u>we define then</u> $\underset{\sim}{g}^O, \ \underset{\sim}{w}^O \in V_O$ <u>by</u>

$$\begin{cases} \alpha \int_\Omega \underset{\sim}{g}^O.\underset{\sim}{z} \, dx + \nu \int_\Omega \underset{\sim\sim}{\nabla g}^O.\underset{\sim\sim}{\nabla z} \, dx = \ <J'(\underset{\sim}{u}^O),\underset{\sim}{z}> \quad \forall \underset{\sim}{z} \in V_O, \\[4mm] \underset{\sim}{g}^O \in V_O, \end{cases} \tag{30}$$

$$\underset{\sim}{w}^O = \underset{\sim}{g}^O, \tag{31}$$

<u>respectively</u>.

<u>Then for</u> $n \geq O$, <u>assuming that</u> $\underset{\sim}{u}^n, \ \underset{\sim}{g}^n, \ \underset{\sim}{w}^n$ <u>are known we obtain</u> $\underset{\sim}{u}^{n+1}, \ \underset{\sim}{g}^{n+1}, \ \underset{\sim}{u}^{n+1}$ <u>by</u>

Step 1: **Descent**

$$\begin{cases} \underline{\text{Find } \lambda^n \in \mathbb{R} \text{ such that}} \\[2ex] J(\underset{\sim}{u}^n - \lambda^n \underset{\sim}{w}^n) \leq J(\underset{\sim}{u}^n - \lambda \underset{\sim}{w}^n) \quad \forall \, \lambda \in \mathbb{R} \end{cases} \tag{32}$$

$$\underset{\sim}{u}^{n+1} = \underset{\sim}{u}^n - \lambda^n \underset{\sim}{w}^n . \tag{33}$$

Step 2: **Calculation of the new descent direction**

$$\begin{cases} \underline{\text{Find } \underset{\sim}{g}^{n+1} \in V_O \text{ such that}} \\[2ex] \alpha \int_\Omega \underset{\sim}{g}^{n+1} \cdot \underset{\sim}{z} \, dx + \nu \int_\Omega \nabla \underset{\sim}{g}^{n+1} \cdot \nabla \underset{\sim}{z} \, dx = \langle J'(\underset{\sim}{u}^{n+1}), \underset{\sim}{z} \rangle \quad \forall \underset{\sim}{z} \in V_O, \end{cases} \tag{34}$$

$$\gamma_n = \frac{\alpha \int_\Omega \underset{\sim}{g}^{n+1} \cdot (\underset{\sim}{g}^{n+1} - \underset{\sim}{g}^n) \, dx + \nu \int_\Omega \nabla \underset{\sim}{g}^{n+1} \cdot \nabla(\underset{\sim}{g}^{n+1} - \underset{\sim}{g}^n) \, dx}{\alpha \int_\Omega |\underset{\sim}{g}^n|^2 \, dx + \nu \int_\Omega |\nabla \underset{\sim}{g}^n|^2 \, dx}, \tag{35}$$

$$\underset{\sim}{w}^{n+1} = \underset{\sim}{g}^{n+1} + \gamma_n \underset{\sim}{w}^n , \tag{36}$$

$$n = n+1, \text{ \underline{go to} } (32).$$

As we shall see in Secs. 4.3.2, 4.3.3, applying algorithm (29) – (36) to solve the least square problem (27) requires the solution at each iteration of several Dirichlet problems associated to the elliptic operator $\alpha I - \nu \Delta$.

4.3.2. *Calculation of J'.*

A most important step, when making use of algorithm (29) – (36) to solve the least square problem (27), is the calculation of $\langle J'(\underset{\sim}{u}^{n+1}), \underset{\sim}{z} \rangle$ at each iteration; owing to the importance of this calculation we shall give it in detail.

Let $\underset{\sim}{v} \in V_g$ and let $\underset{\sim}{\delta v}$ be a __perturbation__ of $\underset{\sim}{v}$ such that $\underset{\sim}{\delta v} \in V_O$ (i.e. $\underset{\sim}{\delta v} = \underset{\sim}{O}$ on Γ); we have for the corresponding variation of $J(\underset{\sim}{v})$

$$\delta J(\underset{\sim}{v}) = \langle J'(\underset{\sim}{v}), \underset{\sim}{\delta v} \rangle . \tag{37}$$

320   GLOWINSKI et al.

Using (26), (28) we also have that

$$J(\underset{\sim}{v}) = \int_\Omega \{\alpha \underset{\sim}{y} . \underset{\sim}{\delta y} + \nu \nabla \underset{\sim}{y} . \nabla \underset{\sim}{\delta y}\} \ dx \tag{38}$$

where $\underset{\sim}{\delta y}$ is the solution of the linear problem

$$\begin{cases} \underset{\sim}{\delta y} \in V_O, \\[2ex] \alpha \displaystyle\int_\Omega \underset{\sim}{\delta y} . \underset{\sim}{z} \ dx + \nu \int_\Omega \nabla \underset{\sim}{\delta y} . \nabla \underset{\sim}{z} \ dx = \alpha \int_\Omega \underset{\sim}{\delta v} . \underset{\sim}{z} \ dx + \nu \int_\Omega \nabla \underset{\sim}{\delta v} . \nabla \underset{\sim}{z} \ dx + \\[2ex] + \displaystyle\int_\Omega ((\underset{\sim}{\delta v} . \nabla) \underset{\sim}{v}) . \underset{\sim}{z} \ dx + \int_\Omega ((\underset{\sim}{v} . \nabla) \underset{\sim}{\delta v}) . \underset{\sim}{z} \ dx \ \forall \underset{\sim}{z} \in V_O. \end{cases} \tag{39}$$

Taking $\underset{\sim}{z} = \underset{\sim}{y}$ in (39) we obtain from (37), (38) that

$$\begin{cases} <J'(\underset{\sim}{v}), \underset{\sim}{\delta v}> = \alpha \displaystyle\int_\Omega \underset{\sim}{y} . \underset{\sim}{\delta v} \ dx + \nu \int_\Omega \nabla \underset{\sim}{y} . \nabla \underset{\sim}{\delta v} \ dx + \int_\Omega ((\underset{\sim}{\delta v} . \nabla, \underset{\sim}{v}) . \underset{\sim}{y} \ dx + \\[2ex] \displaystyle\int_\Omega ((\underset{\sim}{v} . \nabla) . \underset{\sim}{\delta v}) . \underset{\sim}{y} \ dx. \end{cases}$$

Thus $J'(\underset{\sim}{v})$ can be identified with the linear functional from V_O to $\mathbb{R}$ defined by

$$\begin{cases} <J'(\underset{\sim}{v}), \ \underset{\sim}{z}> = \alpha \displaystyle\int_\Omega \underset{\sim}{y} . \underset{\sim}{z} \ dx + \nu \int_\Omega \nabla \underset{\sim}{y} . \nabla \underset{\sim}{z} \ dx + \int_\Omega \underset{\sim}{y} . (\underset{\sim}{z} . \nabla) \underset{\sim}{v} \ dx + \\[2ex] \displaystyle\int_\Omega \underset{\sim}{y} . (\underset{\sim}{v} . \nabla) \underset{\sim}{z} \ dx \ \forall \underset{\sim}{z} \in V_O; \end{cases} \tag{40}$$

<u>it has therefore a purely integral representation</u>, which is of major importance in view of <u>finite element</u> implementations of algorithm (29) - (36).

From the above results, to obtain $<J'(\underset{\sim}{u}^{n+1}), \underset{\sim}{z}>$ we proceed as follows:

(i)   We compute $\underset{\sim}{y}^{n+1}$ from $\underset{\sim}{u}^{n+1}$ through the solution of (25) with $\underset{\sim}{v} = \underset{\sim}{u}^{n+1}$, i.e. we solve the Dirichlet system

$$\begin{cases} \alpha \underset{\sim}{y}^{n+1} - \nu \Delta \underset{\sim}{y}^{n+1} = \alpha \underset{\sim}{u}^{n+1} - \nu \Delta \underset{\sim}{u}^{n+1} + (\underset{\sim}{u}^{n+1} \cdot \nabla) \underset{\sim}{u}^{n+1} - \underset{\sim}{f} \ \underline{\text{in}} \ \Omega, \\[2em] \underset{\sim}{y}^{n+1} = \underset{\sim}{0} \ \underline{\text{on}} \ \Gamma. \end{cases} \qquad (41)$$

(ii)   We finally obtain $<J'(\underset{\sim}{u}^{n+1}), \underset{\sim}{z}>$ by taking in (40)
$\underset{\sim}{v} = \underset{\sim}{u}^{n+1}$ and $\underset{\sim}{y} = \underset{\sim}{y}^{n+1}$.

4.3.3.   *Further comments on algorithm (29) - (36)*.

Each step of algorithm (29) - (36) requires the solution of
several Dirichlet systems for the operator $\alpha I - \nu \Delta$; more precisely
we have to solve the following such systems:

(i)   System (41) to obtain $\underset{\sim}{y}^{n+1}$ from $\underset{\sim}{u}^{n+1}$

(ii)   System (34) to obtain $\underset{\sim}{g}^{n+1}$ from $\underset{\sim}{u}^{n+1}$, $\underset{\sim}{y}^{n+1}$,

(iii) Two systems to obtain the coefficients of the quartic
       polynomial

$$\lambda \to J(\underset{\sim}{u}^n - \lambda \underset{\sim}{w}^n).$$

Thus we have to solve 4 Dirichlet systems for $\alpha I - \nu \Delta$ at each
iteration (or equivalently 4N scalar Dirichlet problems for
$\nu I - \nu \Delta$ at each iteration).

From the above observations it appears clearly that the
practical implementation of algorithm (29) - (36) will require
an efficient (direct or iterative) elliptic solver.

The solution of the one-dimensional problem (32) can be done
very efficiently since it is equivalent to finding the roots of
a single variable cubic polynomial whose coefficients are known.

As a last comment we would like to mention that algorithm
(29) - (36) (in fact its finite element variants) is quite
efficient; when used in combination with the alternating direction methods of Sec. 3 to solve the test problems of Sec. 7,
three iterations suffice to reduce the value of the cost function J by a factor of 10^4 to 10^6.

5. SOLUTION OF "QUASI" STOKES LINEAR SUBPROBLEMS

5.1 *Formulation. Synopsis*

At <u>each</u> full step of the alternating direction methods (8) – (10) and (11) – (14) we have to solve a <u>linear problem</u> of the following type

$$\begin{cases} \alpha \underset{\sim}{u} - \nu \Delta \underset{\sim}{u} + \nabla p = \underset{\sim}{f} \ \underline{in} \ \Omega \\ \nabla . \underset{\sim}{u} = 0 \ \underline{in} \ \Omega, \\ \underset{\sim}{u} = \underset{\sim}{g} \ \underline{on} \ \Gamma \ (\underline{with} \int_{\Gamma} \underset{\sim}{g}.\underset{\sim}{n} \ d\Gamma = 0), \end{cases} \tag{42}$$

where α and ν are two positive parameters and where f and g are two <u>given</u> functions defined on Ω and Γ, respectively.

We recall that if $\underset{\sim}{f}$ and $\underset{\sim}{g}$ are sufficiently smooth, then problem (42) has a <u>unique</u> solution in $V_g \times (L^2(\Omega)/IR)$ (with V_g still defined by (22); $p \in L^2(\Omega)/IR$ means that p is defined only to within an arbitrary constant).

We shall discuss in Secs. 5.1, 5.2 two iterative methods for solving (42), quite easy to implement using finite element methods (other methods are discussed in Glowinski (1982)).

5.2 *A first iterative method for solving (42)*

This method which is quite classical is defined as follows:

$$p^0 \in L^2(\Omega), \ \underline{given}, \tag{43}$$

<u>then for</u> $n \geq 0$, <u>define</u> $\underset{\sim}{u}^n$ <u>and</u> p^{n+1} <u>from</u> p^n <u>by</u>

$$\begin{cases} \alpha \underset{\sim}{u}^n - \nu \Delta \underset{\sim}{u}^n = \underset{\sim}{f} - \nabla p^n \ \underline{in} \ \Omega, \\ \underset{\sim}{u}^n = \underset{\sim}{g} \ \underline{on} \ \Gamma, \end{cases} \tag{44}$$

$$p^{n+1} = p^n - \rho \nabla . \underset{\sim}{u}^n \tag{45}$$

Concerning the convergence of algorithm (43) – (45) we have the following

Proposition 5.1: <u>Suppose that</u>

$$0 < \rho < 2\,\frac{\nu}{N}\,;$$

<u>we have then</u>

$$\lim_{n\to+\infty} \{\underset{\sim}{u}^n, p^n\} = \{\underset{\sim}{u}, p_0\} \ \underline{\text{strongly in}} \ (H^1(\Omega))^N \times L^2(\Omega), \quad (47)$$

<u>where</u> $\{\underset{\sim}{u}, p_0\}$ <u>is that solution of</u> (42) <u>such that</u>

$$\int_\Omega p_0 \, dx = \int_\Omega p^0 \, dx. \qquad (48)$$

<u>Proof:</u> We shall only prove the convergence of $\{\underset{\sim}{u}^n\}$; for a proof of the convergence of $\{p^n\}_{n\geq 0}$ (which is more complicated) see, e.g. Glowinski, (1982). Let $\{\underset{\sim}{u}, p\}$ be a solution of (42); we clearly have

$$p = p - \rho \nabla . \underset{\sim}{u} \ \forall \ \rho > 0. \qquad (49)$$

We define then $\underset{\sim}{\bar{u}}^n, \bar{p}^n$ by $\underset{\sim}{\bar{u}}^n = \underset{\sim}{u}^n - \underset{\sim}{u}$, $\bar{p}^n = p^n - p$, respectively; by subtraction between (44) and (42) (resp. (49) and (45)) we obtain

$$\begin{cases} \alpha \underset{\sim}{\bar{u}}^n - \nu \Delta \underset{\sim}{\bar{u}}^n = - \nabla \bar{p}^n \ \underline{\text{in}} \ \Omega, \\[2ex] \underset{\sim}{\bar{u}}^n = 0 \ \underline{\text{on}} \ \Gamma \end{cases} \qquad (50)$$

(resp.

$$\bar{p}^{n+1} = \bar{p}^n - \rho \nabla . \underset{\sim}{\bar{u}}^n). \qquad (51)$$

Using the notation $|q|_0 = ||q||_{L^2(\Omega)}$, we obtain from (51) that

$$|\bar{p}^n|_0^2 - |\bar{p}^{n+1}|_0^2 = 2\rho \int_\Omega \bar{p}^n \nabla . \underset{\sim}{\bar{u}}^n dx - \rho^2 |\nabla . \underset{\sim}{\bar{u}}^n|_0^2; \qquad (52)$$

on the other hand multiplying (50) by $\underset{\sim}{\bar{u}}^n$ and <u>integrating by parts</u> we obtain

$$\alpha \int_\Omega |\underset{\sim}{\bar{u}}^n|^2 \, dx + \nu \int_\Omega |\nabla \underset{\sim}{\bar{u}}^n|^2 \, dx = \int_\Omega \bar{p}^n \, \underset{\sim}{\nabla} . \underset{\sim}{\bar{u}}^{-n} \, dx. \tag{53}$$

Combining (52) and (53) we finally have

$$|\bar{p}^n|_0^2 - |\bar{p}^{n+1}|_0^2 = 2\rho \int_\Omega \{\alpha |\underset{\sim}{\bar{u}}^n|^2 + \nu |\nabla \underset{\sim}{\bar{u}}^n|^2\} dx - \rho^2 |\underset{\sim}{\nabla} . \underset{\sim}{u}^{-n}|_0^2 . \tag{54}$$

Let $\underset{\sim}{v} \in (H^1(\Omega))^N$; we have

$$\underset{\sim}{\nabla} . \underset{\sim}{v} = \sum_{i=1}^N \frac{\partial v_i}{\partial x_i} ,$$

which implies

$$|\underset{\sim}{\nabla} . \underset{\sim}{v}|^2 = \sum_{i=1}^N (\frac{\partial v_i}{\partial x_i})^2 + 2 \sum_{\substack{1 \le i,j \le N \\ i \ne j}} \frac{\partial v_i}{\partial x_i} \frac{\partial v_j}{\partial x_j} . \tag{55}$$

We also have

$$2 \frac{\partial v_i}{\partial x_i} \frac{\partial v_j}{\partial x_i} \le (\frac{\partial v_i}{\partial x_i})^2 + (\frac{\partial v_j}{\partial x_j})^2 . \tag{56}$$

Combining (55), (56) we clearly obtain

$$|\underset{\sim}{\nabla} . \underset{\sim}{v}|^2 \le N \sum_{i=1}^N (\frac{\partial v_i}{\partial x_i})^2 \le N |\underset{\sim}{\nabla} \underset{\sim}{v}|^2 \le \frac{N}{\nu} \{\alpha |\underset{\sim}{v}|^2 + \nu |\underset{\sim}{\nabla} \underset{\sim}{v}|^2\}. \tag{57}$$

It follows then from (54), (57) that

$$|\bar{p}^n|_0^2 - |\bar{p}^{n+1}|_0^2 \ge \rho(2 - \rho\frac{N}{\nu}) \int_\Omega \{\alpha |\underset{\sim}{\bar{u}}^n|^2 + \nu |\nabla \underset{\sim}{\bar{u}}^n|^2\} \, dx. \tag{58}$$

Suppose that (46) holds (i.e. $0 < \rho < 2\frac{\nu}{N}$); it follows then from (58) that the sequence $\{|\bar{p}^n|_0^2\}_{n \ge 0}$ is <u>decreasing</u>; since it is bounded from below by 0 it converges, implying

$$\lim_{n \to +\infty} \left(|\bar{p}^n|_O^2 - |\bar{p}^{n+1}|_O^2 \right) = 0. \tag{59}$$

Since (46) implies $\rho(2-\rho \frac{N}{\nu}) > 0$ we have from (58), (59) that

$$\lim_{n \to +\infty} \int_\Omega \{\alpha |\bar{\underline{u}}^n|^2 + \nu |\nabla\bar{\underline{u}}|^2\} \, dx = 0.$$

Since $\bar{\underline{u}}^n = \underline{u}^n - \underline{u}$, we have thus proved that

$$\lim_{n \to +\infty} \underline{u}^n = \underline{u} \ \underline{\text{strongly in}} \ (H^1(\Omega))^N.$$

Remark 5.1: It can be proved (see Glowinski (1982)) that $\{\{\underline{u}^n, p^n\}\}_{n \geq 0}$ converges to $\{\underline{u}, p_O\}$ __linearly__ (i.e. the sequences $\{||\underline{u}^n - \underline{u}||_{(H^1(\Omega))^N}\}_{n \geq 0}$ and $\{||p^n - p_O||_{L^2(\Omega)}\}_{n \geq 0}$ converge to zero as fast, at least, as a geometric sequence).

Remark 5.2: When using algorithm (43) - (45) to solve the "quasi" Stokes problem (42), we have to solve at each iteration N __uncoupled__ scalar Dirichlet problems for $\alpha I - \nu \Delta$, to obtain $\underline{u}^n$ from p^n. We see again (as in Sec. 4.3.3.) the importance to have efficient Dirichlet solvers for $\alpha I - \nu \Delta$.

Remark 5.3: Algorithm (43) - (45) is related to the so-called method of __artificial compressibility__ of Chorin-Yanenko; indeed we can view (45), (49) as obtained by a __time discretization__ process from the equation

$$\frac{\partial p}{\partial t} + \nabla \cdot \underline{u} = 0$$

(ρ being the size of the time discretization step).

Remark 5.4: In practice we should use instead of algorithm (43) - (45) a __conjugate gradient__ variant of it whose convergence is much faster and which is no more costly to implement (see Glowinski (1982) for the description of such conjugate gradient algorithm).

5.3 A second iterative method for solving (42)

This second method is in fact a generalization of algorithm (43)-(45), defined as follows (with r a <u>positive</u> parameter):

$$p^0 \in L^2(\Omega) \underline{\text{given}}, \tag{60}$$

<u>then for</u> $n \geq 0$ define $\underset{\sim}{u}^n$ <u>and</u> p^{n+1} <u>from</u> p^n <u>by</u>

$$\begin{cases} \alpha \underset{\sim}{u}^n - \nu \Delta \underset{\sim}{u}^n - r \nabla (\nabla . \underset{\sim}{u}^n) = \underset{\sim}{f} - \nabla p^n \underline{\text{in}} \ \Omega, \\ \underset{\sim}{u}^n = \underset{\sim}{g} \ \underline{\text{on}} \ \Gamma, \end{cases} \tag{61}$$

$$p^{n+1} = p^n - \rho \nabla . \underset{\sim}{u}^n. \tag{62}$$

Concerning the convergence of algorithm (60) - (62) we have the following

Proposition 5.2: <u>Suppose that</u>

$$0 < \rho < 2(r + \frac{\nu}{N}); \tag{63}$$

<u>then the convergence result</u> (47) <u>still holds for</u> $\{u^n, p^u\}$.

The proof of Proposition 5.2 is quite similar to that of 5.1; moreover the convergence of $\{\{\underset{\sim}{u}^n, p^n\}\}_{n \geq 0}$ is also <u>linear</u> (as shown in Glowinski (1982)).

Remark 5.5 (About the choice of ρ and r): In practice we should use $\rho = r$, since it can be proved that in that case the convergence ratio of algorithm (60) - (62) is $O(\frac{1}{r})$, for large values of r. In many applications, taking $r = 10^4 \nu$ we have a practical convergence of algorithm (60) - (62) in 3 to 4 iterations. There is however a practical upper bound for r; this follows from the fact that for too large values of r, problem (61) will be <u>ill-conditioned</u> and its practical solution sensitive to <u>round off</u> <u>errors</u>.

Remark 5.6: Problem (61) is more complicated to solve in practice than problem (44), since the components of $\underset{\sim}{u}^n$ are coupled by the linear term $\nabla(\nabla . \underset{\sim}{u}^n)$. Actually the partial

differential elliptic operator in the left hand side of (61) is
very close to the _linear elasticity operator_, and close variants
of it occur naturally in compressible and/or turbulent viscous
flow problems.

Remark 5.7: Other techniques for solving the "quasi" Stokes
problem (42) are discussed in Bristeau et al (1980a),
Bristeau et al (1980b) and Glowinski (1982).

6. FINITE ELEMENT APPROXIMATION OF THE TIME DEPENDENT NAVIER-
STOKES EQUATIONS

6.1 Generalities. Synopsis

We shall describe in this section a specific _finite element
approximation_ for the time dependent Navier-Stokes equations.
Actually this method which leads to _continuous approximations_
for both pressure and velocity is fairly simple and has been
known for years; it has been advocated for example by Taylor
et al (1973), among other people. Other finite element
approximations of the incompressible Navier-Stokes equations
can be found in Bristeau et al (1980a), Bristeau et al (1980b)
and Glowinski (1982) and also in Girault and Raviart (1979)
(see also the references therein).

6.2 Basic hypotheses. Fundamental discrete spaces

We suppose that Ω is a _bounded polygonal_ domain of $\mathbb{R}^2$. With
T_h a standard finite element triangulation of Ω, and h the
maximal length of the edges of the triangles of T_h, we introduce
the following discrete spaces (with P_k = space of the polynomials
in two variables of degree $\leq k$).

$$H_h^1 = \{q_h \,|\, q_h \in C^0(\bar{\Omega}) \times C^0(\bar{\Omega}),\ q_h|_T \in P_1\ \forall T \in T_h\}, \qquad (64)$$

$$V_h = \{\underset{\sim}{v}_h \,|\, \underset{\sim}{v}_h \in C^0(\bar{\Omega}) \times C^0(\bar{\Omega}),\ \underset{\sim}{v}_h|_T \in P_2 \times P_2\ \forall T \in T_h\}, \qquad (65)$$

$$V_{Oh} = V_O \cap V_h = \{\underset{\sim}{v}_h \,|\, \underset{\sim}{v}_h \in V_h,\ \underset{\sim}{v}_h = \underset{\sim}{0}\ \underline{\text{on}}\ \Gamma\}. \qquad (66)$$

A useful variant of V_h (and V_{Oh}) is obtained as follows

$$V_h = \{\underset{\sim}{v}_h \,|\, \underset{\sim}{v}_h \in C^0(\bar{\Omega}) \times C^0(\bar{\Omega}),\ \underset{\sim}{v}_h|_T \in P_1 \times P_1\ \forall T \in \tilde{T}_h\} \qquad (67)$$

where, in (67), $\tilde{T}_h$ is that triangulation of Ω obtained from T_h by joining the midpoints of the edges of $T \in T_h$ as indicated on Fig. 2.

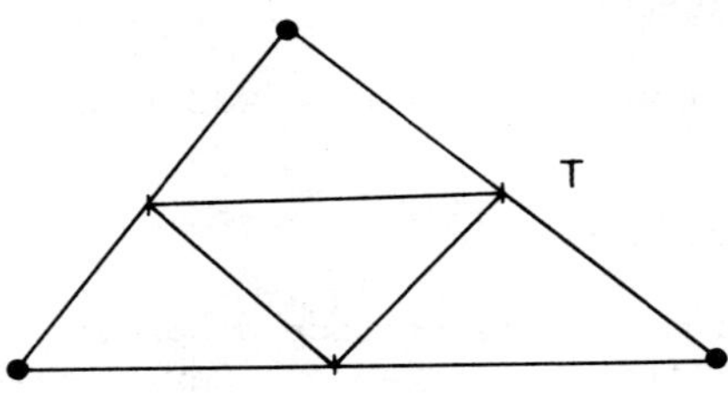

Figure 2

We have the same global number of unknowns if we use V_h defined by either (65) or (67); however the matrices encountered in the second case are more compact.

As usual the functions of H_h^1 will be defined from the values they take at the vertices of T_h; in the same fashion the functions of V_h will be defined by the values they take at the vertices of $\tilde{T}_h$ (resp. the vertices and the midpoints of T_h if V_h is defined by (67) (resp. (65)).

6.3. *Approximation of the boundary conditions*

Suppose that the boundary conditions are still defined by

$$\underset{\sim}{u} = \underset{\sim}{g} \ \underline{on} \ \Gamma, \ \underline{with} \ \int_\Gamma \underset{\sim}{g} \cdot \underset{\sim}{n} \ d\Gamma = 0; \tag{68}$$

for simplicity we suppose that g is $\underline{continuous}$ over Γ. We define now the space γV_h as

$$\gamma V_h = \{\underset{\sim}{\mu}_h | \underset{\sim}{\mu}_h = \underset{\sim}{v}_h |_\Gamma, \ \underset{\sim}{v}_h \in V_h\} \tag{69}$$

i.e. γV_h is the space of the traces on Γ of those functions $\underset{\sim}{v}_h$ belonging to V_h.

Actually if V_h is defined by (65) (resp. (67)), γV_h is also the space of those functions defined over Γ, taking their values in $\mathbb{R}^2$, continuous over Γ and piecewise quadratic (resp. linear) over the edges of T_h (resp. $\tilde{T}_h$) contained in Γ.

Our problem is to construct an approximation g_h of g such that

$$g_h \in \gamma V_h, \quad \int_\Gamma g_h \cdot n \, d\Gamma = 0. \tag{70}$$

If $\pi_h g$ is the unique element of γV_h obtained from the values taken by g at those nodes of T_h (or $\tilde{T}_h$) belonging to Γ, we usually have

$$\int_\Gamma \pi_h g \cdot n \, d\Gamma \neq 0.$$

To overcome the above difficulty we may proceed as follows:

(i)   We define an approximation n_h of n as the solution of the following <u>linear variational problem</u> in γV_h

$$\begin{cases} n_h \in \gamma V_h, \\ \int_\Gamma n_h \cdot \mu_h \, d\Gamma = \int_\Gamma n \cdot \mu_h \, d\Gamma \quad \forall \mu_h \in \gamma V_h; \end{cases} \tag{71}$$

problem (71) is in fact equivalent to a linear system whose matrix is sparse, symmetric, positive definite, and quite easy to compute.

(ii)   Define then g_h by

$$g_h = \pi_h g - \left[\frac{\int_\Gamma \pi_h g \cdot n \, d\Gamma}{\int_\Gamma n \cdot n_h \, d\Gamma} \right] n_h. \tag{72}$$

It is quite easy to check that (71), (72) imply (70).

6.4. *Space discretization of the time dependent Navier-Stokes equations.*

Using spaces H_h^1, V_h and V_{Oh} we approximate the time dependent Navier-Stokes equations as follows:

$$
\text{Find } \{u_h(t), p_h(t)\} \in V_h \times H_h^1 \quad \forall t \geq 0 \text{ such that}
$$

$$
\int_\Omega \frac{\partial u_h}{\partial t} \cdot v_h \, dx + \nu \int_\Omega \nabla u_h \cdot \nabla v_h \, dx + \int_\Omega (u_h \cdot \nabla) u_h \cdot v_h \, dx
$$

$$
+ \int_\Omega \nabla p_h \cdot v_h \, dx =
$$

$$
= \int_\Omega f_h \cdot v_h \, dx \quad \forall v_h \in V_{Oh} , \tag{73}
$$

$$
\int_\Omega \nabla \cdot u_h \, q_h \, dx = 0 \quad \forall q_h \in H_h^1 \tag{74}
$$

$$
u_h = g_h \quad \underline{\text{on }} \Gamma, \tag{75}
$$

$$
u_h(x,0) = u_{Oh}(x) \quad (\underline{\text{with }} u_{Oh} \in V_h); \tag{76}
$$

in (73)-(76), f_h and u_{Oh} are convenient approximations of f and u_O, respectively, and g_h has been defined in Sec. 6.3.

We have thus reduced the solution of the time dependent Navier-Stokes equations to that of a nonlinear system of algebraic and ordinary differential equations.

We observe that the incompressibility condition is approximately satisfied only. The time discretization of system (73) - (76) is discussed in the following Sec. 6.6.

6.5. *Time discretization of (73) - (76) by alternating direction methods.*

We consider now a fully discrete version of the scheme (8) - (10) discussed in Sec. 3.1; it is defined as follows (with Δt and θ as in Sec. 3.):

$$
u_h^O = u_{Oh}, \tag{77}
$$

<u>then for</u> $n \geq 0$, <u>compute</u> (<u>from</u> $\underset{\sim}{u}_h^n$) $\{\underset{\sim}{u}_h^{n+1/2}, p_h^{n+1/2}\} \in V_h \times H_h^1$, <u>and then</u> $\underset{\sim}{u}_h^{n+1} \in V_h$, <u>by solving</u>

$$
\begin{cases}
\displaystyle\int_\Omega \frac{\underset{\sim}{u}_h^{n+1/2} - \underset{\sim}{u}_h^n}{(\Delta t/2)} \cdot \underset{\sim}{v}_h \, dx + \theta\nu \int_\Omega \underset{\sim}{\nabla u}^{n+1/2} \cdot \underset{\sim}{\nabla v}_h \, dx + \int_\Omega \underset{\sim}{\nabla p}_h^{n+1/2} \cdot \underset{\sim}{v}_h \, dx = \\[4mm]
\displaystyle = \int_\Omega \underset{\sim}{f}_h^{n+1/2} v_h \, dx - (1-\theta)\,\nu \int_\Omega \underset{\sim}{\nabla u}_h^n \cdot \underset{\sim}{\nabla v}_h \, dx - \int_\Omega (\underset{\sim}{u}_h^n \cdot \nabla)\underset{\sim}{u}_h^n \cdot \underset{\sim}{v}_h \, dx \;\; \forall \underset{\sim}{v}_h \in V_{0h},
\end{cases}
$$
(78)

$$
\int_\Omega \underset{\sim}{\nabla} \cdot \underset{\sim}{u}_h^{n+1/2} \, q_h \, dx = 0 \quad \forall q_h \in H_h^1 ,
$$
(79)

$$
\underset{\sim}{u}_h^{n+1/2} \in V_h , \;\; p_h^{n+1/2} \in H_h^1 , \;\; \underset{\sim}{u}_h^{n+1/2} = \underset{\sim}{g}_h^{n+1/2} \;\; \underline{\text{on}} \; \Gamma,
$$
(80)

<u>and then</u>

$$
\begin{cases}
\displaystyle\int_\Omega \frac{\underset{\sim}{u}_h^{n+1} - \underset{\sim}{u}_h^{n+1/2}}{(\Delta t/2)} \cdot \underset{\sim}{v}_h \, dx + (1-\theta)\nu \int_\Omega \underset{\sim}{\nabla u}_h^{n+1} \cdot \underset{\sim}{\nabla v}_h \, dx + \\[4mm]
\displaystyle \hspace{3cm} \int_\Omega (\underset{\sim}{u}_h^{n+1} \cdot \nabla)\underset{\sim}{u}_h^{n+1} \cdot \underset{\sim}{v}_h \, dx = \\[4mm]
\displaystyle = \int_\Omega \underset{\sim}{f}_h^{n+1} \cdot \underset{\sim}{v}_h \, dx - \theta\nu \int_\Omega \nabla\underset{\sim}{u}_h^{n+1/2} \cdot \nabla v_h \, dx - \\[4mm]
\displaystyle \hspace{3cm} \int_\Omega \underset{\sim}{\nabla p}_h^{n+1/2} \cdot \underset{\sim}{v}_h \, dx \;\; \forall \underset{\sim}{v}_h \in V_{0h} ,
\end{cases}
$$
(81)

$$
\underset{\sim}{u}_h^{n+1} \in V_h, \;\; \underset{\sim}{u}_h^{n+1} = \underset{\sim}{g}_h^{n+1} \;\; \underline{\text{on}} \; \Gamma.
$$
(82)

Obtaining the fully discrete analogue of the scheme (11)-(14) described in Sec. 3.2 is left as an exercise to the reader.

*6.6. Some brief comments on the solution of the linear and non-
linear discrete subproblems*

The linear and nonlinear subproblems which have to be solved
at each full step of scheme (77)-(82) are the discrete analogues
(in space) of those continuous subproblems whose solution has
been discussed in Secs. 4 and 5; actually the methods described
in these sections apply with almost no modification to the solution of problems (78)-(80) and (81), (82). For this reason they
will not be discussed here (they are however discussed in detail
in Glowinski (1982).

7. NUMERICAL EXPERIMENTS

We illustrate the numerical techniques described in the previous sections by presenting the results of numerical experiments
where these techniques have been used to simulate several flows
modelled by the Navier-Stokes equations for incompressible
viscous fluids.

7.1 Flow in a channel with a step

The first numerical experiment that we have done concerns a
Navier-Stokes flow in a channel with a step, at Re = 191; the
characteristic length used to compute the Reynold's number is
the height of the step. Poiseuille velocity profiles have been
prescribed upstream and quite far downstream.

The alternating direction schemes of Sec. 6.5 have been used
to integrate the time dependent Navier-Stokes equations until a
steady state has been reached. The corresponding stream-lines
are shown on Fig. 3.

We clearly see on Figure 3 a thin separation layer starting
slightly below the upper corner of the step, and separating a
recirculation zone from a zone where the flow is quasi-potential.

The results obtained for this test are in very good agreement
with those previously obtained by the present authors, and also
with those obtained by other authors, using different methods
(see in particular Bristeau et al (1980b) and Hutton (1975)).

7.2 Flow around and inside a nozzle

This experiment concerns an unsteady flow around and inside
a nozzle at high incidence, at Re = 100 (the characteristic
length being the distance between the nozzle walls).

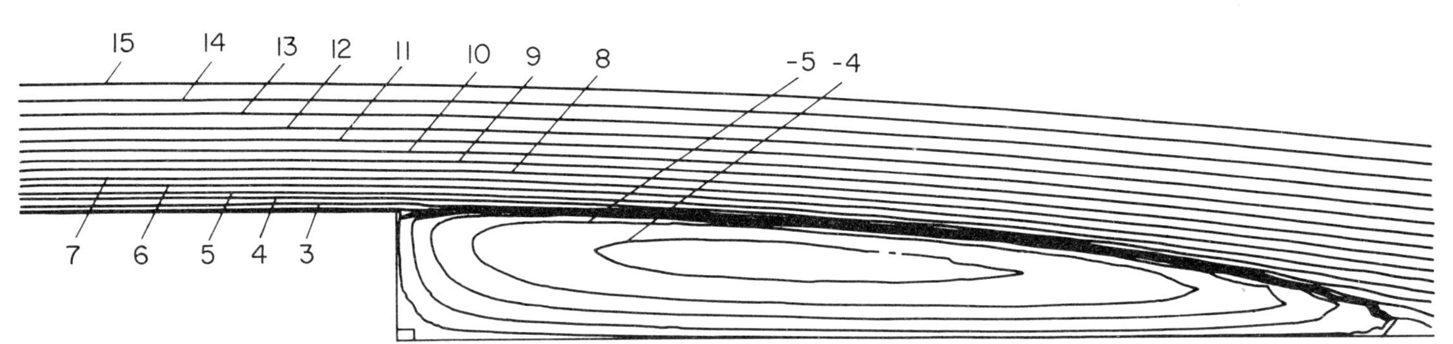

Figure 3 Stream lines for a flow in a channel with a step at Re = 191.

The streamlines shown are those for which the streamfunction assumes values $(n/15)^3$, for integers n between −5 and +15. The stepped (lower) boundary of the channel corresponds to n = 0.

The velocity distribution has been visualised on Fig. 4, showing clearly the creation and the motion of eddies inside and behind the nozzle.

8. CONCLUSION

The methods discussed in this paper combine a time discretization by alternating direction schemes and a space approximation by finite elements. Compared to the methods discussed in Bristeau *et al.* (1980) they produce more accurate results for less computational effort. The main cause of that improvement is the <u>decoupling</u> between nonlinearity and incompressibility obtained through the application of alternating direction schemes.

We are still looking at further improvements and we have the feeling that methods making use of characteristic methods have a good future in order to stimulate viscous flows governed by the incompressible or compressible Navier-Stokes equations.

9. REFERENCES

Adams R.A., (1975) "Sobolev spaces", Academic Press, New-York.

Benque J.P., Ibler B., Keramsi A., Labadie G., (1980), "A Finite element method for Navier-Stokes equations", Proceedings of Third International Conference on Finite Element in Flow Problems, Banff, Alberta, Canada, 10-13 June, pp. 110-120.

Bristeau M.O., Glowinski R., Periaux J., Perrier P., Pironneau O., Poirier G., (1980a) "Application of Optimal control and finite element methods to the calculation of transonic flows and incompressible viscous flows", in "Numerical Methods in Applied Fluid Dynamics", (Ed. B. Hunt), Academic Press, London, pp. 203-312.

Bristeau M.O., Glowinski R., Mantel B., Periaux J., Perrier P., Pironneau O., (1980b), "A finite element approximation of Navier-Stokes equations for incompressible viscous fluids, Iterative methods of solution", in Approximation Methods for Navier-Stokes problems (Ed. R. Rautmann), Lecture notes in Mathematics, **771**, Springer-Verlag, Berlin, pp. 78-128.

Girault V., Raviart P.A., (1979) "Finite Element Approximation of Navier-Stokes Equations", Lecture Notes in Mathematics, **749**, Springer-Verlag, Berlin.

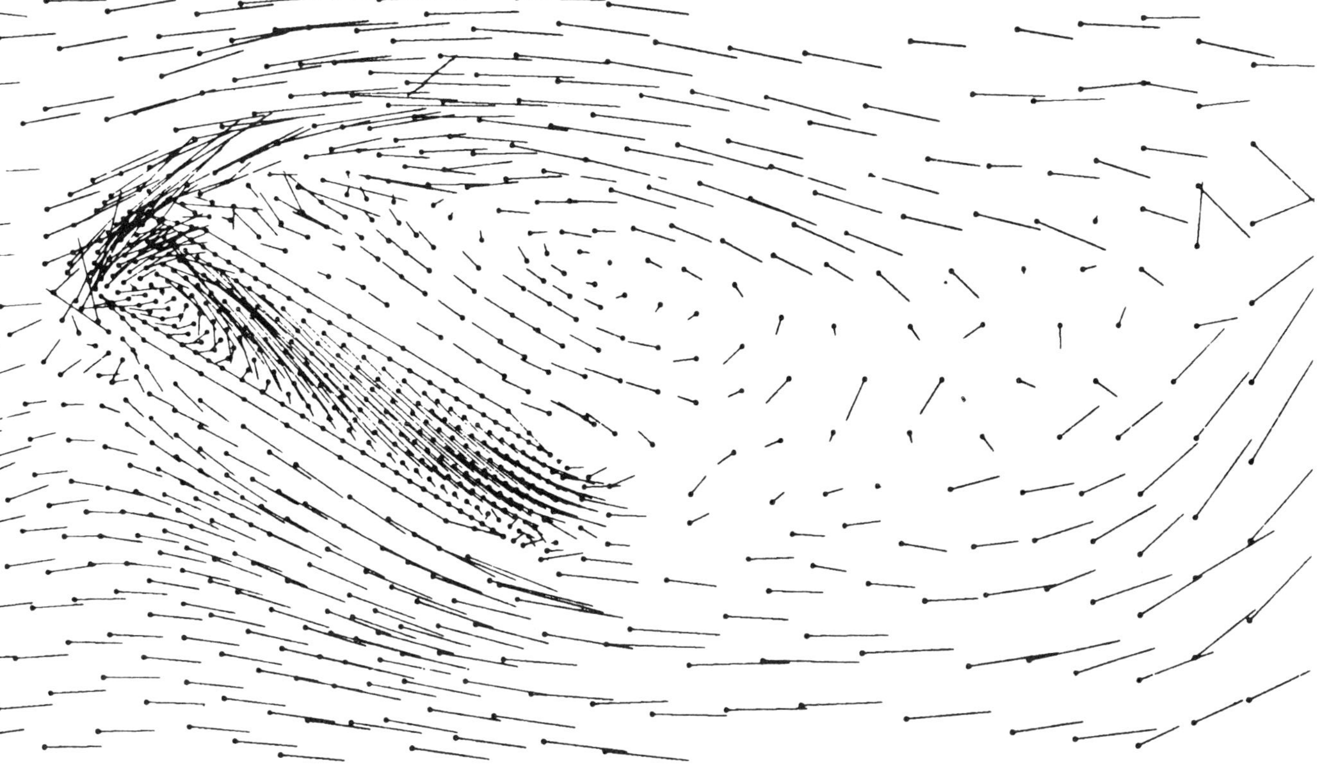

Figure 4 Velocity distribution for a flow around and inside a nozzle at Re = 100.

Glowinski R., (1982) "Numerical Methods for Nonlinear
Variational Problems," 2nd Edition, Springer-Verlag (in press).

Hutton A.G., (1975), "A general finite element method for
vorticity and stream function applied to a laminar separated
flow", Central Electricity Generating Board Report, Research
Dept. Berkeley Nuclear Laboratories.

Ibler B.,"Resolution des équations de Navier-Stokes par une
méthode d'éléments finis", (1981), Thèse de 3ème cycle,
Université Paris-Sud.

Necas J., (1967) "Les méthodes directes en théorie des équations
elliptiques",Masson, Paris.

Oden J.T., Reddy, J.N., (1976), "An introduction to the
mathematical theory of finite elements", J. Wiley and Sons, New-
York.

Polak E., (1971) "Computational Methods in Optimization",Academic
Press, New-York.

Taylor C., Hood P., (1973) "A Numerical solution of the Navier-
Stokes Equations using the Finite Element Technique",*Computers
and Fluids*, **1**, pp. 73-100.

Temam R., (1977) "Navier-Stokes equations" , North-Holland,
Amsterdam.

SURVEY OF TECHNIQUES FOR ESTIMATING VISCOUS EFFECTS IN EXTERNAL AERODYNAMICS

R.C. Lock and M.C.P. Firmin

(Royal Aircraft Establishment, Farnborough)

ABSTRACT

For a wide range of aerodynamic problems concerned with the practical design of aircraft wings, satisfactory results can be obtained by an iterative procedure which involves successive calculations of the external inviscid flow and of the turbulent viscous flow in the boundary layer and wake, linked together by a suitable mathematical model of the interaction between them. The paper reviews recent developments in the subject, in two main parts. In the first part an account is given of the basic theoretical foundations, with particular emphasis on important second-order effects which are normally omitted. A generalisation is described of the mathematical modelling of the displacement effect and of the momentum integral equations for the turbulent boundary layers, from which a simple correction is derived to allow for the effects of normal pressure gradients. There follows an explanation of the ways in which an iterative computational procedure for solving complete interactive problems can be built up, together with a linearised stability analysis of the convergence process. The second part of the paper describes some of the principal methods that are currently available for solving practical problems in viscous aerofoil and wing theory, and assesses their performance by comparison with experiment; examples include single and multiple aerofoils at high lift (low speeds), and aerofoils and wing-body combinations at high (transonic) speeds.

1. INTRODUCTION

Most of the other papers in this volume deal with recent work on the computation of inviscid flows, covering a wide range of problems and of techniques for solving them. But air is undoubtedly a viscous fluid, and it might therefore have been expected that research of this type would be of only limited value in the production of methods that will be of real

---

practical use to the aircraft designer. Fortunately, this is
not really the case, and it is the object of the present paper
to show how methods, of which an external inviscid flow calculation forms a major component part, can in fact be developed
which can be successfully applied to a wide variety of important
practical problems.

All real flows are indeed viscous, but the effects of viscosity can vary greatly in importance from case to case. At
one extreme we have for example the flow over a thin symmetrical
wing or slender body at a small angle of incidence, for which
one can start by calculating the inviscid pressure distribution,
use this to calculate the boundary layer and wake, and continue
if necessary with one further inviscid calculation taking into
account the displacement effect of the viscous layers together
with a final boundary-layer calculation. In such a situation -
sometimes known as a 'weak interaction' problem - this simple
procedure, in which separate inviscid and viscous calculations
are involved almost independently, would lead to estimates of
satisfactory accuracy for both the overall forces and the
detailed pressure distribution. At the other extreme, there are
many important problems, involving gross flow separations, in
which a classical 'attached flow' solution can be completely
misleading, although even in such cases progress can be made by
essentially inviscid methods if appropriate vortex sheets are
introduced (see for example the paper in this volume by J.H.B.
Smith).

In between these two extremes lies a wide range of situations,
of great practical importance, in which separate inviscid and
boundary-layer solutions are still relevant and can provide high
overall accuracy provided that they are combined with sufficient
attention to the mathematical modelling of the interaction
between them. It is essential to arrange the calculation in an
iterative manner, so that each successive inviscid solution
provides the pressure distribution for the next boundary-layer
solution, the displacement effect of which is then used to
modify the inner boundary conditions for a new inviscid calculation; this 'loop' must then be repeated, if necessary a large
number of times, until adequate convergence is obtained.
Methods for solving problems of this 'strong interaction' type
have been under development for over a decade and have now
reached a high level of sophistication in applications to
external aircraft aerodynamics. They are capable of dealing
satisfactorily with low speed flows about wings with high lift
devices ('multi-aerofoils') as well as with transonic flows in
which the interaction between a shock wave and a boundary layer
is likely to be an important phenomenon. In most examples in
which the wing is generating lift it has been found important
to take into account the interactive effect of the near wake,
since this can appreciably affect the pressure distribution

over the whole wing, not just near the trailing edge as might
perhaps have been expected.

This type of approach has had its most consistent successes
in situations where the boundary layers remain attached up to
the trailing edge of the wing, as would normally be the case for
a transport aircraft under cruising conditions. Very recently,
however, the possibility has emerged of extending the technique
to deal with cases where a significant amount of boundary-layer
separation is present, either near the trailing edge or at the
foot of a shock wave. In such cases it is necessary to carry
out the calculation of the boundary layer in an inverse manner,
in which the displacement thickness (or some analogous quantity)
is regarded as the input variable and the pressure distribution
is obtained as part of the output.

A comprehensive idea of the current situation in this topic
can be gained from the proceedings[1] of a recent AGARD Conference
on "Computation of viscous-inviscid interactions". In the
present paper we shall attempt to review briefly what we regard
as the most important features of the subject. After a short
section in which the physical nature of the flows of interest
is described, we give an account of recent developments in the
mathematical modelling of the interaction process and some
associated topics in turbulent boundary-layer theory. This is
followed by an explanation of the ways in which an iterative
computational procedure for solving complete interactive problems
can be built up, together with a simple linearised stability
analysis of the convergence process from which suitable relaxation factors can be derived. In the remainder of the paper we
describe some of the principal methods that are currently available for solving practical problems in viscous aerofoil and wing
theory and assess their performance in comparison with experiment; examples include multi-aerofoils at high lift (low speed),
and aerofoils and wing-body combinations at high (transonic)
speeds.

2. GENERAL FEATURES OF VISCOUS INTERACTIONS FOR WINGS

The general nature of the viscous flow over a lifting aerofoil or wing, at high Reynolds number (say 5 to 50 $\times$ 10^6) and
transonic speeds, is sketched in Fig. 1. The boundary layers
on the upper and lower surfaces will normally (except on a
highly swept wing) be laminar at the leading edge, but at full-
scale conditions transition to turbulent flow will take place
after a relatively short distance, and if this would not occur
naturally in a wind-tunnel experiment it is desirable to stimulate it artificially by means of roughness bands; thus we shall
be dealing primarily with turbulent boundary layers and wakes.
Because the mean adverse pressure gradient on the upper surface

will be higher than on the lower, the upper surface boundary
layer will be the thicker of the two at the trailing edge, and
the net displacement effect will therefore simulate a negative
camber over the rear of the aerofoil, which is in turn mainly
responsible for the overall reduction in circulation, and hence
in lift, caused by viscosity. A second significant result of
the displacement effect is the reduction in pressure at and
near the trailing edge from the stagnation value which it would
have in inviscid flow to the value, just above that of the free
stream, normally observed in experiment; this is caused by the
smoothing of the effective shape, produced by the boundary
layers and near wake.

When the flow - normally on the upper surface of a lifting
aerofoil - is super-critical, a shock wave of appreciable
strength may be present, and its interaction with the boundary
layer at its foot will clearly be of importance. Again, the
boundary layer has a smoothing effect which transforms the discrete pressure jump of the inviscid flow into a rapid, but continuous, pressure rise. As the shock increases in strength it
will eventually cause the boundary layer to separate, and the
consequent separation bubble will then spread towards the trailing edge. Even before this happens the shock wave will thicken
the boundary layer at its foot, an increment which will be
amplified by the additional adverse pressure gradient approaching the trailing edge, thus enhancing the effective negative
camber mentioned above and increasing the overall loss of lift;
so that viscous effects may be expected to be particularly
severe at transonic speeds. The presence of a shock wave (with
or without a local separation bubble at its foot) will exacer-
bate the tendency of the boundary layer to separate before it
reaches the trailing edge, a condition which heralds the onset
of buffeting and limits the usable lift on the wing. It must
be our eventual aim to predict all these phenomena with reasonable accuracy.

Near the trailing edge, on the upper side under high lift
conditions or on both sides if the trailing edge angle is
appreciable, the streamlines in the outer part of the shear
layers will become highly curved, and as a result the change of
pressure across the boundary layer, in a direction normal to the
surface of the wing or to the dividing streamline of the wake,
may no longer be neglected; thus the basic assumption of first
order boundary-layer theory will break down and higher order
effects must be taken into account, at least approximately.
Another region where a similar phenomenon occurs is the interaction of a shock wave with a boundary layer. For both these
regions the consequent variation of pressure is sketched in
Fig. 1. It will be noticed that the pressure at the outer edge
of the boundary layer (or wake) is less than that at the wall

(or dividing streamline) wherever the curvature of the displacement surface is positive (concave outwards), and greater when
the curvature is negative. This qualitative observation can be
put on a quantitative basis, as shown in section 3 below.

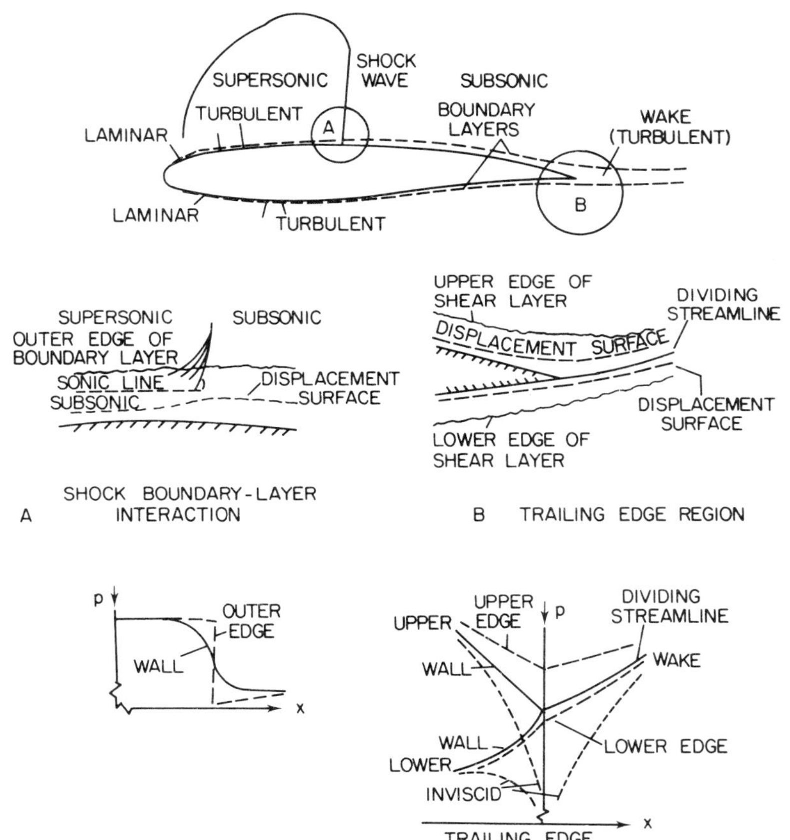

Fig. 1 Viscous flow over an aerofoil at transonic speeds

The cumulative effect of these various types of viscous
interaction on the pressure distribution, for a typical modern
aerofoil close to its buffet boundary, is indicated in Fig. 2.
Compared with the values for inviscid flow, significant changes
of pressure are observed over nearly the whole of both surfaces;
they are particularly marked on the upper surface where there is
a dramatic effect on both the position and strength of the shock
wave. The net reduction in lift can easily exceed 25%, even at
full scale Reynolds numbers.

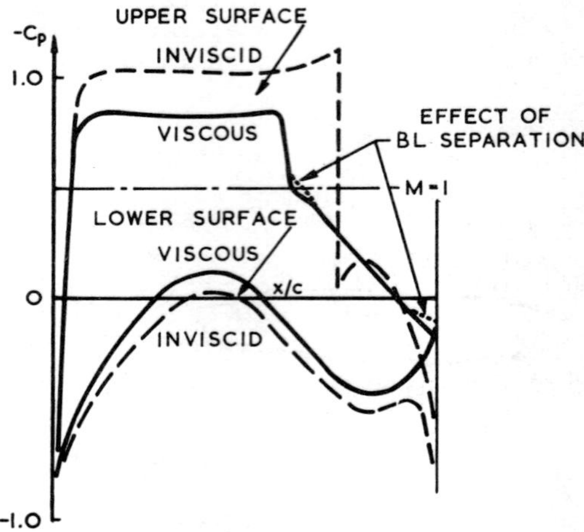

Fig. 2   Viscous effects on pressure

In the case of a wing under high lift conditions at low
speeds, with leading edge slat and trailing edge flaps deployed,
the problems of transonic flow are largely absent*, but additional viscous effects may occur which further complicate the
situation. The chief of these is the interaction of a wake and
a boundary layer, as can be seen from Fig. 3, in which the
general nature of the flow over a three-element aerofoil is
sketched. The wake from the trailing edge of the slat is initially separate from the boundary layer on the main aerofoil
growing beneath it, but the presence of each will affect the
development of the other until eventually they merge to form a
single boundary layer. Usually this will happen before the
trailing edge of the main aerofoil is reached, when the same
process starts again as the flow develops over the flap, while
in the case of a double slotted flap (not shown here) the two
individual wakes, together with the boundary layer on the final
element, may be present as separate entities. Practical design
requirements usually lead to shapes for the under surfaces of
both slat and shroud which give rise to local separation bubbles
in the cavities beneath them, and these must be included in the
mathematical model used to represent the flow. In general, by
the time the flap is reached the viscous layers will be relatively thick (the dimensions shown in Fig. 3 are not exaggerated)
and normal pressure gradients are likely to be appreciable,
while the flow will frequently separate from the upper surface
of the rear flap well before the final stall takes place.

---

*Except on the upper surface of the slat, where a small local
supersonic region is sometimes found.

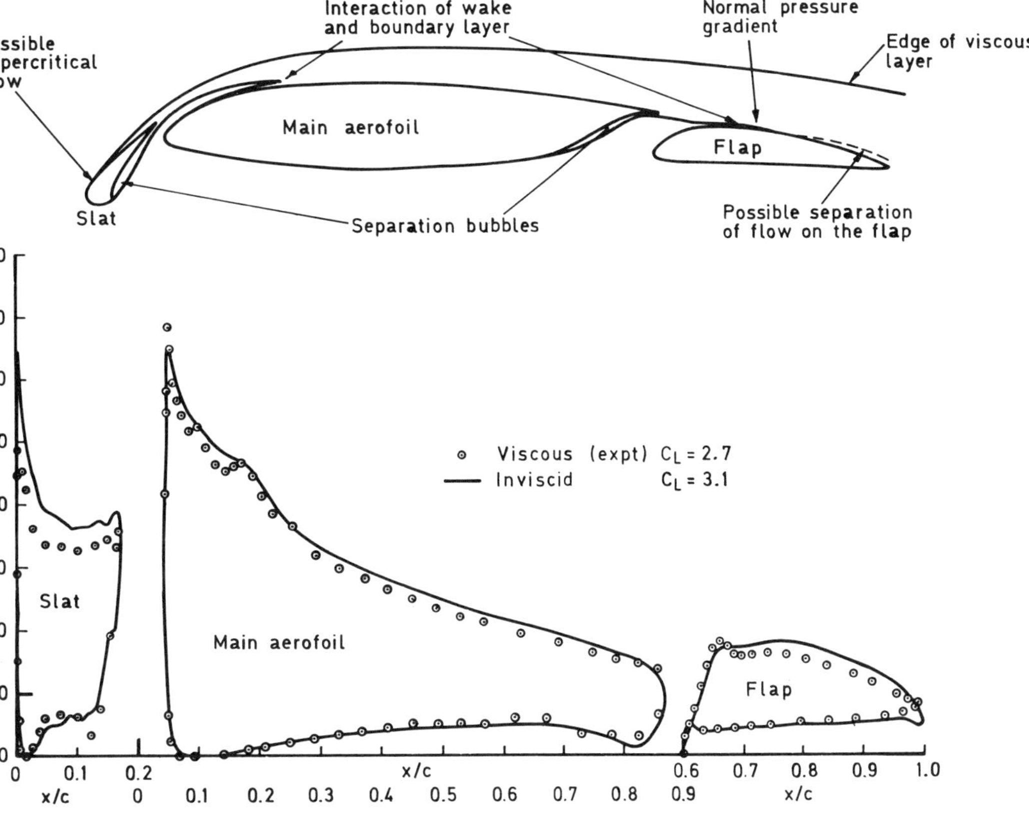

Fig. 3 Viscous flow over a multi-element aerofoil at low speed

Under circumstances of such complexity it might well be
expected that nothing short of a solution of the full Navier-
Stokes equations, at least for the immediate neighbourhood of
the aerofoil system, would yield results of adequate accuracy.
But while this would certainly be desirable, reasonably satisfactory predictions have in fact already been achieved by matching separate inviscid and viscous calculations, together with a
relatively crude modelling of their interaction, and the incorporation of the higher order treatment described in section 3
below should lead to further increases in accuracy.

3. BASIC THEORETICAL PRINCIPLES

The basic mathematical model of the interaction between a
boundary layer and the external inviscid flow, in which the
effect of the former on the latter is simulated by means of an
inviscid calculation of the flow over a 'solid' displacement
surface, is of course of classical familiarity. Only a little
less familiar is the alternative 'transpiration' model, proposed
originally by Lighthill[2], in which the displacement effect
is introduced by means of a distribution of sources on the true
surface of the wing, leading to a non-zero component of normal
velocity (v_{iw}) there, given by

$$v_{iw} = \frac{1}{\rho_e} \frac{d}{ds} (\rho_e u_e \delta*)$$ (1)

(in two dimensions).

Both of these concepts were however originally derived by an
analysis in which the standard approximations of first-order
boundary-layer theory, in particular that the transverse variation in pressure across the boundary layer is negligible, were
assumed - either explicitly or implicitly - to hold. It should
be clear from the physical description given above (section 2)
that these assumptions are likely to become inadequate in some
of the situations which confront us in practice. An important
advance during the past few years has been the development of a
much less restrictive theoretical background to the subject, as
regards both the matching between the inviscid and viscous
elements of the problem and the calculation of the latter:
noteworthy contributions have been made by Myring[3],
Le Balleur[4] and East[5].   The account that follows is based
mainly on the work of East.

It is first necessary to make the fundamental assumption that
to any order of accuracy that could reasonably be required in
practice, the overall flow field can be divided into regions of
two kinds: inner regions (shear layers) in which the Navier-

Stokes equations (or some suitable approximation to them) must
be solved, and outer regions where the inviscid (Euler) equations are sufficiently accurate. Such an assumption is certainly
valid for the flows at high Reynolds number, in which the shear
layers are largely turbulent, that are of greatest interest in
applications to external aerodynamics. Then the essential
feature of the matching between such a pair of inviscid and
viscous calculations is simply that the composite flow field
shall be continuous, as regards both the magnitude and direction
of the velocity, at the interface between the two regions.

Perhaps the most obvious way of performing the matching procedure is to do it at (or near) the outer edge of the shear
layers, and indeed if a 'differential' (rather than an
'integral') method* is being used to calculate the viscous part
of the flow this seems also the most satisfactory procedure.
But in this case it is clearly essential that something better
than the first-order boundary-layer equations should be solved,
in order to take some account of the normal variations in
pressure and other higher order terms that are usually neglected.
Just such a procedure is currently being investigated by Bradshaw and his colleagues at Imperial College. Although the
equations that they use involve few approximations on the full
(time-averaged) Navier Stokes equations, the computational
labour of solving the latter is largely avoided by carrying out
the basic iterations in two steps. In the first it is assumed
that the pressure variation (in both streamwise and normal
directions) is known from a previous iteration, so that equations
of boundary-layer type can be solved in the usual way by marching in the streamwise direction; in the second, the momentum
equation normal to the surface is solved to give a better
approximation to the pressure variation in that direction. The
method has already been successfully applied to problems of flow
in highly curved ducts (Bradshaw and Mahgoub[6]) and an
analogous treatment of the two-dimensional aerofoil problem
(initially at low speeds) is approaching fruition. Any subsequent extension to transonic flow will require (at least in
principle) the ability to deal with cases in which shock waves
may penetrate the viscous layer.

In the majority of current work, however, particularly when
an 'integral' method is being used (for reasons of economy, or
otherwise), it is more convenient to perform the matching by
means of one or other of the two alternative models of the dis-

---

*A 'differential' method is one in which the full governing
partial differential equations are solved in detail; an 'integral' method is one in which only integrated quantities (e.g.
displacement and momentum thicknesses) are calculated.

placement effect mentioned above. This leads either to the
simplest possible (zero normal velocity) inner boundary condition for the inviscid flow calculation (but on a surface whose
position will vary during the course of an iterative process),
or to a relatively simple transpiration condition on the fixed
surface of the body itself. The latter condition has strong
practical advantages; for example if a panel method is being
used, the matrix of influence coefficients can be set up and
inverted once and for all, so that subsequent inviscid calculations become trivial, while with a field calculation method
(finite difference or finite volume) a fixed computational grid
can be used throughout. So this is the approach which will be
followed through most of the present paper.

The need to derive boundary conditions for the inviscid flow
at positions inside the shear layers that are present in the
real viscous flow* leads naturally to the concept of an equivalent inviscid flow*, which can be defined as an analytic continuation, or smooth extrapolation, of the inviscid flow outside
the shear layers in the direction normal to the interface
between them. Such a concept is entirely satisfactory (apart
from possible doubts as regards the mathematical existence or
uniqueness of this continuation) in the framework of an iterative numerical scheme, in which the inviscid flow equations are
being solved over the full region with boundary conditions
appropriate to the current iteration; it is more difficult to
apply to experimental measurements of a real flow, where the
extrapolation required to define the EIF would introduce
uncertainties in the definition of integral quantities (though
these uncertainties would still be less than if first-order
definitions were used).

3.1 *Mathematical models of the displacement effect*

The derivation that follows is restricted to two dimensions,
though an extension to three dimensions is no doubt feasible.
Since we are concerned with the possibility of appreciable
normal pressure variations, and since a likely source of such
variations is surface curvature, it is important to work
throughout in curvilinear coordinates; the simple system that
we use, in which s and n are distances parallel to and normal
to the wall respectively, is sketched in Fig. 4a. Values of the
equivalent inviscid flow are denoted by the suffix i, and values
at the wall by the suffix w. The curvature of the wall is κ_w,
taken to be positive if the wall is concave outwards. The variation across the boundary layer of the streamwise mass flow
rate is shown in Fig. 4b, both for the real viscous flow (ρu)
and for the equivalent inviscid flow ($\rho_i u_i$). Note that we do

---

* We shall sometimes abbreviate these to RVF, EIF respectively.

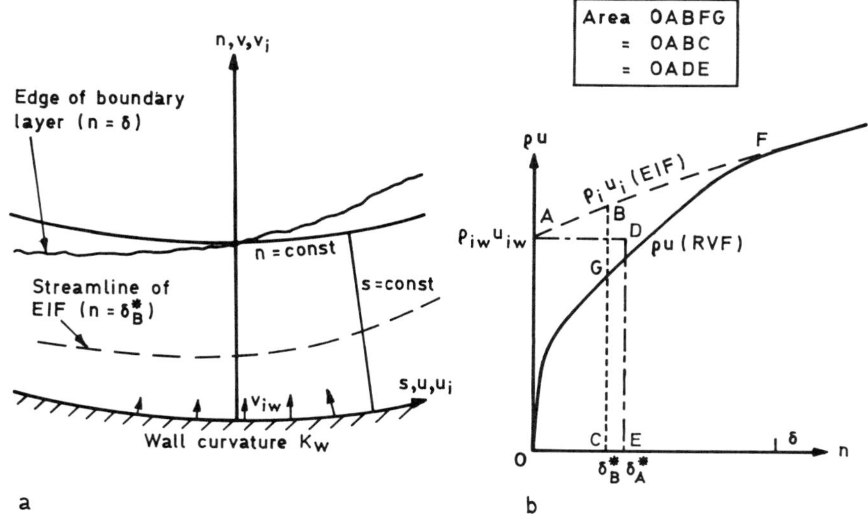

Fig. 4 a and b

<u>not</u> assume that $\rho_i u_i$ is constant.

It is convenient, following Le Balleur[7], to work in terms of the difference between the two flows ('équations déficitaires'). Thus the equation of continuity is written

$$\frac{\partial}{\partial s}\,(\rho_i u_i - \rho u) + \frac{\partial}{\partial n}\,[(1 - \kappa_w n)(\rho_i v_i - \rho v)] = 0. \qquad (2)$$

Integrating across the boundary layer, from n = 0 to δ, we obtain at once

$$\rho_{iw} v_{iw} = \frac{d}{ds}\,\int_0^\delta (\rho_i u_i - \rho u)\,dn, \qquad (3)$$

since both $(\rho_i v_i - \rho v)$ and $(\rho_i u_i - \rho u)$ vanish at n = δ because of the matching condition, while v = 0 at the wall (n = 0). If then we define a displacement thickness, δ_A^* say, by

$$\delta_A^* = \frac{1}{\rho_{iw} u_{iw}}\,\int_0^\delta (\rho_i u_i - \rho u)\,dn, \qquad (4)$$

then

$$v_{iw} = \frac{1}{\rho_{iw}}\,\frac{d}{ds}\,(\rho_{iw} u_{iw} \delta_A^*), \qquad (5)$$

precisely analogous to the first order definition of Lighthill[2] (equation (1) above).

The 'classical' displacement thickness, δ_B^* say, is defined by the condition that the total mass flow in the equivalent inviscid flow shall equal that in the real viscous flow, so that

$$\int_{\delta_B^*}^{\delta} \rho_i u_i dn = \int_O^{\delta} \rho u dn \qquad (6)$$

It is shown by Lock[8] that $n = \delta_B^*$ is indeed a streamline of the equivalent inviscid flow. Unlike equation (4), this is only an implicit definition of δ^*.

The relation between these two alternative definitions of displacement thickness can be obtained by adding

$$\int_O^{\delta_B^*} \rho_i u_i dn$$

to both sides of equation (6). We obtain

$$\int_O^{\delta_B^*} \rho_i u_i dn = \rho_{iw} u_{iw} \delta_A^*. \qquad (7)$$

Thus the transpiration condition (5) is equivalent to

$$\rho_{iw} v_{iw} = \frac{d}{ds} \int_O^{\delta_B^*} \rho_i u_i dn,$$

which provides a physical explanation of it: that the mass flow injected normal to the wall in the equivalent inviscid flow, in order to displace the dividing streamline outwards the required distance δ_B^* from the surface, must equal the streamwise rate of change of the mass flow deficit in the boundary layer.

To investigate the difference between δ_A^* and δ_B^* we can expand $\rho_i u_i$ as a Taylor series from $n = O$. Now

$$\left[\frac{\partial}{\partial n} (\rho_i u_i)\right]_{n=O} \simeq \rho_{iw} (1 - M_{iw}^2) \frac{\partial u_i}{\partial n}\bigg|_{n=O},$$

where M_{iw} is the Mach number at the surface in the equivalent inviscid flow; and since this flow is irrotational,

$$\left.\frac{\partial \mu_i}{\partial n}\right|_{n=0} = \frac{dv_{iw}}{ds} + \kappa_w u_{iw}.$$

Hence the first order expansion of $\rho_i u_i$ is

$$\rho_i u_i = \rho_{iw} u_{iw}\{1 + k(1 - M_{iw}^2)n\} \tag{8}$$

where

$$k = \kappa_w + \frac{1}{u_{iw}} \frac{dv_{iw}}{ds} \tag{8a}$$

and substituting this in equation (7) we find that

$$\delta_A^* \simeq \delta_B^*\{1 + \tfrac{1}{2}k(1 - M_{iw}^2)\delta_B^*\} + O(\delta_B^{*3}). \tag{9}$$

Combining equations (5) and (8a), we have

$$k = \kappa_w + \frac{1}{u_{iw}} \frac{d}{ds} \{\frac{1}{\rho_{iw}} \frac{d}{ds} (\rho_{iw} u_{iw} \delta_A^*)\} = \kappa_w + \frac{d^2 \delta_A^*}{ds^2} + \text{smaller terms}.$$

Thus k, the parameter which determines the difference between the alternative definitions of displacement thickness, is approximately equal to

$$\kappa^*(=k_w + \frac{d\delta_A^*}{ds^2}),$$

the curvature of the displacement surface. This important quantity will be shown later (section 3.3) to govern also the magnitude of the pressure change across the boundary layer.

The relative difference between δ_A^* and δ_B^* is thus

$$\frac{\delta_A^* - \delta_B^*}{\delta_B^*} \simeq \tfrac{1}{2}(1 - M_{iw}^2)\kappa^*\delta_B^*. \tag{9a}$$

To get an idea of the magnitude involved, we note that near the trailing edge of a typical lifting aerofoil (at $Re = 10^7$, say) one might have

$$\delta_B^* \sim 0.02, \quad \kappa^* \sim 1 \text{ (in terms of the aerofoil chord)},$$

so that the two definitions should normally differ by less than
1%.

If however we consider the standard first-order definition
of displacement thickness,

$$\delta^*_{(1)} = \int_O^\delta (1 - \frac{\rho u}{\rho_e u_e}) \, dn \tag{10}$$

(where suffix e denotes values at the edge of the boundary
layer), then we find (Lock 1981) that

$$\delta^*_{(1)} - \delta^*_A \simeq \tfrac{1}{2}(1 - M_{iw}^2)\kappa^*\delta(\delta - 2\delta^*_{(1)}) + O(\delta^3), \tag{11}$$

implying a larger (4-5% say) discrepancy, and one which is
clearly dependent on the precise choice of δ, a somewhat arbitrary quantity under the circumstances envisaged in Fig. 4b.

The arguments given above have provided two alternative
inner boundary conditions which can be used in the calculation
of the equivalent inviscid flow: (A) a transpiration condition of non-zero normal velocity at the surface (equation (5)),
or (B) the condition that the displacement surface is a streamline (equation (6)). It is important to appreciate that either
of these mathematical models is an equally valid, and quite
general, representation of the displacement effect of the boundary layer on the external flow. Normally, the transpiration
model (A) is the more convenient to use, and this is the one
which is chosen for the remainder of the paper, in which the
suffix 'A' will therefore be omitted.

3.2   *The momentum integral equations*

The argument given above has shown how the concept of displacement effect can be generalised without difficulty in situations where the assumptions of first order boundary layer
theory are no longer tenable. While it is not the objective of
the present paper to discuss in detail the corresponding
developments in boundary layer theory itself, it is important
to see how the fundamental equation of all 'integral' methods -
the so-called 'momentum integral' equation - can be generalised
in a similar way; and at the same time we need to investigate
the variation of pressure normal to the wall, which can be done
by considering the equation for the component of momentum in
that direction. The treatment follows the lines of a recent
paper by East[5].

3.2.1 The streamwise momentum equation

As in the discussion of the displacement effect given in section 3.1, we start from the __difference__ between the streamwise momentum equations for the two flows (RVF and EIF), in the coordinate system of Fig. 4a:

$$\frac{\partial}{\partial s}(\rho_i u_i^2 - \rho u^2) + \frac{\partial}{\partial n}[(1-\kappa_w n)(\rho_i u_i v_i - \rho uv)] - \kappa_w(\rho_i u_i v_i - \rho uv)$$

$$= -\frac{\partial}{\partial s}(p_i - p) - \frac{\partial}{\partial n}[(1 - \kappa_w n)\tau] + \frac{\partial}{\partial s}(\overline{\rho u'^2}) \qquad (12)$$

$$- \kappa_w \overline{\rho u' v'},$$

where u' and v' are the fluctuating components of velocity in the turbulent shear flow, and $\tau = \mu \frac{\partial u}{\partial n} - \overline{\rho u' v'}$ is the Reynolds shear stress. For turbulent flows at moderate Mach number this equation can be regarded as exact.

Integrating with respect to n, from O to δ, and applying the matching condition at $n = \delta$ and the inner boundary condition $v = O$ (but $v_i \neq O$) on $n = O$, we obtain

$$\frac{d}{ds}[\rho_{iw} u_{iw}^2(\theta + \delta^*)] - \rho_{iw} u_{iw} v_{iw} - \tau_w = \kappa_w \int_O^\delta (\rho_i u_i v_i - \rho uv)dn -$$

$$- \kappa_w \int_O^\delta \overline{\rho u' v'} dn - \frac{d}{ds}\int_O^\delta (p_i - p)dn + \frac{d}{ds}\int_O^\delta \overline{\rho u'^2}dn,$$

$$\qquad (13)$$

where the momentum thickness θ is defined by

$$\rho_{iw} u_{iw}^2(\theta + \delta^*) = \int_O^\delta (\rho_i u_i^2 - \rho u^2)dn,$$

so that

$$\theta = \frac{1}{\rho_{iw} u_{iw}^2} \int_O^\delta [(\rho_i u_i^2 - \rho u^2) - u_{iw}(\rho_i u_i - \rho u)]dn$$

$$\qquad (14)$$

$$= \frac{1}{\rho_{iw} u_{iw}^2} \int_O^\delta [\rho u(u_{iw} - u) + \rho_i u_i(u_i - u_{iw})]dn.$$

Using equation (5) for v_{iw}, substituting in equation (13) and dividing by $\rho_{iw} u_{iw}^2$, we obtain (after some further reduction) the non-dimensional equation

$$\frac{1}{\rho_{iw}u^2_{iw}} (\rho_{iw}u^2_{iw}\theta) + \frac{\delta^*}{u_{iw}}\frac{du_{iw}}{ds} \overset{(1)}{-} \tfrac{1}{2}C_f \overset{(2)}{=} \kappa_w(\Theta_n + \delta\bar{C}_\tau)$$

$$+ \frac{1}{\rho_{iw}u^2_{iw}}\frac{d}{ds}[- \int_O^\delta \overset{(3)}{(P_i} - p)\,dn + \int_O^\delta \overset{(4)}{\rho u'^2}dn\,], \qquad (15)$$

where $C_f = \dfrac{\tau_w}{\tfrac{1}{2}\rho_{iw}u^2_{iw}}$ is the skin-friction coefficient,

$$\Theta_n = \frac{1}{\rho_{iw}u^2_{iw}} \int_O^\delta (\rho_i u_i v_i - \rho uv)\,dn \text{ may be called the 'normal}$$

momentum thickness' and $\bar{C}_\tau = \dfrac{1}{\delta}\displaystyle\int_O^\delta \dfrac{(\overline{-\rho u'v'})}{\rho_{iw}u^2_{iw}}\,dn$ is the mean Reynolds

shear stress coefficient.

In first-order boundary layer theory the right-hand side of
equation (15) is neglected, and the familiar 'momentum integral'
equation is obtained, usually written

$$\frac{d\theta}{ds} + (H + 2 - M^2)\frac{\theta}{u_e}\frac{du_e}{ds} = \tfrac{1}{2}C_f \qquad (16)$$

where $H = \dfrac{\delta^*}{\theta}$;

while equation (14) for θ then reduces to the standard definition

$$\theta = \int_O^\delta \frac{\rho u}{\rho_e u_e}(1 - \frac{u}{u_e})\,dn \quad . \qquad (17)$$

We must now examine the order of magnitude of the higher order
terms, labelled (1) to (4), in equation (15). East[5] has shown
that adequate numerical approximations to three of these terms
are given by the semi-empirical relations

$$(1) \qquad \theta_n = (\theta + \delta\ast) \frac{d\delta\ast}{ds} \quad \text{(see Appendix A)} \qquad (18)$$

$$(3) \quad \int_0^\delta (p_i - p)\, dn = \tfrac{1}{2}\rho_{iw} u_{iw}^2 \kappa\ast (\theta + \delta\ast)^2 + \int_0^\delta \overline{\rho v'^2}\, dn \quad \text{(see section}$$
$$3.2.2)$$

with

$$\int_0^\delta \overline{\rho v'^2}\, dn \simeq 0.05\, \rho_{iw} u_{iw}^2 (\frac{H-1}{H})\theta \qquad (19)\ast$$

and

$$(4) \qquad \int_0^\delta \overline{\rho u'^2}\, dn = 0.12 \rho_{iw} u_{iw}^2 (\frac{H-1}{H})\theta\ . \qquad (20)\ast$$

Hence the complete momentum integral equation can be approximated by

$$\frac{1}{\rho_{iw} u_{iw}^2} \frac{d}{ds}(\rho_{iw} u_{iw}^2\, \theta) + \frac{\delta\ast}{u_{iw}} \frac{du_{iw}}{ds} - \tfrac{1}{2}\, C_f$$

$$= \kappa_w \underset{(1)}{\Big\{} (\theta+\delta\ast) \frac{d\delta\ast}{ds} + \underset{(2)}{\delta\bar{C}_\tau} \Big\} + \frac{1}{\rho_{iw} u_{iw}^2} \frac{d}{ds} \Big|\rho_{iw} u_{iw}^2 \Big\{ \underset{(3)}{-\tfrac{1}{2}\kappa\ast (\theta+\delta\ast)^2} +$$

$$+ \underset{(4)\ast}{0.07 (\frac{H-1}{H})\theta} \Big\} \Big| . \qquad (21)$$

In this equation, the terms labelled (1) and (3) represent corrections due to the effects of curvature of the mean flow, while (2) and (4) are second-order turbulence terms. For most aerofoils, the surface curvature κ_w will be large (relative to the reciprocal of the chord) only near the leading edge, but here the boundary layer is so thin that all the higher order terms are negligible. Further back along the chord $\kappa_w c$ is generally of order $(\frac{t}{c})$, and then the terms (1) and (2) on the right-handside are of third order (respectively $O(\frac{t}{c} \frac{\delta\ast}{c} \frac{d\delta\ast}{dx})$ and $O(\frac{t}{c} \frac{\delta}{c} \bar{C}_\tau)$) and can certainly be neglected; though there are particular regions (e.g. the "knuckle" of a deflected control

---

\ast Valid for near-equilibrium boundary layers, with $H < 2.3$ (cf Ref. 24)

surface, or the "cove" on the lower surface of a rear-loaded
aerofoil) where $\kappa_w c$ may be $O(1)$ or higher, and where these terms
might need to be taken into account. Normally, however, only
the terms (3) and (4) are likely to be significant; of these
the turbulence terms (4) will have a small but cumulative
effect over the whole chord, while the curvature term (3)
will be very small except near the trailing edge, where
κ^* may be relatively large and growing rapidly. Neither term
is expected to have a significant effect on the calculated drag,
because they both involve differentiation with respect to s and
so their contributions, when the momentum integral equation is
integrated from the leading edge to far downstream, will tend
to be self-cancelling.

A more subtle, but nevertheless important, second order effect
stems from the fact that the velocity appearing on the left-hand
side of equation (15) is u_{iw}, the inviscid velocity at the
surface, rather than the value at the outer edge of the boundary
layer (u_e) as in the standard equation (16). Now the argument
given in section 3.1 shows that

$$\frac{u_e - u_{iw}}{u_{iw}} \simeq \kappa^*\delta,$$

so that in circumstances such that $c\kappa^*$ is $O(1)$ or higher (for
example near the trailing edge), the relative change in the
velocity gradient term $\dfrac{1}{u}\dfrac{du}{ds}$ will be at least $O(\dfrac{\delta}{c})$, and is
therefore likely to be significant, particularly since a boundary
layer approaching separation is sensitive to small changes in
the effective adverse velocity gradient.

To sum up at this stage, it appears that the standard first-
order momentum integral equation (16) is likely to be satisfactory for most purposes, provided that u_{iw} is used in place of
u_e, while the semi-empirical approximations to the higher-order
terms, given above, will allow them to be included without much
difficulty wherever their effects are expected to be significant.
Of at least equal importance is the effect of normal pressure
gradients (already anticipated in term (3) above), and this will
be considered in the following section.

3.2.2 *The normal momentum equation*

The second momentum equation, written in a form analogous to that of equation (12) above, is

$$\frac{\partial}{\partial s}(\rho_i u_i v_i - \rho uv) + \frac{\partial}{\partial n}\left[(1 - \kappa_w n)(\rho_i v_i^2 - \rho v^2)\right] + \kappa_w(\rho_i u_i^2 - \rho u^2)$$

$$= -(1 - \kappa_w n)\frac{\partial}{\partial n}(p_i - p) + \frac{\partial}{\partial s}(\overline{\rho u'v'}) + \frac{\partial}{\partial n}\left[(1 - \kappa_w n)\overline{\rho v'^2}\right]$$

$$+ \kappa_w \overline{\rho u'^2}. \tag{22}$$

This again is exact apart from the omission of the viscous laminar term $\mu \nabla^2 v$.

Integrating across the boundary layer in the usual way, we obtain the following equation for the difference between the values of pressure at the wall for the two flows (EIF and RVF):

$$\Delta p \equiv p_{iw} - p_w = \frac{d}{ds}(\rho_{iw} u_{iw}^2 \theta_n) - \rho_{iw} v_{iw}^2(\theta + \delta^*)$$

$$+ \kappa_w \int_0^\delta (p_i - p)\,dn - \kappa_w \int_0^\delta \overline{\rho u'^2}\,dn + \frac{d}{ds}\int_0^\delta (-\overline{\rho u'v'})\,dn. \tag{23}$$

In deriving this we have made use of the fact that $\overline{v'^2} \to 0$ for both $n = 0$ and $n = \delta$, and have integrated the expression $\int_0^\delta n\frac{\partial}{\partial n}(p_i - p)\,dn$ by parts. In most circumstances the only significant second-order terms are those on the first line. Thus we can write in non-dimensional form

$$\frac{\Delta p}{\rho_{iw} u_{iw}^2} \simeq \frac{1}{\rho_{iw} u_{iw}^2}\frac{d}{ds}\left(\rho_{iw} u_{iw}^2 \theta_n\right) - \left(\frac{v_{iw}}{u_{iw}}\right)^2 + \kappa_w(\theta + \delta^*). \tag{24}$$

Substituting the expression for $\dfrac{v_{iw}}{u_{iw}}$ given by equation (5) and the approximation for θ_n given by equation (18), and using the first-order form of the streamwise momentum integral equation (13), we find that

$$\frac{\Delta p}{\rho_{iw} u_{iw}^2} = \left(\kappa_w + \frac{d^2 \delta^*}{ds^2} \right)(\theta + \delta^*) + E(\delta^*, \rho_{iw} u_{iw}, C_f)$$

$$\text{where } E = \frac{d\delta^*}{ds} \left\{ \tfrac{1}{2}C_f - \frac{\delta^*}{\rho_{iw} u_{iw}} \frac{d}{ds}(\rho_{iw} u_{iw}) \right\} - \left\{ \frac{\delta^*}{\rho u} \frac{d(\rho_{iw} u_{iw})}{ds} \right\}^2 .$$

Of the three terms in the function E, the first will certainly be negligible near separation (because $C_f \to 0$), and since $\dfrac{1}{\rho_{iw} u_{iw}} \dfrac{d}{ds}(\rho_{iw} u_{iw})$ is likely to be $O(1)$ (except at a shock wave) the third term is $O(\delta^{*2})$ and therefore also negligible. The remaining term is $- \dfrac{d\delta^*}{ds} \dfrac{\delta^*}{\rho u} \dfrac{d}{ds}(\rho_{iw} u_{iw})$, and numerical experiments (cf Appendix A) suggests that this should not exceed about $\dfrac{1}{4}\left(\dfrac{d\delta^*}{dx}\right)^2$ which is again negligible (less than 0.005, say) in most relevant cases. Thus the simple numerical approximation

$$\frac{p_{iw} - p_w}{\rho_{iw} u_{iw}^2} = \kappa^*(\theta + \delta^*) \tag{25}$$

should be adequate for most of the situations that we are likely to encounter in practice (see Appendix A for a fuller discussion).

The physical plausibility of this result (due to East[5]) can be indicated as follows. The pressure gradient normal to the streamlines, in both inviscid and viscous flows, is approximately $-\kappa \rho u^2$ where κ is the appropriate streamline curvature, and since the streamline direction does not deviate much from that of the wall, it follows that $\dfrac{\partial p}{\partial n} \simeq -\kappa \rho u^2$ and hence that

$$\Delta_p \simeq \int_0^\delta (\kappa_i \rho_i u_i^2 - \kappa \rho u^2)\,dn .$$

Now for the equivalent inviscid flow a good approximation to κ_i is clearly the curvature of the displacement surface, namely κ^*. For the real viscous flow the streamline curvature will vary from κ_w at the wall (usually small and negative) to a value close to κ^* at the outer edge of the boundary layer. But the major contribution to the integral $\int_0^\delta \kappa\rho u^2 dn$ will in fact come from the outer part because - particularly for flow near separation - u^2 is relatively much smaller over the inner part. Hence it is reasonable to assume that

$$\Delta p \simeq \kappa^* \int_0^\delta (\rho_i u_i^2 - \rho u^2)\, dn \, ,$$

from which equation (25) follows at once. This result, or its near-equivalent with

$$\kappa^* = \kappa_w + \frac{d}{ds}\left(\frac{v_{iw}}{u_{iw}}\right) \tag{26}$$

had in fact been conjectured by Collyer and Lock[9,26] and Le Balleur[10].

Equation (25) tells us that normal pressure gradients will be significant if and only if κ^*, the curvature of the displacement surface, is relatively large when the boundary layer is at the same time fairly thick. Typical situations in which this is likely to occur include: in the region just upstream and downstream of the trailing edge, particularly on the upper surface of a lifting aerofoil; in the hollow (or "cove") on the lower surface of a rear-loaded aerofoil; near the "knuckle" of a deflected flap or control; in the approach to separation on an aerofoil with rear separation (but not in the subsequent separated flow, where both normal and streamwise pressure gradients will be small) and finally in the neighbourhood of a shock-boundary layer interaction (though here equations (18) and (25) may be only qualitatively correct).

It is interesting to note that, for the equivalent inviscid flow, the normal pressure gradient at the wall is given by

$$\left.\frac{1}{\rho_{iw}u_{iw}^2}\frac{\partial p_i}{\partial n}\right|_{n=0} = -\frac{1}{u_{iw}}\frac{dv_{iw}}{ds} - \left.\frac{v_{iw}}{u_{iw}^2}\frac{\partial v_i}{\partial n}\right|_{n=0} - \kappa_w$$

$$= -\left\{\kappa_w + \frac{d}{ds}\left(\frac{v_{iw}}{u_{iw}}\right)\right\} \quad \text{(for incompressible flow)}$$

$$\simeq -\kappa^*.$$

Since the pressure gradient for the incompressible flow will not vary appreciably across the boundary layer, this implies that

$$\frac{p_{iw} - p_e}{\rho_{iw}u_{iw}^2} \simeq \kappa^*\delta \qquad (27)$$

and comparing with equation (25) we see that the pressure change across the boundary layer in the real viscous flow is given by

$$\frac{p_w - p_e}{\rho_{iw}u_{iw}^2} = \kappa^*(\delta - \theta - \delta^*) \ . \qquad (28)$$

For a typical example near the trailing edge of an aerofoil, with $H \sim 2$ and $H_1\left(= \dfrac{\delta - \delta^*}{\theta}\right) \sim 4$ we have $\delta \sim 3\delta^* \sim 2(\theta + \delta^*)$, so that $(p_w - p_e)$ and $p_{iw} - p_w$ are roughly equal. The situation is thus as sketched in Fig. 5, which also suggests that a reasonable approximation to the contribution to the integral $\displaystyle\int_0^\delta (p_i - p)\,dn$ from streamline curvature effects is $\frac{1}{2}\rho_{iw}u_{iw}^2\kappa^*(\theta + \delta^*)^2$, as assumed in section 3.2.1.
We notice also that

$$p_w \simeq \left.p_i\right|_{n=\theta+\delta^*} \qquad (29)$$

a relation which could be used to derive an alternative "Kutta" condition, to fix the circulation for a lifting aerofoil (see 3.4 below).

We note finally that the conjecture made by Myring[3] and Le Balleur[4], that the pressure *difference* $(p_i - p)$ may be considered negligible even though the individual changes in p_i and p across the boundary layer are significant, is only valid for values of H close to 1 (when $\theta + \delta^* \ll \delta$). However the integral $\int_0^\delta (p_i - p)\,dn$ may well be negligible even though $p_{iw} - p_w$ is not.

3.3 Matching conditions in the wake

The conditions that are needed to effect a matching between the viscous and inviscid calculations required to simulate a turbulent wake are slightly more complicated than those over the wing surface, essentially because both the thickness distribution and position of the wake are (in general) unknown and need to be determined as part of the interactive process. When one is using the "solid displacement surface" model of the interaction, it is necessary to introduce a discontinuity in pressure across the wake (related to its curvature) in order to fix its position to better than first order accuracy. When the "surface transpiration" model is used, a dividing line has to be chosen in the wake to separate the upper and lower parts of the inviscid flow, and on this line discontinuities are required in both normal and tangential components of velocity, so that it can be thought of as a combined source and vortex sheet. It is the development of this second model of the wake that we shall consider in the present section.

If we examine the derivation, given above in sections 3.1 and 3.2, of the conditions determining the transpiration velocity v_{iw} and the pressure difference $(p_{iw} - p_w)$, we find that at no stage does the assumption have to be made that $u \to 0$ at the "wall" in the real viscous flow (even though of course this is actually the case), only that $v \to 0$ there. It follows that precisely the same conditions are valid in the wake, provided that we replace the "wall" by the dividing streamline from the trailing edge (in the real viscous flow), so that v_w is still zero on it. The derivation of the corresponding "jump" conditions across this line in the equivalent inviscid flow is then straightforward, as explained below.

A sketch showing the flow in the near wake and explaining the notation used is given in Fig. 6. In most practical applications no distinction is made (or need be made) between the distance along the streamline, s, and the chordwise distance x.

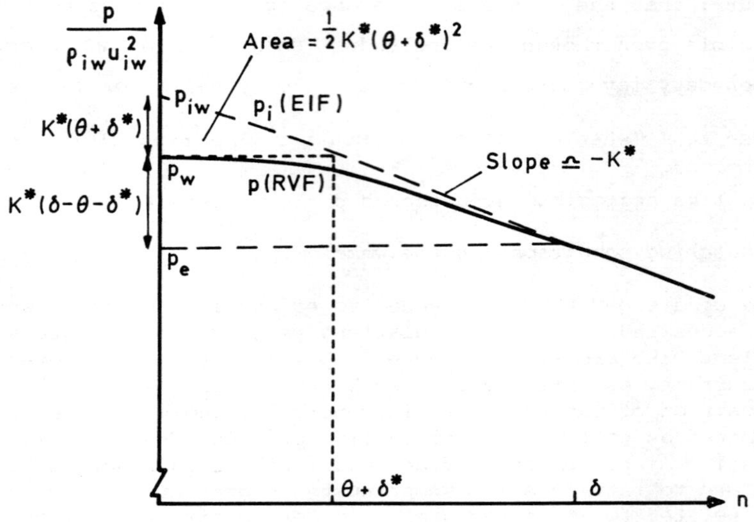

Fig. 5.

3.3.1 Displacement effect

At points C and D on the upper and lower sides of the dividing streamline (see Fig. 6), the components of transpiration velocity, v_{iu} and $v_{i\ell}$ are respectively (see equation (5))

$$\left.\begin{array}{l} v_{iu} = \dfrac{1}{\rho_{iu}} \dfrac{d}{ds}\left(\rho_{iu}u_{iu}\delta_u^*\right) \\[3ex] v_{i\ell} = -\dfrac{1}{\rho_{iu}} \dfrac{d}{ds}\left(\rho_{i\ell}u_{i\ell}\delta_\ell^*\right). \end{array}\right\} \qquad (30)$$

and

Here the sign convention has been used that v is measured positive in the direction of the upward normal to the wake. Hence a jump Δv in the component of velocity normal to the wake is required, given by

$$\Delta v_i \equiv v_{iu} - v_{i\ell} = \frac{1}{\rho_{iu}} \frac{d}{ds}\left(\rho_{iu}u_{iu}\delta_u^*\right) + \frac{1}{\rho_{i\ell}} \frac{d}{ds}\left(\rho_{i\ell}u_{i\ell}\delta_\ell^*\right) \qquad (31)$$

Upper edge of viscous layer
Dividing streamline
$y = Y_u + \delta_u^*$
Upper displacement surface
$y = Y_w + \delta_u^*$
δ_u^*
δ_w^*
A
u
u_{in}
v_{iu}
v_{il}
C
D
B
u_{il}
v_{in}
Y_w
x
S
Y_u
$-Y_l$
δ_l^*
δ_u^*
$y = Y_w - \delta_l^*$
Lower displacement surface
$y = Y_l - \delta_l^*$
Lower edge of viscous layer
y

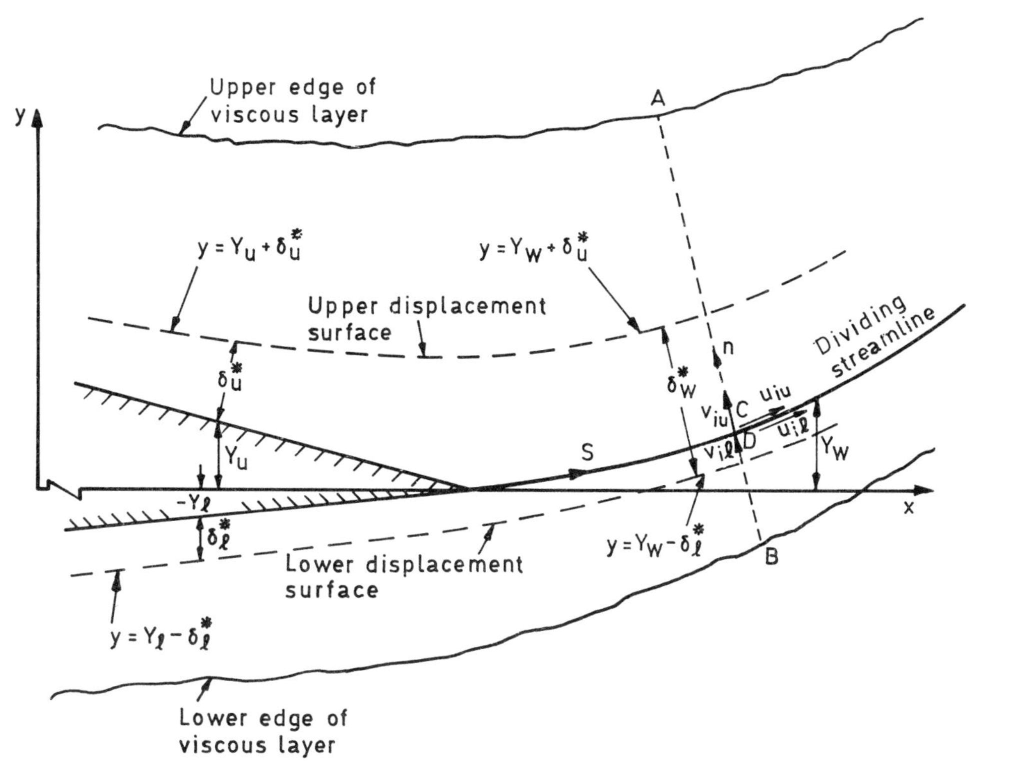

Fig. 6

If desired a simpler equation can be used, obtained by replacing
$u_{iu}, u_{i\ell}$ and $\rho_{iu}, \rho_{i\ell}$ by their mean values (where a capital W
suffix refers to the wake)

$$\bar{u}_{iW} = \tfrac{1}{2}(u_{iu} + u_{i\ell}), \qquad \bar{\rho}_{iW} = \tfrac{1}{2}(\rho_{iu} + \rho_{i\ell});$$

then   (31a)

$$\Delta v \simeq \frac{1}{\bar{\rho}_{iW}} \frac{d}{ds} \left(\bar{\rho}_{iW} \bar{u}_{iW} \delta^*_W \right)$$

where $\delta^*_W = \delta^*_u + \delta^*_\ell$ is the total displacement thickness of the
wake. This is the equation normally used; however the more
accurate equation (31) would be no more difficult to use in
practice, if the development of the upper and lower parts of
the wake is calculated separately, as is often the case.

3.3.2 *Curvature effect*

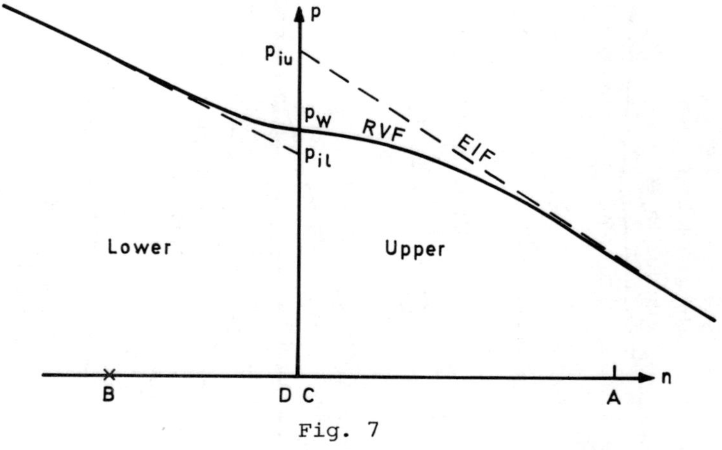

Fig. 7

Consider a section of the wake along the normal AB through
CD (Fig. 6); the variation of pressure along it, for both the
real and inviscid flows, is sketched in Fig. 7. Using the
argument of section 3.2.2 (equation (25)), we have, for the
upper and lower parts respectively

$$P_W - P_{iu} = - \kappa^*_u \rho_{iu} u^2_{iu} \left(\delta^*_u + \theta_u \right) ,$$ (32)

$$P_W - P_{i\ell} = + \kappa^*_\ell \rho_{i\ell} u^2_{i\ell} \left(\delta^*_\ell + \theta_\ell \right) ,$$ (33)

where $\kappa^*_u = \kappa_W + \dfrac{d^2\delta^*_u}{ds^2}$ and $\kappa^*_\ell = \kappa_W - \dfrac{d^2\delta^*_\ell}{ds^2}$ are the curvatures of the upper and lower displacement surfaces (reckoned positive if concave upwards in both cases - hence the difference in sign between equations (32) and (33)), and κ_W is the curvature of the dividing streamline. Subtracting equation (32) from equation (33), we obtain

$$\Delta p_i \equiv p_{iu} - p_{i\ell} = \kappa^*_u \rho_{iu} u^2_{iu}\left(\delta^*_u + \theta_u\right) + \kappa^*_\ell \rho_{i\ell} u^2_{i\ell}\left(\delta^*_\ell + \theta_\ell\right). \quad (34)$$

Again, if the difference between κ^*_u and κ^*_ℓ is neglected, equation (34) can be replaced by

$$\Delta p_i \simeq \kappa_W \bar{\rho}_{iW} \bar{u}^2_{iW}\left(\delta^*_W + \theta_W\right), \quad (34a)$$

but in fact κ^*_u and κ^*_ℓ can differ appreciably near the trailing edge and so equation (34) is preferable.

It is usually more convenient to think of the wake as a vortex sheet producing a jump Δu_i in tangential velocity; this is given to first order by

$$\Delta u_i \equiv u_{iu} - u_{i\ell} \simeq - \left[\kappa^*_u u_{iu}\left(\delta^*_u + \theta_u\right) + \kappa^*_\ell u_{i\ell}\left(\delta^*_\ell + \theta_\ell\right)\right] \quad (35)$$

or the simpler version

$$\Delta_{ui} \simeq - \kappa_W \overline{u_i}\left(\delta^*_W + \theta_W\right). \quad (35a)$$

Equations (31) and (34) (or (35)) provide the jump conditions needed to represent the interaction of the viscous wake with the equivalent inviscid flow. In most practical methods for the problem, the question of determining precisely the shape of the wake has been avoided, and instead the appropriate jump conditions have been applied on some suitable approximation to it; this should not introduce appreciable errors provided that the two shapes diverge only slowly, because most of the 'wake effects' came from the immediate vicinity of the trailing edge. In this case it is convenient to derive the two curvatures, κ^*_u and κ^*_ℓ, required in equation (35), by approximating them as

the streamwise rate of change of the net flow direction (with
respect to fixed axes) in the equivalent inviscid flow just
above and below the chosen dividing line. This is equivalent to

the approximation $\kappa^* = \kappa_w + \dfrac{d}{ds}\left(\dfrac{v_{iw}}{u_{iw}}\right)$ given by equation (26).

3.4 The 'Kutta' condition for viscous flow

The most important of the second-order effects which were
discussed in section 3.2 is the 'pressure correction'
$\Delta p(= p_{iw} - p_w) \simeq \kappa^*\rho_{iw}u_{iw}^2(\delta^* + \theta)$, which is required to adjust
the values of the pressure, calculated at the surface of the
aerofoil or the wake line for the equivalent inviscid flow, to
give the true values for the real viscous flow when normal
pressure gradients are taken into account. The implications of
this on the 'Kutta-Joukowski' condition - the condition which
fixes the value of the circulation in the equivalent inviscid
flow over a lifting aerofoil so that the corresponding real
viscous flow is physically realistic at the trailing edge - are
discussed briefly below.

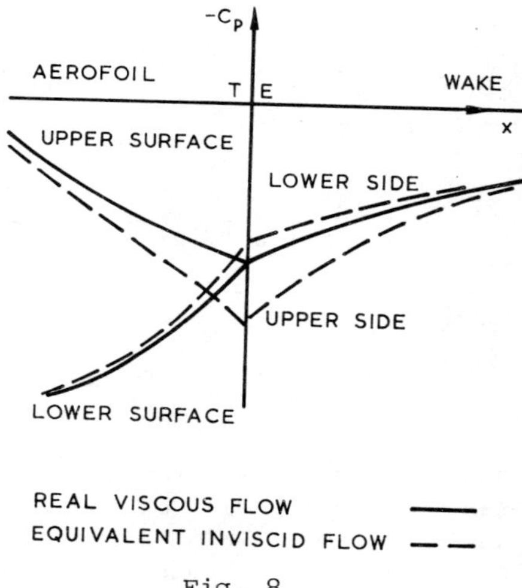

Fig. 8

We start by considering the qualitative nature of the pressure
distribution near the trailing edge of the aerofoil in the real
flow, over the solid surface and along the dividing streamline
in the wake (see Fig. 8, full line). The pressures on the upper
and lower surfaces must approach the same limiting value at the

trailing edge (assuming that this is sharp), which will normally
be slightly higher than the free stream value but well below
stagnation pressure; on the wake line the pressure is of course
single valued. All pressure gradients will be finite and
moderate in value, but may have mild streamwise discontinuities
at the trailing edge on either or both surfaces.

In equivalent inviscid flow (dashed lines in Fig. 8), the
pressures will differ from their corresponding values in the
real flow by an amount given by equation (25), depending on the
thickness of the appropriate boundary layer (or part of the wake)
and the curvature of its displacement surface. As we have seen
in section 3.3.2, this implies that there is now a discontinuity
in the values of p_i on the upper and lower sides of the wake
line, given by equation (34). If we approach the trailing edge
from behind the limiting value of this discontinuity becomes

$$(\Delta p_i)_{TE} = \left[\kappa^*_u \rho_{iu} u^2_{iu}\left(\delta^*_u + \theta_u\right) + \kappa^*_\ell \rho_{i\ell} u^2_{i\ell}\left(\delta^*_\ell + \theta_\ell\right)\right]_{TE} \tag{36}$$

and precisely the same value is found when the trailing edge is
approached from in front, provided that the displacement surfaces
are assumed to be sufficiently smooth as to have continuous
variation of curvature through the trailing edge. Such an
assumption is consistent with the available evidence, both
experimental (cf Ref 11) and theoretical. As an example of
the latter we refer to Fig. 9, taken from Ref 9, which shows
the results of a converged solution for the aerofoil RAE 2822
at $M_\infty = 0.68$, $\alpha = 1^\circ$, obtained by the VGK method which will be
described later (section 5.2.1). The quantity plotted here is
the slope, relative to the chord line of the aerofoil, of the

surface transpiration velocity vector, namely $\left(\dfrac{dy}{dx} + \dfrac{v_{iw}}{u_{iw}}\right)$; the

line '$\theta = 0$' referred to in the key is the line on which the
wake matching conditions are applied. The quantity κ^*, as
defined by the alternative equation (26), is then the streamwise
slope of the appropriate curve in Fig. 9. We can see clearly
how, for the converged solution (particularly when curvature
effects are taken into account (full lines)), the curves are
essentially smooth through the trailing edge on both upper and
lower surfaces; in particular the discontinuity produced by the
finite trailing edge angle of the aerofoil (dashed line) is
completely removed.

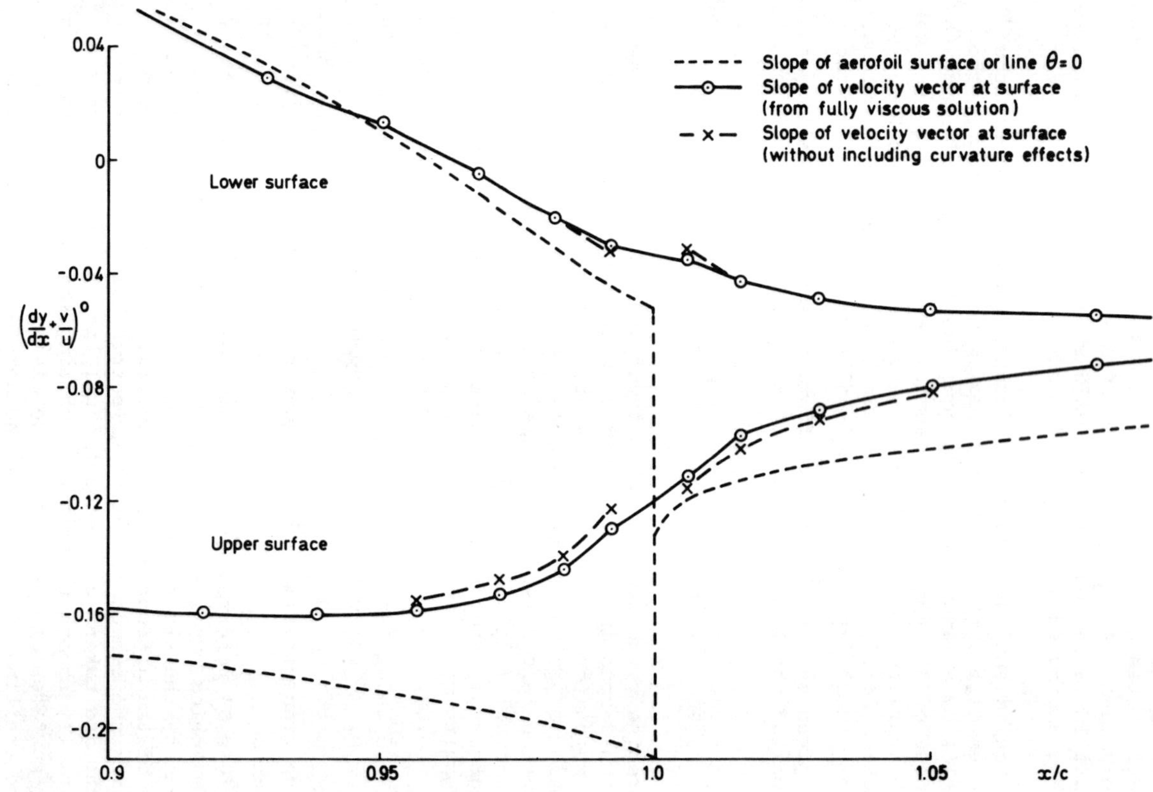

Fig. 9 Slope $\left(\dfrac{dy}{dx} + \dfrac{v}{u}\right)$ near trailing edge for RAE 2822 aerofoil at $M_\infty = 0.676$, $\alpha = 1.06^\circ$

We may therefore regard equation (36) above as an effective
Kutta condition* for our viscous flow problem, which replaces
the condition $(\Delta p)_{TE} = 0$ for a purely inviscid flow. However
this should not be thought of as a condition which needs to be
artificially enforced on the solution. Any iterative method
which incorporates an inviscid element that allows for the
presence of a streamwise-continuous distribution of 'bound'
vorticity in the wake and upstream through the trailing edge,
should converge to a solution which satisfies this condition
automatically. In a sense, the position at which a 'Kutta'
condition needs to be applied has been shifted from the trailing
edge to far downstream in the wake, where the vorticity must
approach zero in the same way as does the wake curvature.

What has been said above should not be taken to imply that
such a solution can be achieved without numerical difficulties.
On the contrary: the importance in second order theory of the
curvature of the displacement surface, together with the rapid
variations in the shape of this surface that may be expected in
the region near the trailing edge, are features to which both
the viscous and inviscid elements of the solution are
particularly sensitive, so that problems of numerical waviness
and slow convergence are almost inevitable if high accuracy is
to be attained in difficult cases. These problems can usually
be overcome by the use of localised numerical smoothing techniques and the choice of appropriate under-relaxation factors in
the iterative procedure, a subject which we shall discuss in
the following section.

There is in fact an important alternative technique, which
has been developed recently by Melnik and his colleagues (see
for example Refs 12 and 13). This involves a special treatment
of the regions, such as the foot of shock wave or the neighbourhood of the trailing edge, where second-order effects are known
to be important, based on the asymptotic theory of turbulent
boundary layers, in the limit as Re $\rightarrow \infty$ (due to Mellor[14]). In
such regions it is found that a 'multi-deck' approach to the
problem is required, the main deck being a layer which the flow
can be regarded as inviscid but rotational, with the vorticity
determined by the shape of the velocity profile in the outer
part of the oncoming boundary layer and subsequently 'frozen'
along streamlines. It is in this outer layer that most of the
transverse variation of pressure - the chief of the 'second-
order' phenomena referred to in section 3.2 above - can be shown

---

* An alternative form of this condition, which may sometimes be
convenient, follows from equation (20). At the trailing edge,
in the equivalent inviscid flow, we must satisfy the relationship
$[p_i\big|_{n = \theta + \delta^*}]_{upper} = [p_i\big|_{n = \theta + \delta^*}]_{lower}.$

to take place. For the trailing edge region, Melnik and Chow[15] have been able to obtain an elegant analytic solution for the outer layer and have shown how the results can be used within the framework of the modelling of the viscous-inviscid matching by surface transpiration. They have developed a powerful iterative computational procedure incorporating these ideas[16], some results of which will be given later (section 5.2).

This approach has many points in common with the recent work of East[5] that we have already described. Perhaps the main difference - apart from the greater analytical complexity - is the dependence on asymptotic theory, which would be expected to become less reliable the closer separation is approached. By contrast, the numerical approximations to the second-order terms used by East were all derived on the assumption of fairly large values of the shape parameter H and therefore become more - rather than less - accurate during the approach to separation. On the other hand there is no doubt that Melnik's technique does obviate the need for artificial smoothing or similar devices which has been experienced with the simpler methods and which inevitably introduces a certain arbitrariness into their numerical results.

3.5 Three-dimensional problems

At the present time there is no three-dimensional counterpart to the second order theory described in sections 3.1 to 3.4 above. But the most important concept - the matching between the viscous and inviscid elements by one of the two alternative models of the displacement effect - is of course still valid. Of these the 'surface transpiration' model remains the more satisfactory, and can be derived without difficulty in the following way:

We choose for simplicity a rectangular coordinate system (x, y, n) with x and y in the wing surface and n normal to it; corresponding velocity components are $(u^{(x)}, u^{(y)}, v)$. If surface curvature is neglected the equation of continuity may be written, again in 'difference' form, as

$$\frac{\partial}{\partial n} (\rho_i v_i - \rho v) = - \frac{\partial}{\partial x} \left[\rho_i u_i^{(x)} - \rho u^{(x)} \right] - \frac{\partial}{\partial y} \left[\rho_i u_i^{(y)} - \rho u^{(y)} \right].$$

$$(37)$$

Integrating this with respect to n from 0 to δ, we obtain for the surface transpiration velocity the expression

$$\rho_{iw}v_{iw} = \frac{\partial}{\partial x}\int_0^\delta \left(\rho_i u_i^{(x)} - \rho u^{(x)}\right)dn + \frac{\partial}{\partial y}\int_0^\delta \left(\rho_i u_i^{(y)} - \rho u^{(y)}\right)dn,$$

which may be written

$$v_{iw} = \frac{1}{\rho_{iw}}\left[\frac{\partial}{\partial x}\left(\rho_{iw}u_{iw}\delta_1^*\right) + \frac{\partial}{\partial y}\left(\rho_{iw}u_{iw}\delta_2^*\right)\right], \tag{28}$$

where $u_{iw} = \sqrt{u_{iw}^{(x)^2} + u_{iw}^{(y)^2}}$ is the surface speed (in the inviscid flow)

and

$$\delta_1^* = \frac{1}{\rho_{iw}u_{iw}}\int_0^\delta \left(\rho_i u_i^{(x)} - \rho u^{(x)}\right)dn,$$

$$\delta_2^* = \frac{1}{\rho_{iw}u_{iw}}\int_0^\delta \left(\rho_i u_i^{(y)} - \rho u^{(y)}\right)dn$$

are the components of the displacement thickness. The form that this equation takes when a more convenient, general non-orthogonal coordinate system is used in the wing surface, has been given by P.D. Smith [25].

If on the other hand it is desired to use the more conventional "solid displacement surface" model of the interaction, it is necessary (cf Refs. 2 and 25) to solve first the partial differential equation

$$\frac{\partial}{\partial x}\left(\rho_{iw}u_{iw}^{(x)}\delta^*\right) + \frac{\partial}{\partial y}\left(\rho_{iw}u_{iw}^{(y)}\delta^*\right) = \rho_{iw}v_{iw} \tag{39}$$

(where v_{iw} is given by equation (38)), to obtain the relevant displacement thickness δ^*. It is clearly inconvenient to have to do this, and since there are in addition even greater advantages in three dimensions than in two for being able to use a boundary condition for the inviscid flow calculation that is applied on the actual, fixed wing surface, it seems that the transpiration model of the interaction is definitely preferable.

There is little doubt that second-order effects, analogous
to those already identified in two dimensions, will be just as
important in three-dimensional flows. Although it is not yet
clear precisely what form they will take it seems probable that
the most important of them will remain the effect of streamline
curvature, and that the governing parameter will be the curvature of the displacement surface in the streamwise direction
(or the streamwise rate of change of v_{iw}/u_{iw}). This implies
that corrections to the pressures calculated for the equivalent
inviscid flow will be required both over the wing surface and
in the wake, where "jump" conditions similar to those applied
in two dimensions will result. Thus it will be necessary to
introduce near the surface of the wake a distribution of spanwise "bound" vorticity (as well as the sinks needed to account
for the displacement effect), which must be considered in
addition to the mainly streamwise vorticity distribution which
is already required to satisfy the wake boundary conditions for
a purely inviscid flow. All this will of course introduce
further complexity into an already complicated computational
problem, but the difficulties should not be insuperable.

Additional problems, that have no apparent analogue in two
dimensions, will occur in corner regions (such as the junction
of a wing with a fuselage or pylon) and near the wing tip, where
the normal assumptions of thin shear-layer theory are no longer
valid, while the cross flow in the wake and its eventual rolling-
up also present extra complications as regards both the inviscid
and viscous elements of the solution. Eventually, we shall have
to face up to such problems, but up to the present they have
been either ignored or overcome by arbitrary means.

4. THE ORGANISATION AND NUMERICAL STABILITY OF AN ITERATIVE
    PROCEDURE

*4.1 Organisation of an iterative computation of viscous-
      inviscid interaction*

In the many numerical methods that have been developed
recently for calculating viscous effects on aerofoils and wings,
the overall scheme of computation usually follows one of two
well-defined iterative patterns, which may be termed "direct"
and "inverse" (or "semi-inverse"). The most common is a direct
procedure which can best be described with the aid of a simplified flow diagram, as shown in Fig. 10. The steps involved
are as follows:

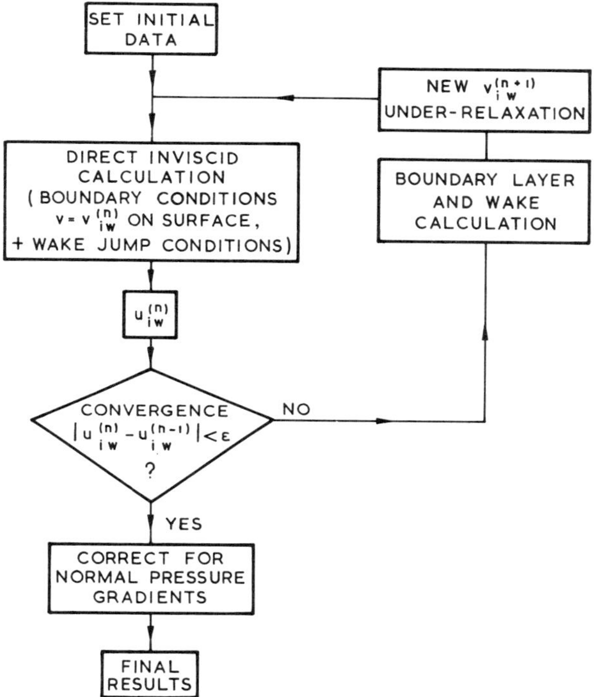

Fig. 10   Flow diagram for direct iterative calculation

(i) After the initial data – wing geometry, Mach number, angle of incidence, Reynolds number and other relevant boundary-layer parameters – have been read in, an initial solution is obtained to the purely inviscid problem, which at this stage need only be approximate. Thus if an iterative procedure is being used to solve a finite difference (or similar) representation of the inviscid flow equations, only a few iterations need be performed.*

(ii) The values of u_{iw} obtained from the inviscid solution, on the aerofoil surface and in the wake, are used as input to the numerical method which has been chosen to calculate the development of the viscous layers.

---

* If a panel method is being used, the inviscid flow calculation will not normally require iteration and can be considered as fully accurate at each stage.

(iii) From this viscous calculation, values are derived of the surface transpiration velocity v_{iw} (or values of $\delta*$ if the solid displacement surface model is being used) over the aerofoil surface and of the jumps in v_i and u_i in the wake (the last requires further output from the inviscid calculation to determine $\kappa*$).

(iv) These values are used, usually with some under-relaxation (see later) to modify the appropriate boundary conditions for a new inviscid flow calculation, a few more iterations of which are then performed*.

(v) The overall convergence is investigated. Usually this is assessed by considering the changes after each successive iteration to the velocity u_{iw}, the boundary layer thickness (or v_{iw}) and the force coefficients C_L and C_D.

(vi) If convergence is judged to be inadequate, steps (ii) to (v) are repeated.

(vii) When satisfactory convergence has been achieved, final corrections to the calculated pressure distribution to allow for the effects of normal pressure gradients may be applied and the values of C_L, C_D etc., can be computed and output.

4.2 *Numerical stability: determination of under-relaxation factors*

The numerical stability of such an iterative procedure has been investigated recently by Le Balleur [17]. His theory assumes that, as suggested in section 3, the coupling between the viscous and inviscid elements is effected by means of the non-dimensional "source strength" $\Sigma = \dfrac{v_{iw}}{u_{iw}}$, the slope of the transpiration velocity vector relative to the aerofoil surface. The basis of the argument is to use a linearised error analysis to derive approximate relations between small changes in Σ and u_{iw}, together with similar relations from boundary-layer (entrainment) theory, and to combine these to study the convergence behaviour of the iterative process and to deduce "optimum" relaxation factors which may be used to avoid instability and accelerate convergence.

---

* See footnote on previous page.

For the inviscid flow, we take coordinates s and n along and normal to the aerofoil surface, and consider small perturbations

$$\sigma^{(n)} = \Sigma^{(n)} - \Sigma^{(\infty)}, \quad u^{(n)} = \frac{u_{iw}^{(n)} - u_{iw}^{(\infty)}}{u_{iw}^{(\infty)}} \quad \text{from the converged}$$

solutions $\Sigma^{(\infty)}$, $u_{iw}^{(\infty)}$, where the superscript (n) denotes the iteration number. Since Σ is effectively the flow direction relative to the wall, and u_{iw} the velocity component along it, we can define a perturbation potential ϕ such that $u = \phi_s$, $\sigma = \phi_n$, and then we have approximately

$$(1 - M^2)\phi_{ss} + \phi_{nn} = 0 \tag{40}$$

where M is the surface Mach number which may be considered as a local "constant". We look for solutions periodic in s, of wavelength λ, proportional to $e^{2\pi i s/\lambda}$. For subsonic flow, the solution which decays exponentially to zero as $n \to \infty$ is $\phi = Ae^{-2\pi\beta n/\lambda}e^{2\pi i s/\lambda}$, where $\beta = \sqrt{1 - M^2}$, so that $u = \frac{2\pi i}{\lambda}\phi(s,0)$

and $\sigma = -\frac{2\pi\beta}{\lambda}\phi(s,0)$ and hence

$$\left.\begin{array}{c} u = -\dfrac{i}{\beta}\,\sigma \\[2em] \dfrac{du}{ds} = \dfrac{2\pi}{\lambda\beta}\,\sigma \end{array}\right\} \quad . \tag{41}$$

or

An essential feature of the analysis is the concept that – say in the late stages of convergence of an iterative process – the disturbance (or error) can be considered as made up of Fourier components of which the minimum wavelength is $2\Delta s$, where Δs is the step length involved in the discretisation process (eg. mesh size or panel length); so that $\lambda \geqslant 2\Delta s$.

The corresponding fundamental relation from first-order boundary layer theory is the equation connecting the source strength Σ with the velocity gradient parameter $\dfrac{\theta}{u_{iw}}\dfrac{du_{iw}}{ds}$. This may be written (see Appendix B) in the form

$$\Sigma = -B \cdot \frac{\theta}{u_{iw}}\frac{du_{iw}}{ds} + C \tag{42}$$

where for incompressible flow

$$B = \frac{(H + 1)(H_1 - HH_1')}{(- H_1')}$$ (43)

and

$$C = \frac{\{\tfrac{1}{2}C_f(H_1 - HH_1') - C_E\}}{(- H_1')} .$$ (44)

Here H_1 is the shape parameter $(\delta - \delta^*)/\theta$, $H_1' \equiv \dfrac{dH_1}{dH}$ and C_E is the entrainment coefficient:

$$C_E = \frac{1}{\rho_e u_e} \frac{d}{ds} \int_0^\delta \rho u \, dn = \frac{1}{\rho_e u_e} \frac{d}{ds} (\rho_e u_e H_1 \theta) .$$

There is no distinction between u_e and u_{iw} in first order boundary layer theory. (The corresponding equation for compressible flow is given in Appendix B.)

In many integral methods based on the idea of entrainment (e.g. Refs. 18 and 19) an empirical relationship between H_1 and H is assumed, for example that of Ref. 18:

$$H_1 = 2 + 1.5 \left(\frac{1.12}{H - 1}\right)^{1.09} + 0.5 \left(\frac{H - 1}{1.12}\right)^{1.09} .$$

From this equation it can be seen that H_1' is negative for $1 < H < 2.85$ and vanishes when $H = 2.85$, a value associated with separation conditions. It follows that for attached flows the coefficient B in equation (42) is positive, but is likely to approach infinity at separation, as indicated in Fig. 11.

From equation (42), we see that the perturbations in Σ and u_{iw} are connected by

$$\left.\begin{aligned}
\sigma &= - B\theta \frac{du}{ds} \\[2ex]
\sigma &= - \frac{2\pi i}{\lambda} B\theta u
\end{aligned}\right\}$$

or   (45)

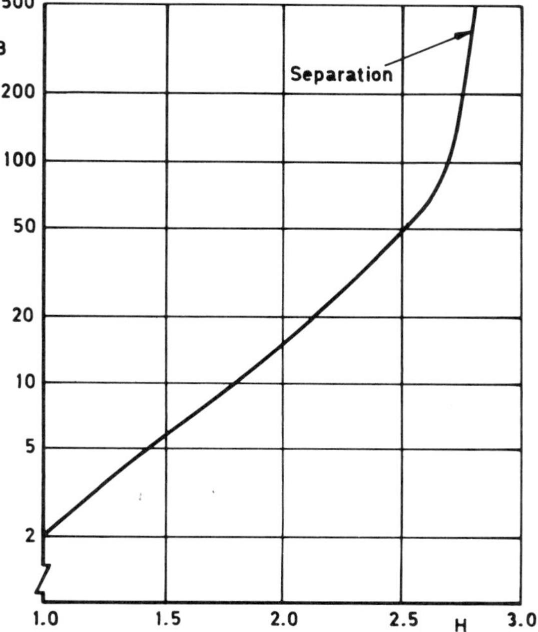

Fig. 11   The boundary layer parameter B

again assuming that the coefficients in equation (42) may be
regarded as "constants" in this locally-linearised analysis.

If equations (41) and (45) are now combined, it follows that
the result of one complete iteration of a "direct" iterative
process gives approximately

$$\left.\begin{array}{c} \sigma^{(n+1)} = \mu\sigma^{(n)} \\[2ex] \mu = -\dfrac{2\pi B}{\beta}\dfrac{\theta}{\lambda} \end{array}\right\} \quad . \tag{46}$$

with

The parameter μ is in effect an amplification factor for the
natural, undamped iterative process. Because μ is negative this
is essentially an oscillatory process, and since $|\mu|$ may be
large - if B is large (near separation), β is small (M near 1)
or $\lambda < \theta$ (disturbance of small wavelength) - the process is
likely to be divergent, and damping in the form of under-
relaxation will therefore be required.

If we denote by $\Sigma*^{(n+1)}$ the new value of Σ predicted *without* damping, then from equation (46)

$$\Sigma*^{(n+1)} - \Sigma^{(\infty)} = \mu\left(\Sigma^{(n)} - \Sigma^{(\infty)}\right).$$

The under-relaxation process, with relaxation factor ω, is equivalent to taking for the new value of Σ

$$\Sigma^{(n+1)} - \Sigma^{(n)} = \omega\left(\Sigma*^{(n+1)} - \Sigma^{(n)}\right) \tag{47}$$

which is equivalent to

$$\Sigma^{(n+1)} - \Sigma^{(\infty)} = [1 + \omega(\mu - 1)]\left(\Sigma^{(n)} - \Sigma^{(\infty)}\right),$$

thus giving a new amplification factor

$$\mu' = 1 + \omega(\mu - 1). \tag{48}$$

To achieve the most rapid convergence we want $|\mu'| \ll 1$, and should therefore choose

$$\omega = \omega_{opt} = \frac{1}{1 - \mu} \tag{49}$$

$$\simeq -\frac{1}{\mu} = \frac{\beta}{2\pi B}\frac{\lambda}{\theta} \qquad \text{if } |\mu| \text{ is large.}$$

For incompressible flow ($\beta = 1$), we see by reference to Fig. 11 that the value of ω_{opt} varies from about $\frac{1}{25}\frac{\lambda}{\theta}$ for $H = 1.3$, through $\frac{1}{100}\frac{\lambda}{\theta}$ for $H = 2$ to zero at separation.

Le Balleur's stability analysis certainly provides a qualitative explanation for many of the features of the under-relaxation process that have been found from experience to be necessary in a direct iterative method: in particular a smaller value of ω is required

(a) with a fine computational grid than with a coarse one (because λ decreases),

(b) at a low Reynolds number than at a high one (because θ increases)

(c)   in a flow that approaches separation than in a milder one
(because B increases).

The main difficulty in applying it quantitatively lies in the
interpretation of the parameter λ, the wavelength of the
disturbance. This could vary from about twice the minimum mesh
size (or panel length) involved in the inviscid flow calculation,
to the full chord of the aerofoil; possibly the use of a
sequence of values covering this range, as in the approximate
factorisation schemes for the inviscid problem described by
Baker[20], might be advantageous. The analysis also suggests
that the use of a relaxation factor that varies from point to
point along the chord, as specified by equation (46), might be
desirable in contrast to the globally constant value that is
normally chosen.

4.3.   Inverse and semi-inverse techniques for separated flows

Both theory and experience are united in indicating that, for
flows which approach separation too closely, an ordinary direct
iterative process will become unstable or at least require such
heavy damping that computing times to ultimate convergence would
become prohibitive. In such cases - which are of course of
great practical importance - it is therefore necessary to have
recourse to some form of inverse approach in which the roles of
input and output are reversed.

A fully inverse technique, in which both the inviscid and
viscous parts of the calculation are inverted and used in
succession, will involve an amplification factor μ equal to the
reciprocal of its value for a direct method, namely

$$\mu = - \frac{\beta}{2\pi B} \frac{\lambda}{\theta} \, . \tag{50}$$

Since $|\mu|$ will be small in the circumstances under which this
technique is required, convergence should be relatively rapid
and under-relaxation is unlikely to be necessary. But although
it is usually straight forward to invert the order of computation
for the boundary layer (see Appendix B or Ref. 21), this may be
difficult - or at least inconvenient - for the inviscid part of
the calculation. For this reason the so-called "semi-inverse"
methods have much to recommend them.

Such a technique can best be explained by reference to the
simplified flow diagram shown in Fig. 12. The basic ideas is
that, at a stage when updated values of the source strength, $\Sigma^{(n)}$

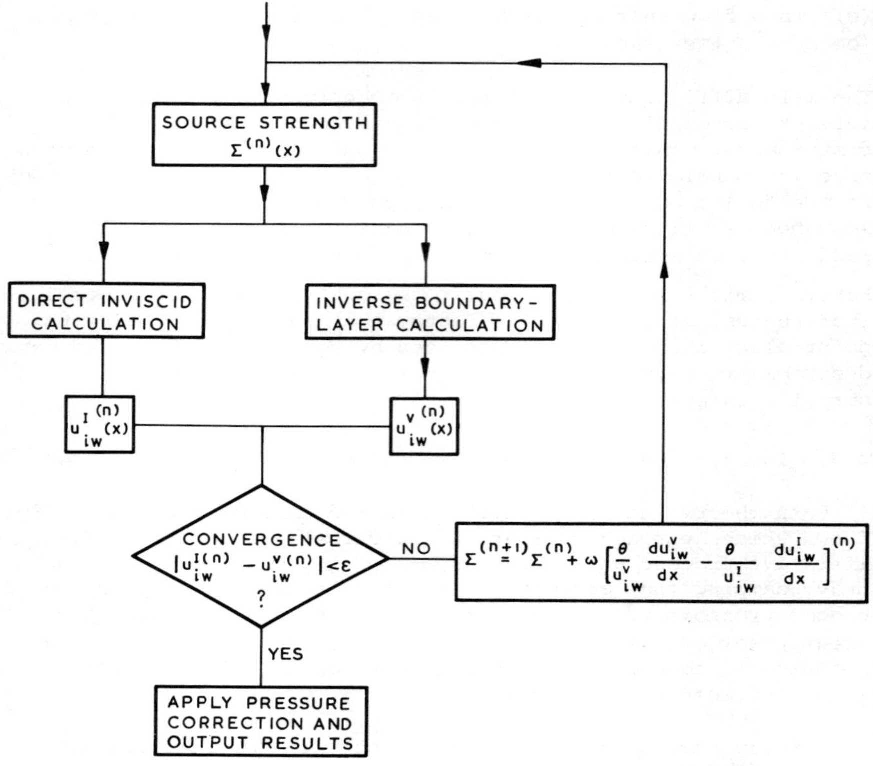

Fig. 12   Flow diagram for semi-inverse scheme

have been determined, two alternative estimations of u_{iw} can
be made with the *same* boundary conditions: (a) a direct inviscid
calculation in the usual way, (b) an inverse boundary layer
calculation, solving for $\dfrac{\theta}{u_{iw}}\dfrac{du_{iw}}{ds}$ with Σ given, for example in
the way described in Appendix B.   Denoting these two solutions by
u_{iw}^{I} and u_{iw}^{V} respectively, we can judge convergence by examining
the difference $|u_{iw}^{I} - u_{iw}^{V}|$.   If this is too large a "corrector"
must be found to provide new values for Σ, and this can be done
in the following way, again making use of the analysis given in
section 4.2 (see Ref. 17).

What we require is, in effect, a simulation of a fully
inverse method for the inviscid part of the calculation.   If
such a method was available, we should use the values u^{V} of

the velocity perturbation, produced by the inverse boundary
layer calculation, to give new values for the source perturbation σ, according to equation (41):

$$\sigma^{(n+1)} = i\beta u^{V(n)} . \tag{51}$$

But the values u^{I} that are actually available from the direct
inviscid calculation are

$$u^{I(n)} = \frac{i}{\beta} \sigma^{(n)} . \tag{52}$$

Comparing equations (51) and (52), we see that
$\sigma^{(n+1)} - \sigma^{(n)} = i\beta \left(u^{V(n)} - u^{I(n)} \right)$, which is equivalent to

$$\Sigma^{(n+1)} - \Sigma^{(n)} = \omega \left(\frac{\theta^{(n)}}{u_{iw}^{V(n)}} \frac{du_{iw}^{V(n)}}{ds} - \frac{\theta^{(n)}}{u_{iw}^{I(n)}} \frac{du^{I(n)}}{ds} \right) \tag{53}$$

where $\omega = \dfrac{\beta}{2\pi} \dfrac{\lambda}{\theta}$.

Qualitatively speaking, this equation tells us that, if the
adverse velocity gradient parameter $\left(-\dfrac{\theta}{u_{iw}^{I}} \dfrac{du_{iw}^{I}}{ds} \right)$ predicted by
the inviscid part of the calculation is greater than the
corresponding value from the inverse boundary layer calculation,
then the source strength must be proportionally increased at the
next iteration. Quantitatively, the presence of the somewhat
arbitrary factor $\dfrac{\beta}{2\pi} \dfrac{\lambda}{\theta}$ in equation (53), together with the other
approximations involved in the analysis, means that there is
bound to be an element of empiricism in using this scheme.
Considerable success has however been achieved in this way, for
example by Le Balleur and Neron[22] in calculating flows with rear
separation on both single and multiple aerofoils at low speeds,
and by Le Balleur[7,17] for some flows involving shock-induced
separation.

A final point of interest here is the relationship between
Le Balleur's scheme and the analogous one due to Carter[23] which

has also been used with success for separated flow problems by
various authors. The latter involves the relation

$$\frac{\delta*^{(n+1)}}{\delta*^{(n)}} = 1 + \Omega \left(\frac{u_{iw}^{V(n)}}{u_{iw}^{I(n)}} - 1 \right), \tag{54}$$

where Ω is another relaxation factor.

Since $\Sigma \sim \frac{d\delta*}{ds}$, differentiating equation (54) gives (after
some further approximations)

$$\Sigma^{(n+1)} - \Sigma^{(n)} = \Omega\delta*^{(n)} \left(\frac{1}{u_{iw}^{V(n)}} \frac{du_{iw}^{V(n)}}{ds} - \frac{1}{u_{iw}^{I(n)}} \frac{du_{iw}^{I(n)}}{ds} \right),$$

which is identical to equation (49) if $\Omega = \frac{\beta}{2\pi} \frac{\lambda}{\delta*}$.

Thus the two approaches are essentially equivalent, but
Le Balleur's - connecting as it does the two essential non-
dimensional parameters representing respectively the source
strength and the velocity gradient - seems to have perhaps the
sounder foundation.

5.   CALCULATION METHODS FOR AEROFOILS

In this and the succeeding section we shall describe briefly
some of the practical methods that are currently available for
performing calculations of viscous inviscid interactions of the
type envisaged in the preceding account of the underlying theory,
and try to assess their accuracy by comparison with experiment.
A reasonably comprehensive review of the subject was given
recently by the first author[8], so in the present paper we shall
be more selective and restrict ourselves to methods in which
(a) the matching is done by the transpiration model of the
interaction and (b) at least some account is taken of the
important effects of the wake. Having said this, in describing
any particular method we need only refer to the way in which the
separate (inviscid and viscous) elements of the calculation are
performed, and to any individual features of the interaction
process that are of interest.

5.1 Methods for low speed flows

A number of workers have used "boundary integral" techniques ("panel methods") to provide the inviscid element of the computation for both single and multiple aerofoils at low speeds: see for example Ref. 1, Papers 11, 19, 25. For illustrative purposes we have chosen two of these methods.

5.1.1 ONERA method[7,22]

The main elements of this method are as follows:

(i) for the inviscid flow, a panel method due to Neron[22] which uses linearly varying distributions of source and vortex strength. Redundancy of equations is avoided by satisfying internal, as well as external, boundary conditions to give improved accuracy near thin trailing edges;

(ii) for the boundary layers and wake, the entrainment method of Michel et al[27] with further refinements due to La Balleur to increase its relevance to separated flows;

(iii) for the matching, the transpiration model with allowance for wake thickness (but not curvature) effects. Use is made of the semi-inverse technique described in section 4.3, whenever the boundary-layer shape parameter H exceeds about 1.8.

Comparisons are shown in Fig. 13 of some theoretical results of this method with experimental pressure measurements made in a low-speed wind tunnel at ONERA on a model of the NACA 0012 aerofoil; the speed is 40 ms (M_∞ = 0.12) and the Reynolds number is 2×10^6. Fig. 13a ($\alpha = 14°$) gives the results of both inviscid and viscous calculations, showing that viscous effects for this uncambered aerofoil are relatively small (compared with Fig. 2, for example) but nevertheless important. Agreement between theory and experiment is excellent, except just upstream of the trailing edge on the upper surface, where the discrepancy could be explained by the neglect of the second-order pressure correction described in section 3.2.2 (the curvature of the displacement surface, κ^*, will be high in this region). In this example separation is predicted on the upper surface at about 95% chord; in the case shown in Fig. 13b, $\alpha = 16°$, $C_L = 1.5$ (close to its maximum value), the separation has moved forward to 75% chord. Here the discrepancy near the trailing edge is of opposite sign, but remains small; and the divergence of trailing edge pressure which signals the onset of serious separation effects is well predicted.

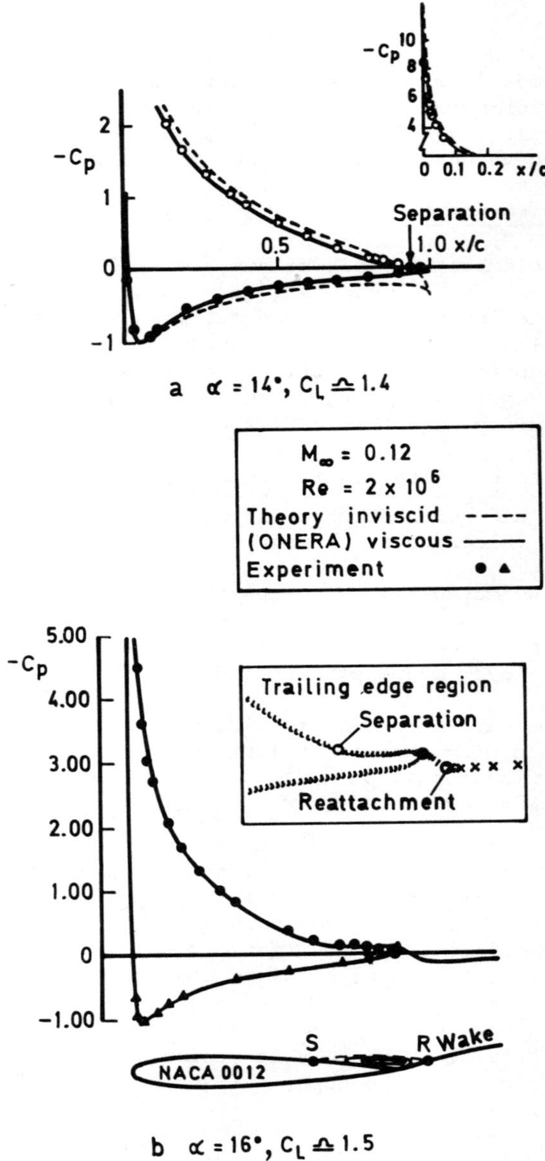

Fig. 13a and b   NACA 0012 aerofoil at low speed

5.1.2 RAE/BAe method 'MAVIS': Butter and Williams[28]

This method, whose acronym stands for 'Multi Aerofoil Viscous Iterative System', was developed to calculate viscous flows over high-lift aerofoil systems, with up to three elements (slat, main aerofoil and flap). Its main constituents are:

(i) for the inviscid flow, the multi-aerofoil panel method of Newling and Butter[29]. Constant source and linearly varying vortex distributions are assumed, arranged so that the strengths are equal on opposing panels on the upper and lower surfaces; these singularity distributions are extended into the wake of each aerofoil element so that both thickness and curvature effects can be included;

(ii) for the boundary layers and wakes (where not affected by the wake of an upstream element), the lag-entrainment method of Green et al[19];

(iii) for confluent boundary layers and wakes (cf Fig. 3), the integral method of Irwin[30]. A sixth order polynomial is used to describe the velocity profile in the mixing region between two shear layers;

(iv) for second-order effects in the matching, the pressure correction given in section 3.2.2 is applied. This is consistent with the use of the wake jump conditions derived in section 3.3.

In its present form, the method operates only in a direct iterative mode (cf section 4.1); thus flows with large separated regions cannot be calculated. The 'bubble' type separations under the 'heel' of a slat or shroud are faired over empirically.

As an example of the use of the method, we show in Fig. 14 results for a three-element aerofoil (similar to that shown in Fig. 3) at an angle of incidence of 18° with a flap defection of 20°; the lift coefficient is 3.9, the Reynolds number 3×10^{6} and the Mach number 0.16. The standard of agreement between theory and experiment is quite good on all three elements (note the different scales). The discrepancy near the leading edge of the slat is unlikely to be due to compressibility effects, since the flow remains well subcritical. On the flap, a separation which is probably spurious was predicted at about 75% chord; this appeared at an early stage in the iterative process and the subsequent 'fixes' failed to remove it during convergence.

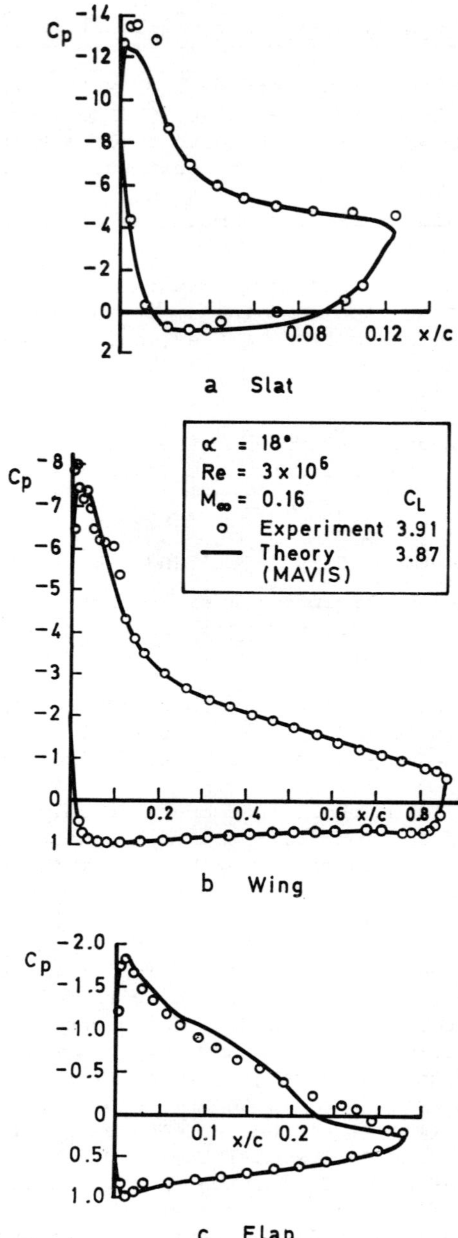

Fig. 14a-c   Comparison with experiment for a three element
aerofoil at 18° incidence

Work is currently in hand to improve the representation and
prediction of separated flows along the lines suggested by
Le Balleur. For this approach to give really satisfactory
results, further experimental information will certainly be
needed about the mean and turbulence structures of such flows,
so that at least a rough estimate can be made of the higher-order
terms in the momentum integral equations (section 3), which may
well be of importance in these situations.

5.2 *Methods for transonic flows*

A large number of methods for the transonic problem have
appeared recently, including even some for multiple (or at least
two-element) aerofoils[31-33]. For present purposes we have chosen
to describe two of them (for single aerofoils) in which some
attempt has been made to allow for the principal higher-order
effects identified in section 3, including in particular the
effects of both thickness and curvature of the wake: namely the
methods of Collyer and Lock[9,26] and of Melnik *et al*[15,16]. Both
these methods rely for their inviscid element on finite-
difference solutions of the full potential (FP) equations using
a coordinate system obtained by conformally mapping the flow
field onto the inside of a circle, as originally suggested by
Sells[34]. Both use the lag-entrainment method of Green *et al*[19]
to calculate the boundary layers and wakes, together with the
transpiration model of the displacement effect, and in neither
is any special treatment attempted of any shock-wave boundary-
layer interactions that may occur. With so many features in
common it might be felt hardly sensible to describe them
separately, but in fact - apart from having been developed
entirely independently - they do have certain important points
of difference, as will appear from the brief account given below.

5.2.1 Method of Collyer and Lock (VGK)

This method was conceived as an extension to viscous flows of
the first satisfactory numerical solution of the full potential
(inviscid) equations, under transonic flow conditions, obtained
by Garabedian and his colleagues in the early 1970s[35,36] (hence
the initials *V*iscous *G*arabedian and *K*orn). In its original
form this involved a treatment of shock waves of non-conservative
(NC) type. A subsequent idea of Murman[37], further developed
by Jameson[38], led to the introduction[39] into the G & K method
of 'quasi-conservative' (QC) differencing at the shock point*
which enabled mass conservation across shock waves to be

---

*The first subsonic point downstream of a shock wave.

(approximately) achieved with a full potential (FP) method in a
simple way, without altering the governing equations to a fully
conservative form. This non-uniqueness of solution with non-
conservative difference schemes has been frequently criticised,
but in the VGK method we have in fact taken advantage of it by
attempting to simulate, in a crude but effective way, both the
position of, and pressure rise through, a shock wave that would
be predicted by an accurate solution of the Euler equations.
The device, which is fully described in Ref. 40, consists of a
simple modification to the finite difference operator at the
shock point involving an arbitrary parameter λ, such that $\lambda = 0$
gives a NC solution, $\lambda = 1$ a QC solution, while a 'partially
conservative' (PC) solution with $\lambda = \frac{1}{4}$ has been found to give
reasonable agreement with several numerical solutions of the

Euler equations obtained recently by Sells[44]. A particularly
relevant (unpublished) example, referring as it does to the
RAE 2822 aerofoil for which comparisons with experiment will be
given next, is shown in Fig. 15. These calculations, at a Mach
number of 0.725, have all been made at the same value, 0.85, of
the lift coefficient C_L. Note first the appreciable discrepancy,

as regards both shock position and strength, between the true
(Euler) solution and either the NC ($\lambda = 0$) or QC ($\lambda = 1$) limits
of the FP solution. Much better agreement is obtained in all
respects with the PC solution ($\lambda = \frac{1}{4}$). But there remains a
discrepancy as regards the angle of incidence required to
achieve this lift coefficient: 2.3° for the Euler solution and
1.9° to 1.75° with the FP solutions. This implies that the
precise value of the lift coefficient (for a given angle of
incidence) predicted by *any* method using the FP equations must
be regarded with suspicion when strong shock waves are present –
and the shock strength in the example, though sufficient to
produce appreciable wave drag ($C_D \simeq 0.004$), would certainly not

be great enough to cause separation in a real flow, so that
this is by no means an extreme case.

Other relevant features of the VGK method include:

(a) in its normal form[26, 41], the only higher order effects
allowed for are those of displacement surface curvature, both
on the aerofoil and in the wake. But the pressure correction
formula[25] is applied *before* entering the boundary layer
calculation, so that p_w (rather than p_{iw} as now proposed) is
used to derive 'u_e' in the first-order momentum integral equation. The effect of various modifications of the type described
in section 3.2 is currently being investigated;

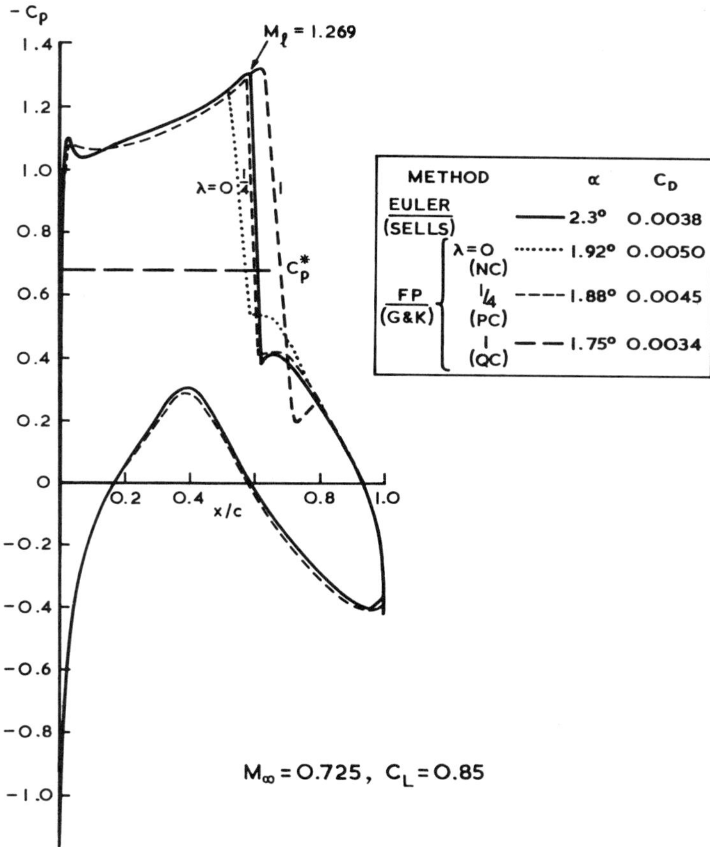

Fig. 15   Inviscid solutions for RAE 2822 aerofoil

(b)   simple smoothing and extrapolation routines (see Ref. 9
for details) are required to define the pressure distribution
in the immediate neighbourhood of the trailing edge, but the
region affected is only from 99% to 101% chord (an interval in
which there are 10 mesh points);

(c)   the boundary conditions that are needed to specify the
equivalent inviscid flow in the wake (section 3.3) are applied,
not as they should be on the dividing streamline of the real
viscous flow, but on the coordinate line ($\theta = 0$) stemming from
the trailing edge as a result of the conformal mapping.   As
indicated in Fig. 16, this approximation to the shape of the
wake should be satisfactory in the region just downstream of the
trailing edge, from which comes the main part of the 'wake
effect'.

(d)   the drag coefficient (C_D) can be calculated in two different ways (for subcritical flow): first, from the value of the momentum thickness far downstream in the wake ($C_{D_\infty} = 2\theta_\infty/c$), and secondly by adding the separate components due to pressure and skin friction ($C_{D_T} = C_{D_P} + C_{D_f}$). With the VGK method, excellent agreement (to better than 5%) has usually been found between these two alternative values, and since any discrepancy appears insensitive to M_∞ or α a suitable correction can easily be applied in supercritical flow, where the value C_{D_T} must of course be used to allow for the presence of wave drag. Thus we usually take

$$C_D = C_{D_T} + \left(C_{D_\infty} - C_{D_T}\right)_{\text{subcritical}} .$$

(e)   the method employs a direct iterative procedure (see section 4.1), and at present constant relaxation factors are used, normally 0.15 with a coarse (80 × 15) grid and 0.075 with a fine (160 × 30) grid. As would be expected from the analysis of section 4.2, lower values are found necessary for flows approaching separation, where the method eventually breaks down because of failure to converge; the use of a semi-inverse procedure (section 4.3) is therefore being explored.

Before we go on to give some comparisons with experiment, it seems appropriate to recapitulate, in Fig. 16, the boundary conditions that are used in the VGK method for the calculation of the equivalent inviscid flow, and the pressure corrections that are applied to the inviscid values.

For illustrative purposes we have chosen to look at three aerofoils typical of modern designs for application to transport aircraft. The first, RAE 2822, has already been mentioned; it is 12% thick and has a moderate degree of rear camber and hence of rear loading. The measurements of pressure shown in Fig. 17 were made in the 8ft × 6ft wind tunnel as RAE[42] at a Mach number of 0.73 and a Reynolds number of 6.5×10^6; transition was fixed by bands of ballotini at $x/c = 0.03$. Comparisons with theory are shown for two values of C_L, 0.66 (case 7 of Ref. 42, Fig. 17a) and 0.79 (case 8, Fig. 17b); the corresponding values for the angle of incidence in the experiment (α_g) (uncorrected for wall interference) are 2.5° and 3.2°. In both cases the agreement between theory and experiment is excellent in most respects. The only appreciable discrepancies are: (a) on the upper surface near the leading edge, probably caused by the roughness band used to fix transition (most marked when the flow

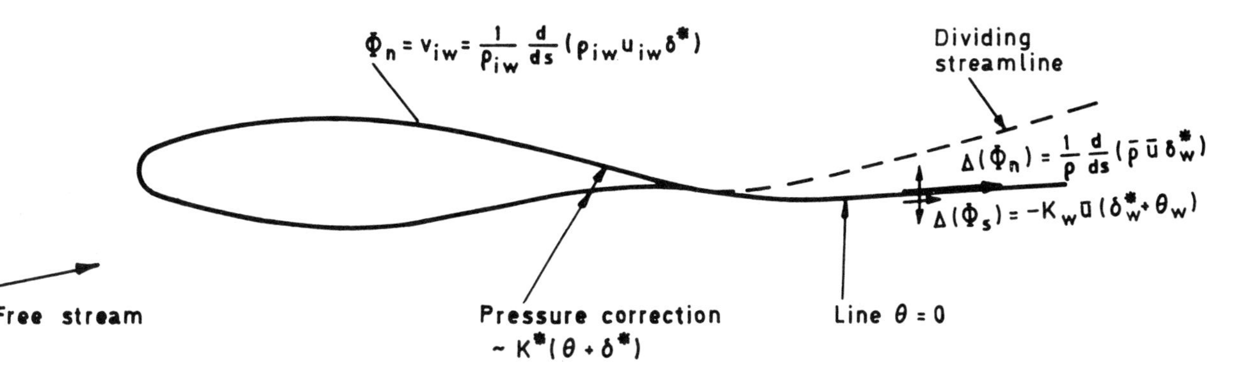

Fig. 16 Boundary conditions and corrections for EIF

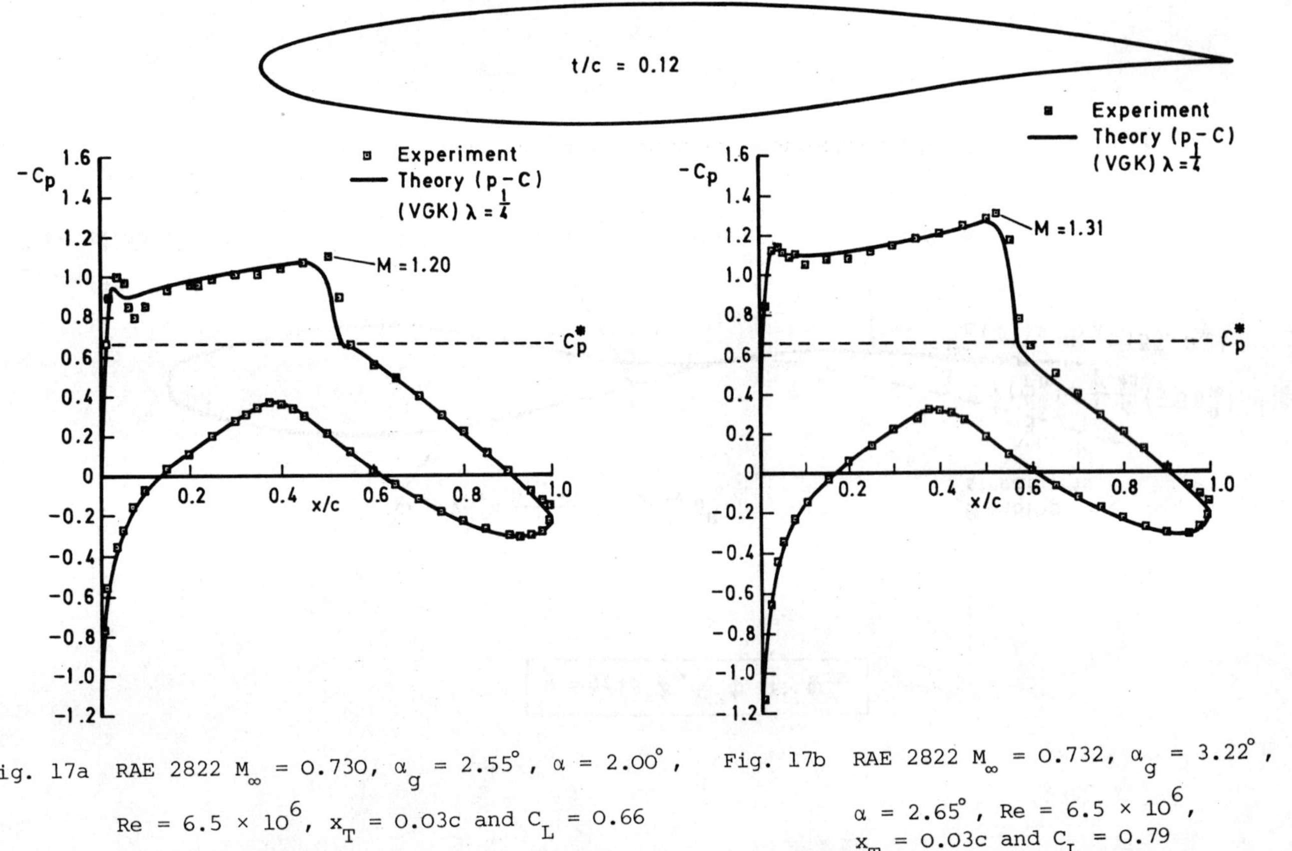

Fig. 17a RAE 2822 M_∞ = 0.730, α_g = 2.55°, α = 2.00°, Re = 6.5 × 10^6, x_T = 0.03c and C_L = 0.66

Fig. 17b RAE 2822 M_∞ = 0.732, α_g = 3.22°, α = 2.65°, Re = 6.5 × 10^6, x_T = 0.03c and C_L = 0.79

is locally just supersonic, Fig. 17a); (b) at the foot of the
shock wave in Fig. 17b, suggesting that the crude modelling of
the interaction is slightly inadequate with this rather strong
shock (M $\simeq$ 1.31); and (c) on the upper surface just ahead of
the trailing edge, where the error is quite small ($\Delta C_p \sim$ 0.05)
in this case but is indicative of probable trouble when
separation is approached more closely (see later).

In this experiment a number of detailed measurements of the
boundary layer were made, allowing a comparison with the
theoretical predictions. Such a comparison is given in Fig. 18
for the first of the two cases (C_L = 0.66), showing that
excellent agreement is obtained as regards the momentum thickness θ, while the rapid growth of the displacement thickness $\delta*$
(and hence of the shape factor $\bar{H}$) towards the trailing edge is
only slightly underestimated. For this aerofoil the agreement
between theory and experiment is also excellent as regards
drag (Fig. 19). Perhaps the least satisfactory feature of the
assessment concerns the prediction of lift, mainly because of the
uncertainty in our knowledge of the corrections that should be
applied for wall-induced upwash in a slotted tunnel such as this.
In the present case the difference between the measured (uncorrected) and theoretical values of α, for given C_L, is about 0.5°
(see Fig. 17), and it is probable that at least half of this
discrepancy can be ascribed to tunnel interference (see further
discussion in section 5.2.2).

We turn now to some results for a so-called 'supercritical'
aerofoil, RAE 5217, which is thicker (t/c = 0.14) and has
substantially more rear loading than RAE 2822, so that for
similar combinations of Mach number and lift coefficient the
boundary layers on both surfaces are subject to more severe
pressure gradients. The experimental measurements were made in
the same wind tunnel (RAE 8ft × 6ft) and at the same Reynolds
number (6 × 10^6).

A general survey of the results, compiled from Ref. 43, is
shown in Fig. 20. Cases (a), (b) and (e) cover a Mach number
range from 0.6 to 0.76 at a constant value of the (uncorrected)
angle of incidence α_g, while cases (c) and (d) are for M_∞ = 0.73
with C_L = 0.79 and 0.03 respectively. The overall impression
is of good agreement between theory and experiment, particularly
as regards position and strength of the shock waves on either
surface, for a wide range of conditions. There is however a
general trend – as noted already for the RAE 2822 aerofoil – for
the pressure to be overestimated towards the end of a long
adverse pressure gradient, wherever the predicted value of the
shape parameter $\bar{H}$ exceeds about 1.8. This is known to be

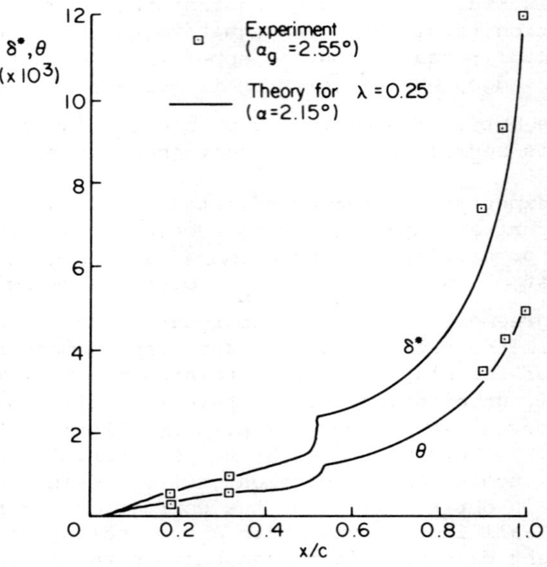

a   Displacement (δ^*) and momentum thickness (θ)

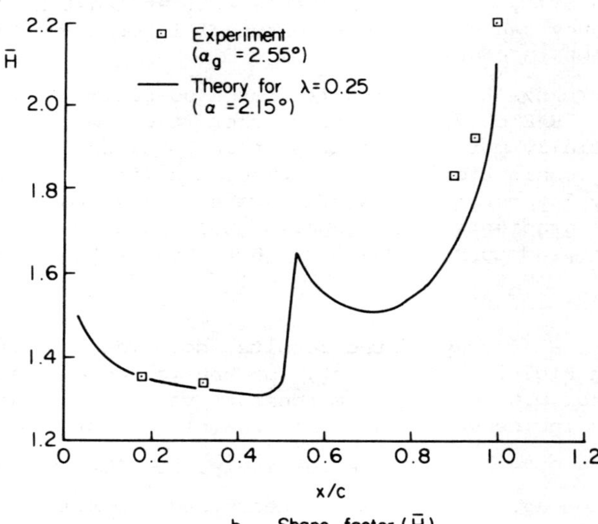

b   Shape factor ($\bar{H}$)

Fig. 18a and b   RAE. 2822 Boundary layer development on the upper surface for $M_\infty = 0.73$, $Re = 6.5 \times 10^6$ and $x_T = 0.03c$

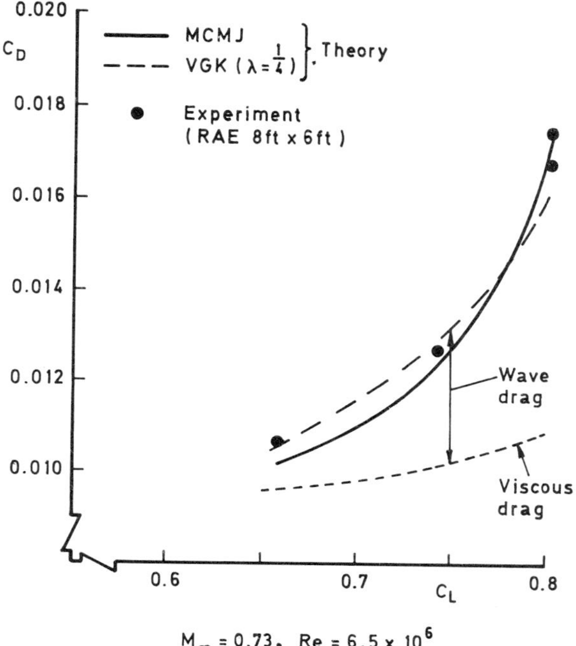

Fig. 19   RAE 2822 – Variation of C_D with C_L

associated with the tendency of the VGK method (in its normal
form) to underestimate the growth of the boundary layer as
separation is approached.

As a result (Fig. 21), the overall drag of this aerofoil is
slightly underestimated, although the general trend of the
variation of drag with Mach number, and in particular the rapid
drag rise at $M_\infty = 0.75$, is very well predicted. Fig. 21 also
includes some calculated results for a representative full scale
Reynolds number, 50×10^6, which suggest that the drag 'creep'
(the slow rise that occurs before the eventual rapid rise)
should be appreciably less under flight conditions than that
measured at Re $= 6 \times 10^6$; and under these circumstances the
predicted values of $\bar{H}$ are considerably reduced so that the
method should be correspondingly more reliable.

Confirmation for this statement can be gained by examining
the results for a third aerofoil, RAE 5225, of the same thickness and similar in other respects to RAE 5217, a model of
which has recently been tested in the RAE 8ft × 8ft wind tunnel

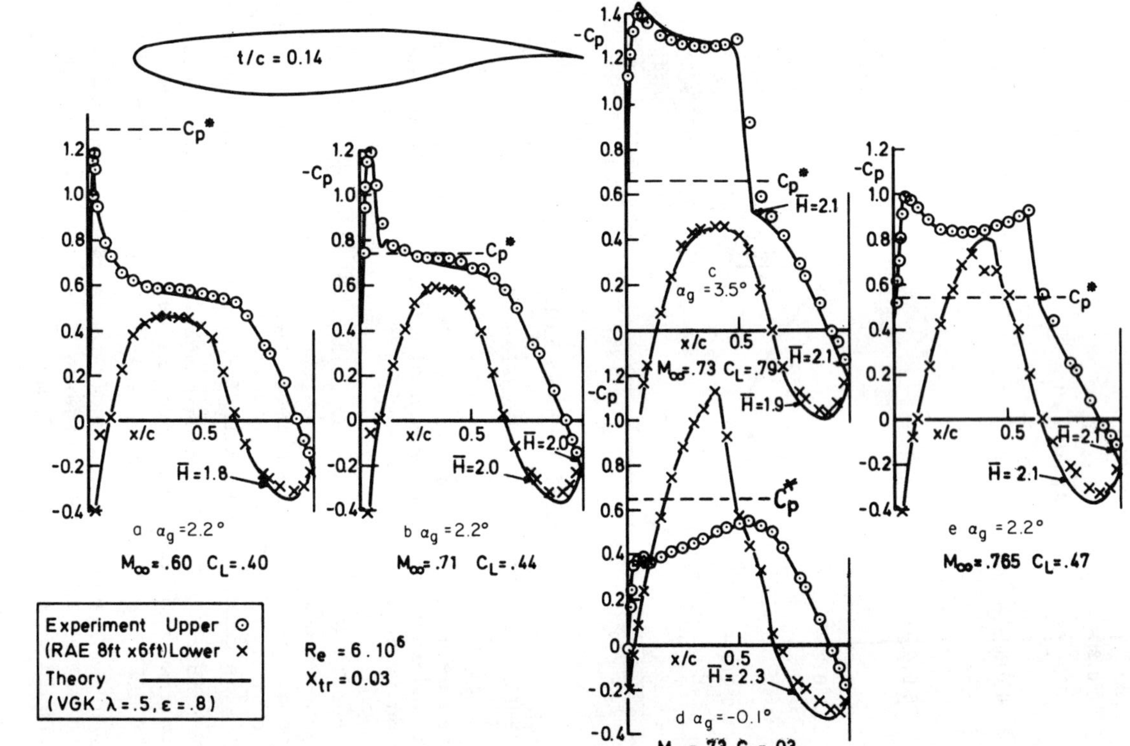

Fig. 20 a – e RAE 5217 – General comparison between theory and experiment

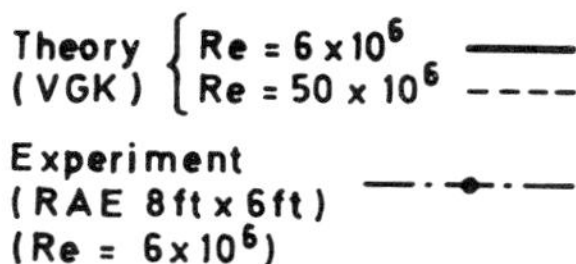

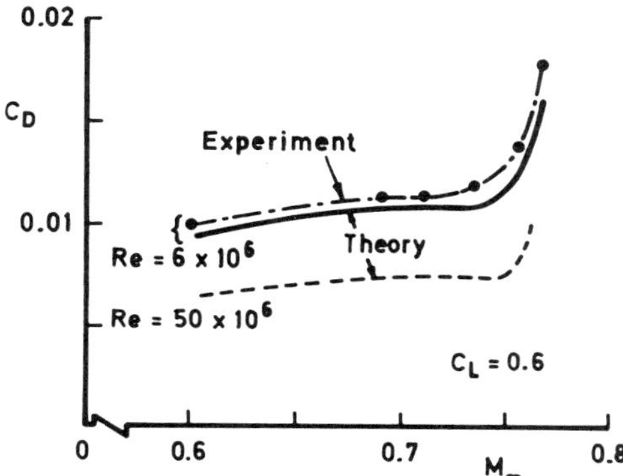

Fig. 21   RAE 5217 - Variation of C_D with M_∞

at Reynolds numbers of 6 and 20 million [47]. These tests have
the additional advantage that, because they were made with solid
walls, a new technique due to Ashill [47] can be used which allows
a more accurate assessment of wind tunnel interference to be
made than is at present possible with slotted walls. The
corrections to be applied comprise changes in the measured
values of M_∞ and α, as well as to the effective camber of the
aerofoil. A comparison between results from the VGK method and
the experimental measurements, for the higher Reynolds number
(20×10^6), is given in Fig. 22. Although the adverse pressure
gradients on the lower surface are just as severe as for
RAE 5217 (cf Fig. 20), the agreement between calculated and
measured pressures is now excellent, at this high Reynolds
number, as indeed it is also on the upper surface, except very
close to the trailing edge. The effect of the induced camber
(compare full and dashed lines) is seen to be negligible for
subcritical, and small for supercritical, cases. The value of
C_D is still under-estimated by about 0.0005, but this appears
to be almost independent of Mach number. The discrepancy
between the theoretical and measured (corrected) values of α is
also approximately constant at about $0.3°$, and of this about
half is thought to be due to the effects of the boundary layers
on the side walls so that the remaining discrepancy is likely
to be very small (say about $0.1°$).

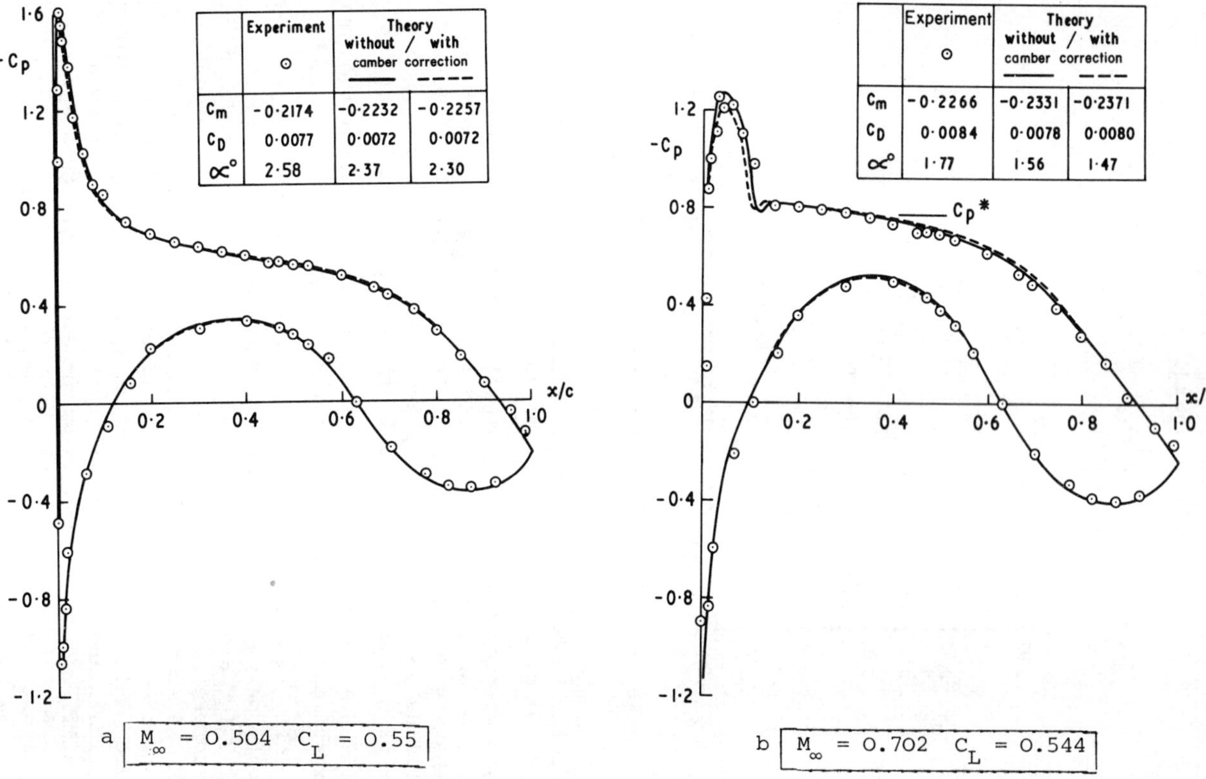

Fig. 22 a and b RAE 5225 Pressure distributions : comparison between VGK theory and measurement.
Re = 20 × 10^6

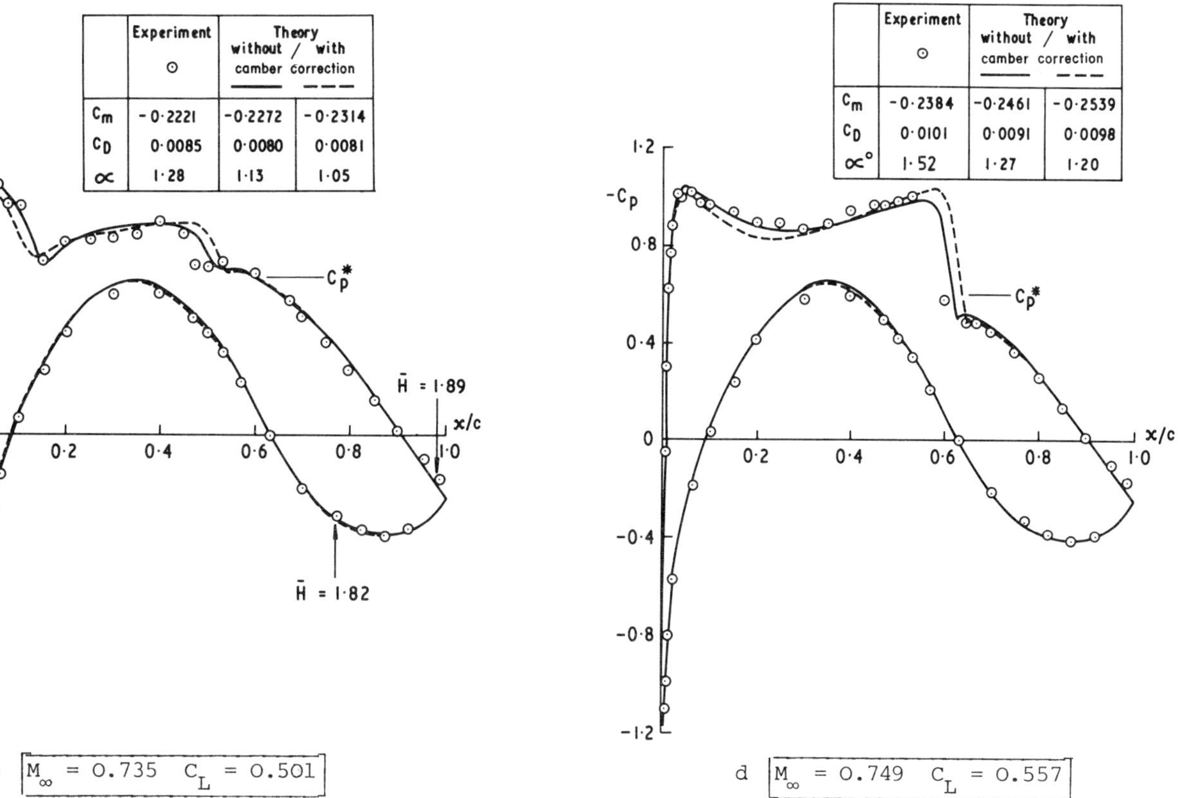

Fig. 22 c and d RAE 5225 Pressure distributions : comparison between VGK theory and measurement

$Re = 20 \times 10^6$

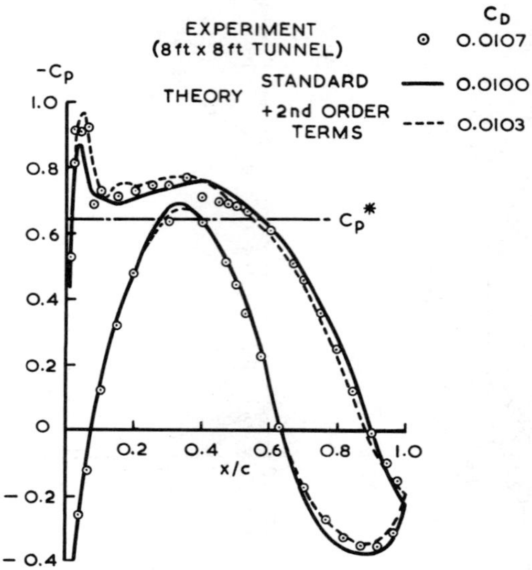

Fig. 23 RAE 5225 : Comparison between VGK theory and measurement: $Re = 6 \times 10^6$

At the lower Reynolds number, 6×10^6 (Fig. 23), the defect noted previously on the lower surface reappears, but the inclusion of second order effects in the boundary layer theory and matching procedure, along the lines discussed in section 3, appears to improve the situation (compare full and dashed lines), while the discrepancy in drag has been reduced from 7% to 4%.

5.2.2 Method of Melnik *et al* (MCMJ)

The most recent and most advanced method for aerofoils in viscous transonic flow has been developed over the past five years by Melnik and his colleagues at the Grumman Aircraft Company; it is known by their initials MCMJ[16]. As mentioned previously, it has a large degree of commonality with the RAE (VGK) method described in section 5.2.1. The main points of difference are

(a) the method for calculating the equivalent inviscid flows involves two options:

(i) a non-conservative (NC) version in which the 'quasi-linear' form of the partial differential equation, namely

$$a^2 \nabla^2 \phi = \underset{\sim}{u} \ \text{grad}\left(\tfrac{1}{2}u^2\right)$$

with

$$\underset{\sim}{u} = \text{grad} \ \phi$$

and

$$a^2 = M_\infty^{-2} + \tfrac{1}{2}(\gamma - 1)\left(1 - \underset{\sim}{u}^2\right),$$

is solved (as in the original G and K method) without special treatment of the shock point; or

(ii)   a fully conservative (FC) version in which the equation to be solved is written explicitly in conservation form, namely

$$\text{div}(\rho \underset{\sim}{u}) = 0$$

with

$$\rho = \left(1 - \tfrac{1}{2}(\gamma - 1)M^2 u^2\right)^{\frac{1}{\gamma - 1}} .$$

The first option corresponds to taking $\lambda = 0$ in the VGK method and the second to taking $\lambda = 1$ (insofar as a QC method is capable of simulating a FC solution). The degree of flexibility which is present in the VGK method is therefore absent - which may or may not be considered desirable. In fact, the FC version is always recommended, with considerable experimental support as we shall see shortly;

(b)   a special treatment is introduced of the 'strong' inter-action process in the trailing edge region, as described briefly in section 3.4 and fully in Refs. 12, 15 and 16 (the analogous treatment of the shock-wave boundary layer interaction [45] is not at present included). This procedure has obvious advantages, particularly as regards the removal of the need for smoothing near the trailing edge and consequent degree of arbitrariness that is present in all other methods. But it does in effect single out the trailing edge as the only region where second-order effects are likely to be of importance, and as we have seen in section 5.2.1 this may not be entirely justified in view of the problems encountered with the VGK method on heavily rear-loaded aerofoils near the 'cove' on the lower surface, which would surely occur also with the MCMJ method.

In this method, the authors did not however consider it appropriate to introduce any special localised treatment (such as their own[45]) for the interaction region at the foot of the shock. Although such a procedure has in fact been used with success recently by Stanewsky and Inger[55], it is not yet clear whether this is really necessary, at least for attached flows.

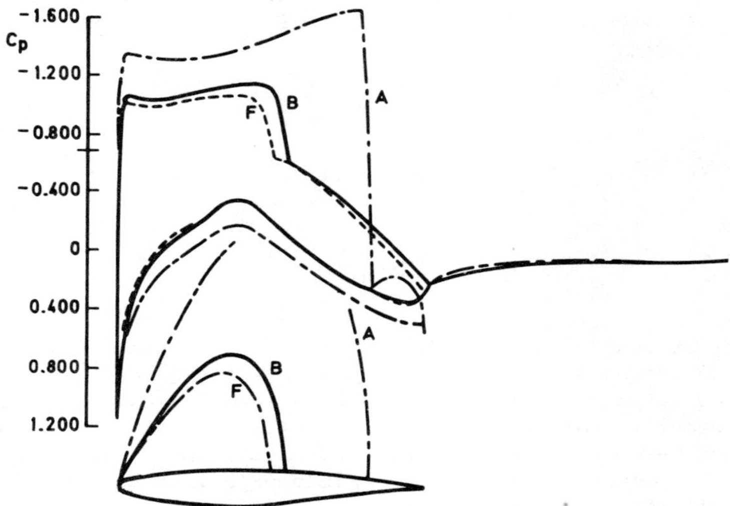

Fig. 24   RAE 2822 -Effect of the wake and trailing edge corrections on the theoretical pressure distribution at $M_\infty = 0.725$, $\alpha = 2.3^\circ$, Re $= 6.5 \times 10^6$ ($X_T = 0.03$)

To illustrate the use of the method we will first discuss some results for the aerofoil RAE 2822 considered previously, taken from Ref. 16. An interesting indication of the magnitude of the viscous effects is provided by Fig. 24, which shows pressure distributions for $M_\infty = 0.725$, $\alpha = 2.3^\circ$ calculated (A) for inviscid flow, (B) with full viscous effects and (F) omitting all effects of the wake (and trailing edge corrections), the last of which is seen to have an appreciable effect on the shock position as well as on the pressure gradient approaching the

trailing edge. The following table gives the values of the lift
coefficient C_L for these three calculations together with one
other (C) in which the effect of the thickness of the wake, but
not its curvature, is included:

A	C	B	F
1.30*	0.76	0.73	0.66

With the viscous solutions, the lowest value of C_L is obtained
(F) with all wake effects omitted. The subsequent inclusion of
wake thickness effects (C) increases the lift because the upper
surface boundary layer is affected more than the lower by a
symmetrical change in pressure; while the final inclusion (B) of
the wake curvature effect naturally reduces the lift once more.
The VGK method predicts a value of 0.72 for this case, in good
agreement with the preferred result (B) above.

   Comparisons with the RAE experimental results are shown in
Fig. 25. The first example (case 7 of Ref. 42, Fig. 25a) is
the same as that of Fig. 17a, and the theoretical pressure
distributions are almost identical; the MCMJ method predicts a
slightly further aft position for the shock wave, as would be
expected since it is fully conservative, and is also in better
agreement with the measured pressures just ahead of the trailing
edge. In the second example (Fig. 25b, case 9 of Ref. 42),
almost identical to case 8, Fig. 17b) the agreement between FC
theory and experiment is again excellent, though the pressure
level between the foot of the shock wave and the trailing edge
is now slightly over-estimated. Note that in both cases the
use of the NC option gives inferior results in the neighbourhood
of the shock wave.

   Turning now to the overall forces, we see from Fig. 19 (p.393
above) that the values of C_D are in good agreement both with the
calculations by the VGK method and with the experimental
measurements. The corresponding picture for the variation of
C_L with α is provided by Fig. 26. This shows that the
predictions of the two methods are in reasonable agreement, with
a small discrepancy that increases with incidence so that when
$C_L = 0.8$ the values of α (for a given value of C_L) differ by
about $0.1°$, as would be expected from a consideration of the
different nature (FC and PC) of the inviscid elements (compare

---

* Note however that a solution by Sells of the Euler equations
(see Fig. 15) gives $C_L \simeq 0.85$ for this case.

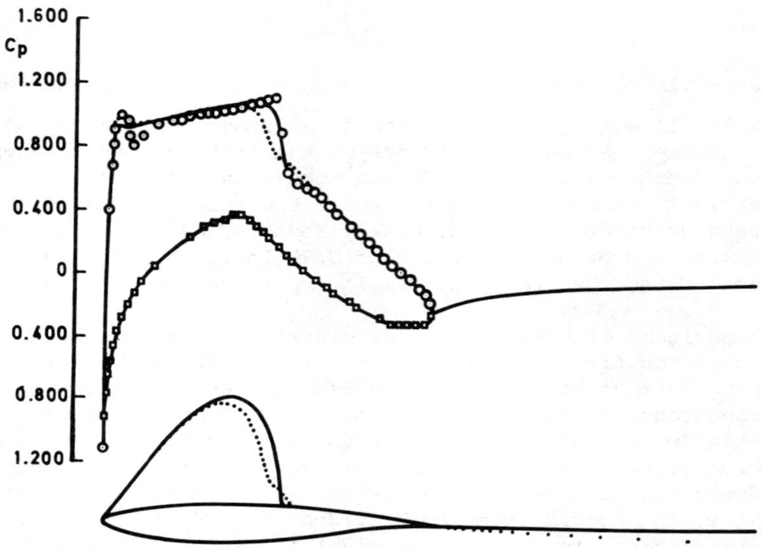

Fig. 25a   RAE 2822 – Pressure distribution at $M_\infty = 0.728$, $C_L = 0.658$, Re $= 6.5 \times 10^6$ ($X_T = 0.03$)

Fig. 15 which refers to inviscid calculations under similar conditions). The uncorrected experimental results differ from the theory by about $0.5°$. There is some uncertainty as regards the correction that should be applied to account for the constraint effect of the slotted walls (1.6% open area ratio); but plausible values* are shown by the arrows in Fig. 26, which suggest that the remaining discrepancy between theory and experiment may be very small.

---

* Obtained by using 'classical' interference theory[46] and assuming that the value of the porosity parameter β/P (3.0) is consistent with the blockage correction ($\Delta M_\infty \simeq 0.004$) which has

been applied in making the pressure comparisons of Figs. 17 and 23. This is consistent with the recommendations made in Ref. 42.

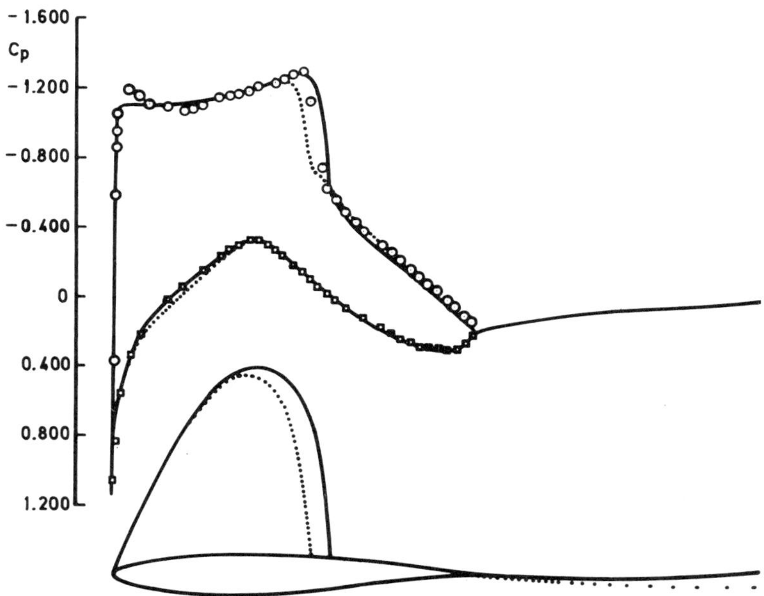

Fig. 25b   RAE 2822 – Pressure distribution at $M_\infty = 0.733$, $C_L = 0.803$, $Re = 6.5 \times 10^6$ ($X_T = 0.03$)

A final example is provided by an aerofoil, designed by Garabedian and his colleagues[36] and known as GK1, a model of which has been tested in the 60in × 15in wind tunnel at NAE, Ottawa[48] at a Reynolds number of 21 million. The aerofoil is 11.5% thick and has only moderate rear camber so that the adverse pressure gradients on the lower (and indeed also on the upper) surface are relatively mild. Two cases have been selected (Fig. 27) in which the shock wave on the upper surface is closer to the leading edge than in most of the examples considered previously. As a result, there is in this case a long region of near-sonic flow downstream of the shock, always a sensitive situation, but the agreement between theory and experiment is nevertheless good, both as regards the pressure rise through

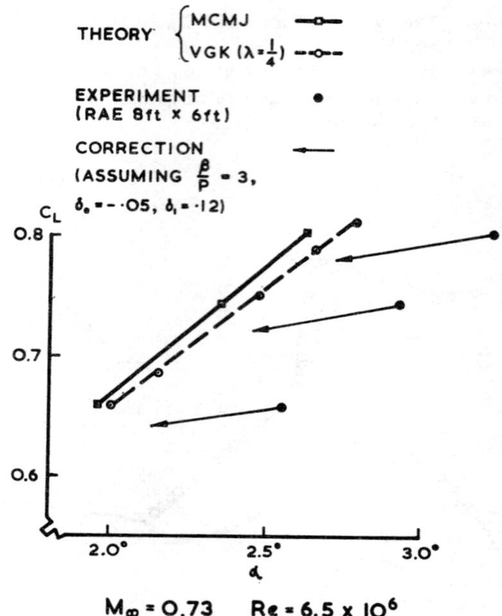

Fig. 26   RAE 2822 : Variation of C_L with α

the shock and the subsequent pressure level - and indeed the whole standard of the prediction of pressure is excellent. The drag, however, is appreciably underestimated, a trend which is confirmed by the drag 'polar' for a higher Mach number, 0.75 shown in Fig. 28. This also contains a curve giving results by the VGK method, which is in slightly better agreement with experiment. The reason for this discrepancy - which persists even in cases where the flow is subcritical - is not yet clear.

The corresponding lift curves are shown in Fig. 29. The two results (MCMJ and VGK) give results in close agreement both with each other and with the corrected measurements, but it must be emphasised that there is again a degree of uncertainty as regards the size of the correction that should be applied.

6.   CALCULATION METHODS FOR WINGS

In this section we shall consider some of the practical methods that are currently available for wings and wing/fuselage combinations. In all the methods described the same concepts are applied to wings as we have already seen to have worked well for aerofoils. The viscous effects are normally neglected entirely for the flow over any part of the structure other than

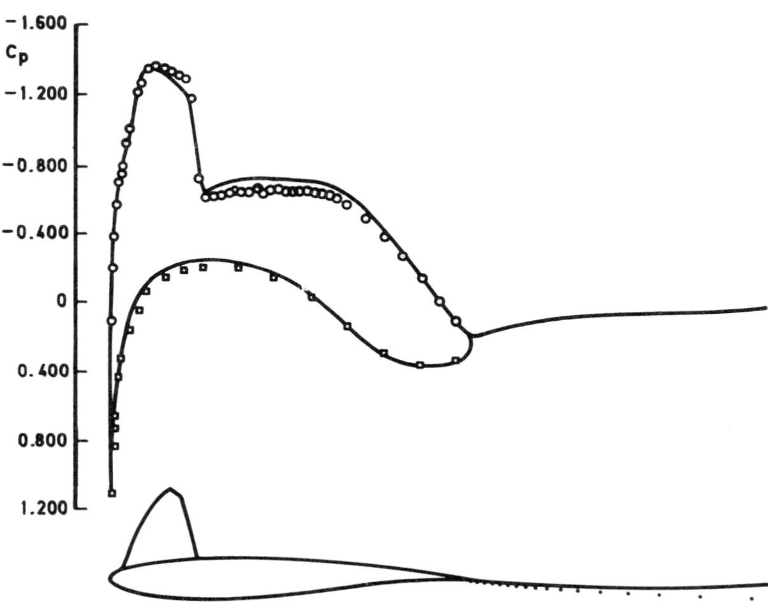

Fig. 27a   GK1 – Pressure distribution at $M_\infty = 0.699$, $C_L = 0.669$, $Re = 21.5 \times 10^6$ ($X_T = 0.10$)

the wing and associated wake. As mentioned in section 3.5, for fully attached flows there are unlikely to be any conceptual reasons, other than the difficulty in modelling the flow in the region of a surface discontinuity, such as the wing-fuselage junction (and wing tip), why the viscous effects cannot be included for all the wetted surfaces, but so far this does not appear to have been done.

6.1 *Methods for low speed flows*

At low speeds, attempts have been made, even for highly separated flows on wings, see Ref. 1, Paper 31, for example, to model the displacement surface for the wing using a 'panel' method with a free vortex sheet at the free-shear layers between the separated flow region and the external flow. This

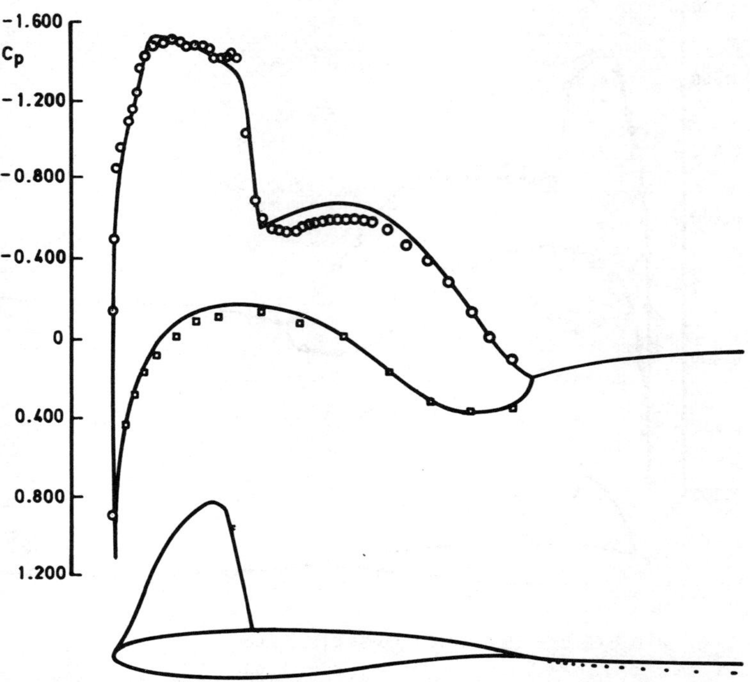

Fig. 27b   GK1 - Pressure distribution at $M_\infty = 0.691$, $C_L = 0.821$, Re $= 21.5 \times 10^6$ ($X_T = 0.10$)

method appears to give results which feature some of the main characteristics of the flow, and the predictions are surprisingly good for the examples presented in spite of the rather empirical approach and conceptual problems. For illustrative purposes we have however restricted ourselves to fully attached flows and have chosen to consider results from one of the first methods into which corrections were included for boundary-layer effects.

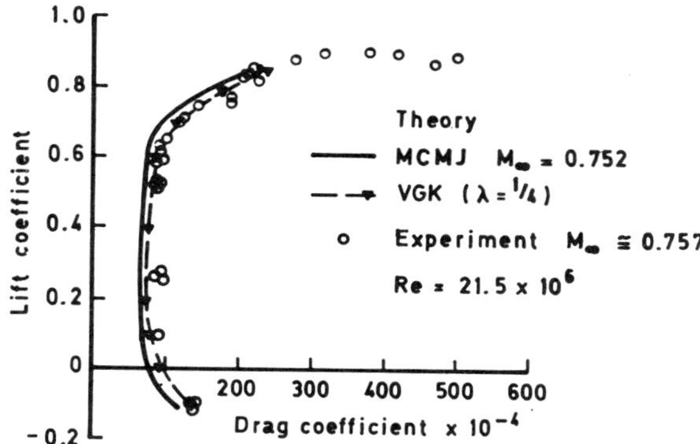

Fig. 28 Drag polar for the G & K 75-06-12 aerofoil (GK1)

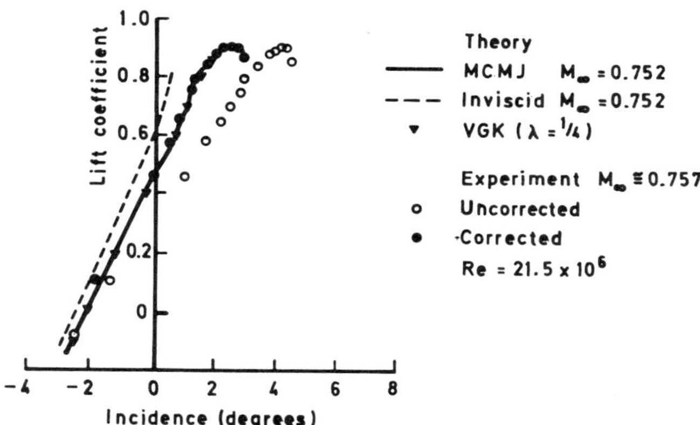

Fig. 29 Lift coefficient for the G&K 75-06-12 aerofoil (GK1)

6.1.1 BAe Spline-Neumann method[50,51]

The main elements of this method are as follows:

(i) for the inviscid flow, a panel method due to Roberts and Rundle[50] is used. In this the novel feature is the use of a double cubic spline system for defining both the shape of the panels and the strength of the fundamental singularity modes. It is expected that considerably greater accuracy will be achieved, for a given number of panels, than with a standard 'plane facet' method, but at the expense of a large increase in computing time;

(ii)   for the boundary layers, the three-dimensional integral method of P.D. Smith[25], based on the entrainment principle, is used. In its original form no account was taken of the 'lag' effect (cf Ref. 19) (although this has now been superseded by a version which does so). For this early (1977) application of the method, no corrections were made for the effect of the wake, but there is no longer any reason why this should not be included;

(iii)   for the matching, the transpiration model is used, with allowance for thickness effects (but not curvature) over the wing only[51].

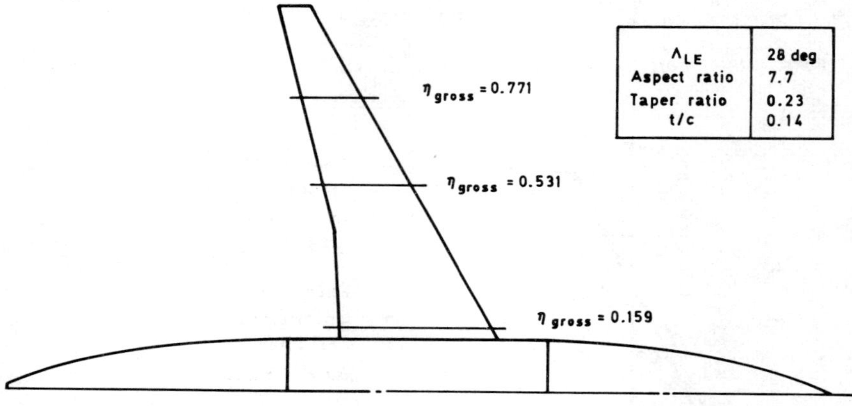

Fig. 30   Wing/body (1) as tested in RAE 8ft × 8ft tunnel

The main details of the wing/fuselage combination considered are given in Fig. 30. The configuration, W/B(1), has a planform as shown, with a trailing edge crank; the leading edge sweep is about 28°, and the aspect ratio is 7.7. The wing is mounted on a circular body in the low position, with fillets to fair the wing to the body over the rear part of the wing. Comparisons are shown in Fig. 31 of some theoretical results from the method with experimental measurements made in the RAE 8ft × 8ft tunnel. The results are for a flow that is fully subcritical ($M_\infty = 0.6$)

and corrections are applied to the results from the panel method to account for compressibiltiy. Boundary layer transition was fixed at 5% chord on both wing surfaces. Calculated results are given for both inviscid and viscous flow, showing that the effect of the viscous corrections can be quite large even for this subcritical example.

The Reynolds number of the wind tunnel tests (3×10^6) is rather low, and this, in combination with a comparatively large

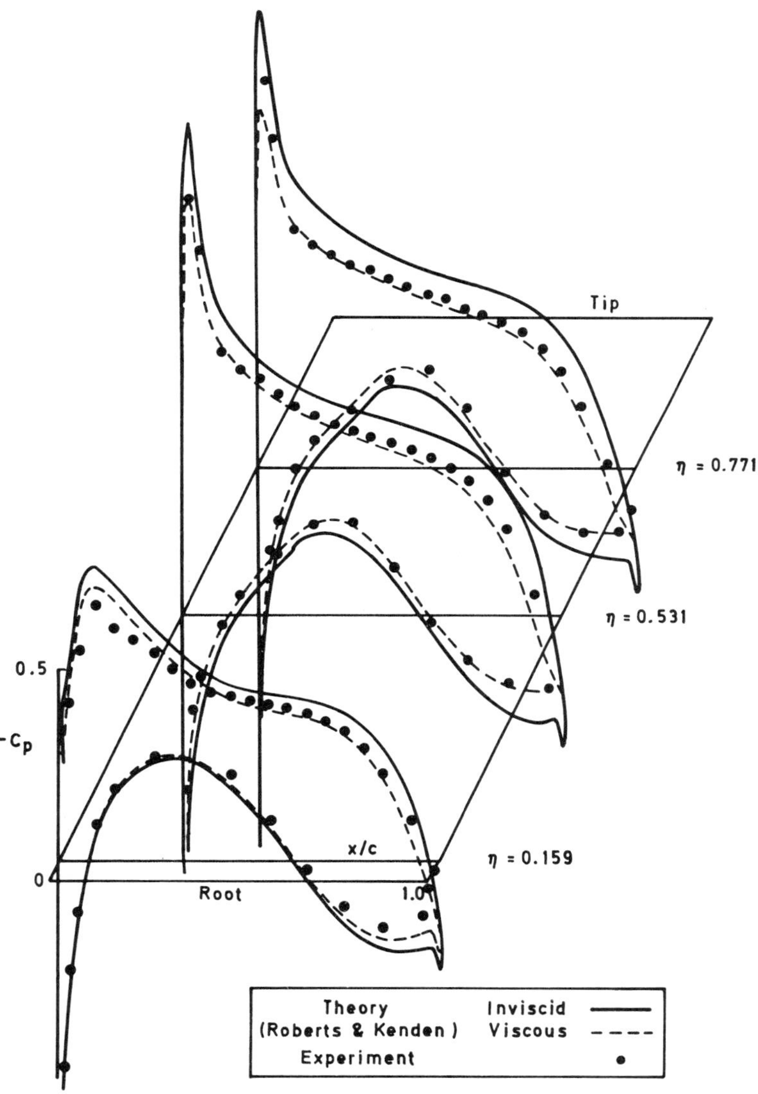

Fig. 31 Distribution of pressure coefficient wing/body (1)
$M_\infty = 0.6$, $Re_{\overline{C}} = 2.85 \times 10^6$, $\alpha_w = 0.3^\circ$

rear loading, results in very significant viscous corrections.
The measurements are in fair agreement with the predictions,
and the remaining differences are of a size likely to be caused
by neglecting the wake thickness effects and the second order
terms which have been shown to be important in the work on
aerofoils (section 5.2.1).

6.2 *Methods for transonic flows*

So far, even for transonic flows with attached boundary
layers, there appear to be no published results from any full
potential (FP) method,[*] for the flow over the wings of a wing/
fuselage combination in which corrections have been made
iteratively for viscous effects using a fully three-dimensional
boundary-layer method, to anything like a comparable standard
to the methods already described in section 5 for aerofoils.
The reasons for this appear to be fourfold;

(i) The full potential (FP) methods have not been available
for as long as the equivalent methods for aerofoils.

(ii) The three-dimensional boundary-layer methods are expensive to run (and this is especially so if a differential method
is used), and of course do not yet include the higher order
terms mentioned in section 3.

(iii) The three-dimensional boundary-layer methods are
susceptible to local disturbances in the pressure distribution
which can cause the calculation to fail even for quite modest
spanwise pressure gradients. Boundary layers are highly sensitive to such gradients but the cause of failure appears to be
connected with the ability to satisfy the domain of dependence
principle. This is a particularly difficult requirement to
satisfy for iterative schemes where unrealistic pressure distributions can occur prior to convergence.

(iv) It is difficult to provide suitable starting conditions
for the boundary layer and wake at the 'side' edges near the
root and tip of the wing, because the flow is not adequately
modelled in such regions when the local flow is entering the
computational region.

Details of several less complete methods were reviewed by the
first author[8], and so we do not propose to do so again here.
From the evidence of that review it appears that the difficulties
outlined can be overcome and we should soon have results from
working methods using a FP method and achieving adequate
convergence. In fact the method proposed by the second
author[52], which uses a TSP scheme rather than an FP one for the

---

[*] (Note added in proof): but see Ref. 56. (issued in JUNE 1981).

inviscid flow and does not include the second order matching or boundary layer terms, gives results which indicate that good predictions can be obtained, and that convergence of the scheme can be achieved in a similar way to that for aerofoils, provided certain precautions are taken to avoid the difficulties mentioned above.

6.2.1   RAE Mk 4W VISTRAN method[52]

The main elements of this method are as follows:

(i)   for the inviscid flow, a TSP method due to Albone, Hall and Joyce[53] is modified to include viscous effects by allowing for a change in the boundary conditions, which are applied in the plane z = O. The boundary conditions are modified for points representing the wing and the wake;

(ii)   for the boundary layers and wake, a 'lag' entrainment integral method is used. This is a three-dimensional turbulent boundary-layer method using a scheme of solution devised by P.D. Smith[25], with the main elements the same as for the 'lag' method[19] used in the VGK program (section 5.2.1) and applied along the external streamlines. It is assumed that the cross-flow may be represented by Mager profiles[54];

(iii)   for the matching, changing the boundary conditions as specified is equivalent to either the transpiration or the solid displacement method to this order of accuracy.

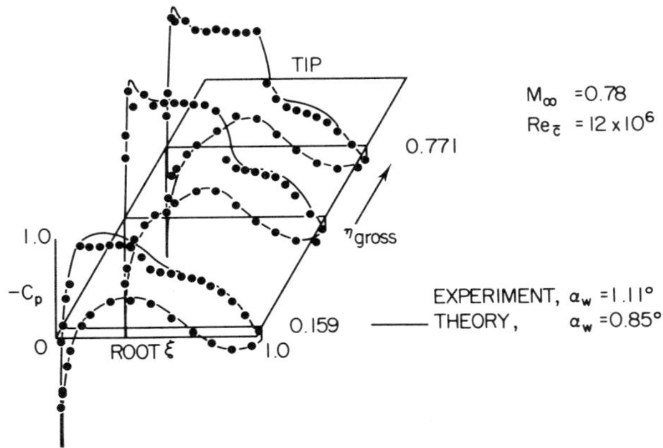

Fig. 32   Distribution of pressure coefficient, wing/body (1), M_∞ = 0.78, $Re_{\bar{c}}$ = 12 × 10⁶

Comparisons are shown in Fig. 32, taken from (52), between some calculated results obtained from the method and measurements from the same series of tests as for the example for subcritical flow (Fig. 31). For this example the flow has a strong shock-wave over the outer part of the wing, just upstream of which a local Mach number of 1.40 is reached near the tip. Although the shockwave is likely to be slightly swept in this region the boundary layer must be close to separation, with the calculated streamwise transformed shape factor reaching a maximum of 2.3 just downstream of the shockwave. Although the results of this comparison are reasonably impressive it is difficult to know whether the remaining differences are due to a failure to model the inviscid flow adequately in the TSP method, to a failure to include details of the fairing at the wing-fuselage junction and of the aeroelastic twist in the specification of the shape of the model, or to inadequacies in the scheme for including viscous effects. Judging by results obtained recently for aerofoils using more advanced matching schemes (section 5.2), it is likely to be a combination of all these causes, although for this example the boundary layers are not particularly close to separation on either surface (except near the tip). We must await results from a FP method to confirm this view.

In the case of W/B(1) the leading edge sweep (28°) is fairly low and the strength of the shock wave does not vary much across the span. As a final example, we consider some results for a further wing-body combination, W/B(3), representative of a long-range transport aircraft, with higher sweep and aspect ratio, on which the flow is more highly three-dimensional and the boundary layers are closer to separation, both on the upper surface near the trailing edge and on the lower surface at the end of the pressure rise (near 80% chord). Details of the model are shown in Fig. 33. The wing is low mounted and has cranked leading and trailing edges. The leading edge sweep is 40° inboard of the crank station and 34.4° outboard, the aspect ratio is 9.0 and taper ratio 0.22. Results are given at $M_\infty = 0.85$, $Re_{\bar{c}} = 3.7 \times 10^6$, for both measured and calculated pressure distributions at eight spanwise stations. On the upper surface the pressure distribution is markedly three-dimensional in form with quite a strong shock wave ($M_{L_{max}} \doteqdot 1.30$) over part of the span. The agreement shown was only achieved by studying effects which were neglected in the previous comparison (modelling of the fillets at the fuselage side, and including aeroelastic twist) and also correcting the inviscid method for finite body length effects.

It is worth perhaps noting that these corrections will still need to be applied when considering more advanced methods, in

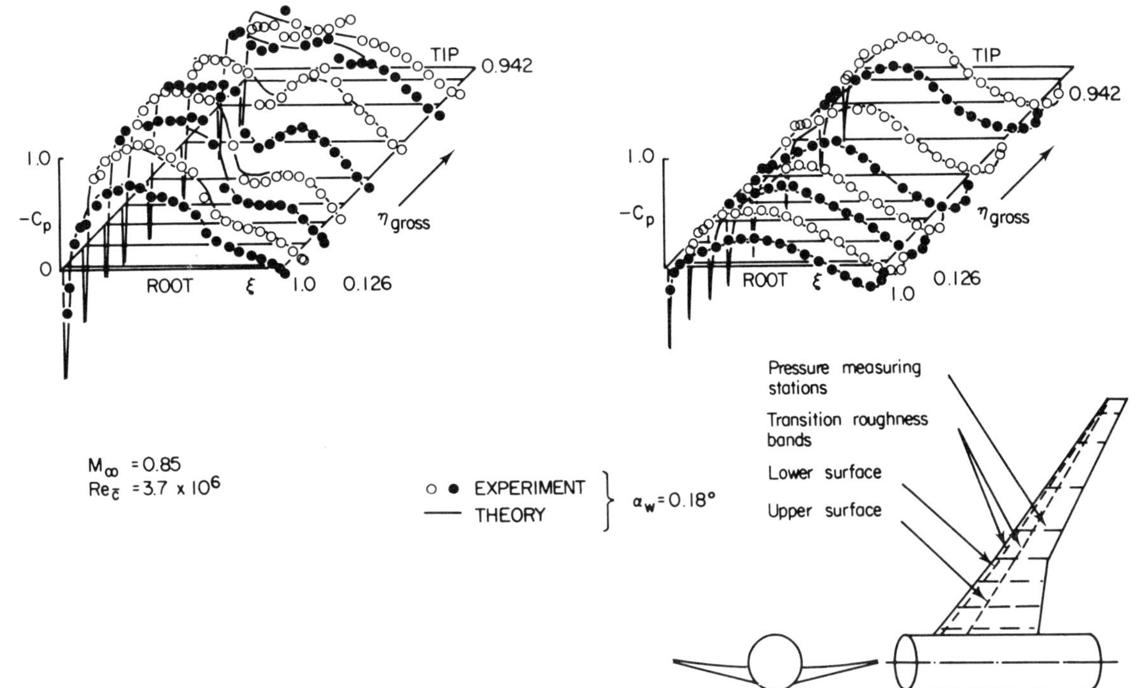

Fig. 33 Distribution of pressure coefficient, wing/body (3), $M_\infty = 0.85$, $Re_{\bar{c}} = 3.7 \times 10^6$, $\alpha_w = 0.18°$

addition to correction of the wind tunnel results for tunnel
wall effects. In order to make an adequate assessment of the
methods it will be necessary to obtain experimental evidence
which includes detailed measurements of the boundary layer
growth on three-dimensional wings in addition to the pressure
measurements upon which our judgements are at present based.

7. CONCLUSIONS

In this final section, we conclude with some remarks on the
current 'state of the art' of this subject, and make some
suggestions about the needs and possibilities for future work
along the same lines.

(i) The generalisation given in this paper (section 3) of the
classical principle of matching between separate inviscid and
viscous elements of the solution through one of the alternative
mathematical models of the displacement effect, has extended the
applicability of this idea to situations in which the viscous
layers are in no sense 'thin' and the assumptions of first-order
boundary-layer theory no longer apply. An essential feature is
the concept of an 'equivalent inviscid flow', defined by analytic
continuation up to the actual surface of the wing, which can be
determined in practice by setting up an iterative procedure for
calculating the combined flow, taking full account of the
viscous-inviscid interaction process.

(ii) In this inviscid flow, the pressures on the upper and
lower sides of the dividing stream surface of the wake are
effectively uncoupled, their difference depending on the local
flow curvature. As a result, the need to ensure that the
corresponding pressures on the two surfaces of the wing should
approach a common limiting value at the trailing edge disappears,
and the condition which fixes the circulation is shifted to far
downstream in the wake.

(iii) For two-dimensional flows over aerofoils, under conditions
well removed from separation onset, the best of the existing
methods which employ this concept of viscous-inviscid matching
have been shown to be extremely successful in predicting the
pressure distribution and overall forces, even when shock waves
of moderate strength are present. This assessment is perhaps
least satisfactory as regards the prediction of lift, where
the main limitation lies in the uncertainties involved in the
determination of wind tunnel interference effects in experiment.

(iv) As the separation boundary is approached and the adverse
pressure gradients on either surface grow stronger, the tendency
arises for methods which are based mainly on the use of first-
order boundary layer assumptions to overestimate the pressure in

regions which are usually rearward-facing, so that the drag is
correspondingly underestimated. Considerable improvement in
this respect has been demonstrated by incorporating the second-
order effects described in section 3, in particular by allowing
for the pressure changes across the boundary layer caused by
high streamline curvature.

(v)   At the same time, however, increasingly severe convergence problems are encountered when a natural, direct iterative
procedure is used, for reasons which can be explained on the
basis of a linearised stability analysis (section 4). Under
these circumstances, therefore, there are strong grounds for
adopting a 'semi-inverse' scheme of the type described in
section 4.2, in which an inverse approach is used for the
boundary layer calculation while the direct form of the inviscid
flow solution is still retained.

(iv)   With such a semi-inverse technique, some success has
already been achieved in predicting two-dimensional flows which
involve a modest degree of rear separation. It does not yet
appear possible to treat cases right up to the maxium lift
coefficient, nor those involving a substantial region of shock-
induced separation. To do so will require both an improvement
in our physical modelling of separated flows and still further
attention to the convergence of the iterative process. When
strong shock waves are a dominant feature, particular care must
also be taken to represent them correctly in the inviscid part
of the flow, preferably by solving the Euler rather than the
full potential equations.

(vii)   Assuming optimistically that all this can eventually be
successfully accomplished, a prediction technique will have been
established for steady two-dimensional separated flows which
appear on the one hand conceptually superior to the use of
over-simplified physical modelling - such as the *a priori* assumption that the pressure is constant throughout a separated region;
and on the other hand computationallly far more efficient than
the attempt to obtain a direct numerical solution of the full
Navier-Stokes equations - an attempt which up to the present
has in any case been rewarded with only qualitative success.

(viii) For three-dimensional flows, there is little doubt that
methods based in a similar way on the concept of viscous-inviscid
matching offer just as much promise as in two dimensions; but
their present state of development is still somewhat rudimentary,
being limited to the use of strictly first-order concepts as
regards both the calculation of the boundary layers and the
matching process. Nevertheless, results have already been
achieved (section 6) which are greatly superior to those obtainable from a purely inviscid calculation, so that current methods

are of considerable practical value to the aircraft designer;
while further refinements along the lines indicated from two-
dimensional considerations can be confidently expected.

(ix) The possibility of an extension to deal with separated
flows, even of the comparatively simple type that could be
called 'quasi-two-dimensional', is less certain than in two
dimensions, since it requires the development of an 'inverse'
technique for three-dimensional boundary layers, a problem which
may not be easily soluble. But the potential rewards are so
great that it is worth making a determined effort to overcome
the inevitable difficulties.

APPENDIX A

EVALUATION OF THE INTEGRAL $\theta_n = \dfrac{1}{\rho_{iw} u_{iw}^2} \displaystyle\int_0^\delta (\rho_i u_i v_i - \rho u v)\, dn$

Because this integral parameter, which occurs in both the
streamwise and normal momentum integral equations (sections 3.2
and 3.3), is a second-order quantity, it is sufficient to
evaluate it using first-order assumptions, so that ρ_{iw}, ρ_i, u_{iw}
and u_i can be replaced by their values at the edge of the
boundary layer, ρ_e and u_e respectively (this is not true of v_i).
We also neglect the variation of ρ across the boundary layer (i.e.
we assume that the flow is incompressible).
Thus

$$\theta_n = \frac{1}{u_e^2} \int_0^\delta (u_e v_i - uv)\, dn. \qquad (A-1)$$

We suppose that the velocity profile $u(n)$ is given by a general
one-parameter expression

$$\frac{u}{u_e} = F(\eta, H) \qquad (A-2)$$

where $\eta = \dfrac{n}{\delta}$

and $H = \dfrac{\delta^*}{\theta}$.

If we write

$$A = \int_0^1 (1 - F)\, d\eta, \quad B = \int_0^1 F(1 - F)\, d\eta$$

then

$$\delta^* = A\delta, \quad \theta = B\delta \text{ and } H = \frac{A}{B}.$$

To determine v and v_i we use the equation of continuity, so that

$$-\frac{v}{u_e} = \delta \int_0^\eta \frac{1}{u_e} \frac{\partial u}{\partial s} \, d\eta. \tag{A-3}$$

Now from (A-2)

$$\frac{1}{u_e} \frac{\partial u}{\partial s} = \frac{F}{u_e} \frac{du_e}{ds} - \eta F_\eta \frac{1}{\delta} \frac{d\delta}{ds} + F_H \frac{dH}{ds}, \tag{A-4}$$

where suffices η and H denote partial differentiation.

Hence

$$-\frac{v}{u_e} = \frac{\delta^*}{A} \left[\frac{1}{u_e} \frac{du_e}{ds} \int_0^\eta F \, d\eta + \frac{1}{\delta} \frac{d\delta}{ds} \left(\int_0^\eta F \, d\eta - \eta F \right) + \frac{dH}{ds} \int_0^\eta F_H \, d\eta \right], \tag{A-5}$$

and so

$$-\frac{1}{u_e^2} \int_0^\delta uv \, dn = \frac{\delta^{*2}}{A^2} \left[\frac{I}{u_e} \frac{du_e}{ds} + (I - I_1) \frac{1}{\delta} \frac{d\delta}{ds} + I_2 \frac{dH}{ds} \right], \tag{A-6}$$

where

$$I = \int_0^1 F(\eta) \int_0^\eta F(\xi) \, d\xi \, d\eta. \tag{A-7}$$

$$I_1 = \int_0^1 \eta F^2(\eta) \, d\eta, \tag{A-8}$$

and

$$I_2 = \int_0^1 F(\eta) \int_0^\eta F_H(\xi) \, d\xi \, d\eta. \tag{A-9}$$

Integrating (A-7) by parts, we find that

$$I = \tfrac{1}{2}(1 - A)^2 . \qquad (A-10)$$

In a similar way we have for the equivalent inviscid flow

$$\frac{v_i}{u_e} = \frac{v_{iw}}{u_e} - \delta\eta \, \frac{1}{u_e} \frac{du_e}{ds}$$

$$= \frac{d\delta^*}{ds} + \frac{\delta^*}{u_e} \frac{du_e}{ds} \left(1 - \frac{\eta}{A}\right) \qquad \text{(using equation (5) of section 3.1);} \qquad (A-11)$$

and so

$$\frac{1}{u_e^2} \int_0^\delta u_e v_i \, dn = \frac{\delta^{*2}}{A} \left[\frac{1}{\delta^*} \frac{d\delta^*}{ds} + \left(1 - \frac{1}{2A}\right)\frac{1}{u_e} \frac{du_e}{ds}\right]. \qquad (A-12)$$

Combining (A-6) and (A-12), and using also

$$\frac{1}{\delta} \frac{d\delta}{ds} = \frac{1}{\delta^*} \frac{d\delta^*}{ds} - \frac{A'}{A} \frac{dH}{ds}$$

where $A' = \dfrac{dA}{dH}$,

we obtain finally

$$\theta_n = \frac{1}{2} \frac{\delta^{*2}}{u_e} \frac{du_e}{ds} + a_1 \delta^* \frac{d\delta^*}{ds} - a_2 \delta^{*2} \frac{dH}{ds} , \qquad (A-13)$$

where the coefficients a_1 and a_2 are given by

$$a_1 = \frac{1}{A^2} \left[\tfrac{1}{2}(1 + A^2) - I_1\right]$$

and

$$a_2 = \frac{1}{A^2} \left[- I_2 + \frac{A'}{A} (I - I_1)\right] .$$

The coefficients a_1 and a_2 have been calculated for three alternative one-parameter representations of the velocity profile of a turbulent boundary layer.

(1)   Power-law profile: $\dfrac{u}{u_e} = \eta^{\frac{1}{2}(H-1)}$:

$$
\left.
\begin{aligned}
a_1 &= \frac{H}{H - 1} \\[2em]
a_2 &= \frac{1}{(H + 1)(H - 1)^2}
\end{aligned}
\right\} .
\qquad\text{(A-14)}
$$

(2)   Straight-line profile: $\dfrac{u}{u_e} = 1 - \dfrac{3}{2}\left(\dfrac{H - 1}{H}\right)(1 - \eta)$:

$$
\left.
\begin{aligned}
a_1 &= \frac{1}{6} + \frac{8H}{9(H - 1)} \\[2em]
a_2 &= \frac{(H + 3)}{9H(H - 1)^2}
\end{aligned}
\right\} .
\qquad\text{(A-15)}
$$

(3)   Coles profile: $\dfrac{u}{u_e} = 1 - \dfrac{2}{3}\dfrac{(H - 1)}{H}(1 + \cos \pi\eta)$:

$$
\left.
\begin{aligned}
a_1 &= \frac{5}{4} - \frac{2}{\pi^2} + \frac{(3/2) - (6/\pi^2)}{(H - 1)} = 1.047 + \frac{0.892}{(H - 1)} \\[2em]
a_2 &= \frac{1}{H(H - 1)^2}\left(\frac{H}{\pi^2} + \frac{3}{4} - \frac{4}{\pi^2}\right) = \frac{(0.101H + 0.345)}{H(H - 1)^2}
\end{aligned}
\right\} .
\qquad\text{(A-16)}
$$

Values of a_1 and a_2 for the three alternative profiles are tabulated below:

	a_1			a_2		
H	(1)	(2)	(3)	(1)	(2)	(3)
1.5	3.0	2.83	2.83	1.60	1.33	1.32
2	2.0	1.94	1.94	0.33	0.28	0.27
3	1.5	1.50	1.49	0.06	0.06	0.05
4	1.33	1.35	1.34	0.02	0.02	0.02

It can be seen that for values of H of 2 or above (which are those for which second order effects are likely to be most important) there is little difference between the results from the three different assumptions about the profile shape, which embrace conditions that are applicable both to the boundary layer over a solid surface and to a wake. A satisfactory numerical approximation can therefore be obtained by taking the simplest of the three results, namely that given by East[5] for power law profiles:

$$\theta_n = \frac{H}{H-1}\,\delta^*\,\frac{d\delta^*}{ds} + \frac{1}{2}\frac{\delta^{*2}}{u_e}\frac{du_e}{ds} - \frac{\delta^{*2}}{(H+1)(H-1)^2}\frac{dH}{ds}\;. \qquad (A-17)$$

In this expression, under the circumstances of greatest interest (boundary layers in adverse pressure gradients approaching separation), the first term is usually the biggest while the remaining two are smaller and are both negative (or zero). A further simplification can be achieved by using the streamwise momentum equation, neglecting the skin-friction term in it and making one of two additional assumptions that cover the extremes of the range of conditions that are of practical importance. These are:

<u>Assumption (a)</u> that in a rapid approach to separation the boundary layer is strongly perturbed from equilibrium, and that as a result the entrainment process responds much less rapidly to the severe adverse pressure gradient than does the displacement thickness, so that

$$\frac{d}{ds}(u_e\delta - u_e\delta^*) \ll \frac{d}{ds}(u_e\delta^*)\;. \qquad (A-18)$$

This assumption, coupled with the momentum equation $c_f \sim 0$, namely

$$\frac{d}{ds}\left(\frac{\delta^*}{H}\right) + \frac{(H+2)}{H}\frac{\delta^*}{u_e}\frac{du_e}{ds} = 0\;, \qquad (A-19)$$

and with the relation $\dfrac{\delta}{\delta^*} = \dfrac{H+1}{H-1}$ for power-law profiles, leads to the approximations

$$\frac{\delta^{*2}}{u_e}\frac{du_e}{ds} \sim -\frac{\delta^*}{(H^2+H+1)}\frac{d\delta^*}{ds} \qquad (A-20)$$

and

$$\delta*^2 \frac{dH}{ds} \sim \frac{H(H^2 - 1)}{(H^2 + H + 1)} \delta* \frac{d\delta*}{ds} \qquad \text{(A-21)}$$

for the second and third terms in (A-17). It follows that

$$\left. \begin{aligned} &\theta_n \sim (\delta* + \theta) \frac{d\delta*}{ds} \left[1 + f_{(a)}(H)\right] \\[2mm] \text{with} \\[2mm] &f_{(a)}(H) = \frac{(- H^2 + 3H + 2)}{2(H^2 - 1)(H^2 + H + 1)} \end{aligned} \right\} . \qquad \text{(A-22)}$$

<u>Assumption (b)</u> that as an opposite extreme the boundary layer remains close to equilibrium conditions. This implies that $\frac{dH}{ds} = 0$, $u_e \sim s^m$ and $\frac{d\delta*}{ds} = \frac{\delta*}{s} = $ constant; so that

$$\frac{\delta*^2}{u_e} \frac{du_e}{ds} \sim m\delta* \frac{d\delta*}{ds} .$$

Equation (16) gives $m \sim - \frac{1}{H + 2}$ when $C_f \sim 0$, so that finally

$$\left. \begin{aligned} &\theta_n \sim (\delta* + \theta) \frac{d\delta*}{ds} \left[1 + f_{(b)}(H)\right] \\[2mm] \text{with} \\[2mm] &f_{(b)}(H) = \frac{4 - H}{2(H - 1)(H + 2)} \end{aligned} \right\} . \qquad \text{(A-23)}$$

Values of $f_{(a)}$ and $f_{(b)}$ are tabulated below

H	$f_{(a)}$	$f_{(b)}$
2	0.10	0.25
3	0.01	0.05
4	-0.003	0

It can be seen that, particularly with assumption (a) which is likely to be the more realistic of the two extremes in cases where second-order effects are relatively large, the approximation (18), quoted in section 3.2, namely

$$\theta_n = (\delta^* + \theta)\,\frac{d\delta^*}{ds}$$

should be entirely adequate.

It remains to investigate the accuracy of equation (25) (section 3.3) which gives the pressure correction due to stream-line curvature effects. A straightforward application of the alternative assumptions (a) and (b) above, in equation (24) of the main text, gives eventually (for $\kappa_w = 0$)

$$\frac{\Delta p}{\rho_e u_e^2} = (\theta + \delta^*)\,\frac{d^2\delta^*}{ds^2}\,(1 + f(H)) + g(H)\left(\frac{d\delta^*}{ds}\right)^2,$$

where f takes the alternative forms given above (but of course $\frac{d^2\delta^*}{ds^2} = 0$ for equilibrium boundary layers) and

$$g_{(a)}(H) = \frac{3H^5 - 4H^4 - 10H^3 - 2H^2 + 2H - 1}{2(H - 1)(H^2 + H + 1)^3}$$

$$g_{(b)}(H) = \frac{(H + 1)}{2(H - 1)(H + 2)}.$$

For $H > 2$ we find that $|g_{(a)}| < 0.08$ and $g_{(b)} < \frac{3}{8}$. Thus the biggest error in equation (25) (in which f and g are both neglected) might be expected for equilibrium boundary layers. But inspection of the values of $\frac{d\delta^*}{ds}$ given in Ref. 24 shows that $\frac{\Delta p}{\rho_e u_e^2} < 0.001H$ for $H < 4$, so that Δp is indeed negligible in such cases, and equation (25) should therefore remain a satisfactory approximation even for separated flows – unless the last term in equation (23), $\frac{d}{ds}\int_0^\delta (-\overline{\rho u'v'})\,dn$, should ever turn out to be appreciable. A rough approximation to this quantity, which may be written $\frac{d}{ds}\left(\rho u_e^2 \overline{C}_\tau \delta\right)$, can be obtained as follows. The measurements of Ref. 24 suggest that, for $H > 1.7$,

$$\bar{C}_\tau \simeq 0.03 \left(\frac{H - 1}{H}\right)^2 ,$$

and this is supported by measurements in strongly separated layers, for example Ref. 49, in which $\bar{C}_\tau$ was found to be about 0.015 with $H \sim 4$. Hence the additional term is approximately

$$0.03 \frac{d}{ds} \left[\rho u_e^2 \left(\frac{H - 1}{H}\right)^2 (H + H_1)\theta\right] ,$$

and this is likely to remain negligible (less than 0.01, say) in most conceivable circumstances.

APPENDIX B

DIRECT AND INVERSE FORMS OF THE TURBULENT BOUNDARY LAYER EQUATIONS

The standard equations for an 'integral' method based on the principle of entrainment are (to first order, with the suffices on ρ_{iw}, u_{iw}, M_{iw} omitted):

<u>Momentum</u>

$$\frac{d\theta}{ds} = \tfrac{1}{2}C_f - \left(H + 2 - M^2\right) \frac{\theta}{u} \frac{du}{ds} , \tag{B-1}$$

<u>Entrainment</u>

$$C_E = \frac{1}{\rho u} \frac{d}{ds} \{\rho u(\delta - \delta^*)\} \tag{B-2}$$

coupled with a third equation for C_E which depends on the details of the particular empirical method, and need not concern us here. Writing $\delta - \delta^* = H_1\theta$, (B-2) becomes

$$C_E = H_1 \frac{d\theta}{ds} + H_1(1 - M^2) \frac{\theta}{u} \frac{du}{ds} + \theta \frac{dH_1}{ds} . \tag{B-3}$$

In the methods of Refs. 18 and 19 it is assumed that H_1 is an empirical function of

$$\bar{H} = \frac{1}{\theta} \int_0^\delta \frac{\rho}{\rho_e} \left(1 - \frac{u}{u_e}\right) dn ;$$

$\bar{H}$ is related to H by

$$(H + 1) = (\bar{H} + 1)\left(1 + \frac{1}{5}\,M^2\right). \qquad (B-4)$$

For use in a direct method, (B-1) and (B-3) are usually combined to give

$$\theta\,\frac{d\bar{H}}{ds} = \frac{d\bar{H}}{dH_1}\left[C_E - H_1\left\{\tfrac{1}{2}C_f - (H + 1)\,\frac{\theta}{u}\,\frac{du}{ds}\right\}\right]. \qquad (B-5)$$

In the direct problem, $\dfrac{1}{u}\dfrac{du}{ds}$ is a given function of s and so equations (B-1) and (B-5) (together with the third equation for C_E) can be integrated simultaneously to give the required solution. When this has been done, the *source strength* parameter Σ can be obtained from

$$\Sigma = \frac{1}{\rho u}\,\frac{d}{ds}\,(\rho u H \theta) \qquad (B-6)$$

$$= H\,\frac{d\theta}{ds} + H\left(1 - M^2\right)\frac{\theta}{u}\,\frac{du}{ds} + \theta\,\frac{dH}{ds}. \qquad (B-7)$$

Now from (B-4)

$$\frac{dH}{ds} = \left(1 + \frac{1}{5}\,M^2\right)\frac{d\bar{H}}{ds} + \frac{2}{5}\,M^2(H + 1)\,\frac{1}{u}\,\frac{du}{ds},$$

and substituting for $\dfrac{d\bar{H}}{ds}$ and $\dfrac{d\theta}{ds}$ from (B-5) and (B-1), we find that

$$\Sigma = -\,(H + 1)\,\frac{\theta}{u}\,\frac{du}{ds}\left[-\left(1 + \frac{1}{5}\,M^2\right)H_1\,\frac{d\bar{H}}{dH_1} + \frac{2}{5}\,M^2\right]$$

$$+ \frac{1}{2}\,C_f\left[H - \left(1 + \frac{1}{5}\,M^2\right)H_1\,\frac{d\bar{H}}{dH_1}\right] + C_E\left(1 + \frac{1}{5}\,M^2\right)\frac{d\bar{H}}{dH_1}, \qquad (B-8)$$

the equation whose form for incompressible flow is quoted in section 4, equation (39).

For flows approaching separation $\dfrac{d\bar{H}}{dH_1}$ may become infinite and then both (B-5) and (B-8) will break down, so that an inverse

method of solution is essential. This can easily be achieved
by writing equations (B-1), (B-3) and (B-7) in matrix form

$$
\begin{pmatrix}
1 & H + 2 - M^2 & O \\[2ex]
H_1 & H_1(1 - M^2) & H_1' \\[2ex]
H & H + \frac{1}{5} M^2 (2 - 3H) & 1 + \frac{1}{5} M^2
\end{pmatrix}
\begin{pmatrix}
\dfrac{d\theta}{ds} \\[2ex]
\dfrac{\theta}{u} \dfrac{du}{ds} \\[2ex]
\theta \dfrac{d\bar{H}}{ds}
\end{pmatrix}
=
\begin{pmatrix}
\frac{1}{2} C_f \\[2ex]
C_E \\[2ex]
\Sigma
\end{pmatrix}
; \quad \text{(B-9)}
$$

here $H_1' \equiv \dfrac{dH_1}{d\bar{H}}$. We can now regard Σ as 'known', $\dfrac{1}{u}\dfrac{du}{ds}$ as
'unknown'. These equations can be solved to give

$$
(H + 1)D\, \frac{d\theta}{ds} = -\tfrac{1}{2}C_f \left\{ H_1 - HH_1' - \frac{1}{5} M^2 (4H_1 + 2H_1' - 3HH_1') - \frac{1}{5}M^4 H_1 \right\}
$$

$$
+ \left(H + 2 - M^2 \right) \left\{ C_E \left(1 + \frac{1}{5} M^2 \right) - H_1'\Sigma \right\} , \qquad \text{(B-10)}
$$

$$
(H + 1)D\, \frac{\theta}{u} \frac{du}{ds} = H_1'\Sigma + \tfrac{1}{2}C_f \left\{ \left(1 + \frac{1}{5} M^2 \right) H_1 - HH_1' \right\} - C_E \left(1 + \frac{1}{5} M^2 \right)
$$

$$
\text{(B-11)}
$$

and

$$
D\theta\, \frac{d\bar{H}}{ds} = H_1\Sigma - \frac{1}{5} M^2 H_1 C_f - C_E \left(H - \frac{2}{5} M^2 \right) \qquad \text{(B-12)}
$$

where $D = H_1 - HH_1' + \dfrac{1}{5} M^2 \left(H_1 + 2H_1' \right)$.

Of these, equation (B-11) is simply a rearrangement of (B-8).
It shows that if $H_1' = O$ (i.e. near separation), the relation

$$
(H + 1)\, \frac{\theta}{u} \frac{du}{ds} = \tfrac{1}{2}C_f - \frac{C_E}{H_1}
$$

must be satisfied. Equations (B-10) to (B-12) (or some suitable
combination of them) can be integrated simultaneously, in a
similar way to the direct equations. The parameter D can never
vanish and so the equations are non-singular (except that
$H_1' \to \infty$ as $H \to 1$).

8. REFERENCES

1. Computation of viscous-viscid   interactions, AGARD CP-291, (1981).

2. Lighthill, M.J. (1958). "On displacement thickness", *J. Fl. Mech.*, **4**, p. 383.

3. Myring, D.F.   (1968). "The effect of normal pressure gradients on the boundary layer momentum integral equation", RAE Technical Report 68214.

4. Le Balleur, J.C. (1978). "Calculs couplés visqueux-non visqueux, incluant décollements et ondes de choc en ecoulement bidimensionnel, AGARD LS-94.

5. East, L.F. (1981). "A representation of second-order boundary layer effects in the momentum integral equation and in viscous-inviscid interactions", RAE Technical Report 81002.

6. Mahgoub, H.E.H. and Bradshaw, P. (October 1979). "Calculation of turbulent-inviscid flow interactions with large normal pressure gradients", *AIAA Journal,* **17**, p. 1025.

7. Le Balleur, J.C. (1981). "Calcul des écoulements a forte interaction visqueuse au moyen de methode de couplage", AGARD CP-291, Paper 1.

8. Lock, R.C. (1981). "A review of methods for predicting viscous effects on aerofoils and wings at transonic speeds, AGARD CP-291, Paper 2.

9. Collyer, M.R. (1977). "An extension to the method of Garabedian and Korn for the calculation of transonic flow past an aerofoil to include the effects of a boundary layer and wake, RAE Technical Report 77104.

10. Le Balleur, J.C. (1977). "Couplage visqueux-non visqueux: analyse du problème incluant decollements et ondes de choc", *La Recherche Aérospatiale,* No.1977-6, p.349.

11. Firmin, M.C.P. and Cook, T.A. (1968). "Detailed exploration of the compressible viscous flow over two-dimensional aerofoils at high Reynolds numbers", Proc. 6th ICAS Conference.

12. Melnik, R.E. (1981). "Turbulent interactions on airfoils at transonic speeds - recent developments", AGARD CP-291, Paper 10.

13. Melnik, R.E. and Grossman, B. (1981). On the turbulent viscid-inviscid interaction at a wedge-shaped trailing edge, Symposium on

numerical and physical aspects of aerodynamic flows", California State University, 19-21 January.

14. Mellor, G.L. (1972). "The large Reynolds number asymptotic theory of turbulent boundary layers", *Int. J. Engr. Sci.*, **10**.

15. Melnik, R.E., Chow, R. and Mead, H.R. (1977). "Theory of viscous transonic flow over airfoils at high Reynolds number", AIAA Paper 77-680.

16. Melnik, R.E., Chow, R., Mead, H.R. and Jameson, A. (1981). "An improved viscid/inviscid interaction procedure for transonic flow over airfoils", NASA CR (to be issued).

17. Le Balleur, J.C. (1978). "Couplage visqueux-non visqueux: méthode numérique et applications aux écoulements bidimensionnels transsoniques et supersoniques", *La Recherche Aérospatiale*, No. 2.

18. Green, J.E. (1972). "Application of Head's entrainment method to the prediction of turbulent boundary layers and wakes in compressible flow", RAE Technical Report 72079.

19. Green, J.E., Weeks, D.G. and Brooman, J.W.F. (1973). "Prediction of turbulent boundary layers and wakes in compressible flow by a lag-entrainment method", ARC R & M 3791.

20. Baker, T.J. (1982). "Approximate factorisation methods", This volume.

21. East, L.F., Smith, P.D. and Merryman, P.J. (1977). "Prediction of separated flows by the lag-entrainment method", RAE Technical Report 77046.

22. Le Balleur, J.C. and Neron,M. (1981). "Calculs d'écoulements visqueux décollés sur profils d'ailes par une approche de couplage", AGARD CP-291, Paper 11.

23. Carter, J.E. (1979). "A new boundary layer inviscid iteration technique for separated flow", AIAA Paper 79-1450.

24. East, L.F., Sawyer, W.G. and Nash, C.R. (1979). "An investigation of the structure of equilibrium turbulent boundary layers", RAE Technical Report 79040.

25. Smith, P.D., (1973). "An integral prediction method for three-dimensional compressible turbulent boundary layers", ARC R & M 3452.

428   LOCK and FIRMIN

26.   Collyer, M.R. and Lock, R.C. (1979). "Prediction of viscous effects in steady transonic flow past an aerofoil", *Aero. Qu.*, **30**, p. 485.

27.   Michel, R., Quemand, C. and Cousteix, J. (1972). "Méthode practique de prévision des couches limites turbulentes bi-et-tri-dimensionelles", *La Recherche Aerospatiale* No. 1972-1.

28.   Butter, D.J. and Williams, B.R. (1981). "The development and application of a method for calculating the viscous flow about high-lift aerofoils", AGARD CP-291, Paper 25.

29.   Newling, J.C. (1977). "An improved two-dimensional multi-aerofoil program", HSA-MAE-R-FDM-0007.

30.   Irwin, H.P.A.H. (1974). "A calculation method for the two-dimensional turbulent flow over a slotted flap", ARC CP-1267.

31.   Grossmann,G. and Volpe, G. (1977). "The viscous transonic flow over two-element aerofoil systems", AIAA Paper 77-688.

32.   Leicher, S. (1981). "Viscous flow simulation of high lift devices at subsonic and transonic speeds", AGARD CP-291, Paper 6.

33.   Rosch, H. and Klevenhusen, K.D. (1981). "Flow computation around multi-element aerofoils in viscous transonic flow", AGARD CP-291, Paper 7.

34.   Sells, C.C.L. (1968). "Plane subcritical flow past a lifting aerofoil", *Proc. Roy. Soc.* (A), **308**, p. 377.

35.   Garabedian, P.R. and Korn, D.G. (1971). "Analysis of transonic aerofoils", *Comm. Pure Appl. Math.*, **24**, p. 841.

36.   Bauer, F., Garabedian, P.R. and Korn, D.G. (1972). "Supercritical wing sections", (Lecture notes in Economics and Mathematical Systems, No. 66), Springer-Verlag, Berlin.

37.   Murman, E.M. (1974). "Analysis of embedded shock waves calculated by relaxation methods", *AIAA Journal,* **12**, p. 626.

38.   Jameson, A. (1976). "Numerical computation of transonic flow with shock waves", Symposium Transsonicum II, Springer Verlag, Berlin, p. 385.

39.   Bauer, F., Garabedian, P.R., Korn, D.G. and Jameson, A. (1975). "Supercritical wing sections II", (Lecture notes in Economics and Mathematical Systems, No. 108), Springer Verlag, Berlin.

40. Lock, R.C. (1981). "A modification to the method of Garabedian and Korn": *in* "Numerical methods for the computation of inviscid transonic flows with shock waves". (A GAMM Workshop), Friedr. Vieweg & Sohn, Braunschweig.

41. Collyer, M.R. and Lock, R.C. (1978). "Improvements to the viscous Garabedian and Korn (VGK) method for calculating transonic flow past an aerofoil", RAE Technical Report 78039.

42. Cook, P.H., McDonald, M.A. and Firmin, M.C.P. (1979). "Aerofoil RAE 2822 - pressure distributions and boundary layer and wake measurements", AGARD AR 138, Paper A6.

43. Collyer, M.R. (1978). "A comparison between theory and experiment for the RAE 5217 aerofoil in transonic flow", RAE Technical Report 78094.

44. Sells, C.C.L., (1980). "Numerical solutions of the Euler equations for flow past a lifting aerofoil", RAE Technical Report 80065.

45. Melnik, R.E. and Grossman, B. (1975). "Analysis of the interaction of a weak normal shock wave with a turbulent boundary layer", AIAA Paper 74-598 (1974), *see also* Symposium Transsonicum II, Springer-Verlag, P. 262.

46. Rogers, E.W.E. (1966). "Subsonic wind tunnel wall corrections", AGARDograph 109.

47. Ashill, P.R. and Weeks, D.J. (1981). "An investigation of viscous effects on a family of advanced aerofoils of 14% thickness; and Techniques developed in Europe for tunnel-wall corrections using measured boundary conditions". Papers given at AGARD FDP TES Subcommittee meeting on Integration of Computers and Wind Tunnel Testing, RAE Bedford, 18-19 February 1981.

48. Kacprzynski, J.J. (1972). "A second series of wind tunnel tests of the shockless lifting aerofoil No.1", Project Rep. 5×5/0062, NRC of Canada, Ottawa.

49. Johnson, D.A., Bachalo, W.D. and Owen, F.K. (1981). "Transonic flow past a symmetrical airfoil at high angle of attack", *AIAA Journal of Aircraft*, **18**, No. 1, p. 7.

50. Roberts, A. and Rundle, K. (1973). "The computation of first order compressible flow about wing-body configurations", BAC Report Aero MA 20, DRIC S & T Memo 14/73 (BR36898).

51. Roberts A. (1977). "Subsonic viscous flow about wing-body combinations", BAC Report Aero MA22, British ARC 37645.

52.   Firmin, M.C.P. (1981) "Calculations of transonic flow over wing/body combinations with an allowance for viscous effects", AGARD CP-291, Paper 8.

53.   Albone, C.M., Hall, M.G. and Joyce, G. (1975). "Numerical solutions for transonic flows past wing-body combinations", IUTAM Symposium Transsonicum II, Springer-Verlag.

54.   Mager, A. (1952). "Generalisation of boundary layer momentum integral equations to three-dimensional flow, including those of rotating systems", NACA Report 1067.

55.   Stanewsky, E., Nandanan, M. and Inger, G.R. (1981). "The coupling of a shock boundary layer interaction module with a viscous/inviscid computation method", AGARD CP-291, Paper 4.

56.   Streett, C.L. (1981). "Viscous-inviscid interaction for transonic wing-body configurations including wake effects", AIAA Paper 81-1266.

ACHIEVEMENTS AND PROBLEMS IN MODELLING HIGHLY-SWEPT
FLOW SEPARATIONS

J.H.B. Smith

(Royal Aircraft Establishment)

ABSTRACT

   After a brief consideration of some kinematic differences
between separated flow in two and three dimensions, the vortex-
sheet model of separated flow originating from a highly-swept
separation line is introduced. Some evidence for the adequacy
of the model is reviewed before a presentation of two recent
applications. The first is a treatment, by staff of the Boeing
Company, of separation from the highly-swept leading edges of a
wing of low aspect ratio in subsonic flow, by means of a panel
method. The second is an attempt, made at RAE, to reproduce the
mechanism of separation from the curved surface of a slender
cone at incidence, including the prediction of the laminar separation line.

1. INTRODUCTION

   The phrase 'boundary-layer separation', or 'separation' for
short, conveys many ideas, some of which it is worth reviewing
briefly. To limit the field a little, only steady flows are
considered. The view most familiar to theoreticians is the one
from classical boundary-layer theory. When a two-dimensional
laminar boundary layer is calculated under a specified adverse
pressure gradient, the displacement thickness increases and the
skin-friction falls to zero. The rate of decrease of the skin
friction becomes infinitely great where the friction itself
vanishes, giving rise to a 'singularity at separation'. It is
now a familiar idea that if, instead of specifying the pressure
distribution, we specify either the skin-friction or the displacement thickness, no singularity will arise at the separation
point. However, we can still identify separation with the vanishing of the skin friction. Lock and Firmin (1982) describe
how such an inverse approach can be used to calculate two-
dimensional flows with relatively thin separated regions, such

---

as arise on aerofoils at moderate angles of incidence.

In a three-dimensional flow the skin-friction has two components, and these only vanish simultaneously at isolated points.
Separation, however, normally takes place along lines. In a
calculation of a three-dimensional boundary layer under a specified external flow, it is found that the limiting streamlines on
the surface converge rapidly close to certain lines and the
boundary layer thickens rapidly there. Similar convergence and
thickening is observed experimentally as the boundary layer
approaches separation. The limiting streamline towards which
the convergence takes place is therefore regarded as a separation line. One result of the convergence is that the boundary
layer calculation cannot be carried across a separation line.
When the separation line forms a closed curve on the surface,
it therefore defines a region of inaccessibility, which cannot
be reached by a boundary layer growing from the upstream stagnation point. In other circumstances, it may be possible for
the boundary layer calculation to approach the separation line
from both sides.

In a two-dimensional flow, a streamline which leaves the
surface at a separation point, called a separation streamline,
divides the flow field into two parts. The part which lies on
the downstream side of the separation point may reasonably be
called a region of separated flow. If the region of separated
flow does not extend to infinity downstream it terminates at a
reattachment point, where the separation streamline reattaches
to the surface of the body or to its wake. The topology of
three-dimensional flows is different. The stream surface leaving the surface of the body along a separation line may reasonably be called a separation surface, but this separation surface
is not usually closed; that is to say, it does not usually
divide the flow field into distinct parts. Even if the separation line forms a closed curve on the body, the separation surface may be open. There is therefore usually no definite region
of separated flow, nor is there a reattachment line.

In many three-dimensional flows it is also possible to recognise attachment lines. These are characterised by the rapid
divergence of the limiting streamlines away from them. They are
important both physically and computationally because they act
as virtual origins for the boundary layers on either side of
them. The stream surface which meets the body surface along an
attachment line may reasonably be called an attachment surface.

Confusion begins to arise when we draw two-dimensional pic-
tures of three-dimensional flows. If we draw a section through
a three-dimensional flow it will not in general contain stream-

lines. In general*, the nearest we can get to a streamline is
a section through a stream surface. Fig. 1 illustrates the
point for a simple geometry. On the left is a representation of
a two-dimensional flow down a step, to be regarded either as
corresponding to a steady flow at a low Reynolds number or to a
mean flow at a higher Reynolds number. On the right is a representation of a three-dimensional flow down a swept step**, in a
plane normal to the step. To enhance the three-dimensionality,
the height of the step increases in the downstream direction.
In both cases the lines drawn are sections through stream surfaces, but only on the left are they also streamlines. On the
left we can identify a region of separated flow, bounded by a
separation streamline which reattaches. On the right there is
no defined region of separated fow. The separation stream surface does not reattach, and the attachment stream surface is
quite distinct from it.

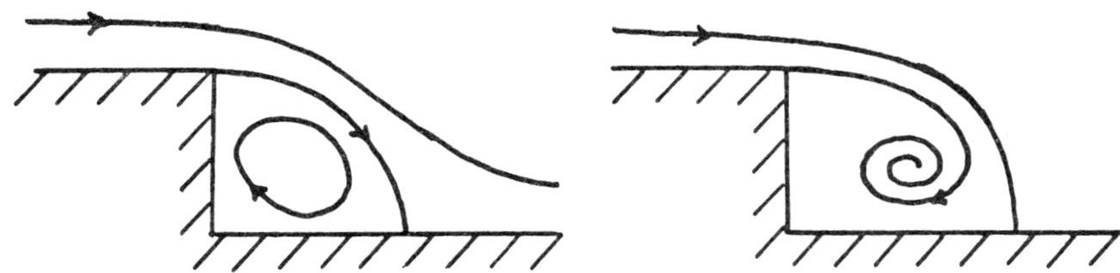

Fig. 1 Flow down a step: planar (left), three-
        dimensional (right)

Having noted the differences between the two parts of Fig. 1,
let us turn to the similarities. In each case fluid which has
acquired rotation during its passage along the upstream wall
leaves the wall at the separation line. It penetrates into the
body of the fluid a distance which, at least for moderate and
large Reynolds numbers, is much greater than the thickness of
the upstream boundary layer. Vorticity is being convected, as
well as diffused, away from the solid surface. This is the
descriptive view of separation. At one time it appeared to
have so little relation to the boundary layer view of separation
that the word 'break-away' was used for it. We are now much
closer to a reconciliation of the two views.

---

*For the case of a conical three-dimensional flow it is possible
to progress further with a study of the conical stream surfaces,
as indicated by Smith (1972), for instance.
**An inviscid model of separated flow has recently been applied
by Billet (1980) to a particular form of swept step, and the
results compared with experimental measurements.

Also, in each part of Fig. 1 we can point to a region of rotational flow, and these regions are not so very different. In particular, if the Reynolds number is not high, the free shear layer which surrounds the spiral separation surface on the right will be quite thick. It could well be thick enough to merge with the boundary layer growing upstream from the attachment line. Indeed, the rotational flow could extend as far as the attachment stream surface, and the regions of rotational flow in the two parts of Fig. 1 would then be very similar.

As the Reynolds number increases, diffusion normal to the stream surface becomes less significant in comparison with convection parallel to it. This matters relatively little in the left-hand picture, because velocities are very low in the separated region. It matters much more on the right, where convection normal to the plane of the picture allows larger velocities in the plane of the picture. Consequently, at larger Reynolds numbers, the rotation in the fluid on the right is confined more and more closely to the neighbourhood of the separation stream surface. In the limit of infinite Reynolds number, the vorticity becomes infinite on the separation stream surface and zero elsewhere. The flow has become irrotational with an embedded vortex sheet. (A vortex sheet is a surface across which the tangential component of the velocity vector is discontinuous, but the pressure and the normal component of the velocity are continuous. In steady flow it is a stream surface. Internal energy, entropy and density may also be different across a vortex sheet, but we shall not be considering the effects of this possibility.) It is not clear that there is any useful steady limit of the two-dimensional flow as the Reynolds number tends to infinity. A trivial limit consists of a stagnant region behind the step separated from a uniform stream above it by a plane vortex sheet of constant strength springing from the salient edge. This is unstable, but would be consistent with the behaviour of the flow past the swept step, if in the latter the Reynolds number is first allowed to tend to infinity and the sweep is then reduced towards zero. Other possible limiting forms might be constructed using a standing-eddy of uniform vorticity, in accordance with the concept of Batchelor (1967).

In any event it is clear that the structure of the three-dimensional flow is different from that of the two-dimensional flow, and that physically significant differences become more apparent as the Reynolds number increases. It is with modelling flows at high Reynolds numbers with highly swept separation lines that we shall be concerned. The model used involves the representation of the rotation in the real flow by vortex sheets and line vortices embedded in inviscid, potential flow. The basis for this model will first be established, and then

accounts will be given of two particular recent developments
which make use of it. The first of these is an extension of the
panel method for attached flow developed over a number of years
at the Boeing Company to the treatment of flows past wings with
sharp, highly-swept, leading edges, in which leading-edge vortices are formed. This provides, for the first time, good predictions of both the vortex configuration and the non-linear
lift at low speeds. The second and more recent development is
the application of the model to describe the vortices shed from
the smooth surfaces of pointed bodies at incidence. This
involves the prediction of the separation line, as well as the
calculation of the inviscid flow. The work, carried out at RAE,
has been limited so far to the study of cones of elliptic cross-
section, treated in the framework of slender-body theory.

These two developments involve essentially the same physical
and mathematical model of the structure of the vortex, so it
seems appropriate to start with a description of the real structure, followed by an account of the model.

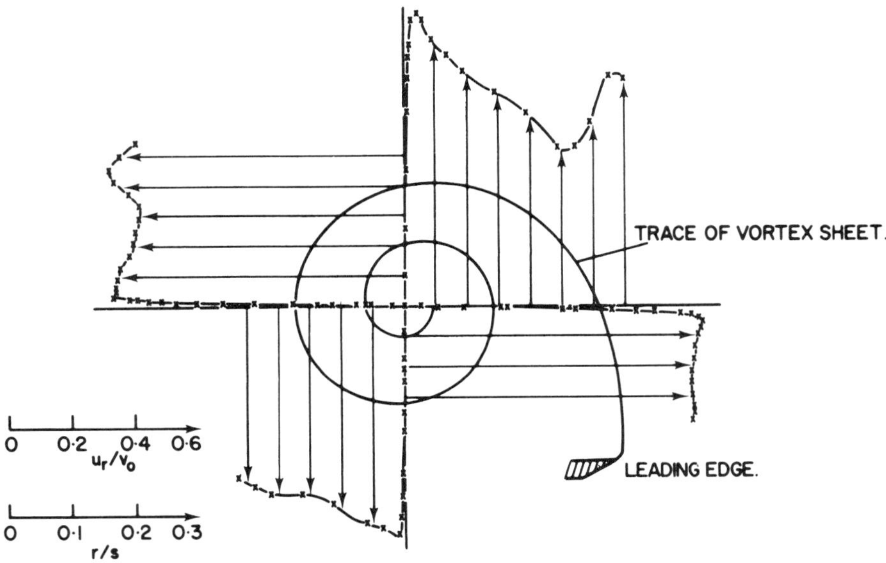

Fig. 2   Circumferential velocity in a leading-edge vortex,
         Earnshaw (1961)

2. VORTEX STRUCTURE AND MODELLING

Fig. 2 shows measured values of the circumferential component
of velocity in a leading-edge vortex on a delta wing of aspect
ratio unity at an incidence of $15°$, at low wind speed, and at a
Reynolds number of 1.5×10^6 (based on distance to the measuring
station). The circumferential velocity has been plotted along
each of four orthogonal radii through the axis of the vortex.
From this we can see first that the region near the axis which
is dominated by viscous diffusion, in which the circumferential
velocity falls to zero, is very small. Surrounding this is a
region in which the circumferential velocity approaches the
free-stream speed and displays little systematic variation. The
flow here must be rotational. At the outside edge of the vortex,
the shear layer is clearly detectable. Closer examination
reveals inflexions in the velocity profiles, which provide evidence of about one more turn of a spiral structure. An attempt
has been made to connect the locations of these inflexions to
produce the spiral shown. We can imagine that, at larger Reynolds numbers, more turns of the spiral would become recognis-
able, with regions of irrotational flow between them; and an
infinite spiral would emerge at infinite Reynolds number.

The mathematical treatment of such an infinite vortex sheet
presents formidable difficulties, even in incompressible flow.
If the shape of the spiral is roughly circular, there is general
agreement about the existence of asymptotic solutions for the
inner part of the spiral (Mangler and Weber 1967, Guiraud and
Zeytounian 1977). However, it is likely that the spiral is
generally somewhat elliptical in form, being flattened by its
proximity to a solid surface. A solution of this kind has been
proposed by Maskell (1964), but no complete analysis is available. Further investigations by Mangler and Sells (1967) and
by Moore (1975) have cast doubt on whether the asymptotic problem is well posed. Effects of compressibility have been included by Brown and Mangler (1967). The alternative of a continuous, rotational core has been examined by Hall (1961), Brown
(1965), and Stewartson and Hall (1963). These solutions are
useful if the overall size and circulation of the core of the
vortex are known and we wish to discover what happens inside it.
We shall be looking at the complementary problem: given the body
and the incident flow, what would be the positions, sizes and
circulations of the vortices.

It turns out that this question can be answered without representing the spiral structure in any detail. Since this
observation is fundamental to the modelling, it is worth spend-
ing some time in justifying it. The essence of the model is
simply the replacement of the inner part of the spiral vortex
sheet by a single line vortex and a cut which connects the line

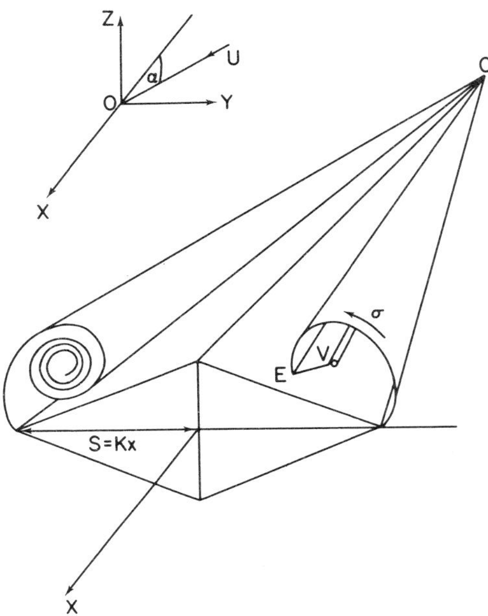

Fig. 3   Model of the infinite spiral vortex sheet

vortex to the free end of the remaining outer part of the sheet.
Fig. 3 shows a thick delta wing with a spiral vortex sheet
springing from its port leading edge, while on the starboard
side the inner part of the spiral has been replaced by a line
vortex OV and a cut OVE. This combination is equivalent to a
doublet sheet OVE whose strength depends only on the axial
coordinate, x. The cut can be regarded in the first place as a
mathematical device to keep the velocity potential single-valued,
but it has a more significant role. This arises because, in the
figure, the circulation of the vortex sheet on the left grows
along its length as vorticity is shed into it from the wing.
Since we do not want the outer part of the sheet on the right to
wrap itself further round the line vortex, the circulation of
the line vortex must be allowed to grow. This is a straightforward violation of Helmholtz theorem, the price of which is
the appearance of a pressure difference across the cut. In
symbols:

$$\Delta p = -\tfrac{1}{2}\rho\Delta\{(U+u)^2 + v^2 + w^2\} \simeq -\rho U\Delta u = -\rho U\Delta\,\frac{\partial\Phi}{\partial x} = -\rho U\frac{d\Gamma}{dx}$$

$$(1)$$

Here u, v and w are the components of the disturbance velocity,
U is the free-stream speed, p and ρ are pressure and density, Δ
denotes the difference across the cut, Φ is the velocity potential and Γ is the circulation of the line vortex. The best we
can do is to eliminate the force on the system by setting the
line vortex at a small angle to the local flow direction so that
it carries a force equal and opposite to that on the cut. All
that can be said at this point is that the approximation will
become closer as the point E moves along the spiral towards its
centre.

This explanation of the model for the core of the vortex is
the most straightforward from a mathematical point of view. It
may help to think of the approximation in two steps. First the
circulation on the inner part of the sheet is pushed along the
sheet into the axis, leaving behind a surface which can be a
streamsurface, but carries a pressure difference. Since the
pressure difference is the same along the whole cross-section,
by equation (1), the force on it depends only on its boundaries
OE and OV, so it can be replaced by a straight cut. Some people
find it more natural to visualize the cut as a series of vortex
segments lying in planes of constant x and connecting OE to OV.
These preserve continuity of circulation at the expense of kinks
in the vortex lines. Use of the Joukowski formula for the
force on each transverse vortex element leads to the same result
for the force on the cut.

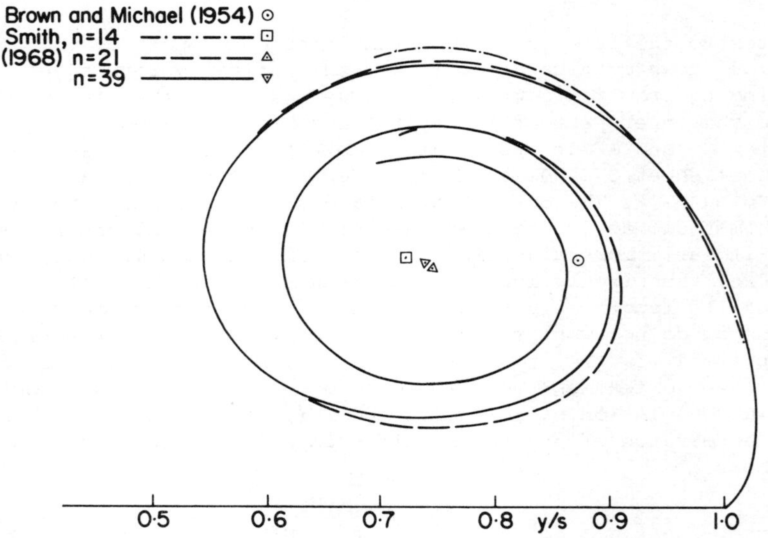

Fig. 4 Shape of vortex sheets of different extent

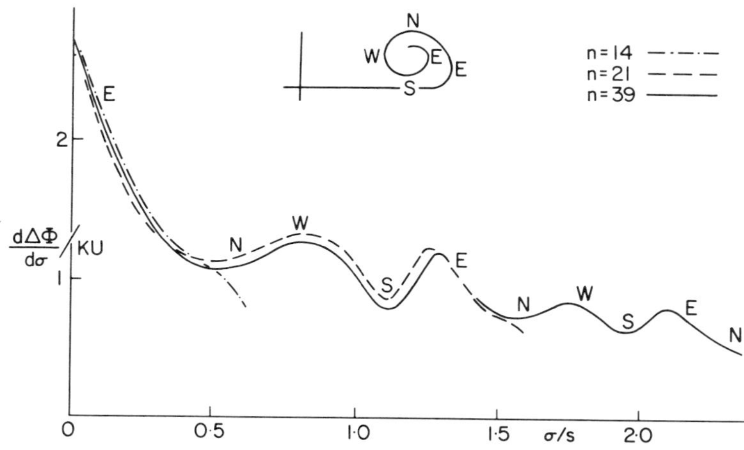

Fig. 5 Strength of vortex sheets of different extent,
        Smith (1968)

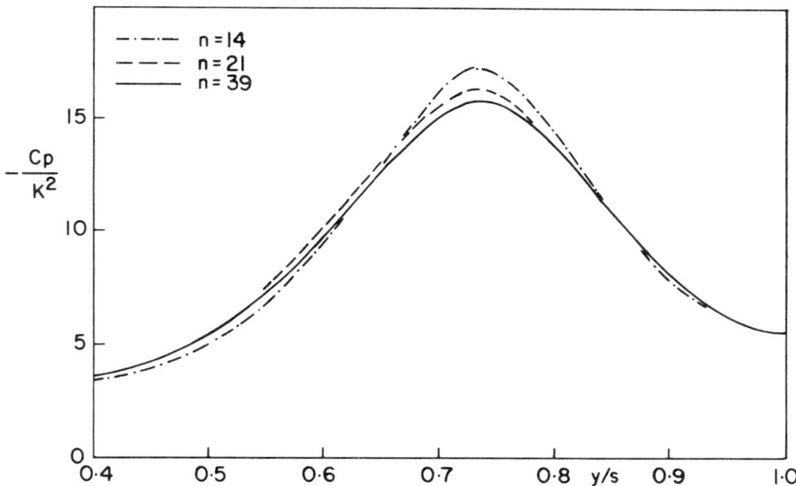

Fig. 6 Suction peaks beneath sheets of different
        extent, Smith (1968)

The justification for this gross simplification comes from the results of calculations using it. For a flat-plate delta wing, using slender-body theory to calculate the velocity field, for the model of the sheet sketched on the right of Fig. 3, we calculate the vortex configurations shown in cross-section in Fig. 4 (Smith (1968)). In slender-body theory, the solution depends only on the similarity parameter α/K, where α is the angle of incidence and K is the tangent of the semi-apex-angle of the wing. These calculations are all for $\alpha/K = 0.91$. The vortex sheet is terminated at three different positions. The shortest sheet extends for between a quarter and half a turn about the line vortex, while the other two extend for one further turn and two further turns. The different solutions are distinguished by the values of the parameter n, the number of points at which the boundary conditions on the sheet are applied in the numerical solution. As can be seen the calculated shapes are very similar, except for a discrepancy where the finite part of the sheet ends. The corresponding positions of the core are also very close. The core position on its own, shown by a circle, is obtained with no outer part of the sheet, i.e. a cut connects it to the leading edge. Clearly some vortex sheet representation is required. To show that the agreement between the three sheet solutions is actually significant, we need to establish that the extra turns do take the model significantly closer to the limit of an infinite spiral. We can do this by presenting the proportion of the total circulation carried on the sheet: the shortest contains 23%, the longer contains 49%, and the longest contains 60% of a total circulation which is virtually the same in all cases.

If we turn to the distribution of the circulation along the sheet, Fig. 5 shows how the calculated jump in tangential velocity varies along the length of the sheet, for the same three extents of sheet. Again, apart from a discrepancy at the free end of the sheet, the distributions are very similar. The waves in the distribution are one aspect of the ellipticity of the vortex, mentioned above. To complete the comparison, Fig. 6 shows how increasing the length of the outer part of the sheet affects the suction peak on the wing under the vortex. Again, the differences are almost negligible, despite the fact that the shortest sheet does not come between the line vortex and the wing. A bigger difference might be found for a smaller value of α/K, for which the vortex would lie closer to the wing.

The comparisons so far have shown the effect of adding a whole turn to the sheet to be very small. This is a slightly favourable view as Fig. 7 shows. Here we see the position of the line vortex varying continuously as the length of the outer part of the sheet is increased from zero. The extent of the outer part of the sheet is described by the angle, θ_E, measured in radians,

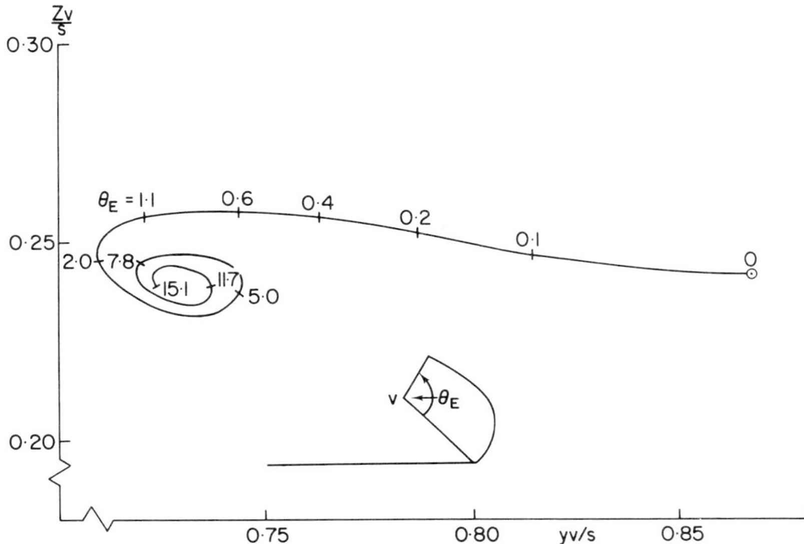

Fig. 7   Position of core vortex for sheets of different
         extent

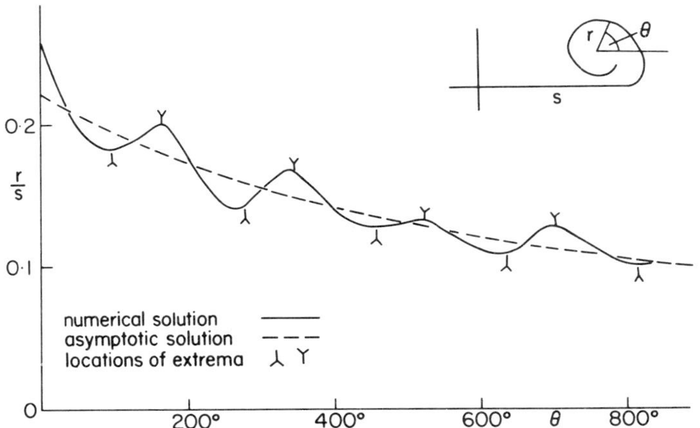

Fig. 8   Shape of vortex sheet compared with asymptotic
         solutions, Smith (1968)

subtended by the sheet at the vortex. Some values of θ_E are
shown along the curve. These calculations are for a delta wing
with $\alpha/K = 1$. It can be seen that the core moves significantly
as the sheet first grows from the leading edge, but its subsequent movement is much smaller. However the change produced by
adding part of a turn of the sheet is larger than that produced
by adding a whole turn.

Finally, let us look at the relation between the numerical
solution and the asymptotic solutions. Fig. 8 shows the radial
distance of the sheet from the line vortex, as a function of
the polar angle, for the longest of the solutions for $\alpha/K=0.91$.
It also shows the variation predicted by the simplest of the
asymptotic solutions (Mangler and Smith (1959)), matched to the
circulation and size of the inner part of the vortex. The general trend agrees much further out along the sheet than could
reasonably be expected. However, the waves, arising from the
ellipticity of the computed sheet, confirm the shortcoming of
the asymptotic solution suggested in relation to Fig. 5. The
marks on the numerical solution are placed at the values of θ
for which the approach by Maskell (1964) predicts a local
extremum of r/s. The measure of agreement achieved appears to
offer some support to the approach.

How is this representation of the vortex to be used to describe the separated flow? There are three further requirements:
the first two are a knowledge of the position of the separation
line and a Kutta condition to be applied there. For the moment
we suppose the separation line is known, either fixed at a salient edge, deduced from experiment, or found in a parallel calculation. The form taken by the Kutta condition is not always
the same. The most obvious form arises when a mapping method is
used for a salient edge. If the edge is regularized by the
mapping, a Kutta condition of a stagnant cross-flow at the
image of the edge in the mapped space ensures a finite velocity
in the physical space. As we shall see, other problems and
other treatments require other conditions. Perhaps for nonlinear governing differential equations no Kutta condition will be
required, as seems to be the case for some aerofoil calculations
in transonic attached flow. For linear governing equations it
seems a Kutta condition is required to produce uniqueness. With
the vortex sheet model its choice is usually straightforward,
based on the elimination of the most blatantly unphysical aspect
of the flow at the separation line. Other models present
greater difficulties. In any case, there is no guarantee that
imposing a Kutta condition does lead to a unique solution. In
particular, for a slender conical wing-body combination,
Levinsky and Wei (1968) found multiple solutions over a range
of moderate angles of incidence. At present we have to rely on
physical insight to choose between multiple solutions if they
arise.

The third requirement is for some mathematical technique to
solve the equations which govern the flow field, in terms of
the boundary conditions. In principle, all the techniques for
inviscid flow that are being described at the conference, as
well as others which are not, are available. In practice,
finite-difference, finite-volume and finite-element methods
have scarcely been applied to flows with vortex sheets. This is
partly because the representation of vortex sheets plays a minor
role in the description of the steady, two-dimensional flows for
which the numerical methods have initially been developed; and
partly because the representation of vortex sheets of unknown
position in mesh schemes is difficult. Capturing methods,
which, as Roe (1982) has shown, can be made to work very well
for shock discontinuities, appear to spread slip discontinuities over many mesh intervals. Fitting, on the other hand, is
complicated, especially when the discontinuity surface rapidly
changes its orientation to the mesh.

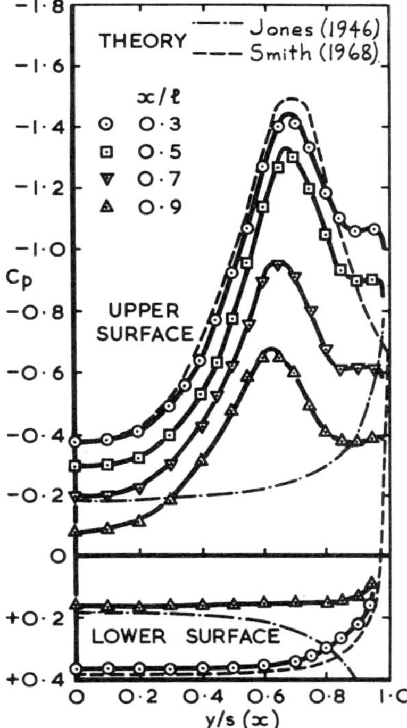

Fig. 9   Pressure distribution on a delta wing; slender-body
theory and experiment: after Hummel and Redeker (1972)

Much of the work that has been done with vortex sheets has been within the framework of slender-body theory, for body cross-sections sufficiently simple for the attached flow to be written in a closed analytic form. Good predictions of vortex shape and position result from this for slender shapes and subsonic speeds. Useful insights have also been gained into the effects of planform, thickness, camber, incidence, side-slip, steady roll, and oscillations in pitch and heave. However Fig. 9 shows that less success is achieved in the basic task of predicting the pressure distribution over the wing. Here the wing is a thin flat delta of aspect ratio one, at $20.5°$ incidence, in a low-speed wind tunnel. Pressures were measured across the local semi-span of the wing at various lengthwise stations defined by the values of x/ℓ listed, and used to define the six solid curves shown, four on the upper surface and two on the lower. It is immediately apparent that the loading falls off rapidly over the rear half of the wing. Slender-body theory predicts the same pressure distribution at each station. The prediction for attached flow appears as the chain line and is not really related to the measurements in any way. The prediction using the vortex sheet model for the separated flow appears as the broken line. It agrees very closely with the measurements at the 30% station on both surfaces, except near the leading edge on the upper surface. Here the boundary layer is being swept outboard under the vortex against a rising pressure, so a secondary separation takes place. The boundary layer is turbulent, otherwise it would have separated earlier and the discrepancy would have been greater.

We can say on the one hand that the vortex model seems to be reflecting the physics quite accurately (except for the secondary separation), but that the slender-body framework is not adequate for quantitative purposes at low speeds. It becomes much better for thin wings as the Mach number approaches one, but deteriorates again at supersonic speeds. Different physical processes intervene, so that leading-edge separation is almost absent when the Mach number of the component of the onset flow normal to the leading edge exceeds about 0.7.

3. PANEL METHOD FOR SHARP-EDGED WINGS

Fortunately there is at least one other framework which is well suited to the representation of vortex sheets. This is the framework of the panel methods, an example of which is described by Butter and Hunt (1982). One particular panel method has been developed over a number of years by staff at the Boeing Company with NASA support, to provide the most elaborate and successful treatment of leading-edge separation on highly swept wings that I know of. The account which follows is based on Johnson, Lu, Tinoco and Epton (1980) and Tinoco, Lu and Johnson (1980).

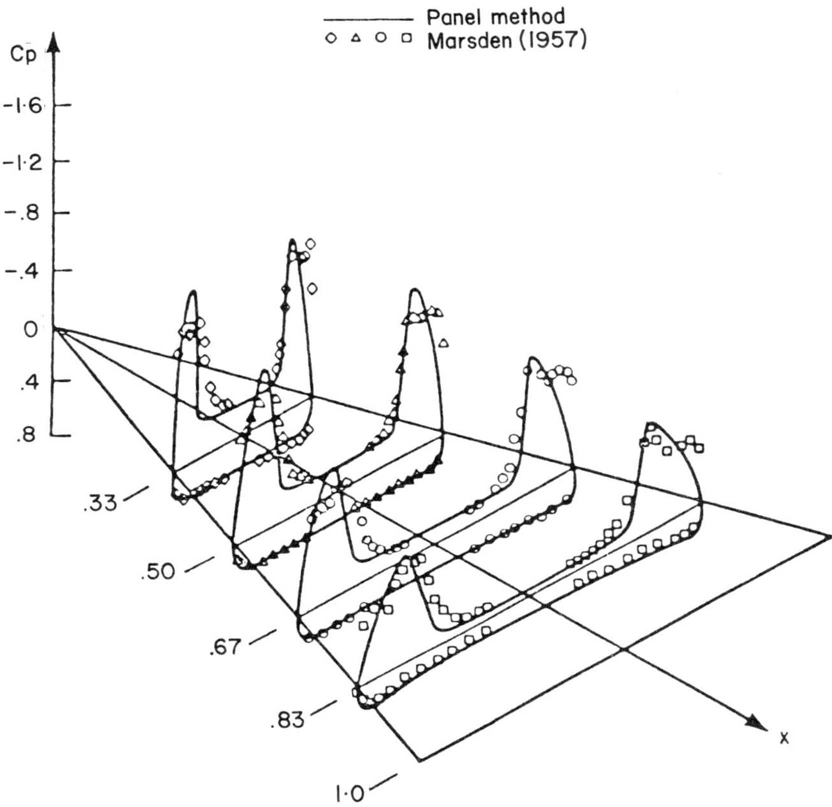

Fig. 10   Pressure distribution on a delta wing, panel method
          and experiment, Weber et al (1975)

Since we have been looking at Fig. 9, which emphasises the
shortcomings of the slender-body framework for the prediction of
pressure distributions, let us contrast it with Fig. 10. Again
we have measurements and calculations for a thin delta wing at
low speed, this time with an aspect ratio of 1.46 and at an
incidence of 14°. Now, the observed reduction in loading from
front to rear is well reproduced by the calculation, both over
the inner part of the wing and beneath the vortices. As would
be expected, there is no improvement in the prediction of the
shape of the suction peak, these measurements being made with
a laminar secondary separation. This particular comparison is
based on an earlier version of the programme, see Weber, Brune,
Johnson, Lu and Rubbert (1975). The main reasons for dissatis-
faction with the earlier programme were difficulties with numerical convergence and doubts about the predicted lift coefficients

446   SMITH

on wings of larger aspect ratio.

It is convenient to describe the model in terms of the ways
in which it differs from the slender-body treatment which led
to Figs. 4 to 9. First, and most important, the governing
differential equation is the Prandtl-Glauert equation for the
small-disturbance approximation to the velocity potential:

$$\beta^2 \phi_{xx} + \phi_{yy} + \phi_{zz} = 0; \quad \beta^2 = 1 - M_\infty^2. \qquad (2)$$

The earlier work used the two-dimensional Laplace equation,
based on the assumption that streamwise variation is slow, an
assumption obviously violated at an unswept trailing edge.
Because equation (2) is linear, solutions can be built up by the
superposition of elementary solutions of the equation. In a
panel method these elementary solutions are distributions of
source and doublet strength over small surface elements, which
may be located either inside or on the surface of solid bodies,
or thin shear layers. In this particular method each panel is
part of a hyperboloidal surface, see Morino et al (1975), so
that the edges of neighbouring panels can always be made to
abut exactly, without gaps. The distribution of singularity
strength can also be made continuous across the panel edges,
and conditions are imposed to ensure that this is so.

On the solid surface and on the vortex sheet a condition of
zero normal flow is imposed. In a panel method for compressible
flow this condition is ambiguous. If the velocity vector is
written in terms of the free-stream speed, U, and the components
of the disturbance velocity, u, v, w,

$$\underline{V} = (U + u)\underline{i} + v\underline{j} + w\underline{k}; \qquad (3)$$

the corresponding first-order approximation to the mass-flux
vector is

$$\rho\underline{V} = \rho_\infty \left[(U + \beta^2 u)\underline{i} + v\underline{j} + w\underline{k} \right] \qquad (4)$$

Consequently, unless $M_\infty^2 u$ can be neglected, the conditions of
zero normal velocity and zero normal mass flux are different.
In the familiar cases of incompressible flow, fully linearised
wing theory, and slender body theory, $M_\infty^2 u$ is either zero or
negligible in comparison with U. However, with a panel method
for compressible flow, a choice has to be made between the two
conditions, and it appears the choice has to be somewhat
arbitrary. In the present problem, the desirability of representing vortex sheets as simply as possible provides a strong

argument for choosing the condition of zero normal mass flux.
A vortex sheet must involve a jump in potential, represented
by a doublet panel. In general, u will be discontinuous across
the sheet; so, unless it is the normal mass flux which is constrained to be zero, a source panel will also be needed to
supply the difference in mass flux across the sheet. Under the
same basic assumption that u, v and w are small and of the same
order, the approximate form of Bernoulli's equation for the
pressure coefficient is

$$C_P = - \frac{2u}{U} - \frac{\beta^2 u^2 + v^2 + w^2}{U^2} . \tag{5}$$

This includes the form used in slender-body theory, in which
$\beta^2 u^2$ is omitted. The presence of quadratic terms is essential
if realistic vortex sheets are to emerge.

In the way the core model was derived earlier, it is implicit
that the line vortex is almost parallel to the stream, so that
equating to zero the component of force in the cross-flow plane
is equivalent to equating to zero the force normal to the vortex.
It is, in fact, the latter which is the fundamental condition,
since there is always an infinite force acting along a line
vortex, corresponding to the infinite energy needed to form it.
In the fully three-dimensional treatment now being described, it
is allowable for the vortices to be inclined at any angle to
the free stream, so the condition to be applied is that the component of the force on the core representation in the plane
normal to the axis of the vortex must vanish.

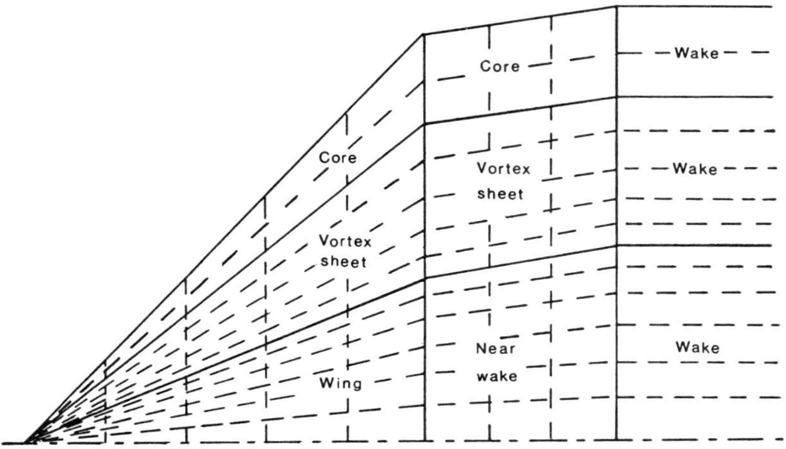

Fig. 11   Panelling of thin delta wing, leading-edge vortex,
          and wake, after Tinoco et al (1980)

Fig. 11 shows a sketch of the panelling arrangement for a
thin delta wing. The vortex sheet and the cut have been
unrolled, and stretched a bit, to allow a planar representation.
Each area enclosed by solid lines represents a network of panels,
and the labels indicate different types of panel used. All the
panels are doublet panels, though source panels could be used
to represent wing thickness. The panels on the wing are of the
'analysis' type, the remainder are of 'design' type, the difference in function relating to whether a boundary condition on
the normal flow or a boundary condition on the pressure is to
be satisfied. The near-wake network remains fixed in position
during the calculation, but the panel strengths vary in order
to satisfy a condition of zero pressure difference. The vortex
sheet networks are allowed to move, so that both zero mass flux
and zero pressure difference conditions can be satisfied. In
the core networks, all panels in the same column in the figure
have the same strength. There are three slightly different
kinds of network labelled 'wake', but they all have the property
that the linearised form of the pressure coefficient is continuous across them: $\Delta u = 0$.

A Kutta condition is applied at all edges of the planform.
For thin wings, on which attention has been concentrated so far,
it works like this. The continuity conditions between the networks already ensure that the potential is continuous across
the edge of the wing on both sides of the doublet surface.
This does not prevent flow round the edge. Flow round the edge,
though, would in principle involve an infinite component of
vorticity in the wing parallel to the edge. It can therefore
be prevented by making the component of the vorticity parallel
to the edge the same on adjacent panels of the wing and the
vortex sheet (or of the wing and the near wake). If the wing
and sheet panels had a common tangent plane, this condition
would mean that the vorticity vector was continuous between the
wing and the sheet, as it should be. In general, only the magnitude of the vorticity vector is continuous and its direction
changes across the leading edge of the wing.

The panel strengths which are to be found may be denoted
collectively by Λ, and the geometrical configuration of the
sheet and the core can be defined by a set of parameters denoted
collectively by Θ. The conditions which these unknown quantities Λ and Θ have to satisfy are grouped into two sets. The
first, denoted by $F(\Lambda,\Theta) = 0$, consists of the condition of zero
mass flux on the wing, the condition of zero pressure difference
across the vortex sheet and the wake, and the Kutta condition.
This provides the same number of equations as there are unknowns
in Λ. The second set of conditions, denoted by $G(\Lambda,\Theta) = 0$ consists of the conditions of zero mass flux on the vortex sheet
and the conditions of zero normal force on the core representa-

tion. This brings the total number of equations up to the
total number of unknowns. The equations are solved by a quasi-
Newton scheme. However, the iterations are nested, with the
equations $F(\Lambda,\Theta) = 0$ being solved for Λ, with Θ held fixed. The
equations $G(\Lambda,\Theta) = 0$ are then introduced in the outer iteration
in order to determine Θ and Λ. The increments which the Newton
algorithm recommends are scaled down, before being applied to
Λ and Θ, by a factor chosen to ensure that the residual actually
reduces at each step.

In some cases a form of instability arises with this method
and so a modification is made to prevent its excitation. The
instability arises because the boundary conditions at a point
of the vortex sheet can be closely approximated by panels with
varying degrees of twist. Inadequate panelling near the apex
can introduce significant amounts of twist into the panels there,
which appears to force even larger twists further downstream,
so that the Newton iteration fails to converge. The instability
is suppressed by introducing supplementary equations which
require the panels to be untwisted. The system is then over-
determined and can only be satisfied in a least-squares approximation. Suitable weighting leads to solutions with only slightly
twisted panels.

The Newton algorithm requires a starting approximation which
is not too far from the solution. Starting approximations
derived from slender-body theory calculations (Smith, 1968) seem
to be adequate, perhaps because it is the starting approximation
to Θ which is most important and the slender body approximation
to the geometry is quite good.

The adequacy of the representation used has been checked by
comparisons of calculations with different panellings. These
are all for delta wings and for combinations of incidence and
aspect ratio for which the vortex is well clear of the wing. The
first question concerns the wing panel arrangement. Two schemes
were applied:the conical arrangement of Fig. 11, and a typical
swept-wing arrangement based on streamwise lines and lines at a
constant proportion of the local chord. With 64 panels on the
half-wing in each scheme, the two schemes yield lift coefficients
less than 0.5% apart and centres of pressure between 1% and 2%
of the root chord apart. The vortex core positions are not significantly different at the trailing edge. It is reasonable
that this change in panel arrangement affects the centre of
pressure rather than the lift.

The second question concerns panel density. Two conical
panelling schemes, one with 30 and the other with 60 panels on
the half-wing, showed a different pattern of discrepancy: about
4% in the lift coefficient and no significant change in the
centre of pressure. The vortex core moved by about 2% of the

semi-span. The type of discrepancy is what would be expected
from a change in the number of conical rays in the scheme, but
no change in the number of panels along each ray.

The third comparison involves roughly doubling the extent of
the sheet which is panelled. Fig. 12 shows sections through the
two sheets for an angle of incidence of 20° and an aspect ratio
of 2. The axis of the vortex has moved about 2% of the semi-
span as a result of the increased length of the outer part of
the sheet. This compares with a shift of 3.5% in Fig. 7 for a
similar extension, using slender-body theory. The change in
lift coefficient between the panellings in Fig. 12 is only 0.5%.
The effect of a similar change in the panelling at smaller
aspect ratios, for the same angle of incidence, is greater however, since the nonlinear lift forms a greater part of the
total. For an aspect ratio of unity, the lift coefficient
changes by 1%, and for the extreme case of the aspect ratio
0.25, the change is 4%. In each case, the longer sheet gives
the larger lift. These results support the conclusion drawn
from the slender-body theory results that between a quarter and
a half turn is enough for the outer part of the sheet.

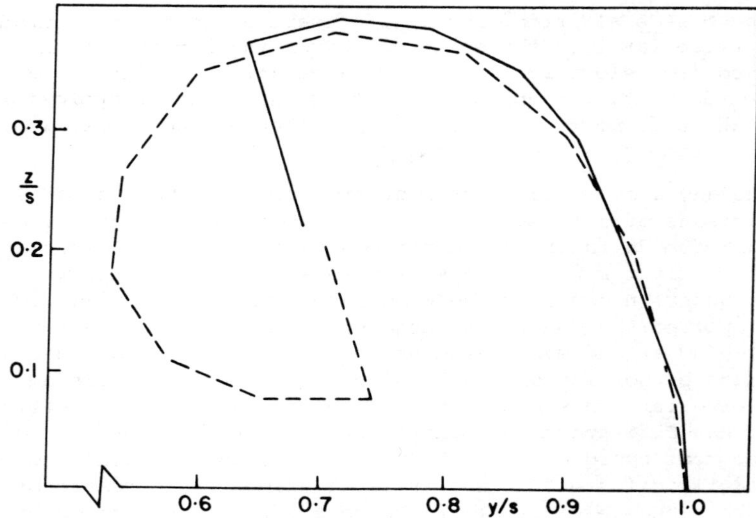

Fig. 12   Standard and extended panelling of the vortex
          sheet, Johnson et al (1980)

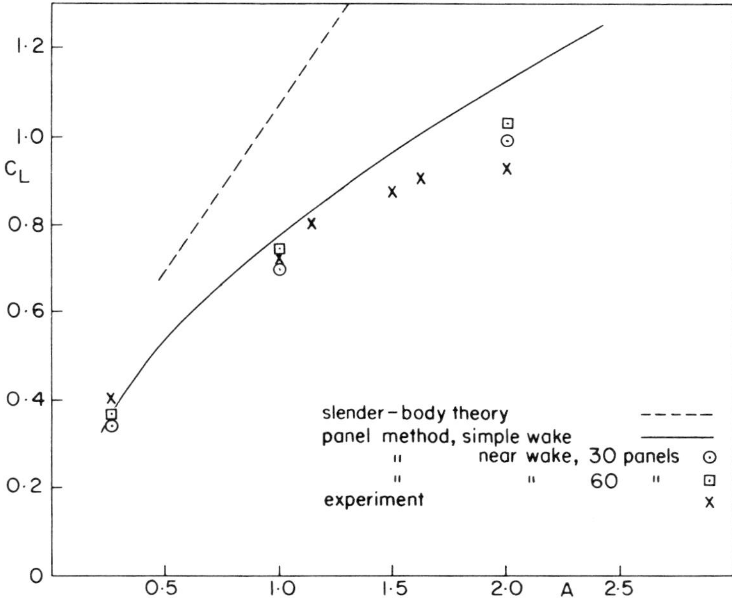

Fig. 13   Variation of lift coefficient with aspect ratio for
delta wings at 20° incidence, after Johnson et al. (1980)

Fig. 12 also shows that the panelling near the leading edge
is not dense enough to reproduce the proper shape of the vortex
sheet in that region.   It might be thought that this would have
a significant effect on the Kutta condition and therefore on
the lift.   However, in another check calculation extra panels
were introduced near the leading edge and produced a change in
lift of less than 0.5%.   The only change in the pressure distribution is very close to the edge.

Finally Fig. 13 shows a comparison between various calculated
and experimental values of the lift coefficient at an incidence
of 20°, as a function of aspect ratio, again for flat-plate
delta wings at low speeds.   The crosses are measured values,
the broken line is obtained from the slender-body theory calculations, and the remaining points and curve come from the panel
method calculations.   It is clear at once that the panel method
is a great improvement over the slender-body theory, even at
aspect ratios below unity where the slender-body assumptions
are less in error.   Comparing the solid curve with the circles,
we see that including the near-wake panelling in the method is
very important, particularly at the higher aspect ratios.   The
effect of the near-wake panelling is to allow the trailing

vortices shed from the wing trailing edge to deviate from the
free-stream direction and align themselves more nearly with the
local flow direction. Comparing the circles and squares, we
see again the effect of the number of wing panels. Lastly, the
squares should be compared with the crosses to assess how the
best results from the model compare with experiment. At first
sight, this is disappointing, with an 11% overprediction at
$A = 2$ and a 6% underprediction at $A = 0.25$. However, there is
little doubt that the vortices in the experiment on the $A = 2$
wing had burst, and there is a possible asymmetry in the vortices in the experiment on the $A = 0.25$ wing. All the other
experimental points line up very closely with the trend
established by the three squares.

From the published account of the method it appears that convergence difficulties arise for small angles of incidence. We
may guess that these difficulties arise from the close approach
of the panels representing the leading-edge vortex to the panels
representing the wing and wake. Situations in which the vortex
sheet is close to the wing are common in cruising conditions,
when it is important to be able to design for low drag in the
presence of vortices. It may be that some form of sub-panelling
like that used by Maskew (1975) would be helpful.

There is also a lack of flexibility in vortex sheet models,
in the sense that the structure of the flow needs to be known
in advance. For instance, a double-delta wing may produce two
vortices on each side at small angles of incidence and only one
vortex on each side at large angles of incidence. The panel
method at present only deals with the single-vortex case, but
it could be developed to deal with the pair of vortices. However, it is difficult to see how it would predict which system
would occur at a given angle of incidence, or at what incidence
the two vortices would amalgamate. For problems like this a
more flexible model is needed, in which the evolution of shed
vorticity is traced in space or time. Such models are easy
enough to construct in slender-body theory, and an interesting
fully three-dimensional time-dependent method has recently been
produced by Rehbach (1978a and b) for incompressible flow.

4. SEPARATION FROM THE SMOOTH SURFACE OF A POINTED BODY

It is relatively easy to show that, if vorticity is being
shed into a free vortex sheet from a separation line on a smooth
body, the sheet is tangential to the surface of the body along
the separation line (J.H.B. Smith, 1977). We can then call the
side of the separation line on which the sheet lies close to the
body surface, the 'downstream' side, and the other side the
'upstream' side. Since both the surface of the body and the
vortex sheet are stream surfaces, the velocity vector on the

downstream side of the separation line must be tangential to the
separation line. On the upstream side, the only constraint is
that the velocity vortex lies in the common tangent plane of the
body and the vortex sheet.

Along the separation line, the velocity is already finite in
the attached flow. When we make the pressure continuous across
the vortex sheet, it is continuous across the separation line.
However, unless it is specified as an extra condition, the
velocity vector on the downstream side will not be along the
separation line: there could be a flow 'sliding' between the
body and the vortex sheet. This condition therefore serves as
the Kutta condition for separation from a smooth surface.

It may sometimes be awkward to evaluate the velocity between
the sheet and the body. An indirect approach may then be useful.
At a point P on the separation line, let us denote by u, v and
w the components of velocity along the separation line, normal
to the separation line but tangential to the surface, and normal
to the surface. Denoting the upstream side of the separation
line by suffix 1 and the downstream side by 2, we can write

$$u_1^2 + v_1^2 + w_1^2 = u_2^2 + v_2^2 + w_2^2,$$

by the condition of continuity of pressure. Moreover

$$w_1 = w_2 = 0,$$

by the body boundary condition;

$$v_2 = 0,$$

by the Kutta condition; and

$$u_2 - u_1 = \frac{\partial}{\partial \sigma}(\phi_2 - \phi_1) = \frac{d\Gamma}{d\sigma},$$

where σ is the arc length along the separation line and Γ is
the total circulation about a contour surrounding the sheet and
crossing it at P. Then

$$v_1^2 = \frac{d\Gamma}{d\sigma}(\frac{d\Gamma}{d\sigma} + 2u_1).   \qquad (6)$$

This expresses the Kutta condition in terms of overall circulation and upstream quantities. A similar formulation involving
the mean, or convective, velocity components is obviously
possible.

Local analysis of the behaviour of a vortex sheet leaving a
smooth surface (J.H.B. Smith (1977), F.T. Smith (1978)), reveals
two quite different possibilities. Looking at a section through
the surface and the sheet normal to the separation line, we find
that the distance of the sheet from the wall, z, is given by

$$z = a_1 (y - y_s)^{n+\frac{1}{2}} + \dots \qquad \text{for } y > y_s, \qquad (7)$$

where $y - y_s$ is the distance along the wall downstream from the
separation line, the dots denote higher order terms, and n = 1,
2, 3 ... . Associated with this shape of the sheet is a similar
form for the pressure upstream of the separation line:

$$p = p_s - a_2 (y_s - y)^{n-\frac{1}{2}} + \dots \qquad \text{for } y \leqslant y_s, \qquad (8)$$

where a_2 is proportional to a_1. The case n = 1 corresponds to
a sheet with infinite curvature at S and an infinite pressure
gradient at S. For n > 1 the curvature and pressure gradient
are finite. It seems likely that the occurrence of n > 1 only
arises through the vanishing of a leading term with n = 1; so
that for a general position of the separation line, S, on the
body, n will be unity, with n greater than unity only at special
positions of S. Calculation confirms this. The positions of S
for which n is greater than unity are called positions of smooth
separation, since the infinite curvature and pressure gradient
are avoided. These results represent a generalisation of the
behaviour of the Kirchhoff free-streamline solution which produces a constant-pressure wake behind a circular cylinder in
plane flow.

Their significance is, that if S is not a position of smooth
separation, a boundary layer calculation will predict singular
separation upstream of S, because of the infinite adverse pressure gradient (8). The only positions of separation consistent
with a regular boundary layer growth are therefore positions of
smooth separation, with the paradox that the pressure gradient
is usually favourable and the boundary layer does not separate
there. Moreover, calculation shows that, for both a circular
cylinder in plane flow and a slender cone at incidence, the position of smooth separation is well upstream of the position of
the laminar separation line observed in real flows. An improvement in the classical boundary layer model is therefore needed
if the position of the separation line is to be predicted
rationally or successfully.

The rational approach to problems of this kind in plane
laminar flow is the asymptotic analysis of the Navier-Stokes

equations for large Reynolds numbers. This leads to the emergence of a triple-deck structure which surrounds the separation
point. With ε a small quantity equal to $Re^{-1/8}$, the streamwise
extent of the triple deck is of order ε^3. Within this length,
a free interaction between pressure and displacement takes place
which enables separation to occur in a regular fashion. It is
the lower deck, whose thickness is of order ε^5, that dictates
the behaviour of the whole. The middle deck is essentially the
main part of the upstream boundary layer, with thickness of
order $\varepsilon^4 = Re^{-1/2}$, displaced outwards a distance $\varepsilon A(X)$, where X
is a scaled streamwise coordinate. The upper deck has a thick-
ness which is of the same order, ε^3, as the streamwise extent
of the region. In it, the flow is inviscid, so that, for an
incompressible flow, the pressure P is related to the displacement A by the familiar integral

$$A'(x) = \frac{1}{\pi} \int_{-\infty}^{\infty} \frac{P(\xi)d\xi}{x - \xi} . \tag{9}$$

The lower decks are governed by the boundary layer equations
and the pressure is constant across the whole thickness of the
layer. F.T. Smith (1977) has shown, by a numerical calculation,
that a structure of this kind does exist, and that it appears to
be unique.

The outcome is that the displacement and pressure have the
forms indicated in equations (7) and (8), with n = 1; but that
the coefficients a_1 and a_2 are themselves dependent on Reynolds
number, being proportional to $\varepsilon^{1/2}$. Thus as $Re \to \infty$ and $\varepsilon \to 0$
separation becomes smooth. At finite Reynolds number separation
takes place regularly at a distance proportional to $\varepsilon^{1/2}$ further
downstream. The role of the triple deck is to permit the growth
of the viscous layer upstream of the separation point, so weakening the severity of the adverse pressure gradient sufficiently
to allow a regular form of separation. The second parameter
which governs the scale of the triple deck is the skin friction
at its upstream end, which is determined by the upstream boundary layer. The larger the skin friction, the more energetic is
the approaching boundary layer, and the larger is the pressure
rise along the length of the triple deck.

It has also been shown by F.T. Smith (1978) that the same
structure describes the behaviour of a laminar three-dimensional
separating flow, in a plane normal to the separation line. This
is a very useful result, because for highly swept separation
lines the vortex sheet model already discussed can be expected

to describe the flow adequately, whether the separating shear
layer springs from a salient edge or a smooth surface. We
therefore have available both a model of the separated flow and
a model of the separation process, both appropriate to flows at
large Reynolds numbers. These two aspects have been brought
together by Fiddes (1980) to study the body vortices on cones of
elliptic cross-section at incidence. His treatment and some of
his results will now be described.

The framework for the inviscid vortex-sheet calculation is
slender-body theory. For supersonic flow this leads to a velocity
field which is conical. For subsonic flow, only the cross-flow
velocity components are constant along rays through the apex of
the cone, but this is enough to ensure the consistency of a conical vortex-sheet configuration. The question of the consistency of the conical inviscid flow with the viscous elements will
be taken up later. For the inviscid calculation it is simply
assumed that there is a straight separation line, an assumption
which is in agreement with experimental observations. The position of the separation line is assumed to be known at this stage.

For the attached flow past a slender elliptic cone at incidence, the velocity field can be expressed in closed form. In
order to avoid any numerical irregularities in the velocity
field of the vortex sheets, and to reflect the known analytic
form (7) near the separation line, a very smooth representation
was developed for the vortex sheets. The first step is to perform the sequence of conformal mappings illustrated in Fig. 14.

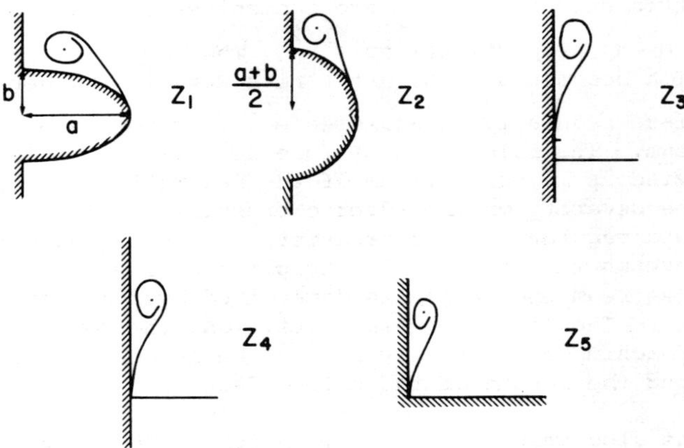

Fig. 14 Conformal mappings leading to a vortex sheet with
         regular curvature, Fiddes (1980)

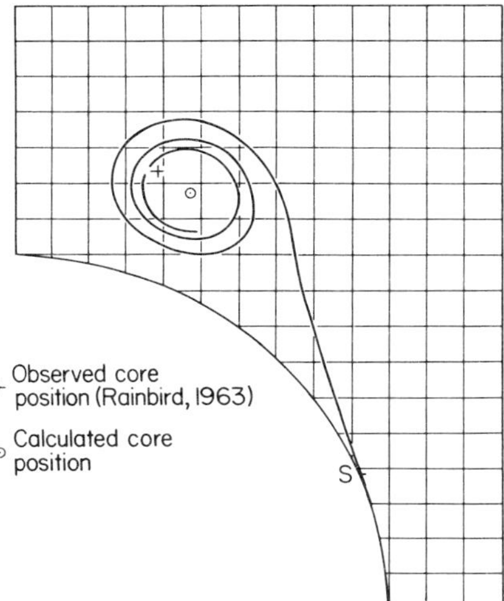

Fig. 15   Vortex sheet from observed laminar separation
line, Fiddes (1980)

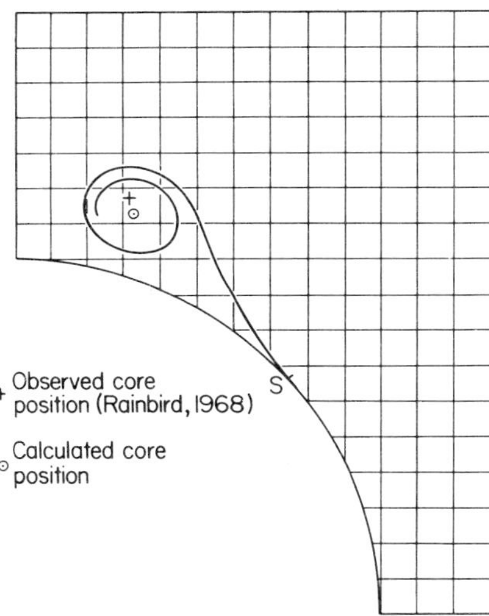

Fig. 16   Vortex sheet from observed turbulent separation
line, Fiddes (1980)

The effect of this is that the body boundary is simplified and
the infinite curvature of the sheet at its base becomes finite
in the final transformed plane Z_5. In this plane, the sheet is
then constructed out of circular arcs, for which the velocity
field is known analytically, Mangler and Smith (1959). The
circular arcs touch at their end points, and the singularity
strength is forced to be continuous at the end points, so that
the resulting induced velocity field is continuous everywhere,
except for the required jump in tangential velocity across the
sheet, and the remaining singularities at its free end and at
the line vortex. The body boundary condition is preserved by
introducing image sheets in the Z_5 plane. The vortex-sheet
boundary conditions are applied at the junctions of the circular
arcs. Together with the Kutta condition described above and the
usual conditions of zero total force on the line vortex and cut,
these conditions provide enough equations to determine the configuration. They are solved by a Newton iteration.

A typical result is shown in Fig. 15, for a circular cone at
an angle of incidence three times its semi-apex-angle. The
separation position is taken from an experiment in a water tunnel
by Rainbird et al (1963), at 113° from the windward generator,
corresponding to laminar flow upstream of separation. The
observed position of the vortex core is shown for comparison
with the prediction. The discrepancy is about 10% of the cone
radius. An error of similar size in solutions for thin delta
wings with laminar secondary separation has been attributed to
failure to model the secondary separation (Smith, 1968). It is
likely that the secondary separation is at least as significant
in the experiment with the circular cone. Fig. 16 shows the
case of an incidence of 2.52 times the cone semi-angle, with the
separation line at 132°, taken from further tests by Rainbird
(1968), this time in a wind tunnel at a Mach number of 1.8. The
boundary layer at separation is now turbulent and the discrepancy in core position is smaller. Rainbird also measured the
total pressure over the region of rotational flow and Fig. 17
shows the contours of the total pressure relative to its free-
stream value. Superimposed on them are two calculated vortex
sheet shapes, one by Fiddes and one by Zacharov (1976).
Zacharov calculates the steady flow as the limit of a flow which
evolves in time, and it is suggested that the discrepancy between
the two calculations might be reduced if the evolution of his
solution were followed to a later time. The secondary separation
shows up in the measurements, which also show air from the free
stream being swept in beneath the vortex. These comparisons
suggest that the inviscid model is performing as well on smooth
bodies as it does on sharp-edged wings, so it is reasonable to
proceed to a viscous model in the hope of being able to calculate the position of the separation line. Since a triple-deck

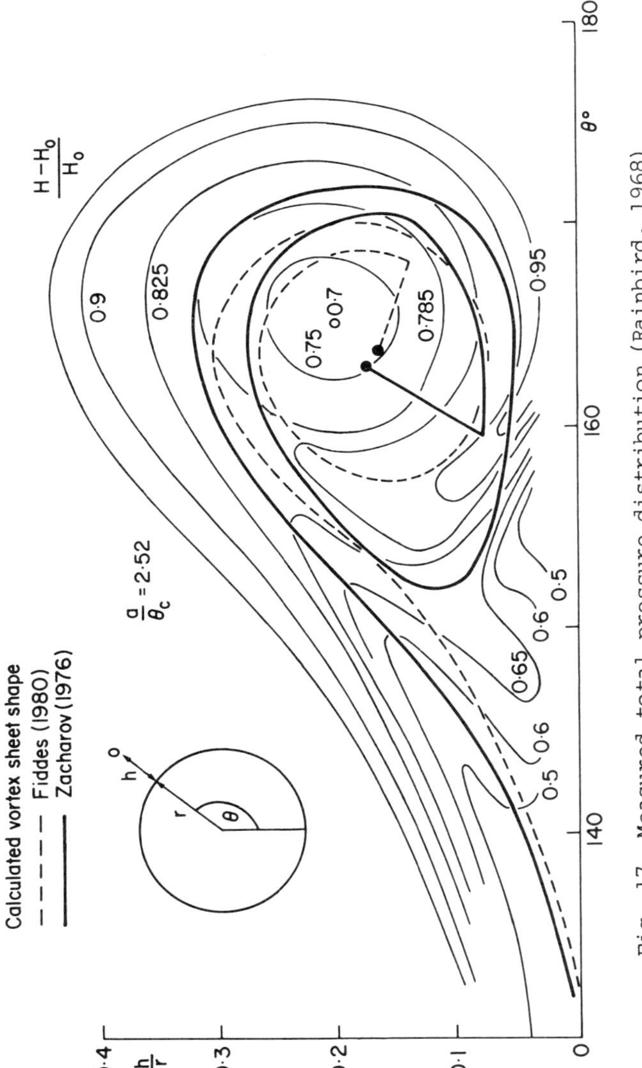

Fig. 17 Measured total pressure distribution (Rainbird, 1968) compared with calculated vortex sheets

treatment is only available for laminar flow, we confine our attention to this case.

We consider first the calculation of the boundary layer on the cone upstream of separation. For a conical external flow, there is a similarity solution of the three-dimensional laminar boundary-layer equations in which the growth of the boundary layer along the conical rays takes a simple form. The non-conical element in the external flow is therefore neglected, to exploit this simplification; with the partial justification that streamwise gradients are small over most of the length of the cone, even in subsonic flow. The similarity form requires a numerical solution of two coupled partial differential equations for the velocity components in a typical cross-flow plane, as functions of the angular distance round the cone and the distance from the wall. For this, an unpublished Keller box method by C.F.M. Boyd (RAE) was used, with the external flow calculated using the vortex-sheet model.

The determination of the position of the separation line at finite Reynolds number starts with the determination of the position of smooth separation, which is the correct position for infinite Reynolds number. From the inviscid calculation, with the separation line at a (parametric) angular position $\theta = \theta_s$, the coefficient of the singularity in the curvature at the base of the vortex sheet can be found. In particular, the quantity

$$A_1(\theta_s) = \lim_{\sigma \to 0} \left(\frac{\sigma}{a}\right)^{\frac{1}{2}} \frac{\rho}{a} \tag{10}$$

is calculated; where σ is the arc length along the cross-section of the sheet, measured from the separation line, a is the semi-major-axis of the cross-section of the cone, and ρ is the radius of curvature of the cross-section of the sheet. A_1 is also proportional to the coefficient of the infinite pressure gradient upstream of the separation line. A_1 then varies with θ_s, as shown, for example, in Fig. 18 for several axis ratios, t, of the elliptic cross-section, all for an angle of incidence equal to three times the semi-apex-angle of the planform of the cone. As would be expected, the flatter the cone, the steeper is the variation. Smooth separation is found where $A_1 = 0$, and we see that this is always upstream of the end of the major axis, where $\theta_s = 90^\circ$. The next step is to calculate the boundary layer which develops under the inviscid flow corresponding to smooth separation. This provides the skin-friction which is the input to the triple-deck.

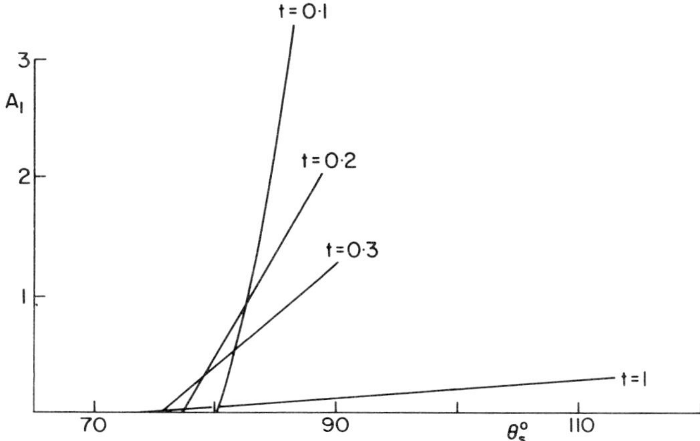

Fig. 18   Variation of inviscid singularity at separation
with position of separation, Fiddes (1980)

When the special features of the conical inviscid flow and
the quasi-conical boundary layer are introduced into the three-
dimensional form of the triple-deck solution of F.T. Smith (1978),
it is possible to deduce the following relation:

$$A_1(\theta_s) = 2c\left(\frac{Ue}{Ve}\right)^{7/4}(k^2 Re_x)^{-1/16}\tau^{9/8}. \tag{11}$$

A_1 is the coefficient defined in (10); c = 0.44 is the constant
calculated by F.T. Smith (1977); Ue and Ve are the inviscid
velocity components along and normal to the smooth separation
line; x is the length from the apex of the cone; k = a/x is a
slenderness parameter; Re_x is the Reynolds number based on x
and the freestream speed; and τ is a scaled skin-friction normal
to the separation line:

$$\tau = \frac{Ve}{kUe}\left(\frac{\partial}{\partial z}\frac{V}{Ve}\right)_{z=0},$$

where V is the circumferential velocity component in the boundary layer and z is the boundary-layer coordinate normal to the
wall. With A_1 given by (11) in terms of the viscous calculation
and by a relation like those shown in Fig. 18 in terms of the
inviscid calculation, a matching of the two determines the

value of θ_s, which specifies the position of the separation line.

To summarise, the procedure is, for a given cone and angle of incidence;

(a) find A_1 as a function of θ_s by performing a number of inviscid calculations, each for a specified value of θ_s;

(b) deduce the value of θ_s for which $A_1 = 0$ (smooth separation) and for this find the inviscid solution, giving Ue and Ve in (11) and the input for the next step;

(c) calculate the boundary layer up to the smooth separation line, giving τ in (11);

(d) for a particular Reynolds number, calculate A_1 from (11)

(e) use the results of step (a) in an inverse sense to find the value of θ_s corresponding to the value of A_1 found in (d);

(f) this is the position of the separation line for this cone, incidence and Reynolds number; calculate the inviscid flow corresponding to it.

Before looking at the results, we need to discuss an inconsistency in the treatment. For the inviscid calculation, the flow is assumed to be conical and, in particular, the separation line is assumed to be straight. However, the procedure described leads to a value of θ_s which depends on x. Fortunately, the x dependence expressed by (11) is very weak, except for very small values of x, i.e. very near the apex of the cone. Many effects make the model unreal at the apex, so that its relevance already depends on the ability of the real flow to 'forget' what happened there.

Coming finally to the results, we see in Fig. 19 the variation of the separation position, θ_s, with a reduced Reynolds number, for a circular cone at an angle of incidence three times its semi-angle. The crosses denote the positions obtained by Rainbird et al (1963) from visualization of the surface flow in a water tunnel. The solid line represents Fiddes (1980) calculations, which clearly represent both the level and trend of the observations quite successfully. The broken line is the position of smooth separation and is therefore an asymptote of the solid line for Re $\to \infty$. The chain line, which gives the general level of the experimental points so successfully, is obtained relatively easily. Cooke (1965) used the attached flow past the circular cone as input to a quasi-conical laminar boundary layer calculation. The boundary layer calculation eventually broke down, but careful extrapolation enabled him to estimate where the square-root singularity at separation would

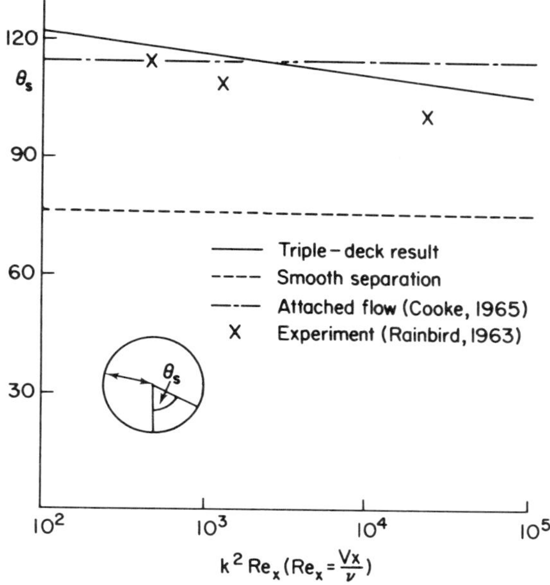

Fig. 19   Variation of separation position with Reynolds
number, Fiddes (1980)

arise. The chain line is the result. It has taken a great
deal of effort to do better.

At present, the method is being extended to asymmetrical
flow and the inviscid part of the calculation seems to be working successfully. It looks straightforward to extend it to
non-conical pointed bodies, still in the framework of slender-
body theory, calculating the inviscid flow and the boundary
layer and the separation line step-by-step from the apex. It
should also be possible to produce a panel method for separation
from smooth pointed bodies, though some care would be needed in
developing an appropriate panelling scheme for the neighbourhood
of the separation line.

A very useful extension would be to a treatment of turbulent
interactions. Here there seem to be difficulties at two levels.
First, the turbulent flow at a particular Reynolds number is
essentially further from the limiting asymptotic behaviour than
a laminar flow would be at the same Reynolds number. Simple
manifestations of this are that the eddy viscosity is larger

than the molecular viscosity and, in the present context, that
the turbulent separation line is further downstream than the
laminar one. It may also be the case that the mechanism of
turbulent separation is different, involving a greater length
scale; so that the main part of the upstream boundary layer
plays a more active role, as a curved shear layer with a variation of pressure across it.

5. CONCLUSIONS

   From the work presented we can conclude that, for the modelling of large-scale separations at high Reynolds numbers from
highly-swept separation lines:

(a) we have a good inviscid model for flows in which the
    separation originates at a pointed apex;
(b) the model has been successfully implemented for flows
    governed by linear partial differential equations, in
    particular, flows past slender bodies and flows which
    are well subcritical; and
(c) under these conditions, the triple deck model of viscous-
    inviscid interaction appears to provide a means of
    determining the position of laminar separation on a
    smooth body.

The remaining problems concern:

(d) the representation of the inviscid model in numerical
    schemes for solving the nonlinear equations of fluid
    flow;
(e) the inviscid modelling of the origins of separated
    flows which do not start in a conical form;
(f) the treatment of the interaction between the inviscid
    flow and a separating turbulent boundary layer; and
(g) the provision of a satisfactory model for high Reynolds
    number flows separating from lines which are swept only
    a little or not at all.

6. REFERENCES

Batchelor, G.K. (1967) Introduction to fluid dynamics, CUP,
pp. 538-9.

Billet, G. (1980) Simulation numérique d'un décollement
tridimensionelle pariétal, Rech. Aerosp. 1980-4, pp. 229-240.

Brown, C.E. and Michael, W.H. (1954) Effect of leading-edge
separation on the lift of a delta wing, *J. Aero. Sci.* **21**,
pp. 690-644 and 706 (NACA TN 3430 (1955)).

Brown, S.N. (1965) The compressible inviscid leading-edge vortex, *J. Fluid Mech.* **22**, pp. 17-32.

Brown, S.N. and Mangler, K.W. (1967) An asymptotic solution for the centre of a rolled-up conical vortex sheet in compressible flow, *Aero. Quart.* **18**, pp. 354-366.

Butter, D.J. and Hunt, B. (1982) Survey on boundary integral methods, in these proceedings.

Cooke, J.C. (1965) The laminar boundary layer on an inclined cone, RAE Technical Report 65178.

Earnshaw, P.B. (1961) An experimental investigation of the structure of a leading-edge vortex, *ARC R & M* **3281**.

Fiddes, S.P. (1980) A theory of the separated flow past a slender elliptic cone at incidence, *in* AGARD CP-291.

Guiraud, J.P. and Zeytounian, R. Kh. (1977) A double-scale investigation of the asymptotic structure of rolled-up vortex sheets, *J. Fluid Mech.* **79**, pp. 93-112.

Hall, M.G. (1961) A theory for the core of a leading-edge vortex, *J. Fluid Mech.* **11**, pp. 209.

Hummel, D. and Redeker, G. (1972) Experimentelle Bestimmung der gebundenen Wirbellinien sowie des Stromungsverlaufs in der Umgebung des Hinterkante einer schlanken Deltaflugels, *Ab. Braunschweig Wiss. Ges.* **22**, pp. 273.

Johnson, F.T., Lu, P., Tinoco, E.N. and Epton, M.A. (1980) An improved panel method for the solution of three-dimensional vortex flows, Vol. I - Theory document, NASA CR-3278.

Jones, R.T. (1946) Properties of low aspect ratio pointed wings at speeds below and above the speed of sound, NACA Rep. 835.

Levinsky, E.S. and Wei, M.H.Y. (1968) Nonlinear lift and pressure distribution on slender conical bodies with strakes at low speeds, NASA CR-1202.

Lock, R.C. and Firmin, M.C.P. (1982) Survey of techniques for estimating viscous effects in external aerodynamics, in these proceedings.

Mangler, K.W. and Smith, J.H.B. (1959) A theory of the flow past a slender delta wing with leading-edge separation, *Proc. Roy. Soc.* **A251**, pp. 200-217.

Mangler, K.W. and Sells, C.C.L. (1967) The flow field near the centre of a slender rolled-up conical vortex sheet, RAE Technical Report 67029.

Mangler, K.W. and Weber, J. (1967) The flow field near the centre of a rolled-up vortex sheet, *J. Fluid Mech.* **30**, pp. 117-196.

Marsden, D.J., Simpson, R.W. and Rainbird, W.J. (1957) The flow over delta wings at low speeds with leading-edge separation, C.O.A. Cranfield Report 114.

Maskell, E.C. (1964) On the asymptotic structure of a conical leading-edge vortex, IUTAM Sym. on Vortex Motions, Ann Arbor (see *Progress in Aeronautical Sciences* **7**, pp. 35-51, Pergamon Press (1966)).

Maskew, B. (1975) A subvortex technique for the close approach to a discretised vortex sheet, NASA TM-X-62-487.

Moore, D.W. (1975) The rolling-up of a semi-infinite vortex sheet, *Proc. Roy. Soc.* **A345**, pp. 417-430.

Morino, L., Chen, L. and Suciu, E.O. (1975) Steady and oscillating subsonic and supersonic aerodynamics around complex configurations, *AIAA J.* **13**, pp. 368-374.

Rainbird, W.J., Crabbe, R.S. and Jurewicz, L.S. (1963) A water-tunnel investigation of flow separation about circular cones at incidence, NRC (Canada) Aeronautical Report LR385.

Rainbird, W.J. (1968) The external flow field about yawed circular cones, AGARD CP-30, paper 19.

Rehbach, C. (1978a) Numerical calculation of unsteady three-dimensional flows with vortex sheets, AIAA Paper 78-111.

Rehbach, C. (1978b) Calcul instationnaire de nappes tourbillonnaires émises par des surfaces portantes fortement inclinées, *in* AGARD CP-247.

Roe, P.L. (1982) The numerical modelling of shockwaves and other discontinuities, in these proceedings.

Smith, F.T. (1977) The laminar separation of an incompressible fluid streaming past a smooth surface, *Proc. Roy. Soc.* **A356**, pp. 443-463.

Smith, F.T. (1978) Three-dimensional viscous and inviscid separation of a vortex sheet from a smooth non-slender body,

RAE Technical Report 78095.

Smith, J.H.B. (1968) Improved calculations of leading-edge separation from slender, thin, delta wings, *Proc. Roy. Soc.* **A306**, pp. 67-90.

Smith, J.H.B. (1972) Remarks on the structure of conical flow, *Progress in Aerospace Sciences* **12**, pp. 241-272, Pergammon Press.

Smith, J.H.B. (1977) Behaviour of a vortex sheet separating from a smooth surface, RAE Technical Report 77058.

Stewartson, K. and Hall, M.G. (1963) The inner viscous solution for the core of a leading-edge vortex, *J. Fluid Mech.* **15**, pp. 306-318.

Tinoco, E.N., Lu, P. and Johnson, F.T. (1980) An improved panel method for the solution of three-dimensional vortex flows, Vol. II - User's guide and programmer's document, NASA CR-3279.

Weber, J.A., Brune, G.W., Johnson, F.T., Lu, P. and Rubbert, P.E. (1975) Three-dimensional solution of flows over wings with leading-edge vortex separation, NASA CR-132709 and CR 132710, also *AIAA J.*, **14**, pp. 519-525 (1976).

Zacharov, S.B. (1976) The calculation of inviscid separated flow about a slender cone with large angles of attack, Uchenye Zapiski TsAGI VII, 6, also RAE Library Translation 2009 (1979).

A SURVEY ON BOUNDARY INTEGRAL METHODS

D. J. Butter*, B. Hunt** and G. R. Hargreaves*

*(British Aerospace, Woodford)
**(British Aerospace, Warton)

1. INTRODUCTION

The boundary integral method has been available in a variety
of guises for some considerable time for external aerodynamic
calculations, under the assumptions of inviscid, irrotational,
"slightly" compressible flow. Early forms of the method included
the widely popular lifting surface theories and vortex lattice
methods, (Labrujere, 1972). The limitations in speed and capacity
of the computers available when these techniques were initially
developed, and the fact that the method originators did not have
access to the advanced mathematical and numerical schemes subsequently developed, led to certain simplifying assumptions being
adopted. These assumptions (e.g. 'linearised' boundary conditions, 'simplified' definition of the relationship between velocity and pressure, etc.) immediately limited the range of application of these methods, for example a very limited representation
of body effects and a restricted range of useful output - such as
estimates of overall forces only.

These 'simplifying' assumptions also led to numerous artificial theoretical difficulties resulting directly from the approximations used; the treatment of the leading edge 'singularities'
which appear as a consequence of using thin wing theory provides
one example where probably as much effort has been expended on
attempting to derive physically meaningful and useful results
from totally artificial computations, as has been expended on the
implementation of the artificial model itself.

A somewhat more sinister problem associated with the use of
'simplified' methods relates to the demand placed upon the program user to have some grasp of the mathematical/numerical concepts upon which the model is based; this understanding may
relate to abstract concepts not even remotely connected with the
physics of the aerodynamic problem in question. Without such an
understanding the user is very likely to attempt to apply the
method outside the range of problems for which the simplifying
assumptions have been empirically (and frequently fortuitously)

established. For example, quite convincing pseudo-theoretical
arguments can be put forward to explain why two-dimensional
vortex lattice models are capable of producing surprisingly
accurate estimates of total lift on an aerofoil (James, 1969).
This does certainly not imply that a 3-dimensional version of the
model is capable of equal accuracy in predicting the lift on a
complex 3-dimensional configuration (at least without paying very
careful attention to the geometrical arrangement of the lattice -
for which an acceptable layout is strongly configuration dependent - and this requires mathematical/numerical knowledge not
possessed by the typical user). Even less does it imply that the
model can generate meaningful estimates for the other aerodynamic
properties; in inexperienced hands the vortex lattice method can
yield totally misleading results - when used to simulate wind
tunnel wall effects, for example.

The above generalisations are intended to indicate that until
recently the theoretical methods available for aerodynamic
predictions were tricky to use, limited in application, and
usually capable of yielding only approximate numerical solutions
of the idealised inviscid, irrotational, 'slightly' compressible
flow over an approximation of the configuration in question. However, the rapid growth in availability of computers has led, over
the past two decades, to the development of increasingly sophisticated theoretical methods of which the so-called panel method is
undoubtedly the most advanced and widely applied for theoretical
aerodynamic predictions. It can be said, in principle, that the
development of the boundary integral theory allows restrictions
related to the approximation of the configuration geometry and of
the associated boundary conditions to be lifted completely: it is
possible to generate nominally exact numerical solutions of the
idealised flow over arbitrary 3-dimensional lifting configurations.

The fact that this is possible in principle does not mean that
it is easy to achieve in practice. The numerical errors inherent
in the numerical implementation - and this problem is common to
all types of computerised theoretical methods - can, and
frequently do render the numerical solutions practically useless,
or at least, make them erroneous.

In common with the earlier generations of methods, the problem
also remains that the 'idealised' flow itself may not give a
realistic simulation of an actual flow. Certain phenomena which
can have a crucial effect on the detailed flow are governed by
properties not represented by the 'idealised' model; such phenomena include shock effects and viscous dominated features such as
flow separation or even the less obvious problem of how in a
practical configuration the lift or circulation induced by a wing
extends across the fuselage. Making an adequate allowance for

such deficiencies continues to pose difficulties for the user;
such difficulties may be expected to be removed at a much slower
rate, even as more sophisticated 'idealised' models are
developed.

As far as the practical aerodynamicist - the 'user' - is concerned, the panel method is by far the easiest to use of all the
computational fluid mechanics techniques under development for
aerodynamic prediction; the data supplied generally comprises
very little more than a description of the surface geometry of
the aircraft, though for a complex configuration this definition
itself can, of course, involve a substantial amount of work.
Apart from the relative ease of use, the user should also be
encouraged by the fact that recent developments are starting to
allow complete aircraft problems to be tackled with a considerable degree of realism; coupling with 3-dimensional boundary
layer calculations; modelling of leading edge and body vortex
flows, prediction and modelling of wing separation allowing estimates of maximum lift coefficient to be obtained; more realistic
estimation of interference between the wing wake and downstream
surfaces; allowance for wind tunnel wall interference; the
possibility of modelling intake flows and jet entrainment effects
and so on.

This paper outlines the major classes of flow for which solutions can be obtained with panel methods and describes the techniques used to extend the solutions beyond the normal linear
inviscid results. There is also a description of the way that a
correctly formulated panel method can be used to produce solutions for a wide variety of nonaerodynamic problems efficiently
and economically.

2. POTENTIAL FLOW CAPABILITIES

2.1 Subsonic method

Currently the panel method is most advanced for the case of low
subsonic, steady flows. Fifteen years development has resulted in
a number of programs which are efficient, accurate, and capable
of dealing with very complex configurations. Figure 1 illustrates
a typical application in the military field which it must be
stressed is for a routine production design calculation and not
for a 'one-off' demonstration case. The development has taken
place in both America and Europe. Their basic formulations are
reviewed by Hunt (1978) and a comparison of their accuracy by
Sytsma *et al* (1979).

The continued development of panel methods has been spurred by
the requirements of users, who, having applied the earlier methods

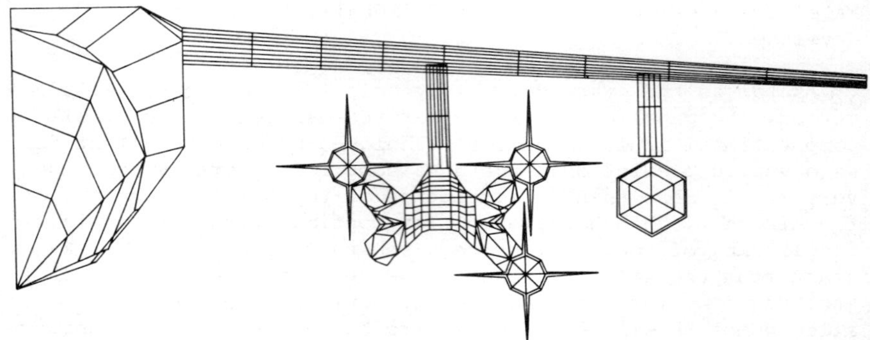

Fig. 1   Typical fighter aircraft configuration

successfully as a routine design procedure, have demanded increased applicability of the methods to more complex flows and
geometric configurations. We have now reached the position where,
as a response to these requirements, what may be termed a second
generation of methods is about to come into use, of which the BAe
Warton Division Mk II method, (Semple (1978)), and the Boeing
PANAIR, (Johnson *et al* (1976)), are the two major programs at
present. These programs, having built on the experiences with the
methods written around the early 1970s, have improved modelling
and accuracy but what distinguishes them most from the earlier
methods is their considerably increased flexibility. The methods
have been developed into complete systems with the addition of
data handling techniques to the basic panel method programs to
facilitate the input of complex configurations, and featuring a
program architecture which will allow an enormous class of problems to be analysed, as will be shown later. In these methods the
formation of the influence coefficients and solution of the
matrix equations make up only a small part of the total coding.

The main difference between the two 'second generation'
methods mentioned above, the BAe Mk II and the Boeing PANAIR,
centres on the order of representation of the singularity distribution and surface geometry, and their efficiency in terms of
computing time to achieve a given accuracy. The Mark II method
employs a first order singularity and surface geometry representation, whilst PANAIR uses a higher order representation, which
involves a much higher computing cost per panel but should
require fewer panels to achieve a given accuracy. The relative
merits of the two approaches could form the basis of a lecture in
itself and we do not intend to deal with this here. However, it
is our opinion that, bearing in mind that there is a minimum
number of points required to define a pressure distribution sensibly, adequate accuracy can be achieved at quite low computing

cost by a well formulated first order panel method, and the BAe
Mk II method, employing a technique known as 'pseudo-error-
minimisation' of the source and doublet gradients, satisfies
these accuracy requirements.

The accuracy of the first order methods can be enhanced by
simply placing the collocation points on the true surface rather
than at the panel centroids and also by a method of local flow
improvement which will be described in Section 3.

Description of the major methods are given by Kraus (1970),
Labrujere *et al* (1970), Morino and Luo (1974), Rubbert *et al*
(1967), Johnson *et al* (1976), Hunt and Semple (1976), Semple
(1978), Hess (1972, 1973), Petrie (1978), and Roberts and Rundle
(1972).

Numerous comparisons of the accuracy of various panel methods
have been made and many of them are presented in Sytsma *et al*
(1979). Just two will be considered here. Figure 2 shows the
chordwise and spanwise velocities over a wing calculated by a
first order method ('H-S sheets') employing piecewise constant
source and vortex singularity distributions. These are compared
with the results obtained using a higher order method (Roberts
and Rundle, (1971)) whose geometry and singularity distributions
are modelled with bi-cubic splines and which for this case can be
regarded as an exact solution of the inviscid flow problem. It
can be seen that virtually complete agreement can be achieved
using a first order formulation. (The first order method used in
this comparison was in fact one of the earlier BAe panel methods
but it is known that the BAe Mk II method will give the same
accuracy at a much reduced computing cost).

The other results shown ('NLR') were produced by an earlier
generation method in which an even simpler vortex singularity
distribution (line vortices) was employed. The increase in accuracy through the use of a piecewise-constant vorticity distribution is quite apparent; this does not involve significantly
higher computational costs.

Figure 3 shows the results for the more difficult case of a
free flow nacelle and it can be seen that there is not even complete agreement between the two higher order methods used for
this case. The BAe Mk II method does, however, achieve comparable accuracy with a smaller number of panels and at considerably
less cost.

It is clear from the literature that panel methods are well
established for subsonic applications and are being used rout-
inely for very complex configurations. The methods are many and
varied, but it is the authors' opinion that a well formulated

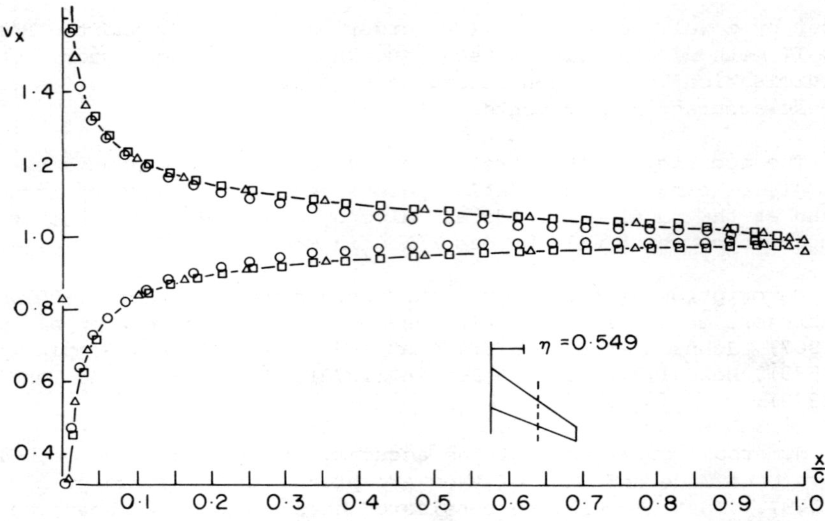

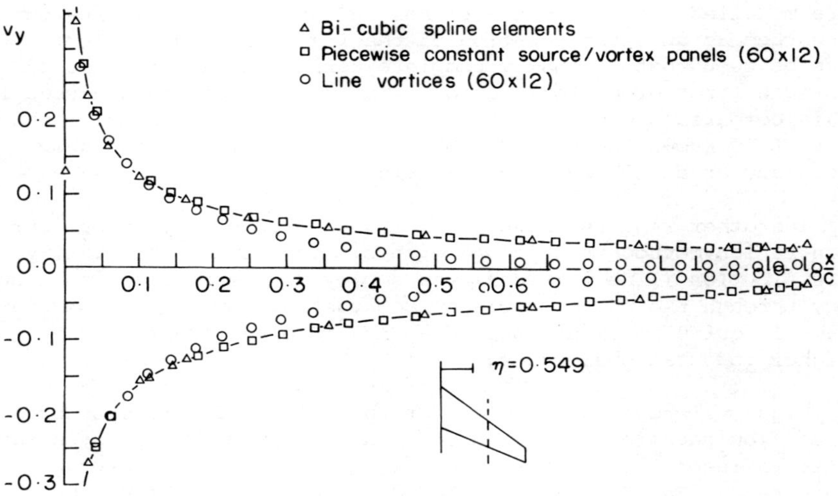

Fig. 2. Comparison of chordwise and spanwise velocity components predicted by three different panel methods

first order method can give the required accuracy for most cases at quite low computing costs, and that their usefulness is greatly

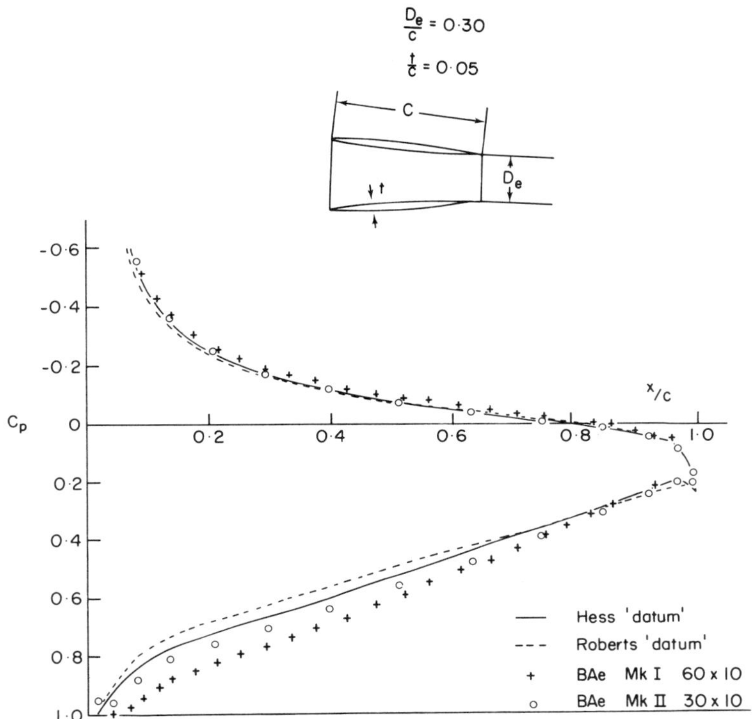

Fig. 3 Pressure distribution on the Euromech nacelle

enhanced by the incorporation of sophisticated data handling
techniques.

2.2 Supersonic methods

Based to a large extent on the success and efficiency of solving linearised potential flow equations with an integral equation
formulation for subsonic flows, the requirement to be able to
compute the supersonic flow (again linearised) around complete
aircraft configurations had led to the development of the supersonic panel method. Probably the first in this field was developed by Woodward (1973). This method has been subsequently modified and in its present form, known as USSAERO/C, is in use by
several design teams.

However, in this country an evaluation of the method revealed
several unsatisfactory aspects of the program, mainly concerned
with singularity modelling deficiencies and also a lack of development potential. This and the problems of using the program led
to the start of an exercise at BAe Warton Division to develop a

new method. At the outset it was decided that this must be compatible with the architecture of the BAe Mk II subsonic method, retaining the user orientated input and output facilities, but obviously it would be necessary to replace the subsonic singularity modelling with a robust supersonic one. This work has now reached an advanced stage and is described by Hewitt *et al* (1980). Results for a body of revolution are shown in Fig. 4 and for a delta wing in Fig. 5. These latter show perturbation velocities per unit incidence.

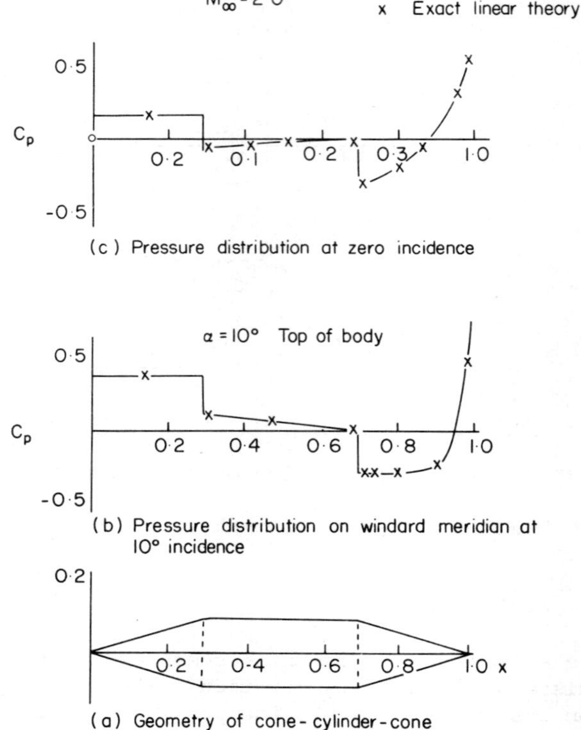

Fig. 4   Results of supersonic panel program for a body of revolution at M_∞ = 2.0

2.3 Compressible Flows

2.3.1 Exact compressible potential flow   If we ignore the entropy changes which in real flows arise due to the presence of shocks, then the flow may be considered to be a potential flow and the equation governing this isentropic compressible flow is

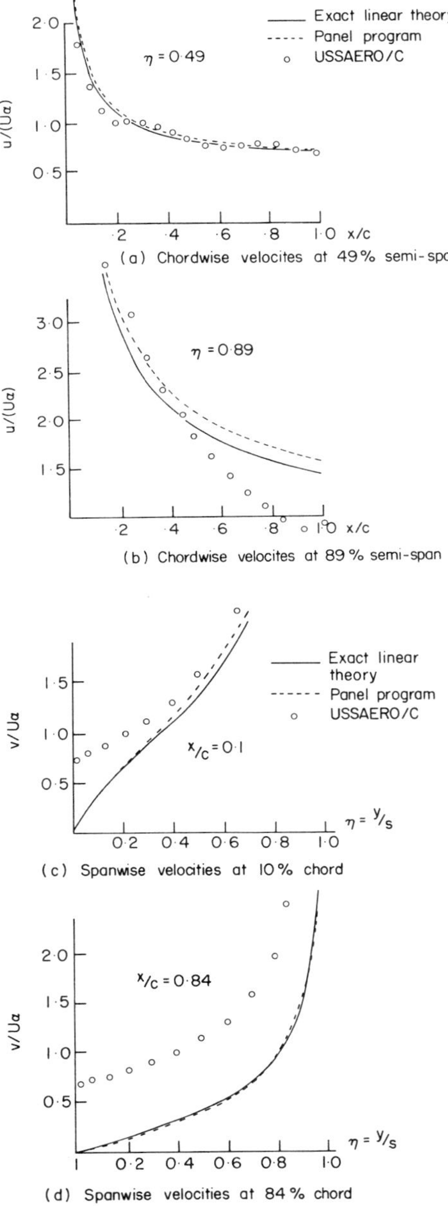

Fig. 5 Results from supersonic panel programs for a 45° delta wing at $M_\infty = \sqrt{2}$

$$\nabla^2 \Phi = M^2 \frac{\partial V}{\partial s} \tag{1}$$

where the local Mach number M is related to the free stream Mach Number M_∞ and the local speed V by the classical isentropic relations; the quantity $\partial V/\partial s$ expresses the local rate of change of the flow speed along the local streamline.

Normally finite-difference, finite-element or finite-volume techniques are applied for a solution of this equation but it is possible - though in the present state of the art, computationally expensive - to obtain exact solutions by splitting Φ in a variety of ways. One possible scheme consists of setting $\Phi = \phi + \varphi + \Phi_\infty$ where ϕ is the potential due to surface singularities and φ is a residual potential which may be treated separately by a field method. The potentials ϕ and φ may be treated individually in turn, coupling them iteratively. The boundary condition for the panel method part of the problem is

$$\frac{\partial \phi}{\partial n} = - \frac{\partial \varphi}{\partial n} - U_\infty \cdot n \tag{2}$$

for a solid boundary S; here φ is the currently known value from the previous iteration of the field calculation. The residual potential φ must then satisfy the condition in the field that

$$\nabla^2 \varphi = M^2 \frac{\partial V}{\partial s} - \nabla^2 \phi \tag{3}$$

with $\nabla^2 \phi = 0$ if a Laplacian panel method is used, and with M and $\partial V/\partial s$ both functions of $(\phi + \varphi + \Phi_\infty)$ and of the associated derivatives. Here φ is the current unknown function to be determined during the field calculation and ϕ is the currently known function from the previous iteration of the panel method.

One possible way of treating the field problem employs an explicit field-integral approach in which the currently computed value of $M^2 \partial V/\partial s$, obtained from the total potential $(\phi + \varphi + \Phi_\infty)$ is considered to be produced by a volume source distribution of local density

$$\sigma_v = M^2 \frac{\partial V}{\partial s} \tag{4}$$

throughout the entire domain. This source then modifies the boundary condition which the next iteration of the panel method calculation must satisfy in order to make the resultant normal velocity $\partial \Phi/\partial n$ equal to zero.

Such approaches are described by Luu *et al* (1972), Piers and Slooff (1979), and Hewitt and Marchbank (1976) and hold except-

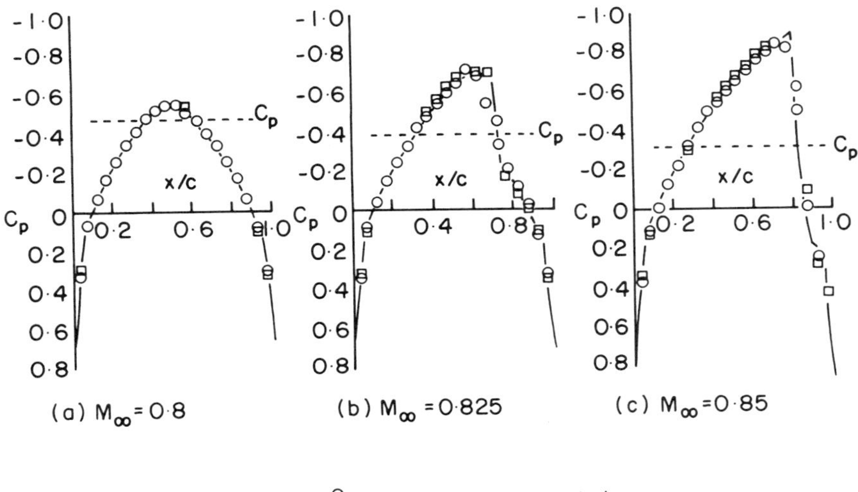

Fig. 6   Results from a transonic panel program for a 10% thick
         parabolic arc aerofoil

ional promise for the exact computation of compressible flows,
though at present certain difficulties remain, primarily associated with shocks. Figure 6 shows the preliminary but very
encouraging results achieved by Piers and Slooff for a parabolic arc
aerofoil; in this case the field equation solved was the so-
called transonic-small-perturbation approximation to the full-
potential equation.

2.3.2. Simple Compressibility Correction A simple approximation
to the solution of the exact compressible flow equations described above has been developed by BAe Warton division, (Hunt,
1980) and has been in use for some time in some of the UK panel
methods. This approximation is explicit and is applied to an
existing solution of the Prandtl–Glauert equation subject to
correct real space normal velocity boundary conditions; the correction is exceptionally simple and economical to apply. The
effect of this scheme is illustrated in Figs. 7 and 8 and in
practice the correction appears to work well even for slightly
supercritical flows. In principle, it is possible to devise a

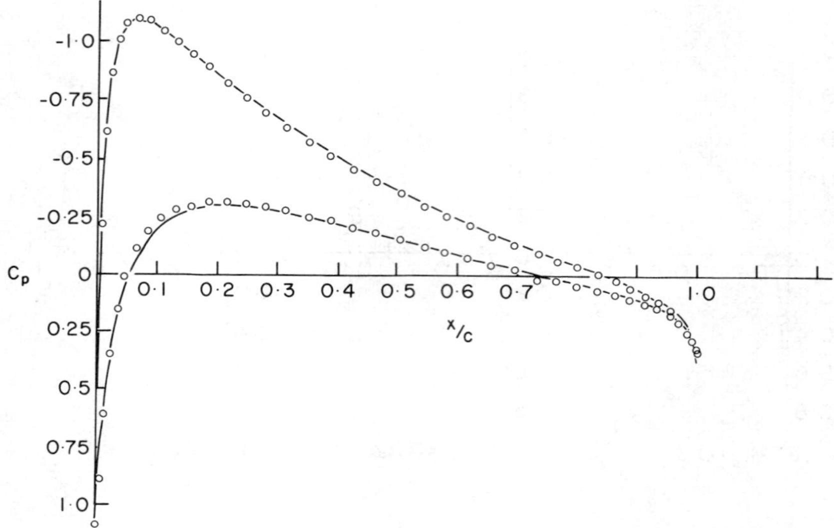

Fig. 7   Calculated pressure distributions on a two-dimensional aerofoil (RAe 100) at M_∞ = 0.6 and α = 2°. The points are obtained from a full-potential solution. The line is obtained by applying Hunts compressibility correction to an incompressible flow solution. Boundary layer effects are allowed for in both cases.

corresponding correction for the solution of the supersonic Prandtl-Glauert equation and work is continuing in this area.

2.4   Unsteady flow

Generalised unsteady flow is in principle amenable to the boundary integral technique and methods already exist for the unsteady aerodynamic input to flutter calculations for subsonic flow, (see Roos *et al* (1977) and Rubbert (1978)). Similar methods are currently being developed for supersonic flows and Morino (1974), has reported a method for unsteady, subsonic and supersonic flow which is widely available in the USA.

However, these methods are written for solutions in the frequency domain and yield results in both the real (amplitude) and imaginary (phase shift) planes. This is not always useful and there are problems which require a general time dependent flow solution such as store release, unsteady stability derivatives, control response, etc. A scheme for calculating these flows with panel methods has been developed by Hunt for incompressible flows. The problem obviously becomes more complex for non-

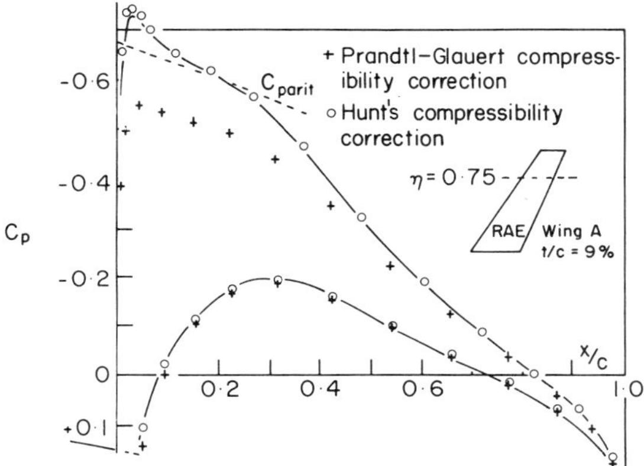

Fig. 8   Calculated pressure distribution at a particular spanwise station on a 3-dimensional wing at $M_\infty = 0$ and
$\alpha = 2°$.   The solid curve shows the results of a second-
order lifting surface theory, due to Hewitt BAe Warton.

zero Mach numbers, with account now having to be taken of the
time it takes for signals to propagate from one panel to another.

There is, however, a class of unsteady problems which are
easily amenable to calculation.   These are the quasi-steady aircraft motions such as steady rate of roll, pitch, yaw, etc., as
seen in uniform aircraft motion and also isolated propellers and
helicopter rotors.   In these cases the flow is actually 'steady'
relative to the body.   (In the case of 'non-lifting' motions not
involving vortex wakes, even this limitation is unnecessary, and
'instantaneous' solutions can be derived for a series of bodies
in arbitrary relative motion.)   In all cases, care needs to be
exercised in interpreting results for flows with non-zero Mach
number (i.e. finite speed of sound).

Solutions to such problems can be obtained from the BAe Mk II
program;   the modifications to the basic steady-flow method
involve only a few trivial extra lines of code.

An analysis to be published by Hunt considers a general motion
of a body, comprising an arbitrary steady rotational velocity $\overline{\Omega}$
and a translational velocity $\overline{V}_i$, located in a uniform freestream
U_∞ , all of these motions being defined relative to a fixed absolute reference system.   In a coordinate system fixed in the body,
the equivalent 'unperturbed' onset flow $\overline{U}_{FS}$ at a point $\overline{r}$ (relative
to the centre of rotation) is given by $\overline{U}_{FS} = \overline{U}_\infty - \overline{V}_i - \overline{\Omega} \times \overline{r}$.

If the perturbation velocity $\overline{V}_p$ due to the body is then calcul-

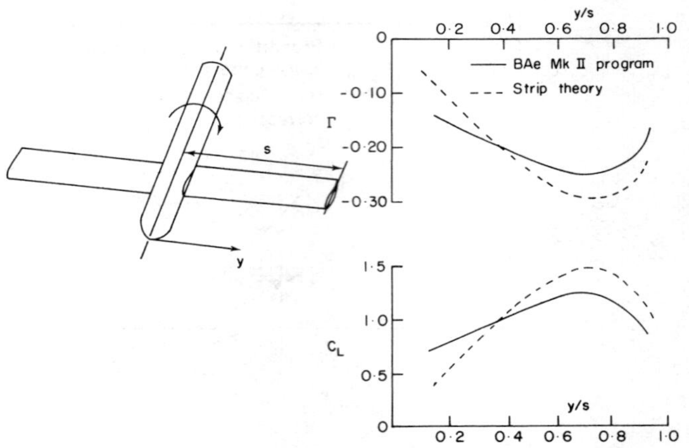

Fig. 9   Circulation (Γ) and lift distribution across the span
         of a wing attached to a body in steady rolling motion.

ated (in the usual way), as if the body were solid, stationary,
and in the non-uniform flow $\overline{U_{FS}}$, then the pressure field is given
by

$$\frac{p - p_\infty}{\tfrac{1}{2}\rho_\infty} = U_{FS}^2 - (\overline{U_{FS}} + \overline{V}_\rho)^2 \tag{5}$$

which has the same form as for steady flow (in which case $\overline{U_{FS}}$
degenerates to being uniform and equal to $\overline{U}_\infty$).   This equation
comprises a generalised form of Bernoulli's equation. Expression
in terms of a pressure coefficient requires a more general definition of a reference speed, V_{REF}.   Note that the above generalised form *automatically* contains the $\partial\phi/\partial t$ term which normally
appears in the unsteady Bernoulli equation.

This simple addition to a standard panel method gives the
capability of calculating a range of quasi-steady flows, both
linear and angular, and *any* combination of these as illustrated
by a rolling wing in forward flight, a helicopter rotor in climb-
ing flight or a propellor in forward flight. Figure 9 shows the
circulation and lift distributions obtained by this method for a
rolling wing installed in a uniform flow along the axis of rotation. The dotted lines in these graphs show the results given
by the traditional 'strip-theory' approximation.

As mentioned above the generalised unsteady case involving
time-dependent circulation and finite signal propagation speed is
more complex but it basically only requires a sequence of potential influence matrices to be calculated and stored as well as a

sequence of velocity influence matrices, which although it is
time-consuming is straightforward.

2.5 Internal flows

Most of the panel methods that have been developed for aeronautical applications have concentrated on external aerodynamics,
where for a first order method the effects of leakage errors for
a well posed problem tend to be small and are reduced by being
of opposite sign to the errors in representing a curved geometry
by plane panels. However, for internal flows the leakage errors
are much more important. For the accurate flow in a pipe, for
example, it is essential that the mass flow leaving the downstream station is the same as that entering the upstream one.

Solutions to these problems require that, for first order
methods, a very large number of panels is used (and this is not
always successful), or that higher order formulations are used.
Hess, (1973), has formulated an axisymmetric higher order method
for internal flows and Figure 10 illustrates how successful this
method is for two fairly severe test cases. Fig. 10a shows the
flow through an axisymmetric duct with a contraction ratio of 16,
with values of the mass flow calculated at various stations. The
large leakage errors present in the base method have been almost
eliminated by the higher-order method. Fig. 10b shows the "semi-
internal" flow inside a forward-facing cavity. Again the higher-
order method is much more accurate. The plot shows surface velocity.

Although the topic has not received much attention by the
aeronautical fluid dynamicist in the past, it is becoming increasingly more important. As complete aircraft are modelled,
there is pressure to be able to model the flow in the intakes
with varying intake mass flows. There is also an increasing
interest in internal flows for non-aeronautical applications
which can benefit from these analyses.

2.6 Design methods

At first glance the so-called design method, which gives the
body shape for a given pressure distribution, should be more useful than the normal analysis method. However, although there
have been many attempts at this both in 2 and 3 dimensions, most
notably by Bristow (1980) and Fray and Slooff (1980), no really
successful method has been developed. Even in 2 dimensions,
there are many problems in this approach, not least those of
uniqueness, existence and closure conditions. Added to these in
3-dimensions, there is the fundamental problem that a three component velocity distribution cannot be uniquely defined by a
given pressure distribution.

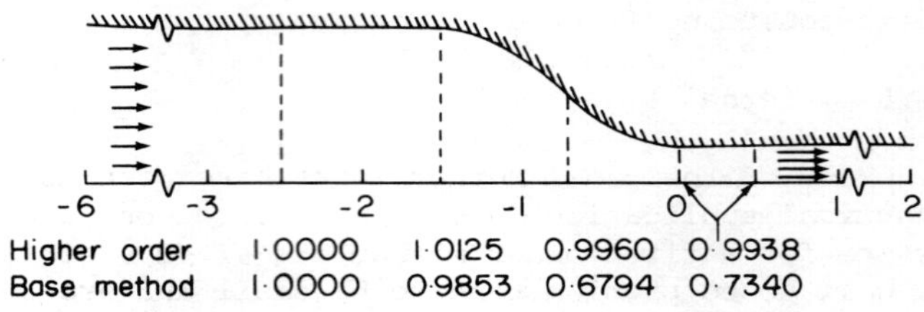

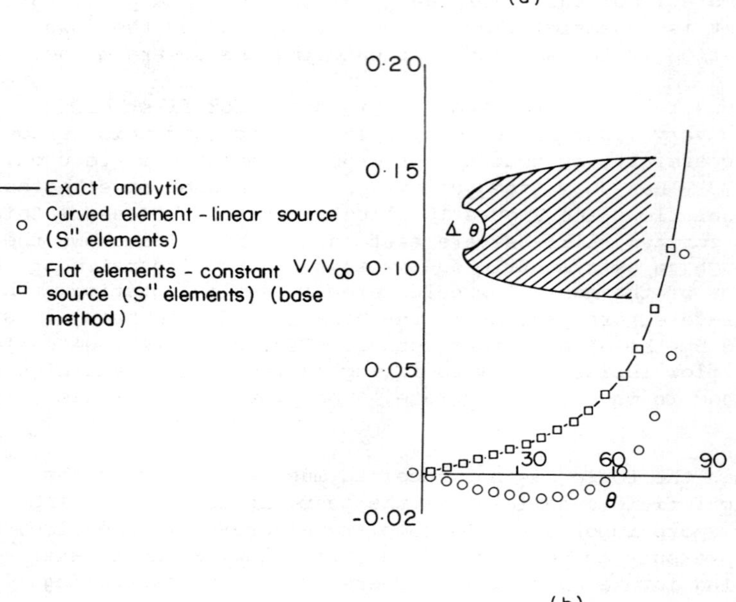

Fig. 10   Axisymmetric imcompressible flows calculated by a higher-order panel method and by a base method.

Fortunately, in terms of designing aircraft the lack of a design method does not really cause an embarrassment to the designer who very successfully uses an analysis method and experience to arrive at the correct geometry generally with very few iterations.

3.   REAL FLOW CORRECTIVE PROCEDURES

It is becoming possible to obtain solutions to a less idealised approximation of the flow by coupling the calculation of the inviscid flow with the calculation of some aspect of the real

flow, such as boundary layer modelling, leading edge and body
vortex flows, flow separation and the improved treatment of compressibility. However, before discussing any of these in particular, it must be realised that if any practical advantage is to
be gained by applying such corrections to the inviscid flow
problem, then the numerical solution of the inviscid problem must
be as accurate as possible. It is a fact that the numerical
errors inherent in many of the early panel methods were such that
the program underestimated the inviscid wing lift to such an
extent (particularly for wings of less than about 8% thickness-
chord ratio unless very many surface panels were used) that the
computed inviscid results actually predicted *less* lift than the
experimental results. The application of boundary layer or compressibility 'corrections' to such inaccurate inviscid predictions would therefore inevitably move these predictions further
away from experiment even if the coupling process were effected
with the utmost care.

Also the models used in many of the corrective procedures
were devised some considerable time ago when the only problems
for which it was possible to contemplate calculating the idealised flow related to very simple geometries and to very simple
flows. For such problems the corrective procedures could yield
indications which 'engineering know-how' converted to useful
estimates of real physical effects pertaining to this class of
problem. One example of this is provided by the artificial concept of the boundary layer displacement thickness. This is based
on two-dimensional arguments but can usefully be extended in an
approximate fashion to large regions of flow over a high aspect
ratio wing. Unfortunately the model begins to break down just
when it is most needed - when the real flow and its idealised
inviscid counterpart are strongly 3-dimensional and contain significant gradients; such is the case near the trailing edge of
a wing. In such regions, of course, the very concept of a
boundary layer itself becomes questionable; the same is true of
many other flow regions of vital interest in computational fluid
mechanics.

Other simple models which start to become physically unrealistic and/or difficult to couple properly with an inviscid flow
computation, when applied to realistic configurations and realistic flow conditions, include the widely-used vortex model
based on the assumption of conical flow or slender-body theory,
and various simple models of trailing edge wakes (particularly
in the case of compressible flow or when the wake is interacting
with other components of the configuration). For many flows of
interest, the basic models and assumptions in such 'real flow
corrective' procedures, which have long enjoyed considerable
success when applied to the simple problems which hitherto were
the only problems amenable to computation of any kind, require

complete revision if they are to serve any useful purpose when
applied to the classes of problems now amenable to inviscid computation.

Once the accuracy of the particular panel method formulation
for the inviscid part of the solution has been confirmed and once
the accuracy of the real flow corrective procedure has also been
confirmed, it becomes a meaningful exercise to compare the
results of coupled calculations with experiment. If such comparisons demonstrate discrepancies, then this casts doubt on the
validity of the coupling procedure or on the suitability of the
corrective model for the application concerned or on the accuracy of the experimental data.

Having outlined the overall problems it is worth now looking
at how some of the techniques are applied in more detail.

3.1 Viscous effects

Most current methods of representing the boundary layer effects
are based on the displacement surface concept, either explicitly
by adding the displacement thickness to the body, or implicitly
by the computationally more efficient way of surface transpiration with artificial boundary conditions on the real body, as
described first by Lighthill and developed by Lock and Firmin in
these proceedings. Both of these formulations have produced satisfactory results for many cases both in 2 and 3 dimensions. Two
examples, one for a high lift multi-component aerofoil and the
other for a three-dimensional wing are shown in Figs 11 and 12.

However, successful as these comparisons seem, it must be
pointed out that in general, the formulations used for the coupling procedure are not sound, and errors are introduced in the
trailing edge region which effectively mean that the Kutta condition is being applied incorrectly, as described by Hunt (1980).
Considerable care must be taken over this point as it is the
trailing edge conditions which control the circulation and lift
on the section and a small error can produce a significant error
in the overall lift. Figure 13 shows the effect different imple-
mentations of the Kutta conditions and wake models can have on
the pressure distribution in the trailing edge region of an aerofoil. And here it must be remembered that if we do not get the
trailing edge pressure correct we will not be able to predict the
profile drag accurately which is of fundamental importance in
aircraft design.

However, considering the problems with the normal simple
approach, it is difficult to see that it will ever yield sensible
results for cases other than the single aerofoil or wing at moderate incidences. A different, and possibly more reliable, form-

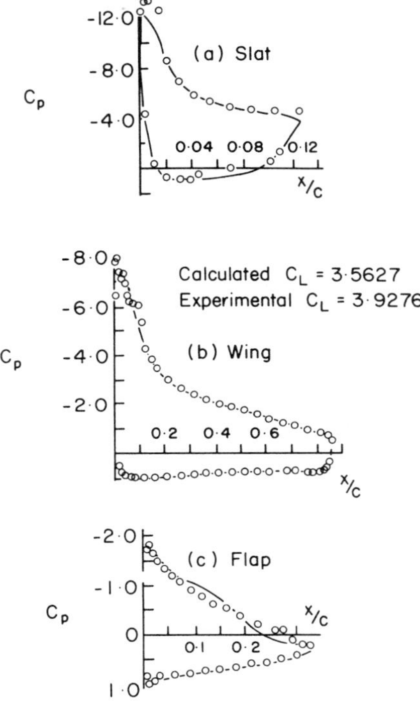

Fig. 11   Comparison between viscous flow model and experiment
for a three element aerofoil at 18° incidence.

ulation of the problem is that derived in America by Melnik
(1980) where the trailing edge region is treated as a region of
'strong interaction' and a triple deck type of formulation is
used.

An alternative method is being developed by Bradshaw (see
Maghoub and Bradshaw (1979)), which does not use the idea of
boundary layer displacement thickness and returns to the basic
concept of viscous/inviscid coupling. The method matches both
the u and v velocity components on an arbitrary surface, not
necessarily a stream surface, around the body, normally just
outside the edge of the viscous layer. Boundary layer equations
are solved between the body and the matching surface and a panel
method is used to compute the inviscid flow outside the matching surface. The drawback of this approach is that it requires a
finite difference boundary layer method and will therefore in
general use a lot of computing time.

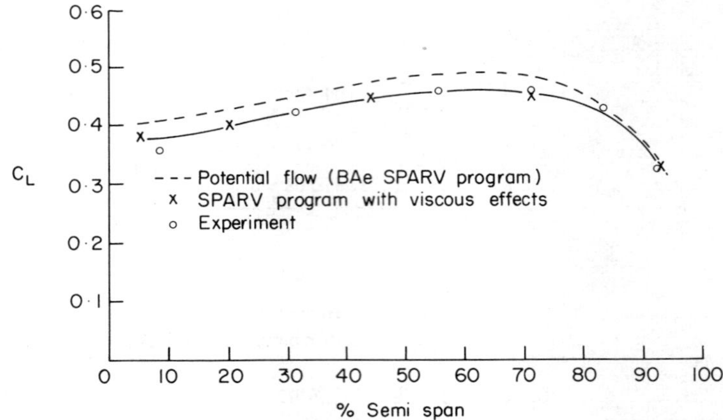

Fig. 12   Sectional lift coefficient on a three-dimensional wing;
aspect ratio = 3.0, incidence = $8°$,   Reynolds number
= 4 x 10^6.

3.2 Separated flow

Having developed a method for including the effects of viscosity into a panel method, it is not long before the users of the program are asking for it to be further extended to include the effects of separated flows.

Separation can be broadly classed into two types - a 'closed bubble' type of separation such as that seen on aerofoils and in the trailing edge region of wings, and an open free shear layer type which is generally seen either on delta wings and on slender bodies, or in junctions such as that of wings and bodies where the flow changes rapidly and a vortex flow is formed. We shall confine our attention here to the closed type of separation. The open separations are very difficult to model analytically and the only real success has been on the leading edge separations from strakes and delta wings, as described later in Section 5.3.

The two dimensional modelling of the closed type of separation has received considerable attention and several distinct models have been developed, falling broadly into two categories.

1. One class of method makes some attempt to predict the actual nature of the separated flow using a boundary layer method, then utilises a viscous/inviscid matching procedure.

2. Some are only concerned with the external flow, to predict accurately the pressure distribution and hence lift on the aerofoil. Because they do not model the detail of the separated flow they all require some degree of empiricism.

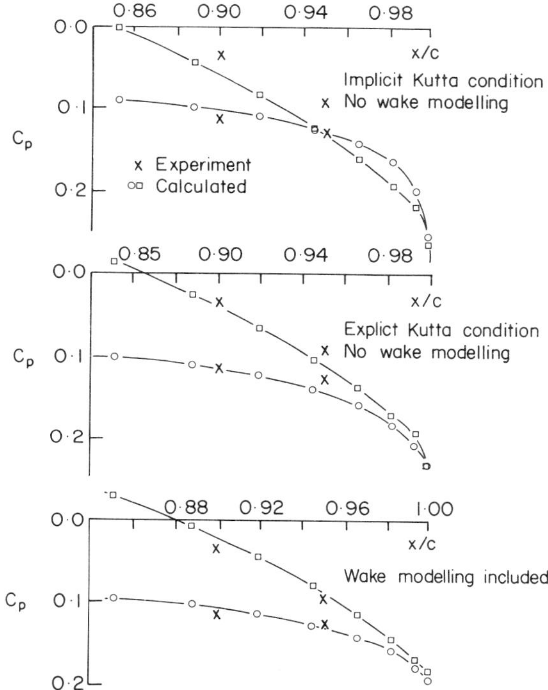

Fig. 13   Distribution of pressure coefficient near the trailing
          edge of an aerofoil at 6° incidence.   Re = 1.78 x 10^6.

The class of methods which attempt to model and predict the
detailed separated flow generally uses an inverse boundary layer
technique. This approach has been shown to be successful but
will inevitably be restricted to small areas of separation because
of the difficulties of the viscous/inviscid matching procedure,
and cannot easily be extended to three dimensions because of the
inverse nature of the method.

So we are left with the second class of methods that predict
only the pressure distribution on the body. In this field there
have been many developments - all with some degree of success.
Initially they were concerned only with predicting the flow round
bodies with limited separations, but more recently some have been
extended to predict the maximum lift. Figure 14 illustrates the
results for a two-dimensional method based on that developed by
Maskew and Dvorak (1978).

This model has been extended to 3-dimensions by Maskew and
Dvorak (1980), and has produced encouraging results. However an
alternative way of extending the 2-dimensional model for the cal-

490   BUTTER *et al.*

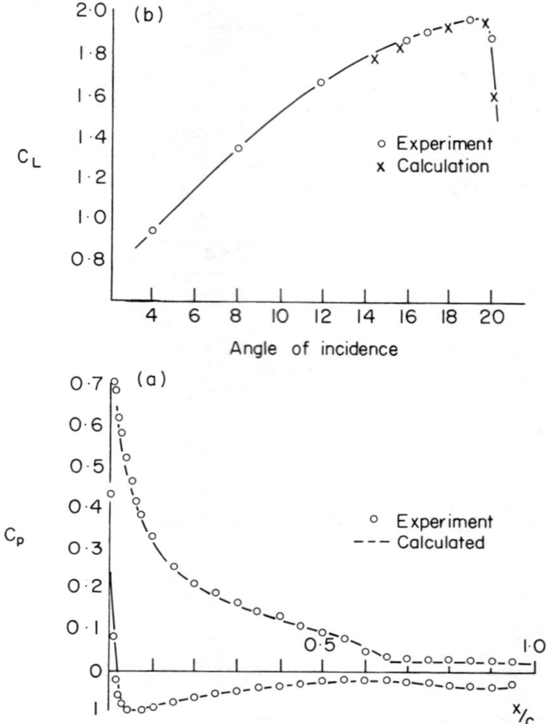

Fig. 14   Comparison between theory and experiment for separated
flow over the GA(W)-1 aerofoil at Re = 6.3 x 10^6.

culation of maximum lift on a wing is to use the hybrid 2-dimensional/3-dimensional method described in section 5.3.

3.3.   Wake relaxation

This subject was treated in depth by Jepps (1978) in the VKI
lecture series.   The formulation of a simple wake and the calculation of the way it deforms under loading has been attempted by
many people with encouraging results.   In most cases the wake is
modelled by discrete vortex filaments trailing from the wing
trailing edge downstream to infinity.   Initially a solution is
obtained for a flow over the wing with the wake in some arbitrary
fixed position, usually in the free stream direction.   With this
solution, the downwash field of the wing and the wake can be computed at discrete points on the wake vortices.   Each wake vortex
is discretised into a number of segments and each segment is
rotated to lie in the local stream direction, the last segment
remaining fixed in the freestream direction.   With these angular

positions of each segment a new shape for the wake is derived by
integrating from the wing trailing edge. The process is iterated
until no further movement of the wake occurs, which is generally
only two or three iterations. Obviously, to avoid numerical
problems some special allowance has to be made where two or more
filaments roll up into the core of the tip vortex. Figure 15
demonstrates the method of the BAe Warton Division for the interesting case of the wake behind a lifting nacelle.

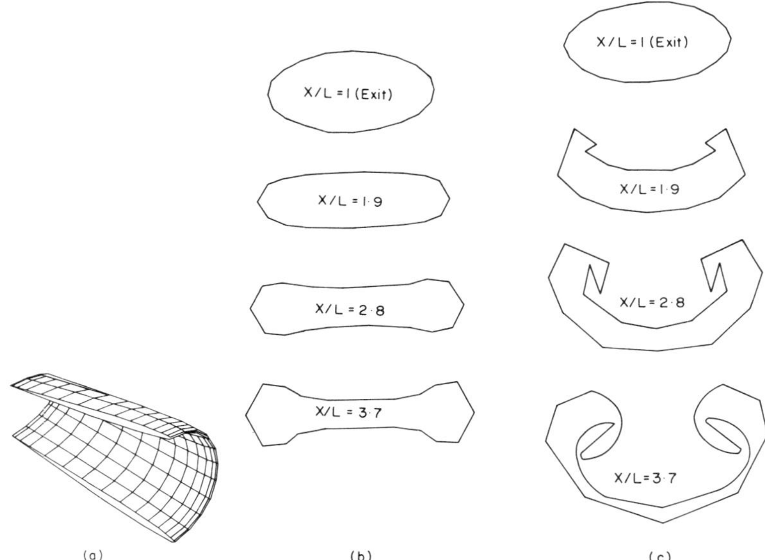

Fig. 15   Modelling the wake behind a flattened nacelle; (a) panel
          distribution, (b) evolution of wake shape at zero
          incidence, (c) evolution of wake shape at 10° incidence.

However, the importance of modelling the wake roll-up is in
predicting the flow field downstream of the wing, and generally a
fully relaxed wake will not significantly alter the wing pressure
distribution from that computed with a rigid wake. The differences are small and usually confined to the wing trailing edge and
tip regions and their effect on the integrated pressure forces is
negligible. But if the effects of a boundary layer are included
as described in Section 3.1, then in order to obtain the correct
trailing edge pressure, and hence, drag, it will be necessary to
relax the wake.

The basic model will give a good estimation of the flow field
at points off the wake, but because it assumes a physically
unrealistic representation of a vortex filament which has an
infinite velocity at its centre, the model becomes incorrect when

considering cases where the wake passes close to or impinges on
another body such as the tailplane or fuselage. Developments at
BAe Warton Division are aimed at overcoming this by using a vortex sheet representation of the wake and also by developing a
model of a discrete vortex which does not have the infinite velocity at its centre, such as the Rankine vortex.

4. APPLICATION

In line with the achieved and anticipated improvements in flow
modelling there is an associated widening in the range of possible aerodynamic applications; predictions of the whole range of
steady and rate derivatives for stability and control of complete
aircraft (Hargreaves (1980), Roos (1978)); store release calculations including the full 3-dimensional and interference effects
and the possibility of high incidence and relative motion effects
on the released store; pylon and nacelle installation design to
minimise drag and aerodynamic loads; aerodynamic input to flutter
calculations with full aircraft modelling, and so on. Without
trying to describe all the applications in great detail, we
should like to consider the first two on this list as they illustrate significant properties of a panel method which make them so
flexible and useful.

4.1 *Derivatives*

When considering the stability and control derivatives used in
normal aircraft design it must be remembered that they are defined
as the first term in a Taylor series expansion, and so for example
$\partial C_L/\partial \alpha$ is the derivative of lift with incidence at zero incidence.
Hence, we do not have to calculate the forces on the aircraft at
large angles of incidence when non-linear effects would become
important. With that in mind it can be shown that all the normal
derivatives for all the steady *and* quasi steady, or rate, motions
can be calculated at little extra computing cost than that for
the basic incidence runs.

The derivatives are calculated directly at the particular
incidence or sideslip required (and not by numerical differentiation between two points) as follows. For two-dimensional flow, we
have simply

$$\frac{\partial C_L}{\partial \alpha} = -\frac{1}{C}\int \frac{\partial C_p}{\partial \alpha}\, d(x/c)$$

$$= -\frac{1}{C}\int \left(1-2v\frac{\partial v}{\partial \alpha}\right) d(x/c) \tag{6}$$

where Cp is the pressure coefficient, V the local velocity and
x/c the chordwise ordinate. Now at any incidence the second
integral can be simply expressed in terms of the two basic solutions at incidences of 0° and 90° - described later in Section 5.
This analysis is described in full by Hargreaves (1980) and can
be extended to all the steady and quasi-steady derivatives, which
will include a contribution to the derivative from the $\partial\phi/\partial t$ term
in the pressure coefficient expression.

Results from this method are shown in Figure 16 and compared
with experimental measurements. This method is still under development and there are plans to extend it to give the unsteady
derivatives as well.

Although agreement from certain derivatives is not as good as
required because of the important effects of viscosity which are
neglected from the theoretical calculations the theory can be used
very successfully to predict the effect of small geometric
changes, such as adding stores or nacelles under wings, which cannot be obtained from the standard data sheets.

4.2 *Store carriage and release*

This is probably one of the most important design areas for
military aircraft and there has been a lot of effort put into it
both experimentally and theoretically. Recently the application
of boundary integral methods to all aspects of this problem has
been producing very successful results. Probably the most well
known is the NEAR method due to Goodwin *et al* (1974) for calculating store loads and trajectories. However this is based on a
very simple modelling of the aircraft and store and it is now
possible to improve the modelling by using a panel method such as
the BAe Mk II method.

In the most frequently applied form of store release calculation the flow field is computed at a regular mesh of off-body
points lying beneath the 'clean' aircraft. Then the forces on the
store in isolation, installed in this non-uniform field, are computed independently with the store fixed in space at a series of
estimated positions along the store trajectory. This trajectory
is estimated as the calculation proceeds by integrating the equations of motion in accordance with the store's inertial characteristics and the most recently computed forces and moments on the
store as it proceeds through the mesh of off-body points.

This approach suffers from three main drawbacks:

1. The interference between aircraft and store is ignored.

2. The flow and hence the forces on the store estimated by a

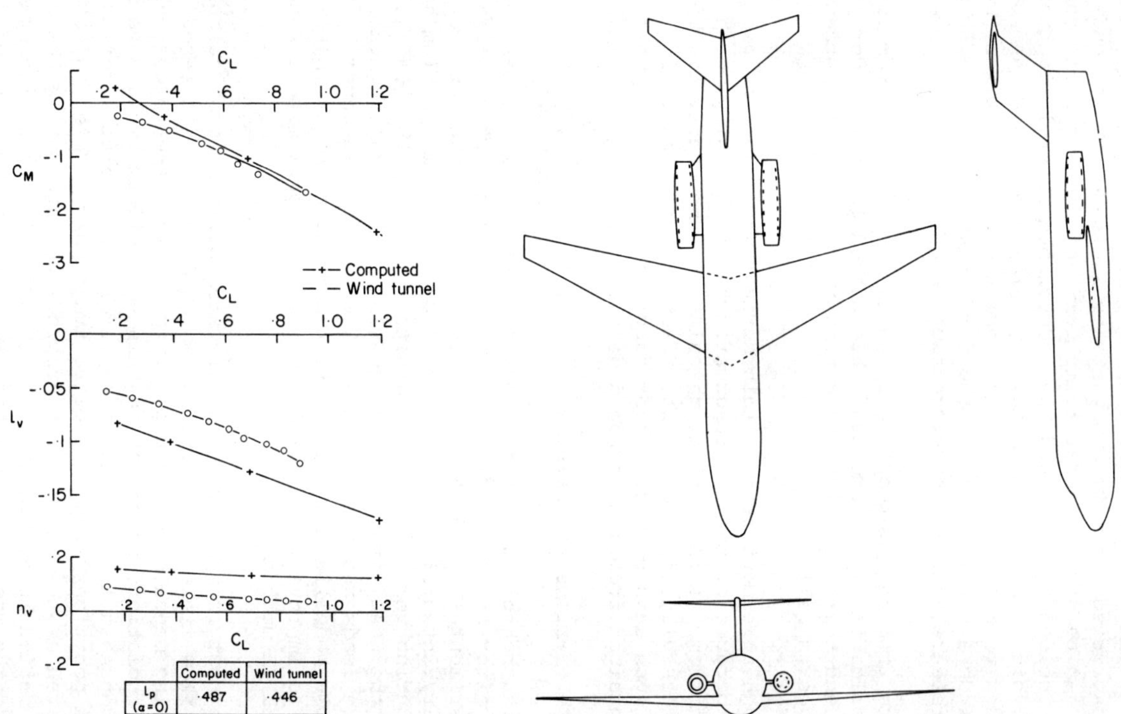

Fig. 16 Aerodynamic derivatives predicted by the method of Section 4.1, compared with experimental measurements

potential flow calculation are unrealistic since the real
flow is viscous and will contain body vortices due to flow
separation from the store.

3.   There is in fact relative motion between the store and the
     aircraft, which can generate significant differences from
     the quasi-static calculation, particularly in the case of
     rapid tumbling motions which provide the greatest danger in
     connection with store release.

Although there are more severe problems at transonic speeds
and with small aircraft-store separation distances, it is possible to improve all these deficiencies by a slight modification
to the standard panel method. A panel method calculation in
which both the aircraft and the store are simultaneously modelled
will of course include the full interference effects for the
inviscid flow. However a set of such complete calculations with
the store in a series of different positions relative to the
aircraft, and with the store modelled in sufficient detail to
allow an accurate calculation of the inviscid forces acting on
it, would be computationally expensive. This expense can be
reduced by computing only once the aircraft-on-aircraft and
store-on-store influence matrices; only the aircraft-store
influence coefficients, which vary with store position, need be
calculated for each store position and these will be, in general,
'far field' and hence relatively cheap to compute.

A presentation at the recent ICAS conference (Deslandes (1980))
shows how successfully the trajectories of stores can be predicted theoretically.

Figure 17 illustrates results due to Ostojic (1980) for store
carriage. In this example, apart from the comparison of the
inviscid potential flow method with experiment, the results are
also shown for an economic and potentially very useful approach
to include the 'real' effects on the store. This method is still
under development but, briefly, it uses the panel method to compute the flow field due to the complete configuration around the
store, which can then be used as input to a more sophisticated
method, including the effects of viscosity and body vortices, for
calculating the loads on the store. In a standard panel method
this still requires the expensive accurate modelling of the
store, but the method can be greatly enhanced by the 'ghost'
facility in the BAe Mk II method described in Section 5.4.

The example shown in Figure 17 is for a store fixed under a
wing but the method can be clearly extended to the trajectory
problem.

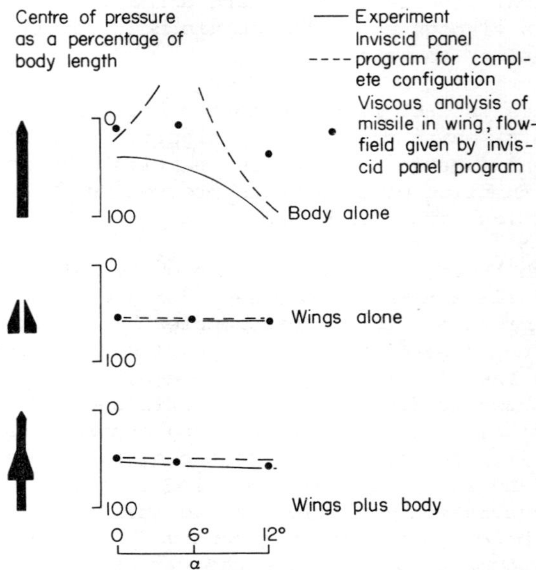

Fig. 17   Behaviour of longitudinal centre of pressure with
incidence for a fixed store under an unswept wing,
calculated by two different methods and compared
with experiments.

4.3 *Wind tunnel effects*

There is certainly a growing awareness that for many flow
problems of practical interest, different experimental results
can sometimes be produced in different wind-tunnels with the same
model. This is particularly true in the case of the various types
of flow involving separation. Obviously this casts doubt for
such cases on the validity of extrapolating wind tunnel tests to
full scale Reynolds numbers, but more pertinently in the present
context it highlights the difficulties in attempting to validate
the success of coupling procedures by means of comparison with
experimental results.

It has become almost traditional to blame the lack of agreement between theory and experiment on such vagaries as 'wind
tunnel wall effects', and a close inspection of many published
comparisons have been effected at equal values of measured and
predicted lift rather than at equal values of incidence. With
panel methods it is possible to apply an adequate modelling of
the wind tunnel walls - even perforated walls in principle - and
this should be done as a matter of course in connection with
effecting comparisons with any experiments for which wall cons-

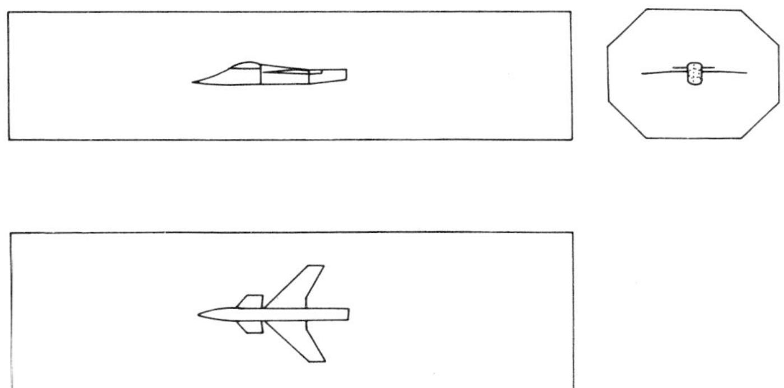

Fig. 18(a)   Sketch of canard aircraft model in wind tunnel

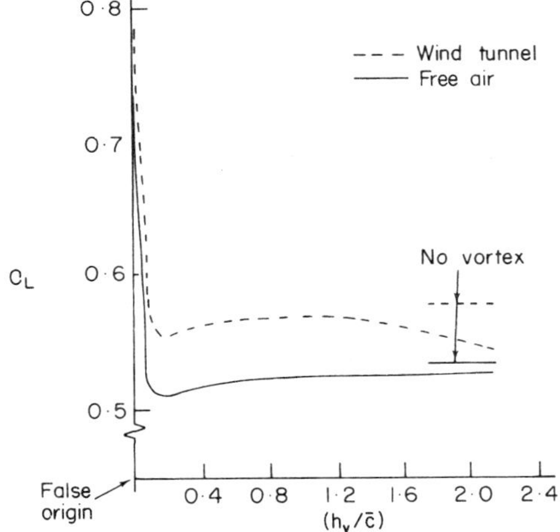

Fig. 18(b)   The predicted lift coefficient of the configuration
            in a wind tunnel or in free air. The incidence is
            7° and the ratio (vortex height above wing)/(wing
            chord) is the independent variable.

traint effects are other than totally negligible.

Figure 18 illustrates how the effects of a tunnel on the overall lift of a model can be predicted for the case of a canard configuration. Here the trailing vortex from the canard over the
wing is being represented by an arbitrary discrete vortex in the
calculation at various heights above the wing. Were this model to
be tested in a wind tunnel, normal corrections would not predict
the effects of tunnel constraint on the position of the canard
trailing vortex and could clearly give difficulties in comparing
theory and experiment.

Apart from the normal tunnel constraints of lift interference
and attached flow blockage, it is also possible to calculate the
tares and strut interference for normal and abnormal models. By
modelling the model *and* its suspension system in the tunnel, it
is possible to obtain accurate estimates of the tares and interference.

5. FORMULATION AND IMPLEMENTATION

We should now like to consider the philosophy of the current
generation of panel methods and to attempt to illustrate the
fundamental concepts behind their implementation, such as multiple solutions, matrix editing, hybrid solutions and 'ghosting',
which give them their efficiency for a variety of flows.

Returning to Figure 1, to give an indication of the power of
a panel method program, having once computed the flow around this
configuration, we can, for example, calculate the flow for a
different aircraft incidence with the outer store removed and
with the aircraft rolling at a steady rate of roll at very little
extra cost. This type of flexibility is inherent in the basic
formulation of the panel method and in the ability to superpose
solutions.

5.1 *Multiple solutions*

The principal equations in a panel method can be written in
simplified form in matrix notation as

(a) to calculate the singularity distribution, m

$$[m] = [A]^{-1} [-\overline{V}_\infty . \overline{n}] \qquad (7)$$

and

(b) to calculate the tangential velocity and hence pressures

$$[\overline{V}_T] = [B] [m] \qquad (8)$$

In this sequence the major investment in computing time is
principally in forming the influence matrices, $[A]$ and $[B]$. The
remaining time, about 20% of the total, is largely used in solving the equations to yield the singularity strengths, with the
other operations, such as forming the panel geometry and calculating the pressures, being in general trivial.

In order to exploit this we make use of the well known fact
that because we are solving Laplace's equation, or the Prandtl-
Glauert transformation of it, then if we have two solutions to
equation (7) any linear combination of the two is also a solution.

This is, for example, how any number of incidences can be treated extremely efficiently. The program calculates a solution for $0°$ incidence (m_0) and $90°$ incidence (m_{90}) and any intermediate incidence is calculated by a simple combination of these two. In computing terms all we have to do is to combine the singularities so that

$$[\overline{V}_{T}]_{\alpha} = [B][m_0\cos\alpha + m_{90}\sin\alpha] \qquad (9)$$

which is clearly a trivial operation.

A further and very important point is that if the two basic solutions are then stored in an output file on the computer we can return to them at a later date and *interactively* produce answers at a variety of incidences at very little cost. This retrospective manipulation of the results is not restricted to pressure distributions only but can be extended to generate other analyses such as streamline tracing.

It can also obviously be extended to include a wider variety of right hand sides to equation (7) such as those for slideslip, rate of roll, rate of yaw and rate of pitch. We have now 6 basic onset flows, the results of which can be linearly combined in any combination to yield the surface pressure for an aircraft under-going any steady motion (remembering that the pressure coeffic-ient for the rate terms will include a $\partial\phi/\partial t$ term as outlined in Section 2.4.). Hence we can write

$$[\overline{V}_{T}] = [B][K_1 m_{\alpha=0} + K_2 m_{\alpha=90} + K_3 m_{\beta=90} + K_4 m_p + K_5 m_q + K_6 m_r] \qquad (10)$$

Admittedly at this point we are only dealing with attached flow, but nevertheless, referring to Figure 1 again, it means we can compute the basic flows for this aircraft on one day and obtain results for a given condition of say $6°$ incidence. Then, provided that we store the solutions, we can subsequently compute the results, at any time, for the yawed configuration at a steady rate of pitch.

This concept can be taken even further and to the ability to compute both steady translational and rotational motions we can add the onset flows of a specified vortex or source distribution on the surface or in the field. Also we can add the effects of non uniform onset flows such as those that arise from boundary layer calculations or, as will be seen later, from the hybrid and ghost techniques.

Although not precisely part of the concept of mutiple solu-tions, it is also possible to change the applied boundary condi-

tions from external to internal, Neumann or Dirichlet or even to
mixed boundary conditions with only minor changes to the calculation of the influence matrix.

All these facilities, it must be stressed, are obtained basically for the cost of forming the influence matrices once only and
of deriving the several solutions for the various basic right
hand sides, which will in general be little more than that for the
single incidence point.

5.2 *Matrix editing*

Another very important feature of a panel method is the ability to change, add or subtract parts of the configuration without
having to recompute the entire influence matrix. For example, if
we were to add the fourth store to the aircraft shown in Figure 1,
it is only necessary to compute the influences of the store on
itself, the store on the aircraft and the aircraft on the store
which in general is only a small part of the total, particularly
as most of the influences in the last two groups will be of the
'far field' type allowing the simplest approximation to be used.
Removing a part is obviously even easier and much cheaper as it
only requires the various influence coefficients to be identified
and removed from the matrix.

Clearly this requires that we store the basic influence matrix
and geometry but if we are prepared to do so, once having computed
a basic aircraft, we have the ability to generate solutions of
modified geometries and applications at very little cost. Over
the life of a project this could represent a very significant
cost saving. This is a major advantage of the integral approach
over field methods where the required grid is generally configuration dependent and many of the cases outlined above would
require a completely new solution.

5.3 *The Hybrid Method*

This subject was treated in depth in the paper by Jepps, in the
1978 VKI lecture series, covering the problems of the trailing
edge wake relaxation, leading-edge vortices and body vortices.

The term 'hybrid' reflects the fact that the concept is partly
based on a standard 3-dimensional (3-D) panel program and partly
on 2-dimensional (2-D) techniques, combined in such a way that a
nominally exact solution to the 3-dimensional problem is achieved
via iteration: the 2-dimensional contribution to the total field
reduces very rapidly as the iterations proceed. The basic iteration algorithm may be summarised schematically as:-

$$(3\text{-}D)_n = \{(3\text{-}D)_{n-1} - (2\text{-}D)_{n-1}\} + (2\text{-}D)_n \tag{11}$$

One iteration consists of applying the algorithm in turn to each 'slice' of the 3-dimensional problem. The nth iteration of the algorithm may be interpreted as computing the 2-dimensional singularities in that slice $(2\text{-D})_n$, which, in the presence of a non-uniform onset flow $\tilde{U}\{\tilde{U}=(3\text{-D})_{n-1}-(2\text{-D})_{n-1}\}$, satisfy the 3-dimensional boundary conditions for that slice $(3\text{-D})_n$. The non-uniform flow U is the 3-dimensional flow induced at that slice by the entire set of 3-dimensional singularities computed during iteration (n-1) minus the 2-dimensional flow induced at that slice by the 2-dimensional singularities computed in that slice during iteration (n-1). The 3-dimensional singularity *strengths* in each slice are generally taken to be the 2-dimensional values computed in the iteration (n-1) for that slice, but their 3-dimensional *influence* is computed by applying these values on the 3-dimensional panels. Thus both a 3-dimensional influence matrix and N 2-dimensional influence matrices need to be generated, where N is the number of slices treated in this way.

This scheme and its numerous variants can also be used to advantage for fixed geometry 3-dimensional calculations: the 3-dimensional equation solution can be broken down into a series of much smaller "2-dimensional" solutions. By employing more 2-dimensional panels than 3-dimensional panels in a particular slice, moreover, (in conjunction with an interpolation process for the singularity strengths and boundary conditions), it is possible to gain a substantial improvement both in the local flow and in the accuracy of a wing calculation, for example, particularly in connection with drag estimation.

The primary advantage of the scheme, however, is gained during vortex calculations in which the position and strength of free vortices is computed as part of the "2 dimensional" calculation. Although work is underway (in particular at Boeing, Johnson *et al* (1979)) where the 3-dimensional flow is treated as an explicit unknown, this approach is prone to computational instabilities and is probably prohibitively expensive for routine applications and for complex aircraft configurations. Using the hybrid method, described by Hunt (1980), pilot versions of this scheme, which were based on the use of a simple vortex lattice scheme instead of a full 3-dimensional panel method and a simple line vortex model of the vortex sheet, lead in general to a very rapid convergence of the solution. Work is currently in hand to apply the technique to trailing edge wake relaxation and to body vortex problems, though the latter suffers from the practical difficulty that the separation line on a smooth body is not known in advance and in the present state of the art cannot be satisfactorily predicted by boundary layer-calculations.

An example of the application of the hybrid scheme is shown in Figure 19. In this calculation a 'semi-relaxation' of the trailing

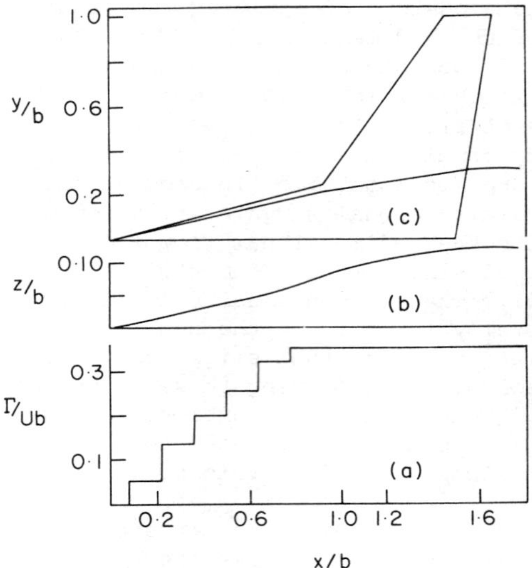

Fig. 19 "Hybrid" computation of vortex flow over a straked wing.
       $M_\infty = 0$, $\alpha = 14.32°$. (a) strength of vortex core,
       (b) height of core above wing, (c) lateral position of
       core.

edge wake was applied: during the computation this wake was held
in the same plane, $Z = 0$ but the direction of the vorticity within
this wake was interactively adjusted to satisfy a zero-pressure-
jump condition at the control points of all the panels in the
wake: the usual type of Kutta condition applied simply near the
trailing edge did not work satisfactorily. The leading edge
vortex was fed only from the strake leading edge: it then convected at fixed strength over the main part of the wing.

It is possible to see that this hybrid method can be extended
to many other classes of flow including the transonic flows
mentioned in Section 2.3, the separated flows described in Section
3.2, intake flows and possibly even as an interim means of calculating the flow over multi-aerofoil configurations, incorporating
viscous effects in the 2-dimensional calculation only.

5.4 *"Ghost" modelling*

Another very important feature is the concept of "ghost"
modelling, which is a particular case of the hybrid approach for
fixed geometry calculations.

Consider for example the case of an aircraft with a nacelle.
The influence felt at the aircraft due to the presence of the

nacelle, in a composite aircraft-plus-nacelle calculation, will
in general be relatively insensitive to the level of detail used
in modelling the nacelle; for this influence a coarse panelling
of the nacelle is sufficient and moreover will give a reasonable
approximation of the influence due to the nacelle if the latter
generates body vortices. Suppose such a calculation is performed
with singularities 1 to N representing the aircraft and singularities N + 1 to M modelling the nacelle. The velocity V_p at a
point P will then be given by:

$$V_p = \sum_{j=1}^{N} V_{p_j} m_j + \sum_{j=N+1}^{M} V_{p_j} m_j + U_\infty \qquad (12)$$

where V_{p_j} is the velocity induced at the point P by a unit singularity density on panel j, and m_j is the generalised singularity
density. Once the system of equations for m_j has been solved, V
will be tangential to the aircraft surface and to the nacelle
surface at every collocation point. However, if we now compute
the velocity $\tilde{V}_p$

$$\tilde{V}_p = \sum_{j=1}^{N} V_{p_j} m_j + U_\infty \qquad (13)$$

with only the N *aircraft* singularities, the nacelle will appear
as a 'ghost' through which the flow $\tilde{V}$ passes freely.

If $\tilde{V}$ at the original nacelle collocation points were used as a
non-uniform onset flow for a new calculation of the same nacelle
in isolation (i.e. if $\tilde{V}$ were used instead of U_∞), then the resultant solution for that nacelle would be exactly as if that solution
were part of the original aircraft-plus-nacelle calculation: i.e.
full potential flow interference terms would be included. However,
a more accurate potential flow solution for the nacelle could now
be obtained by replacing the original coarse panelling by a more
refined panelling of the nacelle; the onset flow $\tilde{V}$ at the collocation points of this different panelling would have to be interpolated from those of the coarse panelling. The interpolation
could be simplified by using the ghost flow $\tilde{V}$ computed not at the
original coarse collocation points but at the points of a regular
mesh set up round the entire nacelle.

Results for such a calculation of a nacelle under a wing are
shown in Figure 20. The results plotted correspond to the vertical component of the velocity at points on a line which passes
longitudinally through the walls of the nacelle. Most of the
results for the complete configuration thus lie in a fictitious
flow region within the section profile of the nacelle. The velo-

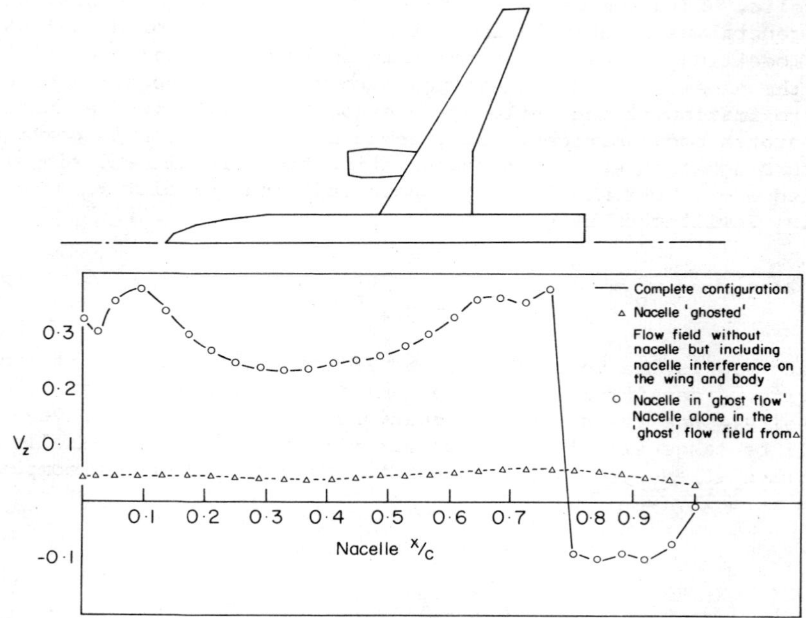

Fig. 20   Application of the "ghost flow" concept to a wing/body/
nacelle configuration. V_z is the vertical component of
velocity at points on a line partly inside the nacelle.

city discontinuity at about three-quarters chord, where the line
emerges into the real flow, can clearly be seen. The corresponding 'ghost' flow, however, is very regular and amenable to interpolation.

A useful variant of this scheme is offered by replacing the
calculation of the finely panelled nacelle 'in isolation' in the
field $\tilde{V}$, by a calculation in which the viscous effects on the
nacelle are modelled. This could comprise an empirically-based
scheme or a purely theoretical scheme including boundary layer
calculations.

This concept can be extended to other configurations, probably
the most obvious of which is that of store carriage and release.
Here the scheme could be extended to a superposition of a set of
'ghost' fields $\tilde{V}_i$, computed for the coarsely panelled store located at a set of representative positions beneath the aircraft.
The finely panelled store at any position in this mesh of off-
body points would feel a $\tilde{V}$ corresponding to near locations of the
original coarse panelled store.

6. USE AND EVALUATION

6.1 *Basic Concepts*

Before concluding this paper, having briefly described the
range of potential and 'real' flow problems to which panel
methods can be applied, we should like to issue a word of warning
to the potential users of these methods by discussing the somewhat more philosophical problem of what exactly is a potential
flow about a complete aircraft configuration.

To illustrate the problem, let us first consider two-dimensional flow. An aerofoil has a sharp trailing edge from which,
for attached potential flow, we know that the flow will always
leave smoothly, and as the incidence increases the lift also
increases in a unique way. But if we have an ellipse this is no
longer true; here there is no unique point defining the rear
stagnation, and at a given incidence the lift and circulation can
assume a range of values depending on where that point is taken,
which gives us an infinite number of potential flows at any incidence. One of these will correspond to the classical potential
flow with zero circulation and zero lift and drag; for this condition the rear separation point will be symmetrically disposed
relative to the front attachment point. Yet we know from experiment that at given Reynolds and Mach numbers there is a unique
(finite) lift curve. Obviously, in this case it is the viscous
effects which fix the separation point and hence control the
circulation. We clearly cannot, therefore, expect a potential
flow method to reproduce the flow around an ellipse and we will
in general require some form of viscous flow modelling to obtain
a sensible answer. We could possibly modify the potential flow
method and include some empirical tricks to obtain the 'right'
answer, but this then brings the basic validity of the solutions
into question and is usually too far from general to be reliable.

This is more complex for 3-dimensional geometries such as fuselage, where the rear separation controls the circulation on the
body, or a wing-body combination where the way the viscous layers
interact in the junction dictates the lift carry-over, etc. For
example, there are currently several different methods of modelling the lift carry-over between a wing and a fuselage in panel
methods, and provided each formulation is a properly posed boundary value problem then each solution is a valid potential flow.
But, in general, each will give different answers, of which one
or all may agree sensibly with the experimental results. That
agreement will in general be *fortuitous*, not *axiomatic*.

The problem becomes more severe for a full aircraft configuration with all its interacting components. We have, for example,
no way of knowing what is the correct potential flow around the

aircraft shown in Figure 1 and we can only use our intuition and
experience as to whether the results we obtain are sensible and
useful. In all applications such as these, it is essential that
we *calibrate* (and we use that word advisedly) the method against
a known, *reliable* set of experimental results on a configuration
as close as possible to that in which we are interested.

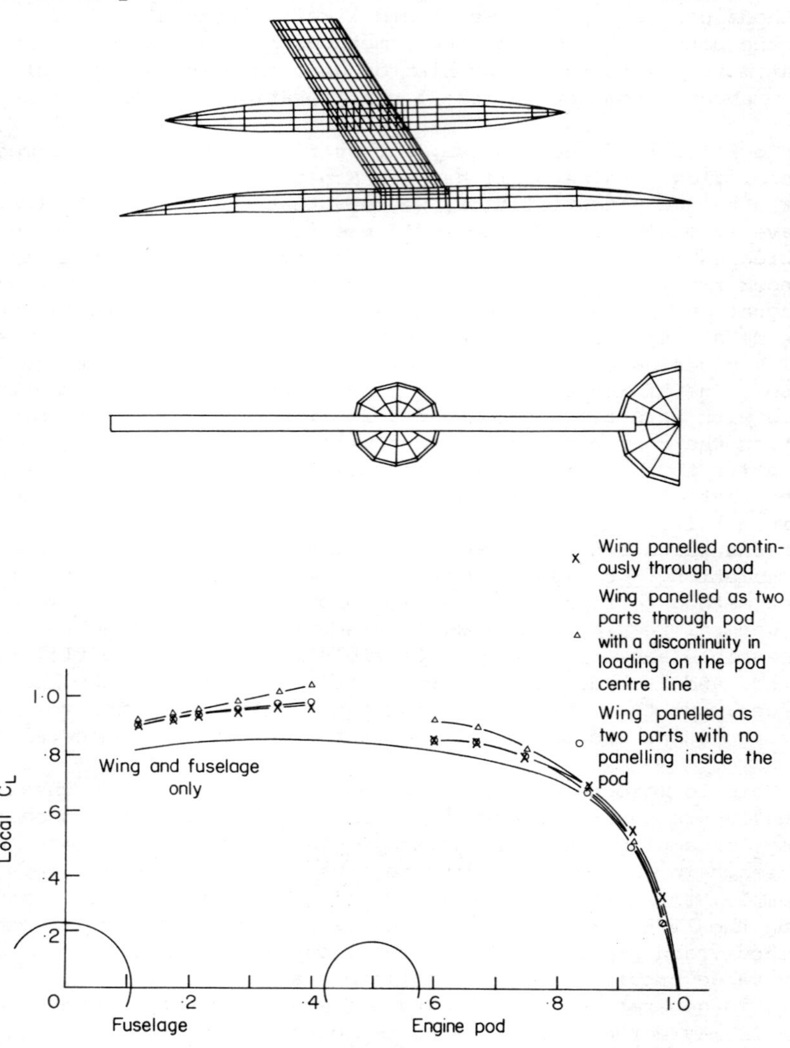

Fig. 21 Effect on local lift coefficient of different
         assumptions regarding carry-over through an
         engine pod.

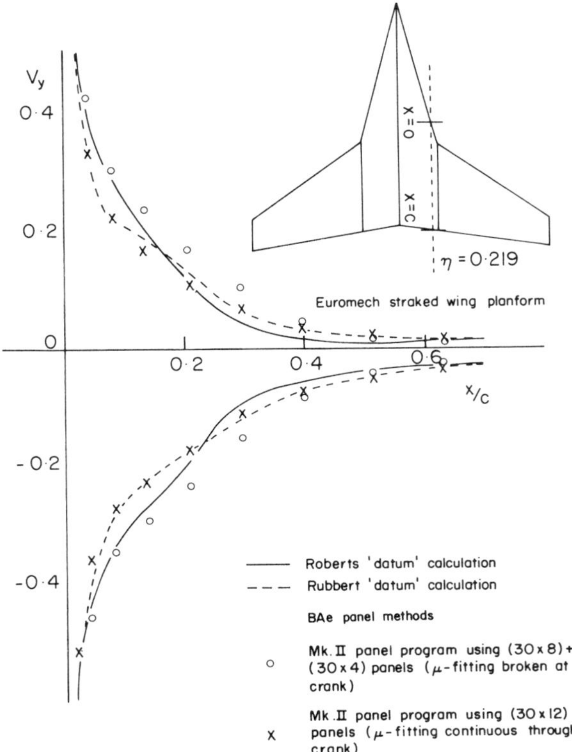

Fig. 22   Spanwise velocity component along one chord of a
straked wing with thickness/chord ratio equal to
0.92, at 5° incidence.

Figure 21 illustrates the problem for a nacelle mounted on the
wing. Different modelling of the lift carry-over gives significantly different answers. The problem of uniqueness is not confined solely to bluff bodies and interactions; Figure 22 shows
the results of potential flow calculations with several methods
for a straked wing where even with the advantages of a sharp
trailing edge, different formulations of the continuity of the
doublet density μ across the kink can produce different answers.

There is an interesting corollary to this when attempting to
model the viscous flow using a viscous/inviscid matching procedure. It is no use taking an existing viscous flow method and an
existing potential flow method and simply combining them, because
if, for example, we take two potential flow methods for the wing/
body, with different lift-carry-over modelling, and the same

boundary layer method, then in general we would get a different
answer. But we believe that, subject to the more esoteric problems of free stream turbulence, etc., at a given Reynolds number
the viscous solution is unique. Therefore in this case it
becomes very important that care is taken in selecting models of
the potential flow and the boundary layer which are compatible
and which will combine to give the 'correct' answer.

So to summarise it is important that the user is aware of the
problems of calculating the potential flow around complex 3-
dimensional geometries. He must also be aware of how the program
that he is using is formulated and should calibrate it against
known test cases. The responsibility for using a program must be
with the user, but at the same time the developer must avoid
'tricks' in his method to improve answers for particular cases.

6.2 *Accuracy*

In this paper we have tried to show the wide range of potential
and 'real' flow problems that can be solved with existing standard
panel methods. The outstanding problem is that of accuracy, not
numerical accuracy which can be assessed either by analysis or by
panel refinement techniques, but how well does the method predict
the 'correct' answer, and this is probably one of the most difficult problems. We are faced with two problems; firstly that for
the complex 3-dimensional geometries, we do not have an analytic
potential flow solution; and secondly that currently even with our
'real' flow models, their representation of the true viscous flow
can only be approximate, to a lesser or greater extent.

Consequently we have to exercise great care when comparing our
results with experiment. For the potential flow solutions comparison with higher order methods can yield useful information but
again care must be taken because the higher order solution will
depend on the basic formulation of the method, which may make
assumptions about the potential flow, such as wing lift carry-
over. However, for the complex geometries we are left with comparison with experiment and here the comparisons should be treated
qualitatively rather than quantitatively. We have all seen results
showing excellent agreement between theory and experiment which
probably only really serve to show that either they are both wrong
or the method has been specially tuned to that case, but certainly
with care panel methods can be used to predict trends and rates of
exchange very successfully on complex geometries, as is illustrated
in Figure 23 for some recent calculations by BAe Brough Division on
a typical fighter aircraft. Here a complex wing/body geometry is
analysed including the flow through the intakes, and, as can be
seen from the results, excellent agreement is obtained with experiment. What is important in this example is not the absolute level
of agreement, which is somewhat fortuitous here, but that the

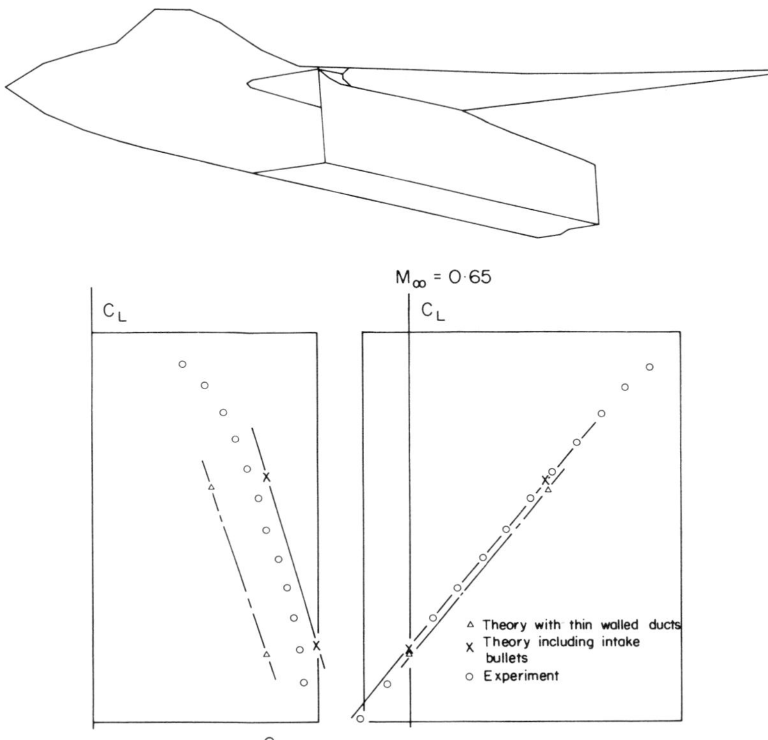

Fig. 23   Lift coefficient plotted against pitching moment and
          incidence for a wind-tunnel model of a fighter aircraft
          with flow through ducts representing engine nacelles,
          at M_∞ = 0.65.

calculation is producing answers of the right magnitude despite
the complex geometry, and the trends of the various parameters
are predicted sensibly. The results also show the significant
effect of the intake bullets.

It must also be stressed that great care should be taken when
comparing pressure distributions; it is more informative to use
the velocity in its cartesian components. Obviously this is very
difficult for experimental comparisons but should always be used
for theoretical comparisons. In two dimensions there is no
problem but in three dimensions the pressure is derived from the
total vector velocity, which is calculated from the sum of the
squares of the three components. Consequently not only is it
possible to have incorrect velocity components compounding to
give a 'correct' pressure but some of them may even be of the
wrong sign.

As mentioned in Section 4.3, it is possible to model the effects of the wind tunnel with the standard panel method. This could be done as a matter of routine for comparisons with experimental results but in general it will not be necessary and the traditional tunnel interference corrections will be adequate. However, for complex geometries, if it is suspected that tunnel interference effects are significant, especially the high lift cases, the tunnel should obviously be included in the modelling.

7. CONCLUDING REMARKS

The difficulties in taking the predictions of theoretical methods closer to physical reality, for problems of practical interest in aircraft aerodynamics, may be likened to the difficulty in approaching a temperature of absolute zero. Each step nearer the goal requires a quasi-exponential increase in the effort applied. The tools and techniques which were satisfactory at an earlier stage break-down or become inappropriate and a new level of technology is required to achieve further progress. It is the writers' opinion, however, that the future of computational fluid mechanics does *not* depend entirely on the use of new technology in the form of enormous computers for solving the Navier-Stokes equations for realistic flows over complete configurations. Apart from the total impossibility, without some form of time-averaging and empirical guess work, of writing down the equations governing the microscopic and macroscopic behaviour of a fluid, there remains the very real barrier to *full*-aircraft calculations of generating the computational mesh in space around the aircraft. A suitable mesh will not only be configuration dependent; it may also vary with incidence, flight Mach number and Reynolds number. The likelihood that self-contained Navier-Stokes solutions or even Euler or full potential solutions will ever be able to simulate such flows for full scale real flight conditions is extremely unlikely.

In view of this anticipated inapplicability of advanced field solution techniques for those areas where the need is most pressing, the boundary integral method is very likely to attract increasing interest throughout the 1980s as the basis of advanced computational schemes for addressing a large subset of the problem. It should be added that although numerous and daunting theoretical difficulties remain in most aspects of the flow modelling, such advances will almost certainly be paced more by the level of effort applied to the method development rather than by the theoretical difficulties themselves. Advances in computer capacity and speed and in the efficiency and accuracy of the numerical processes involved in boundary integral schemes have brought the associated computer times down to acceptable levels for routine aerodynamic computations and these times can no longer be realistically considered as barriers to the use of the computer program.

It is also an inescapable fact that the boundary integral
method is inherently far better suited to dealing with the boundaries of realistic configurations than any of the field methods
used in isolation. It is noteworthy that boundary integral methods are currently being developed and applied for the equally
difficult problems involving elastoviscoplastic behaviour in
highly stressed structures, see Chaudonneret (1978). The kernel
functions of the various surface integrals which appear in the
methods for stress analysis are fundamental solutions of the
equation governing the stress/strain problem rather than of the
more familiar idealised fluid mechanics problem; at the simplest
level, however, both equations reduce to relatively simple forms,
whereas when 'higher-order' versions of the governing equations
are used the respective kernel functions become more complicated.
The differing natures of the problems naturally lead to different
physical interpretations of the basic functions ('tractions' and
displacement for the stress problem rather than the more familiar
velocity components, velocity potential, pressure, etc. for the
fluid mechanics problem) and have required mathematical formulations based on tensor analysis rather than vector analysis. The
central mathematical identity exploited in the stress analysis
formulations is the so-called Somigliana Identity, (see Jaswon
and Symm (1977)) rather than Green's identities and their off-
shoots, which are used in the aerodynamic formulations; nonethe-
less, all these identities spring from the same basic mathematical
concepts and may be considered equivalent. At the same time the
physical problems themselves bear some striking similarities; for
small perturbations both problems reduce to solving relatively
simple field equations (i.e. Laplace's equation in the fluid
mechanics case) whereas larger perturbations lead to non-linear
terms; both problems require special treatment for sharp edges
(e.g. Kutta conditions for aerodynamic problems); both problems
can involve surfaces of discontinuity, cracks being partly analogous to wakes and partly to shocks.

Clearly the specialists in each field - as well as mathematicians - can derive substantial benefit from a study of the techniques developed for other fields. For example, it seems to be
the case that at present boundary integral methods for fluid
mechanics have the lead over those for stress analysis as far as
numerical sophistication and level of implementation is concerned.
On the other hand, the stress methods are more advanced as far as
the complexity of the governing equation is concerned and as far
as the coupling with other types of method (e.g. finite element
methods) is concerned.

We may conclude by observing that frequently the opinion is
offered to the present writers that the panel method for aerodynamic computations seems to have reached its limit of application
and that there is little point in developing it further. This

opinion is certainly premature and whilst substantial work
remains to be done, it would appear that the generalised boundary
integral method is far from having reached its useful limits not
only for aerodynamic applications but also for a very diverse
range of other engineering problems.

REFERENCES

Bristow, D.R. (1980). Development of panel methods for subsonic
analysis and design. NASA CR 3234.

Chaudonneret, M. (1978). Calcul des concentrations de constrainte
en elastoviscoplasticité. ONERA Publications - 1.

Deslands, R. (1980). Evaluation of aircraft interference effects
on external stores at subsonic and transonic speeds. AGARD CP 285
Paper 6.

Fray, J.N.J. and Sloof, J.W. (1980). A constrained inverse method
for the aerodynamic design of thick wings with given pressure
distribution in subsonic flow. AGARD 285 Paper 16.

Goodwin, F.W., Dillenius, H.F., & Nielsen, J.N. (1974). Prediction of six-degree-of-freedom store separation trajectories at
speeds up to the critical speeds. Vol.I. Theoretical method and
comparison with experiment.   U.S. Air Force Flight Dynamics Laboratory TR-72-83.

Hargreaves, G.R. (1980). The use of panel methods for predicting
stability derivatives of transport aircraft. BAe Manchester FDM-
0060.

Hess, J.L. (1972). Calculation of potential flow about arbitrary
3-D lifting bodies. Final Technical report. McDonnell Douglas
Report MDC J5679-01.

Hess, J.L. (1973). Higher order numerical solution of the integral
equation for the two-dimensional Neumann problem. Computer methods
in Applied Mechanics and Engineering No. 2, 1-15.

Hewitt, B.K., Crowley, J.A. & Marchbank, W.R. (1980). Development
of a supersonic panel program. BAE Warton AE 416.

Hewitt, B.L. and Marchbank, W.R. (1976). Unpublished work on
transonic-integral-equation formulation involving the artificial
viscosity concept. BAe Warton.

Hunt, B. and Semple, W.G. (1976). Economic improvements to the
mathematical model in a plane, constant strength panel method.
Euromech Colloquium 75 Branschweig.

Hunt, B. (1978). The panel method for subsonic aerodynamic flow;
A survey of mathematical formulations and numerical models and an
outline of the new BAe scheme. VKI Lecture Series.

Hunt, B. (1980). Recent and anticipated advances in the panel
method. Appendix 1 Viscous flow modelling in panel methods. VKI
Lecture Series.

Jaswon, M.A. and Symm, G.T. (1977). Integral equation methods in
potential theory and elastostatics. Academic Press.

James, R.M. (1969). On the remarkable accuracy of the vortex
lattice discretisation in thin wing theory. McDonnell Douglas
Report DAC 67211.

Jepps, S.A. (1978). The computation of vortex flows by panel
methods. VKI Lecture Series - 4.

Johnson, F.T., Ehlers, F.E., & Rubbert, P.E. (1976). A higher
order panel method for general analysis and design applications
in subsonic flow. Proc. of 5th int. Conf. on Numerical
Methods in Fluid Dynamics, Twente University, Holland. Lecture
notes in Physics, 59, Springer-Verlag.

Johnson, F.T. *et al.* (1979). Recent advances in the solution of
three-dimensional flows over wings with leading edge vortex separation. AIAA Paper 79 - 0282.

Kraus, W. (1970). Das MBB-Unterschall-Panelverfahren. Teils 1 - 3.
MBB-UFE-633-70.

Labrujere, Th.E., Loeve, W. & Sloof, J.W. (1970). An approximate
method for the calculation of the pressure distribution on wing
body combinations at subcritical speeds; AGARD CP No.71 Paper 11.

Labrujere, Th.E. (1972). A survey of current collocation methods
in inviscid subsonic lifting surface theory. VKI LS.44.

Lock, R.C. and Firmin, M.C.P. (1982). Survey of techniques for
estimating viscous effects in external aerodynamics. (This vol.)

Luu, T.S., Coulmy, G. & Dulieu, A. (1972). Calcul de l'ecoulement
transonique autour d'un profil en admettant la loi de compressi-
bilité exacte. ATMA Session.

Maghoub, H.E.H. and Bradshaw, P. (1979). Calculation of turbulent-
inviscid flow interactions with large normal pressure gradients.
AIAA Journal, Vol. 17.

Maskew, B. & Dvorak, F.A. (1978). The prediction of C_{Lmax} using a separated flow model. J. Amer. Hel. Soc. April, 1978.

Maskew, B., Rao, B.M. & Dvorak, F.A. (1980). Prediction of aerodynamic characteristics for wings with extensive separations.
AGARD CP 291 Paper 31.

Melnik, R.E. (1980). Turbulent interactions on airfoils at transonic speeds - recent developments. AGARD CP-291 Paper 10.

Morino, L. (1974). A general theory of unsteady compressible
potential aerodynamics. NADA CR - 2464.

Morino, L. & Luo, C.C. (1974). Subsonic potential aerodynamics
for complex configurations: a general theory. AIAA Journal, 12,
No. 2.

Ostojic, P.L. (1980). BAe(DG) Experience in the use of a subsonic
panel methods to study aircraft/store aerodynamic interactions.
BAe Filton

Petrie, J.A.H. (1978). SPARV - the BAe-Brough panel method. BAe
(Brough) Note YAD 3348.

Piers, W.J. & Slooff, J.W. (1979). Calculation of transonic flow
by means of a shock capturing field panel method. AIAA Paper 79,
1459.

Roberts, A. & Rundle, K. (1972). Calculation of incompressible
flow about bodies and thick wings using the spline mode system.
BAC(CAD) Report Aero MA19.

Roos, R., Bennekers, B. & Zwann, R.J. (1977). Calculation of
unsteady subsonic flow about harmonically oscillating wing/body
configurations. Journal Aircraft, 14, No.5.

Roos, R. (1978). The use of panel methods for stability derivatives. NLR MP 78012U.

Rubbert, P.E. & Saaris, G.R. *et al.* (1967). A general method for
determining the characteristics of fan inwing configurations.
USAAVLABS Technical Report 67-61A.

Rubbert, P.E. (1978). AIAA Professional study series - panel
methods.

Rubbert, P.E. *et al.* (1978). An improved higher order panel
method for linearised supersonic flow. AIAA Paper 78-15.

Semple, W.G. (1978). A pseudo-error minimisation technique for
determining the source/vortex mix on wing type components in 3-D

surface singularity methods. BAe (WARTON) Report Ae/A/555.

Sytsma, H.S., Hewitt, B.L. & Rubbert, P.E. (1979). A comparison
of panel methods for subsonic flow computation. AGARDograph No.241.

Thomas, J.L. & Miller, D.S. (1979). Numerical comparisons of panel
methods at subsonic and supersonic speeds. AIAA Paper 79-0404.

Woodward, F.A. (1973). An improved method for the aerodynamic
analysis of wing-body-tail configurations in subsonic and supersonic flow. Part 1. NASA CR 2228 Pt. 1.

Woodward, F.A. (1979). The supersonic triplet - a new aerodynamic
panel singularity with directional properties. AIAA paper 79-0273.

REQUIREMENTS AND DEVELOPMENTS SHAPING A NEXT GENERATION OF INTEGRAL METHODS

J.W. Slooff

(National Aerospace Laboratory NLR Amsterdam, The Netherlands)

ABSTRACT

A discussion is given of various "push" and "pull" factors affecting the development of a next generation of integral or panel methods for aerodynamic applications.

Amongst the "push" factors to be discussed are increasing utilization and increasing requirements with respect to range of applicability, accuracy, ease and cost of application and integration with other disciplines. Improving these aspects will, in turn, increase the level of utilization, as will be made plausible.

Factors "pulling" towards such improvements are being generated by developments in both "hardware" and "software". A discussion will be given of the possibilities of increased computational power as well as of recently evolved concepts for improved and extended mathematical models and more efficient, multi-grid type algorithms.

Finally, it is argued that, for efficient application and integration with other disciplines, the techniques of modern informatics are indispensable. A "blue-print" will be sketched of how a next-generation integral method (or any other fluid dynamics solver) could be embedded in an aerodynamic information system in a computational/experimental research/design environment.

1. INTRODUCTION

Boundary Integral or "Panel" methods for subsonic as well as for supersonic aerodynamic applications have been in use in the aerospace community since the 1960s and have become indispensable tools in aerodynamic analysis and design. This position has arisen by virtue of the following characteristics:

1) the capability to provide accurate solutions for incompressible potential flow about complex configurations

2) their dimension-lowering property, leading, at engineering
   accuracy, to:
   a) acceptable computational effort
   b) acceptable effort in grid (panel) generation.

The latter feature is probably also the most important one.
It is directly related to the fact that the surface panelling of
complex 3-D configurations is an order of magnitude simpler than
the volume discretization of the space around such configurations
which is needed for finite difference and finite element methods.

Because of the large number of possibilities in the basic
formulation of boundary integral methods as well as in the
numerical implementation thereof, a large number of different
methods can be found in the literature. They can be classified
according to any of the following criteria:

1) the type of (surface) singularity distribution, i.e. doublet,
   source, vorticity or combinations thereof
2) the "order", in terms of power of the panel dimension, of the
   truncation error (1st, 2nd or 3rd order method)
3) the method for solving the resulting system of linear algebraic equations (direct matrix inversion, iterative, etc).

In spite of their differences in formulation and implementation all these methods have in common that, generally speaking,
they are capable of producing results of acceptable accuracy at
acceptable cost for element or panel numbers of the order of
1000—2500. Above the latter number the computational effort and
costs deteriorate rapidly as a result of the fact that the computational effort is proportional to N^2 (or even N^3), if N is the
number of panels. Hence, the currently available panel methods
become rapidly less attractive when changing requirements with
respect to accuracy or resolution ask for larger numbers of
panels.

A typical "first generation" boundary integral method for
3-D lifting flows, featuring piecewise constant surface source
and fixed internal "bound" vortex distributions (Labrujere et
al., 1970), has been operational at NLR since early 1969. It
has been used and abused in several versions, has acquired a
respectable amount of peripheral programs, but, stretched to
its limits, is no longer capable of standing up to the changing
requirements with respect to accuracy, resolution, range of
applicability etc.

It is interesting to note at this point that, at least for
geometrically fairly simple configurations, allowing automated
spatial grid generation, the current successful SLOR-type

F(inite) D(ifference) M(ethod)s for three-dimensional transonic potential flow (Caughey and Jameson, 1979) are also quite competitive in low speed application when we compare the computational effort required with that of the boundary integral method for the same resolution.)*

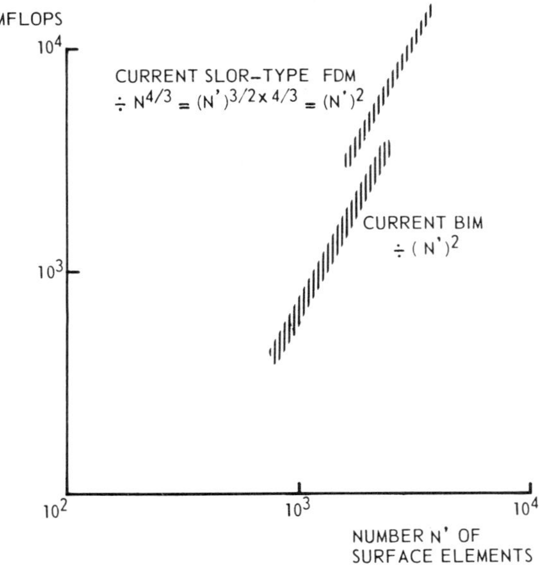

Fig. 1 Computational effort required by current boundary integral and finite difference potential flow methods.

The point is illustrated by Fig. 1 which contains a comparison of the computational effort required by the current BI and FD methods as a function of the number of discretization elements on the surface of the (wing) configuration.

In the context of what was just mentioned the question is sometimes raised whether integral methods still have a future; in particular in view of the current progress in grid generation (Anon, 1980) grid embedding (Boppe, 1980) and fast iterative techniques for FDMs (Jameson, 1979). This paper contains several arguments for a positive answer to this question. At NLR these arguments have led to the decision to start the development of a next generation integral method. The paper will discuss some of the factors that are shaping this new generation

---

*Whether or not the same resolution implies also the same accuracy is not well known; there has been little or no systematic investigation on the subject.

panel method (Fig. 2). In doing so we will distinguish between
factors *pushing* towards increasing capabilities from the experience and requirements of the user of panel methods on the
one hand, and, on the other, advances in CFD-related disciplines,
such as numerical mathematics, computer technology and informatics. The latter are *pulling* CFD methods to higher levels of
sophistication. The discussion, it is hoped, will illustrate
that with new developments at the horizon, integral methods are
worth of further investments, in particular when freed from
the restriction to treat only linear, i.e. boundary integral
problems.

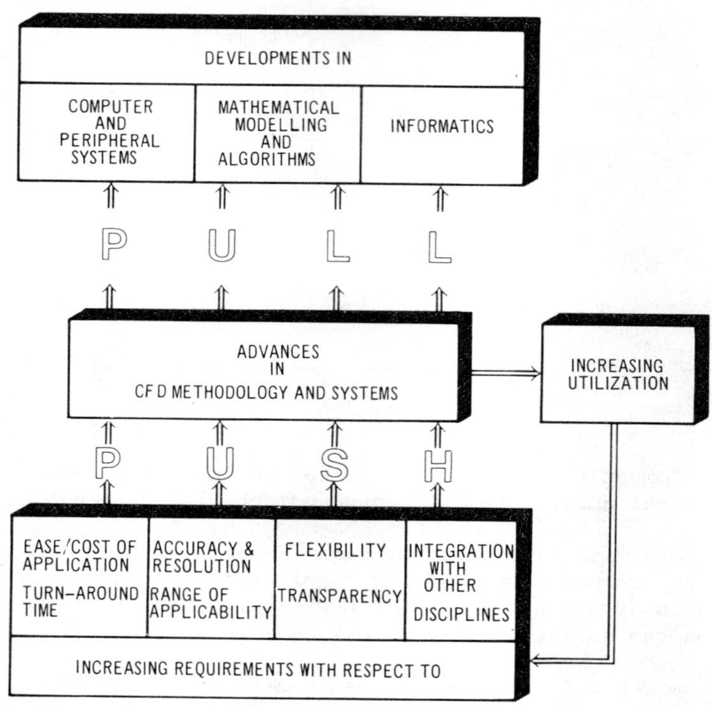

Fig. 2 Factors shaping a next generation of integral methods.

2.   USER-RELATED FACTORS AND REQUIREMENTS "PUSHING" ADVANCES IN CFD IN GENERAL AND BOUNDARY INTEGRAL METHODS IN PARTICULAR

2.1 General Factors

   Amongst the user-related factors that push advances in
integral methods and their computer implementation many are
equally applicable to CFD software systems in a more general
sense. The most important of these general factors have been
summarized in the lower block of Fig. 2. In arbitrary order

these factors are:

1) ease and cost of application
2) integration with other, in particular the experimental, aerodynamic disciplines
3) turn - around time
4) flexibility and transparency of the code
5) accuracy/resolution of results and range of applicability.

Ease of application requires highly automated input preparation and run initialization in the form of an interactive session
on a graphics terminal. It also requires automated means for the
presentation of results including facilities for comparison with
results from other computations and interfacing with other disciplines. We will return to this point in section 5.

Codes that do not exhibit the features just mentioned, or not
to a sufficient extent, are doomed to be used only by the code
designer. There are two, related, reasons for this. First of
all, the number of different CFD codes available to the user/
aerodynamicist is increasing steadily and he will tend to use
only those that are easy to work with. Secondly he will try
to reduce, as much as possible, the time and cost associated
with input preparation, data presentation etc. In NLR experience the cost of human labour constitutes usually between
55% and 75% of the total cost of CFD applications, the remaining
25—45% being computer cost.

Although the computer costs are certainly not insignificant,
a larger impediment to large-scale CFD applications is usually
formed by the appreciable turn-around times required by 3-D
computations in combination with limitations to the accessibility
of computer systems for long-running batch jobs. Hence the user/
aerodynamicist, not surprisingly, is also interested in faster
algorithms and faster computers, both factors, of course, also
tending to reduce the pure computation cost. Note, however,
that a high productivity has little sense when the rate at which
the results can be digested is not matched. We will return to
this point in section 5.

Another important aspect of CFD code design is flexibility and
transparency. NLR experience is that once a code has been accepted and adopted the user/aerodynamicist will, inevitably, use
the program with other objectives and in other ways than it was
originally intended. For that purpose he may even want to change
or replace certain parts of the program himself, provided that
the code is sufficiently transparent and documented. If the
latter is not the case he will either turn to the code designer,
who may be busy or who may even have left the firm, or he will
abandon the code. In either case an(other) investment in software is likely to be lost.

Extension of the range of applicability of existing CFD codes
or the introduction of new ones is generally pushed as much by
the user/aerodynamicist as by the code designer. It is important to realize that the user/aerodynamicist often feels quite
comfortable with his conventional tools (codes) and analysis/
design procedures. Doing the same thing fundamentally better
may not be a sufficient reason for adopting a new tool, in particular when familiarization requires a substantial effort.
Also a new tool may not fit into his existing procedures or,
at best, is not used optimally in the existing procedure. This
is a very important point: *for optimal functioning, CFD software systems and aerodynamic analysis design procedures must be
tuned to each other*. The computational fluid dynamicist/code
designer must be prepared to look upon his product as a small
part of a much bigger system, and tailor his product accordingly.
The user/aerodynamicist on the other hand must be prepared to
adapt his analysis and design procedures in order to be able
to make optimal use of the new possibilities that are being
created by the advances in CFD (Slooff, 1976).

Concluding the discussion on general "push factors" we note
that extending the range of applicability of CFD by improved
modelling of the physics not only tends to increase the computational power required per data point, i.e. per angle-of-
attack/Mach number combination. It also tends to increase the
number of α-M_∞ combinations for which a flow computation has
practical sense. This is due to the fact that by extending the
representation of the physics in the mathematical model the area
in the α-M_∞ plane that is covered by CFD is also increased. The
point is illustrated by Fig. 3, which, schematically, indicates
the coverage of the α-M_∞ plane for several types of mathematical
modelling, the shaded area representing what is being covered
by the current boundary integral methods.

In general it can be stated, that each of the user-related
factors mentioned above will, when properly met and particularly
in combination, in turn increase the level of utilization of
CFD. Hence the feed-back loop in Fig. 2, which implies a substantial rate of amplification. It is important to recognize
this effect in the planning of future computer capacity requirements in aerodynamic analysis/design environments.

2.2 *Specific Requirements for a Next Generation of Integral Methods*

Apart from the general factors mentioned there are a number
of requirements to be met specifically by a next generation of
integral methods. Some of these are, very specifically, related
to the NLR situation. Others apply also to boundary integral
methods more in general. These requirements can be listed as
follows:

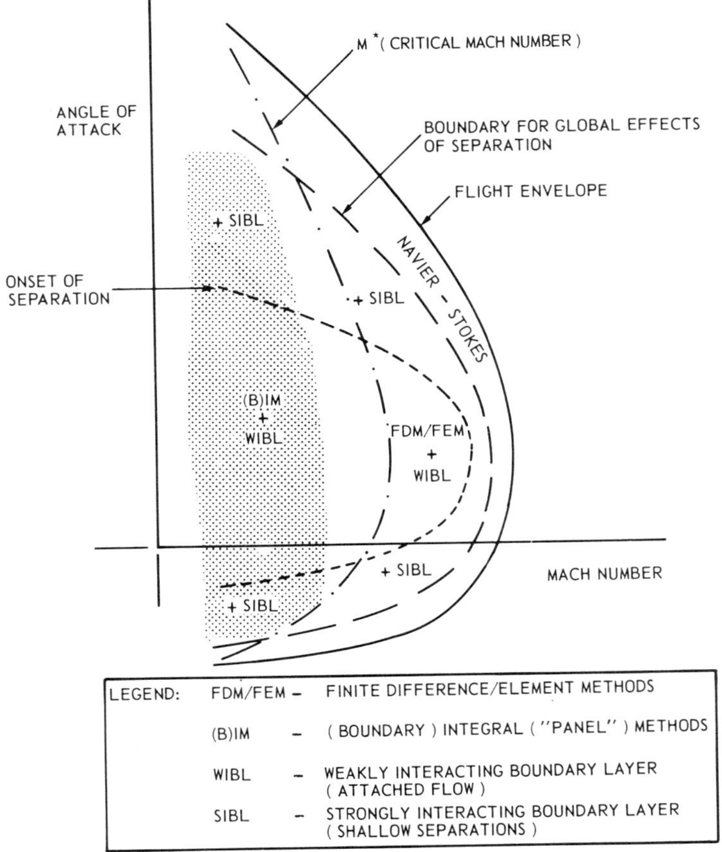

Fig. 3   Schematic of CFD coverage of angle-of-attack-versus-Mach-
number plane (subsonic, transport type aircraft)

1) Increased accuracy and resolution in view of anticipated
   future incorporation of boundary layer effects. Elimination
   at the same time of the excessively large truncation errors
   occurring with (very) thin wings and internal flows, typical
   for several first-generation panel methods (Sytsma et al.,
   1978).
2) Improved modelling of the physics by replacing the linear or
   semi-empirical compressibility correction by a full, numerical representation of compressibility effects.
3) Increased computational efficiency through the use of more
   efficient algorithms.

Note that the latter is needed, amongst others, in view of the

additional computational effort associated with requirements
1) and 2).

In the following chapter we will discuss some developments
that may enable us, in the near future, to satisfy these requirements.

3. "PULL" EFFECTS FROM ADVANCES IN MATHEMATICAL MODELLING AND ALGORITHM DEVELOPMENT

In this section we will consider some recent advances in
mathematical modelling and algorithm development associated with
panel methods. In the past 10 to 15 years there has been relatively little progress in this field. However at present some
new concepts for extended modelling and efficient algorithms are
evolving that hold promise for an order of magnitude increase
in algorithm efficiency as compared with the current generation
of panel methods.

3.1 Surface Discretization Schemes

Although the principles of (boundary) integral methods are
well established in Green's theorems, there is, as mentioned,
a large degree of freedom in the choice of the (surface) singularity distributions. Depending on the type of boundary
conditions, we may use, or may have to use, either or both
doublet/vortex and /or source distributions (see e.g., Hess
et al., 1978). Anticipating boundary conditions which may be of
the Neumann type with non-zero net outflow, Dirichlet type
(flows with lift) as well as, eventually, "design" type boundary
conditions (tangential velocity) we choose to use both types of
singularity distributions at the surface as in the later Hess
(1972), Morino et al. (1974), and Boeing (Johnson et al., 1975)
methods. With this choice the additional complexity of performing integrations over both the external and camber surfaces of
a lifting configuration, typical for the first generation of
panel methods, is avoided.

Having chosen to use both doublet/vortex and source distributions on the bounding surfaces of the configuration, the next step
is to select a means for rendering uniqueness to the formulation. For example one may either satisfy boundary conditions
on both sides of the bounding surface or fix, a priori, one of
the two distributions. A careful study of the numerical errors
resulting from various formulations for two dimensional flow
with lift (Oskam, 1980) has shown that satisfying both external
and internal boundary conditions leads to a lower error level
than formulations based on either external or internal boundary
conditions (see Fig. 4).

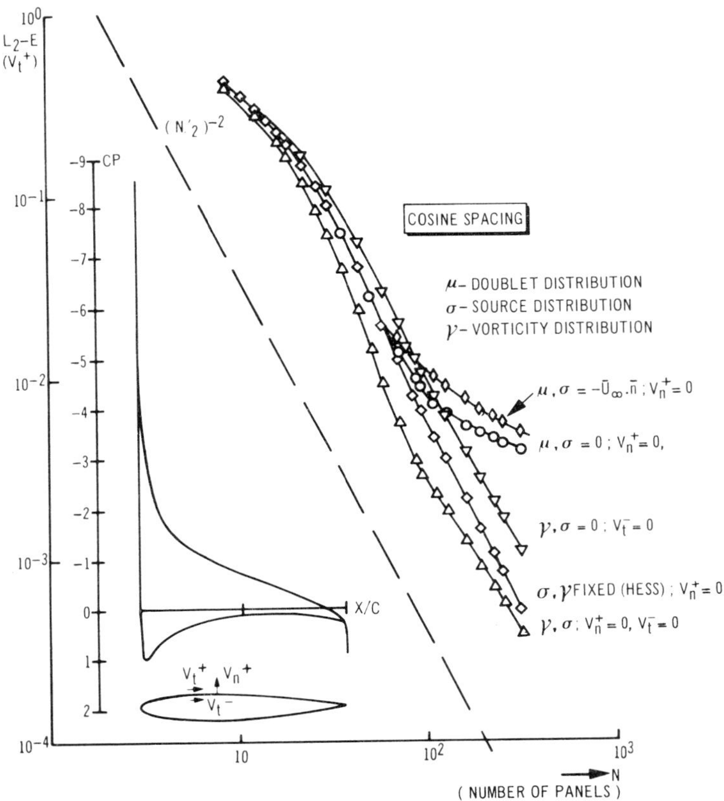

Fig. 4   Accuracy of various (2nd order) BIM formulations for 2-D
lifting flow (12% Von Karman-Trefftz, α=10°), (Oskam,
1980).

It might be argued that satisfying explicitly, both external
and internal boundary conditions would be disadvantageous for
the efficiency of the method because it doubles the number
of unknowns and hence of equations to be solved. This is true,
of course, if the system of equations is solved by means of
direct matrix inversion. However when an iterative solution
procedure is used one can fix one of the two distributions while
solving for the other and after a certain number of iterations
readjust the fixed distribution to better approximate the other
boundary condition. For instance one can solve first for the
source distribution by satisfying an external Neumann boundary
condition while keeping the vorticity distribution fixed and
subsequently update the local vorticity distribution in order to
better approximate the condition of zero internal tangential
velocity (Oskam, 1980).

A very important question is that of the "order" of the truncation error of the numerical scheme to be chosen. The choice
is determined primarily by the requirement to minimize the computational effort for given accuracy and resolution. If N is
the number of unknowns (panels), determining the resolution,
the computational work can asymptotically be expressed as

$O[c(q)N^p]$, $p \geqslant 1$, $q > 0$ and the truncation error as $O[N^{-q}]$. The
coefficient $c(q)$ increases with the order q of the method. Anticipating geometries and solutions exhibiting a low degree of
continuity, we assume that accuracy and resolution are not
necessarily directly related to each other. It then follows
that:

1) requiring increased accuracy (smaller truncation errors),
   for a given type of solver (given p) tends towards higher
   order methods (increasing q) possibly allowing decreasing
   resolution (N).
2) requiring increased resolution (N) for a given type of solver
   tends towards lower order methods (decreasing q) through the
   effect on c(q).
3) for given accuracy and computational effort the introduction
   of fast solvers (decreasing p) tends towards lower order
   methods (decreasing q, increasing N).

In engineering practice one requires *both* sufficient accuracy
and sufficiently high resolution. In the current NLR opinion
this means a normalized L_2-error for the surface velocity of

0.1—1.0% and between 20 and 200 panels per chordwise section.
Note that these figures imply higher requirements than is
currently customary in panel method applications. This is
motivated by the requirement that both accuracy and resolution
should allow the coupling with boundary layers as well as sufficiently sharp modelling of shock waves. (The latter subject
will be treated in more detail in section 3.2). Note that, in
principle, these combined requirements completely determine
the required "order" of the method, irrespective of the solution
algorithm.

The NLR impression is that the required figures for accuracy
and resolution can generally be met by a second order formulation (Fig. 5), and probably in many cases also by a first order
formulation. Because of the relatively small difference in computational effort involved when using consistent curvature and
far-field expansions in formulating the aerodynamic influence
coefficients a second order formulation is preferred.

An important point to be noted in Fig. 5 is the sensitivity
of the error to the type of grid spacing used (compare Figs. a)
and b). This illustrates that from the point of view of accuracy
solution-adaptive grids are highly attractive.

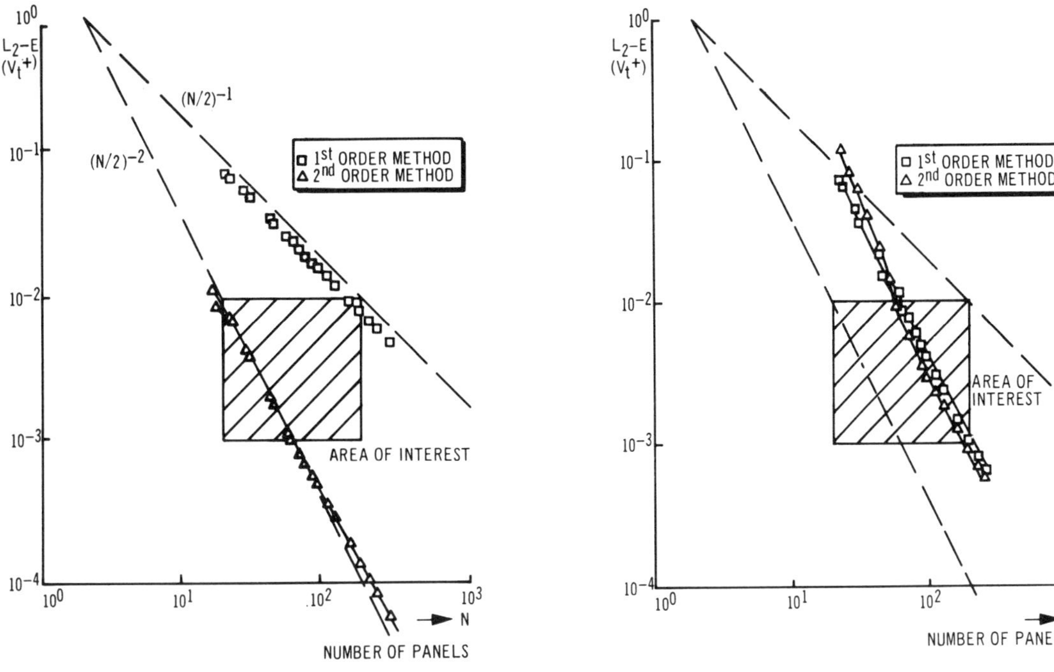

a) near-optimal (curvature adapted) spacing   b) non-optimal (cosine) spacing

Fig. 5 Accuracy of 1st and 2nd order BIM formulations for 2-D lifting flow (12% Von Karman-Trefftz), (Oskam, 1980).

Another aspect of discretization is the (spline-type) stability
of the numerical representation of the singularity distributions
in relation to the choice of collocation point location as used
in evaluating the aerodynamic influence coefficients. The
subject is treated to some extent in Hess et al. (1978). In
agreement with the known criteria for stability both the unknown
source and vortex strength as well as collocation are defined
at panel mid-points in Oskam (1980). The (linear) variation
over a panel is expressed as a two-term Taylor series with the
derivative determined by mean of central differences.

Sensitivity (or the lack of it, rather) to locally singular
behaviour of the solution is yet another aspect of practical
importance. Singular behaviour of the solution does occur, for
instance, at the trailing edge of wings, at planform kinks,
wing-body intersections, etc. The precise character of such
singularities is, in general, not known. If they were, their
inclusion in terms of special mode functions would still be
highly undesirable from the point of view of simplicity of the
code. Hence it is desirable to use a discretization scheme
that, within the requested levels of accuracy and resolution,
does not suffer from significant loss of accuracy at singular
edges.

The characteristics in this respect of the discretization
scheme outlined above can be judged, to some extent, from Fig.
6. The figure presents the maximum norm of the errors in the
surface velocity as a function of the number of panels for the
same application (12% Von Karman-Trefftz aerofoil) as considered
before. The position of the point which determines L_{max} is
indicated. It is clear that in this application the trailing-
edge singularity begins to hamper further reduction of the norm
of the errors at an accuracy and resolution level beyond that
of engineering interest.

3.2 Field distributions

One restriction, if not the most important, of the current
boundary integral methods for aerodynamic applications is that,
in principle, they are limited to incompressible flow. Although
subsonic compressibility effects are usually approximated by
some form of Goethert rule, supplemented in some cases with
semi-empirical nonlinear corrections (Labrujere, et al., 1970;
Hunt, 1980) this situation is far from satisfactory. Not to
mention the fact that such approximations break down completely
when the flow becomes supercritical.

A natural way to extend the capability of boundary integral
methods to compressible potential flows is to represent non-
linear compressibility effects by means of field source distribu-

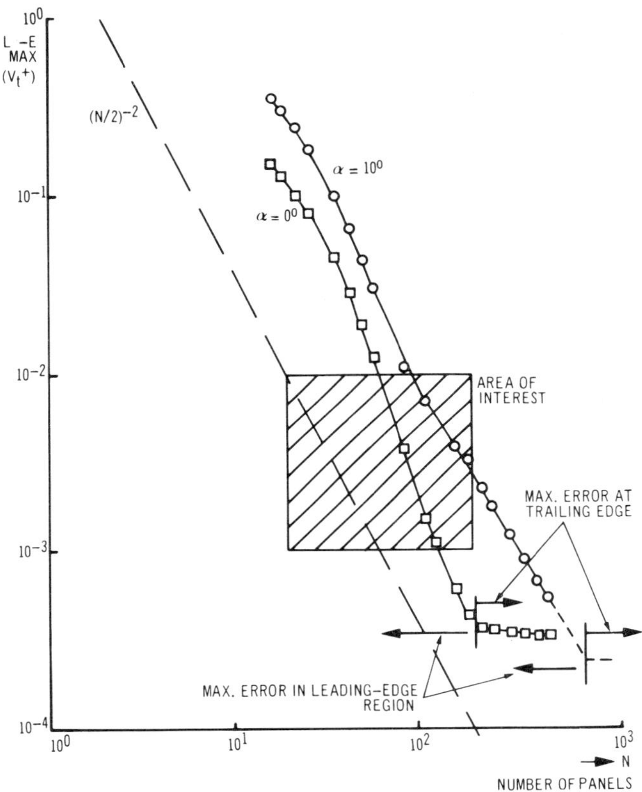

Fig. 6   Maximum error norm showing effect of trailing edge singularity (12% Von Karman-Trefftz, cosine spacing)
(Oskam, 1980).

tions. Utilizing the transonic small perturbation equation it
was demonstrated by Piers and Slooff (1979) for 2-D flow that by
adding artificial viscosity to the governing partial differential
equation it is possible to model transonic flows with shock
waves by means of field source distributions in a way very
similar to the current finite difference methods for transonic
potential flows. Results for the 2-D non-lifting flow about a
10% thick parabolic arc aerofoil from a "quasi-conservative"
and a "fully conservative" formulation have been reproduced
in Fig. 7. They are compared to and in good agreement with
results from a finite difference formulation.

Extension of the formulation just mentioned to the full
potential equation in conservation form is probably most con-

veniently realized if, instead of discretizing the conservative differential form

$$\frac{\partial}{\partial x}(\rho\Phi_x) + \frac{\partial}{\partial y}(\rho\Phi_y) + \frac{\partial}{\partial x}(\rho\Phi_z) = 0 \qquad (3.2.1)$$

at a finite number of grid points, we apply the equivalent integral formulation

$$\iint(\rho\underline{V}\Phi).\underline{n}\ dS = 0 \qquad (3.2.2)$$

to the surface bounding each elementary volume. The velocity vector $\underline{V}\Phi$ and density ρ in (3.2.2) can both be expressed in terms of boundary and field singularity distributions. For transonic flow the artificial viscocity concept in the artificial density form (Hafez et al., 1979) can be introduced to guarantee uniqueness and convergence. The resulting system of non-linear equations for the unknown singularity strengths could be solved by Newton iteration, as in Piers and Sloof (1979).

It is emphasized that a F(ield) I(ntegral) M(ethod) of the type outlined above is just as well suited to complex geometries as are finite element or finite volume methods. However the field integral method has the additional advantage that the field source distribution and application of the mass flux integral (3.2.2) can be limited to the part of the flow field where non-linear compressibility effects are important. In the remainder of the flow field the linear Laplace or Prandtl-Glauert equation and also the boundary conditions at infinity are implicitly satisfied in the field integral formulation. The point is illustrated by Fig. 7 which also presents the boundary of the field source distribution. Extending the field distribution beyond this boundary had no noticeable effect on the solution.

This partial dimension-lowering property of (field) integral methods lowers the number of unknowns in actual calculations as compared with finite element and finite difference methods. The effect is particularly important for moderate and low speed flows involving small pockets with high subsonic or supersonic flow. Perhaps the most important consequence is that the problem of spatial mesh generation is simplified considerably and limited to only a part of the flow field.

It was mentioned in Section 1 that with the asymptotic operations count proportional to at least the square of the number of panels, the computational effort becomes rapidly prohibitive when the number of panels increases. This is particularly true for field integral methods (Piers and Slooff, 1979). In the next section we will discuss some very recent developments that are expected to remove most if not all of this decisive restriction.

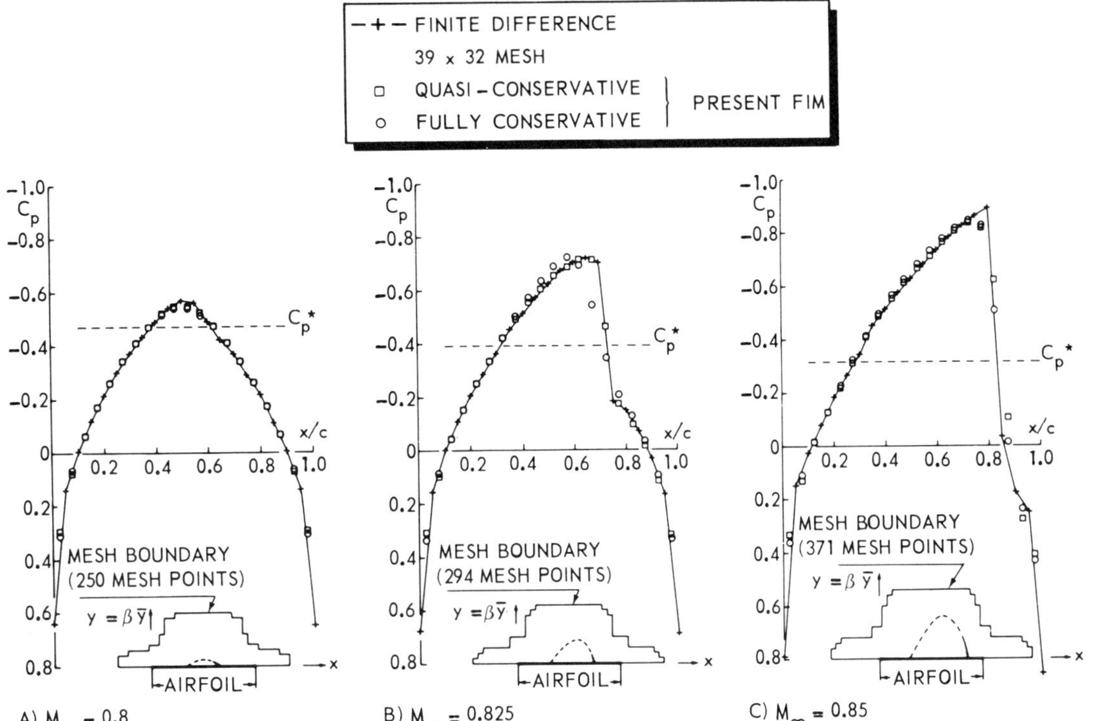

Fig. 7 Results of a shock-capturing field integral method for a 10% parabolic arc airfoil (Piers and Slooff, 1979)

3.3 Fast Solvers

3.3.1 General considerations In order to facilitate the following discussions we write the system of linear equations to
be solved with integral methods as

$$A \underline{x} = \underline{b} \qquad (3.3.1)$$

A is a full N×N matrix (N being the number of panels). In aerodynamic applications the structure of the matrix A depends on
the particular integral equation considered and the discretization scheme used. In general we are dealing with a Fredholm
integral equation of either the second or the first kind.

In some of the panel methods that can be found in the literature the system (3.3.1) is solved by means of a direct method
while in others it is solved by means of iterative techniques.
The iterative techniques usually employ some form of block
Gauss-Seidel or block Jacobi relaxation. Because A is a full
matrix, the asymptotic operations count of the direct methods
is $O[N^3]$ while it is $O[pN^2]$ for the iterative methods, where p
is the number of iterations. As a result the computational
effort for the direct methods becomes prohibitive for N > ~1000.
The iterative methods usually require between 10 and 30 iterations. In this case the admissible number of panels is determined
by the computational effort involved with the aerodynamic influence coefficients.

In the following two sections we will discuss some recent
developments that hold the promise of reducing:

1) the number of iterations needed to obtain an accuracy level
   corresponding with the level of the truncation error,
2) the operations count of a single iteration cycle ($O[N^2]$ for
   the current iterative methods).

3.3.2 Multi-grid techniques In recent years there has been considerable progress towards reducing the number of iterations
needed for solving the sparse matrix equations that result from
finite difference and finite element methods. One of the most
promising techniques in this respect is the multi-grid (MG)
method (see e.g. Brandt, 1977). Although the MG technique is
also applicable to integral equations, (Schippers, 1978), the
number of (reported) applications in this field is extremely
small.

One way to look at multi-grid methods is to consider them
as a residual correction process. In the classical residual
correction approach the solution of the problem say

$$A \underline{x} = \underline{b} \tag{3.3.1}$$

is found by splitting the coefficient matrix A into a form

$$A = \tilde{A} - R \tag{3.3.2}$$

where $\tilde{A}$ is a nonsingular matrix approximating A. Then we may define an iteration process by

$$\tilde{A} \underline{x}^{(n)} = \underline{b} + R \underline{x}^{(n-1)} \tag{3.3.3}$$

starting with some arbitrary vector say $\underline{x}^{(O)}$. This iteration process implies a residual correction matrix, M_r, which operates on the residual $\underline{r}^{(n)} = \underline{b} - A \underline{x}^{(n)}$ like

$$\underline{r}^{(n)} = M_r \underline{r}^{(n-1)} \tag{3.3.4}$$

where

$$M_r = I - A\tilde{A}^{-1} \tag{3.3.5}$$

In this way a sequence of residual vectors is defined which in principle should converge to zero. The rate of convergence depends on how well $\tilde{A}$ approximates A.

In multi-grid algorithms the relaxation process (3.3.4), (3.3.5) is executed on various grid levels. The role of the relaxation process is to efficiently liquidate, on the particular grid considered, those residue components which cannot be re-solved on the next coarser grid. The remaining long Fourier components of the residual vector may be resolved and liqui-dated efficiently on the coarser grids. Utilizing a sequence of grid levels k results in a residual correction process with a high rate of convergence.

In the multi-grid method the operator M_r is often called a smoothing operator. From a given level k the smoothing process is continued either on the coarser grid G_{k-1} after, say p iterations or on the finer grid G_{k+1}, after, say q iterations. Switching to the coarser grid requires restriction of the resi-dual $\underline{r}^{k,(p)}$ from G_k to G_{k-1}. This can be denoted by

$$\underline{r}^{k-1} = R_k^{k-1} \underline{r}^{k,(p)} \tag{3.3.6}$$

Switching to the finer grid, in terms of the formalism (3.3.4) requires:

1) prolongation of correction $\underline{\delta x}$ from G_k to G_{k+1}, denoted by

$$\underline{\delta x}^{k+1} = P_k^{k+1} \; \underline{\delta x}^k , (q) \tag{3.3.7}$$

2) transformation to increment residual $\underline{\delta r}^{k+1}$ through

$$\underline{\delta r}^{k+1} = -A^{k+1} \; \underline{\delta x}^{k+1} \tag{3.3.8}$$

and 3) adding to $\underline{r}^{k+1}$

$$\underline{r}^{k+1} : = \underline{r}^{k+1} + \underline{\delta r}^{k+1} \tag{3.3.9}$$

In principle the switching sequence, i.e. the parameters p and q, should be chosen such that the amount of work to obtain the solution is minimized. How this should be realized is not always evident from the literature; one can find both fixed switching strategies (i.e. fixed p and q) and strategies in which the switching is determined by the smoothing rate and the level of the residual.

Early examples of application of the multi-grid technique to a Fredholm integral equation of the second kind are the two-level methods described by Atkinson (1973). Atkinson considers the integral equation

$$\lambda f(x) - \int_0^1 K(x,\xi) f(\xi) d\xi = g(x) \tag{3.3.10}$$

with continuous kernel $K(x,\xi)$, which with some integration rule is discretized on a fine grid G_2 into

$$\lambda f(x_i) - \sum_{j=1}^{N} w_j \; K(x_i,x_j) f_j \; h_j = g(x_i) \tag{3.3.11}$$

where w_j is a weight function depending on the integration rule chosen and h_j is the mesh width. Rewriting (3.3.11) as (3.3.1) we have

$$\left. \begin{aligned} A = A_{ij} &= I - w_j K_{ij} h_j \\ \underline{x} &= f(x_i) \\ \underline{b} &= g(x_i) \end{aligned} \right\} \tag{3.3.12}$$

Atkinson solves the system (3.3.1) through the following algorithm:

1) Estimate $\underline{x}^2$ on the fine mesh G_2

2) Determine residual $\underline{r}^2$

3) Perform p Jacobi relaxations on the fine mesh

4) Make the restriction

$$\underline{r}^1 = R_2^1 \, \underline{r}^2 \tag{3.3.13}$$

5) Solve correction $\underline{\delta x}^1$ from

$$A^1 \, \underline{\delta x}^1 = \underline{r}^1 \tag{3.3.14}$$

6) Prolongate

$$\underline{\delta x}^2 = P_1^2 \, \underline{\delta x}^1 \tag{3.3.15}$$

by means of natural (Nyström) interpolation i.e. by means of the generalized form

$$f(x) = \frac{1}{\lambda} \left[g(x) + \sum_{j=1}^{N} w_j \, K(x,x_j) f_j h_j \right] \tag{3.3.16}$$

of (3.3.11).

7) Correct $\underline{x}^2$ and return to 1).

Assuming that the number of mesh points on the coarse mesh is much smaller than that on the fine mesh Atkinson proposes to solve (3.3.14) by means of a direct method. One of the variants he considers implies p=0. This choice is possible as a result of the fact that the Nystrom interpolation process itself has smoothing properties similar to those of Jacobi relaxation.

Extensions of the schemes of Atkinson to more than two grids have been considered recently by Schippers (1978) and Hemker and Schippers (1979). A similar technique was studied by Piers (1979) in the aerodynamic model problem of two-dimensional in-compressible lifting potential flow about a flat plate. The problem is described by the Fredholm integral equation of the 1st kind with singular kernel

$$\int_O^1 \gamma(\xi) \, \frac{d\xi}{x-\xi} = \Pi \tag{3.3.17}$$

with the additional (Kutta) condition $\gamma(1)=0$. In order to avoid the singular kernel equation (3.3.17) is written as

$$p(x)\gamma(x) - \int_0^1 \frac{\gamma(x)-\gamma(\xi)}{x-\xi}\, d\xi = \Pi \qquad (3.3.18)$$

where

$$p(x) = \ln \frac{x}{1-x} \qquad (3.3.19)$$

The integral in (3.3.18) is discretized on an equidistant mesh by means of the trapezoidal rule, yielding

$$p(x)\gamma(x) - \sum_j w_j \frac{\gamma(x)-\gamma_j}{x-\xi_j}\, h_j = \Pi \qquad (3.3.20)$$

or,

$$q(x)\gamma(x) + \sum_j w_j \frac{\gamma_j}{x-\xi_j}\, h_j = \Pi \qquad (3.3.21)$$

where

$$q(x) = p(x) - \sum_j w_j \frac{1}{x-\xi_j}\, h_j \qquad (3.3.22)$$

Note that although the original integral equation (3.3.17) is one of the 1st kind, the discretized form (3.3.21) is more like one of the 2nd kind, allowing Nyström interpolation.

Applying mid-point collocation the resulting system of equations is solved by means of three different grid levels using a fixed switching strategy. This strategy is depicted in Fig. 8, showing the sequence of one M(ulti) G(rid) cycle. Figure 8 also presents the convergence history in terms of the error $\Delta\Gamma$ in the total circulation $\Gamma = \int_0^1 \gamma\, dx$ around the plate. Starting from $\gamma = 0$ convergence to truncation error level is reached in 4 MG cycles with a convergence factor of about 0.25.

An interesting adventitious circumstance with the example given above is that Jacobi-relaxation on the finest grid only diverges. This illustrates the power of multi-grid methods in speeding-up the convergence of iterative schemes!

Multi-grid schemes for integral equations utilizing relaxation and prolongation operators different from those implied by the Nyström method have been studied by Hemker and Schippers (1979) and Hackbusch (1978). These studies indicate that for Fredholm integral equations of the 2nd kind, under certain conditions, only 1 or 2 MG-cycles are needed for convergence to truncation error level.

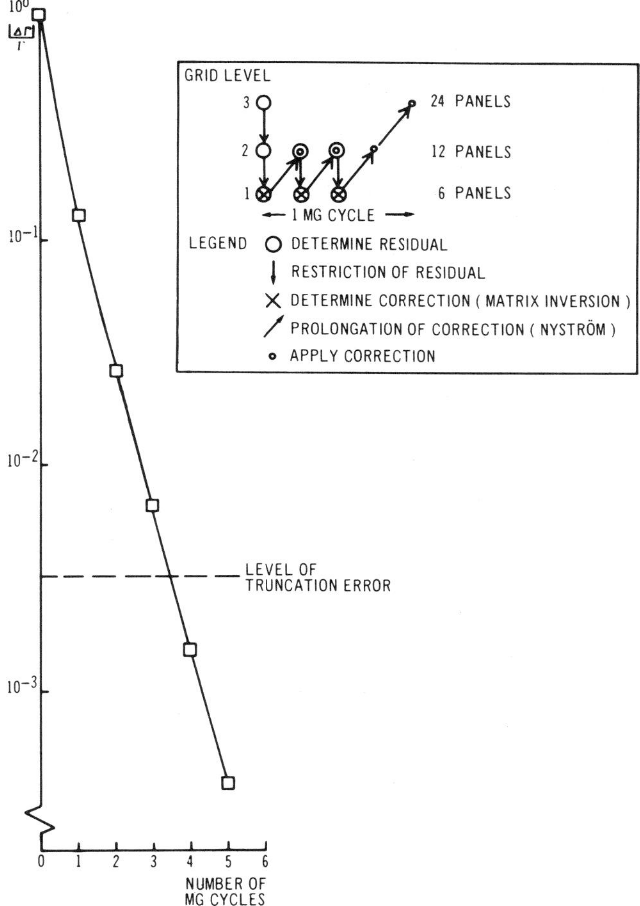

Fig. 8   Characteristics of a multi-grid method with Nyström
interpolation as applied to the integral equation associated with the lifting flat plate problem (Piers, 1979).

The application of such techniques to the integral equation
of the 2nd kind associated with the 2nd order panel method described in section 3.1 is currently being pursued by Oskam and
Fray of NLR.

3.3.3 "Clustering" schemes In spite of the prospects for rapid
convergence sketched above, MG methods as such cannot prevent
that at least one matrix x vector multiplication must be executed on the finest grid. For integral methods with their full

matrices this means that the asymptotic operations count remains
$O(N^2)$. The computational implication of this becomes clear if
we realize that matrix formation and the preceding computation
of $O(N^2)$ aerodynamic influence coefficients (AIC's) requires
about 75% of the total computational effort of the current boundary integral methods (see, e.g., Sytsma et al., 1978). This
illustrates that reduction of the computational AIC work is considerably more important than improvement of the current iterative solution methods.

Methods to reduce the work involved with AIC computation and
to indeed lower the asymptotic operations count are currently
being studied at NLR. The basic concept underlying these
studies is that the influence of distant panels experienced at
a given point can be evaluated, without significant loss of
accuracy, on coarse grid levels. This technique may be interpreted as a "clustering" of distant portions of the singularity
distribution.

In terms of multi-level techniques a general "clustering"
algorithm can be formulated as follows:

Consider the matrix x vector multiplication

$$\underline{v}^k = A^k \, \underline{x}^k \tag{3.3.23}$$

on a grid G_k. We approximate (3.3.23) by the following sequence:
1) splitting $\underline{v}^k$ into

$$\underline{v}^k = \underline{v}_s^k + \underline{v}_d^k \tag{3.3.24}$$

where

$$\underline{v}_s^k = A_s^k \, \underline{x}^k \tag{3.3.25}$$

$$\underline{v}_d^k = A_d^k \, \underline{x}^k \tag{3.3.26}$$

and

$$A^k = A_s^k + A_d^k \tag{3.3.27}$$

A_s is a sparse, banded matrix containing only "near-field"
information. A_d is a dense matrix containing only "far-field"
information with zeros on a finite number of bands.
2) Execute (3.3.25)

3) In order to obtain a (cheap) approximation to $\underline{v}_d^k$ make the restriction

$$\left.\begin{array}{l} A_d^{k-1} = R_k^{k-1} \; A_d^k \; P_{k-1}^k \\[2ex] \underline{x}^{k-1} = R_{k-1}^k \; \underline{x}^k \end{array}\right\} \qquad (3.3.28)$$

to the coarser grid G_{k-1}. R_k^{k-1} and P_{k-1}^k are restriction and prolongation operators respectively, as in (3.3.6), (3.3.7).

4) Execute the matrix x vector multiplication

$$\underline{v}^{k-1} = A_d^{k-1} \; \underline{x}^{k-1} \qquad (3.3.29)$$

by repeating the sequence 1), 2) on the level k-1, with A_d replacing A. On the lowest level G_0 (3.3.29) is executed directly.

5) Do the prolongation

$$\underline{v}_d^k = P_{k-1}^k \; \underline{v}^{k-1} \qquad (3.3.30)$$

6) Add to $\underline{v}_s^k$, i.e. do (3.3.24).

It is important to realize that by the clustering process described above we obtain an *approximation* to $\underline{v}^k$. Hence it implies a loss of accuracy. If A is an N×N matrix, truncation error accuracy can, in principle, be obtained if the matrix A_s contains a certain number m of diagonal bands $1 \leq m < N$. The precise value of m required for truncation error accuracy will depend on the type of integral equation considered and N. Note that the asymptotic operations count of the algorithm is O[mN].

Algorithms of the type outlined above are currently being studied at NLR. A provisional result of an estimate for the asymptotic number of diagonal bands in A_s, obtained by Piers (1981), is that $m = O(N^{1/2})$ if N is increased to reduce the truncation error. Note however that m is independent of N if N is increased without reducing the truncation error. The latter would be the case, for instance, if N is increased in order to treat a more complicated configuration. Another example would be an increase in free stream Mach number for the field integral method of section 3.2.

Very recent information, derived from application of a "clustering" algorithm to the downwash integral (3.3.17) for the flat plate suggests that m=3 is sufficient for engineering accuracy. Note that m=3 means that only the effect of the immediate neighbours is considered on the same grid and that of

all other panels on coarser grids. For the surface distribution
of a 3-D lifting wing this would, presumably, mean
$m = 2 \times (3 \times 3) = 18$, the factor 2 coming from the interaction of
upper and lower surface. Similarly, a 3-D field distribution
would require $m = 3 \times 3 \times 3 = 27$.

In the present author's opinion the multi-level techniques
mentioned above hold the promise that the next generation of
boundary integral methods could be equipped with extremely efficient iterative solvers. When combined with Newton iteration
this should hold also for non-linear field integral methods of
the type discussed in section 3.2.

4. COMPUTATIONAL REQUIREMENTS AND POWER AVAILABLE

On the basis of the, admittedly, scarce information available
on multi-level techniques for integral equations, a tentative
estimate has been made of the computational requirements of
future integral methods for three-dimensional incompressible
(BIM) and compressible (FIM) potential flow. The estimates
are based on the following assumptions:

1) for number m of bands in iteration and AIC matrices:
   $2 \times (3 \times 3) \leq m \leq 2 \times (7 \times 7)$ for BIM
   $3 \times 3 \times 3 \leq m \leq 7 \times 7 \times 7$ for FIM
2) 3 multi-grid cycles for solution of linear problems, one
   cycle requiring twice the mount of work of one relaxation
   sweep on the finest grid.
3) 3 Newton iterations on the finest grid for the non-linear FIM.

The results of these estimates, in terms of number of (mega)
floating point operations (MFLOPs), total and central memory
requirements are presented in Figs. 9, 10 and 11, respectively,
and compared with the corresponding requirements of the current
boundary integral and finite difference methods.

As indicated by Fig. 9, it must be expected that, for the
same resolution and on the same computer, future boundary integral methods will be an order of magnitude faster than the
current BIMs. Depending on the free stream Mach number M_∞, the
number of MFLOPs required by a FIM for compressible potential
flow will range between that of the future BIM and that of the
current FDM. This suggests that integral methods, at least when
equipped with fast solvers, are indeed worthy of further investment!

It is interesting to project the MFLOPs requirements against
the CPU-speed that will, typically, be available to the aerospace community in the 1980s. As indicated by the scales on the
right side of Fig. 9, it must be expected that a typical BIM

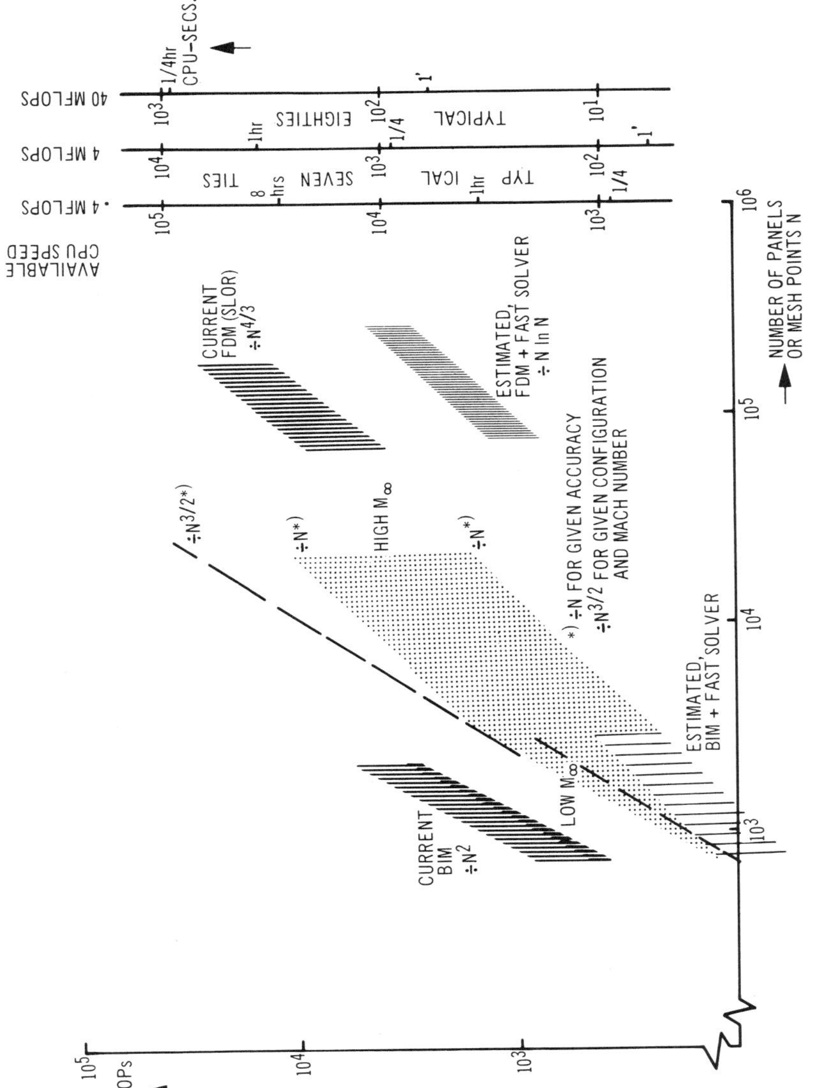

Fig. 9 Computational requirements for potential flow: floating point operations

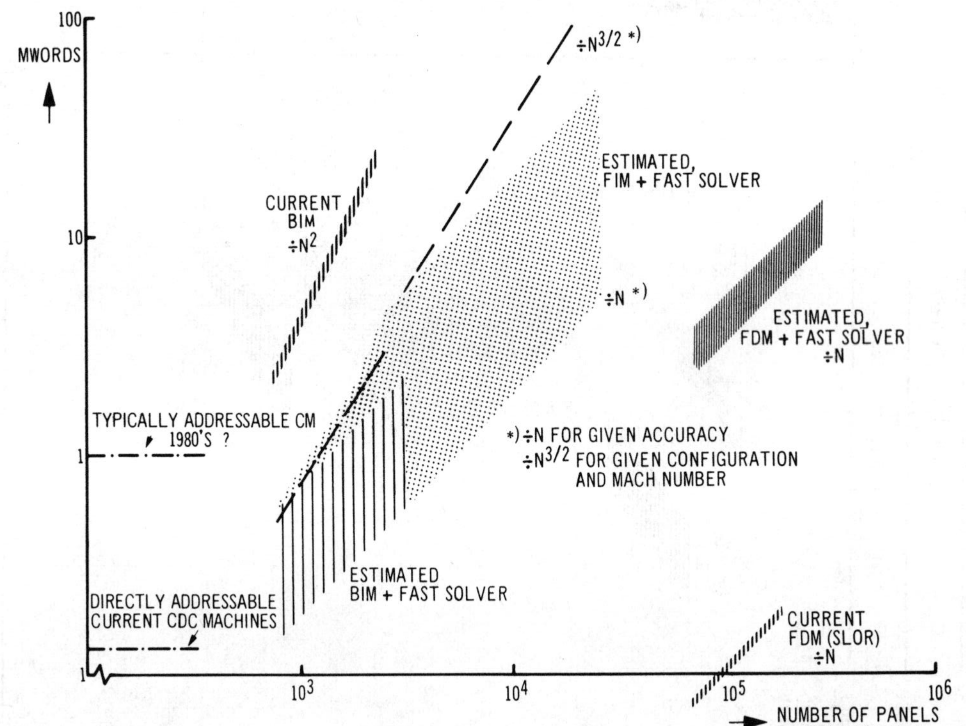

Fig. 10 Computational requirements for potential flow: (total) memory

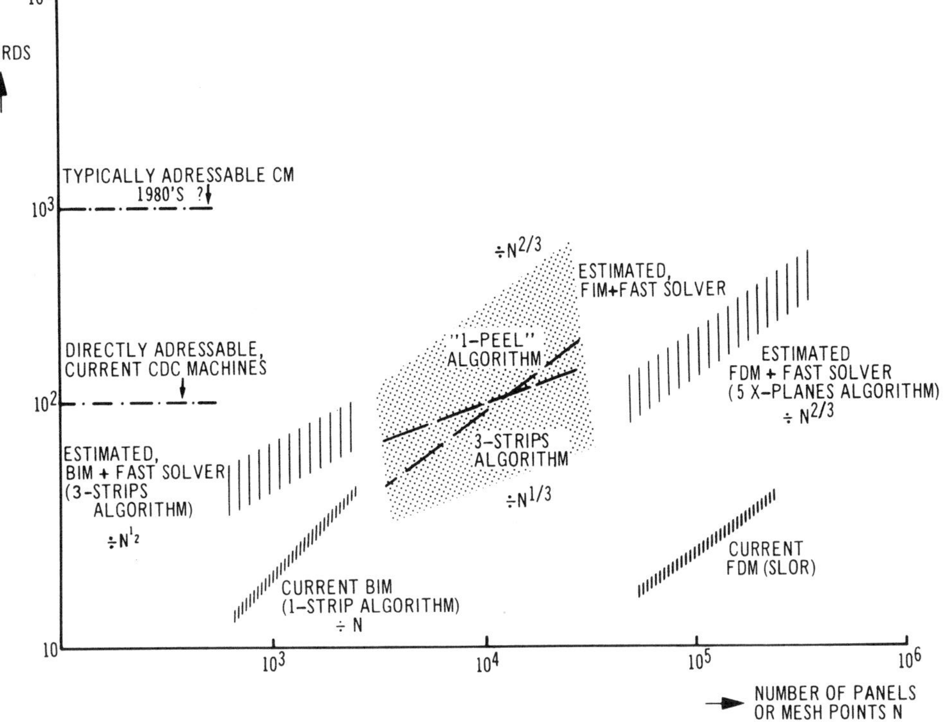

Fig. 11 Computational requirements for potential flow: central memory

run, around say, 1985, will take somewhere between 5 and 50
CPU-seconds. The corresponding figures for the seventies are
typically between 500 and 5000 CPU-seconds. In other words we
may expect an *increase in computational productivity* of roughly
two orders of magnitude! As indicated by Fig. 10 this can be
obtained with a total memory capacity of roughly one order of
magnitude less than required by the current BIM. With field
distributions added the required capacity may be of the same
order of magnitude as that of the current BIM.

Figure 11, finally, provides estimates for the central
memory requirements with varying amounts of data in core. Note
that "1-strip algorithm" denotes a situation in which all the
data related to one chordwise strip of panels around the wing
is stored in central memory. In case of a "1-peel algorithm"
all the data pertaining to a surface completely enclosing the
wing is packed in core. Figure 11 illustrates that for central
memories approaching 1 MWORDS (possibly a typical value for the
1980s) most "algorithms" are feasible, suggesting that I/O will
not be a too constraining factor.

5. CFD AS PART OF AN AERODYNAMICS INFORMATION SYSTEM

As indicated in the preceding section an increase in CFD productivity of roughly a factor 100 may be expected to be realized
in the 1980s. The question must be raised if and how this can
be utilized. One obvious implication is that rather than running
for a given configuration one or at best a few angle-of-attack/
Mach number combinations in overnight batch-mode we will in the
relatively near future be able to cover the complete α-M_∞ plane
in one night. The amount of data produced during such a session
is immense (probably of the order of several MWORDS).

A very important if not decisive point is the answer to the
following question: How long does it take the aerodynamicist
to digest such an incredible amount of information? Hand-
plotting of the surface pressures only will take him about four
months. For obvious reasons this is not acceptable. Equally
unacceptable, of course, would be a situation in which a similar
amount of time would be required for input preparation (see also
section 2).

In order to solve such and related problems we will have to
resort to the techniques of informatics. Figure 12 presents a
"blue-print" for an aerodynamics information system directed
towards solving both the front-end (input generation) and rear-
end (data post-processing) problems as well as integration of
the computational and experimental disciplines. The main ingredients of the system are (Posthuma de Boer et al., 1980):

1) a geometric data base with an appropriate set of data base

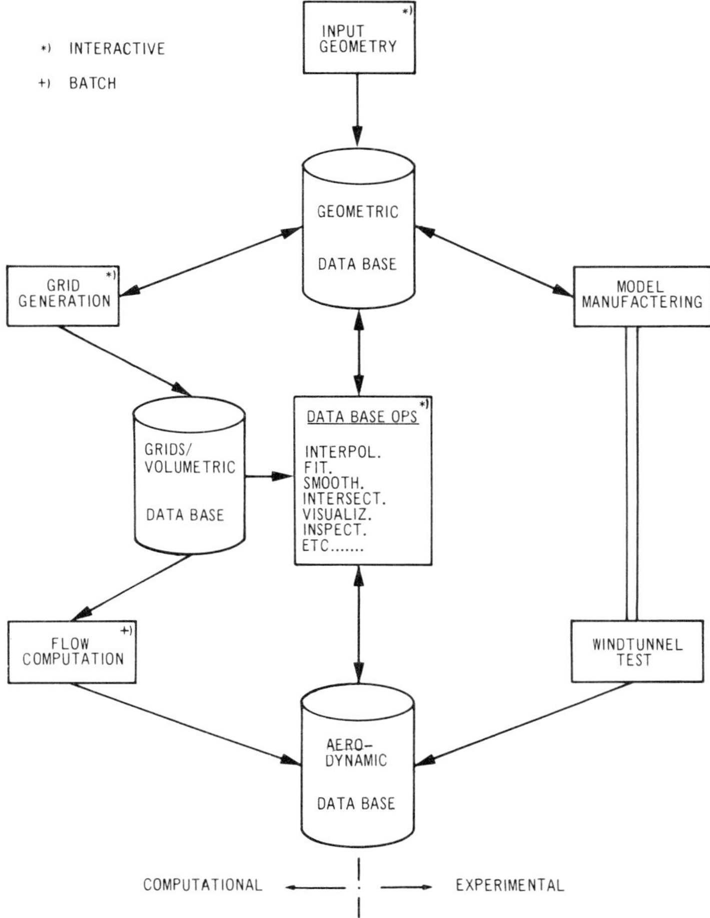

Fig. 12   CFD as part of a Computational/Experimental Aerodynamics
          Information System

    operations (geometry package)
2) an aerodynamic data base with associated set of post-processing operations (not necessarily identical to the set of geometric data base operations)
3) a grid generator with associated data base
4) a fluid dynamics solver
and, on the experimental side
5) model manufacturing
6) wind tunnel testing

Construction of the system is under way with some sub-systems
operational in provisional versions. The final objective is

that geometry input and grid generation are executed in a few
interactive sessions on a graphics terminal during normal working hours. Most of the actual flow computations will be run
in over-night batch jobs, the results of which will be stored
in the aerodynamic data base. Post-processing and visualizing
of the results, including, if required, comparisons with results
of earlier computations or windtunnel tests are again executed
interactively on the graphics terminal.

In the author's opinion facilities of the type just described
are a prerequisite for integration of the computational and experimental disciplines in an aerodynamic research/design environment.

6. CONCLUDING REMARKS

A discussion has been given of factors shaping the development of a new generation of integral or "panel" methods at
NLR. A distinction has been made between "push-factors" (requirements) originating from the user-aerodynamicist and "pull-
factors" from advances in mathematical modelling and algorithm
development.

Amongst the requirements for improvements (relative to the
current NLR panel method) are:

1) higher level of automation in input preparation and presentation of results; better integration with other disciplines.
2) higher accuracy and resolution; in general because of anticipated inclusion of boundary layer effects; in particular
   for thin wing and internal flow applications.
3) better representation of compressibility effects, in particular for low speed, high lift situations.

and, in order to allow the realization of 2) and 3) and to
reduce turn-around times:

4) improved computational efficiency.

On the basis of current progress in mathematical modelling
and algorithm development it is expected that the new panel
methods of the 1980s, as compared with the panel methods of the
1970s, will be characterized by:

1) an order of magnitude increase in algorithm efficiency
2) "exact" representation of compressibility effects, including
   shock waves, by means of field source distributions.

It is further noticed that improvements in computer technology will provide another order of magnitude increase in
computational productivity. In order to enable the user-aerodynamicist to cope with the latter, future CFD system should

be embedded in an efficient computational/experimental aerodynamics information system.

ACKNOWLEDGEMENTS

The author is indebted to his colleagues of the Theoretical
Aerodynamics Department and Informatics Division of NLR for fruit-
ful discussions and comments. Thanks are due in particular to
Messrs Oskam and Piers for providing early results of their
current research.

REFERENCES

Anon (1980). Numerical Grid Generation Techniques, NASA CP-2166.

Atkinson, K.E. (1973). Iterative variants of the Nyström method
for the numerical solution of integral equations. *Numerische
Mathematik* **22**, 17—31.

Boppe, C.W. and Aidala, P.V. (1980). Complex configuration
analysis at transonic speeds, *AGARD CP* No. 285, paper 26.

Brandt, A. (1977). Multi-level adaptive solutions to boundary-
value problems, *Mathematics of Computation* **31**, 333—390.

Caughey, D.A. and Jameson, A. (1979). Recent progress in finite
volume computation for wing-fuselage combinations, *AIAA* paper
79—1513.

Hackbusch, W. (1978). Die schnelle Auflösung der Fredholmschen
Integral-gleichung zweiter Art, Report 78-4, Universität zu
Köln.

Hafez, M., South, J., and Murman, E.M. (1979). Artificial compressibility methods for numerical solutions of the transonic
full potential equation, *AIAA J.*, Vol. **17**, no. 8, 838—844.

Hemker, P.W. and Schippers, H. (1979). Multiple grid methods
for the solution of Fredholm integral equations of the second
kind; Mathematisch Centrum, Amsterdam, NW75/79.

Hess, J.L. (1972). Calculation of potential flow about arbitrary
three-dimensional lifting bodies, final technical report, Mc-
Donnell Douglas Report No. MDC J5679-Ol.

Hess, J.L., Johnson, F.T., Rubbert, P.E. (1978). Panel Methods,
AIAA Professional Study Series.

Hunt, B. (1980). Recent and anticipated advances in the panel
method: the key to generalized field calculations? VKI Lecture
Series 1980—5.

Jameson, A. (1979). Acceleration of Transonic Potential Flow
Calculations on Arbitrary Meshes by the Multiple Grid Method,
Proceedings AIAA Computational Fluid Dynamics Conference, paper
79—1458.

Johnson, F.T. and Rubbert, P.E. (1975). Advanced panel-type
influence coefficient methods applied to subsonic flows, AIAA
paper 75—50.

Labrujere, Th.E., Loeve, W. and Slooff, J.W. (1970). An approximate method for the calculation of the pressure distribution on
wing-body combinations at subcritical speeds, AGARD CP No. 7.

Morino, L. and Kuo, C.-C. (1974). Subsonic potential aerodynamics for complex configurations: a general theory, *AIAA J.*,
Vol. **12**, No. 2, pp. 191—197.

Oskam, B. (1980). Unpublished work at NLR.

Piers, W.J. (1979). Beschrijving van eeen multi-rooster tech-
niek voor een panelenmethode, NLR AT-79-006 I, (unpublished
memorandum).

Piers, W.J. (1981). Unpublished work at NLR.

Piers, W.J. and Sloor, J.W. (1979). Calculation of transonic
flow by means of a shock-capturing field panel method, Proceedings AIAA Computational Fluid Dynamics Conference, paper
79—1459.

Posthuma de Boer, U., Hameetman, G.J., Heerema, J. and van der Vooren,
J. (1980).   Functionele analyse van de automatiseringsbehoeften
en technisch concept van de verdere ontwikkeling van de com-
puter-infrastructuur bij het NLR, NLR TR 80105 L, (unpublished
report).

Schippers, H. (1978). Multi-grid techniques for the solution
of Fredholm integral equations of the second kind, MC-Syllabus
41, Mathematisch Centrum, Amseterdam.

Slooff, J.W. (1976). Windtunnel tests and aerodynamic computations, thoughts on their use in aerodynamic design, AGARD
CP No. 210, paper 11.

Sytsma, H.S., Hewitt, B.L. and Rubbert, P.E. (1978). A comparison of panel methods for subsonic flow computations,
AGARDograph No. 241.